P9-CKX-707

METHODS IN CELL BIOLOGY

VOLUME 24

The Cytoskeleton

Part A. Cytoskeletal Proteins, Isolation and Characterization

Advisory Board

METHODS IN CELL BIOLOGY

Prepared under the Auspices of the American Society for Cell Biology

VOLUME 24

The Cytoskeleton
Part A. Cytoskeletal Proteins, Isolation and Characterization

Edited by

LESLIE WILSON
DEPARTMENT OF BIOLOGICAL SCIENCES
UNIVERSITY OF CALIFORNIA, SANTA BARBARA
SANTA BARBARA, CALIFORNIA

1982

ACADEMIC PRESS
A Subsidiary of Harcourt Brace Jovanovich, Publishers

New York London
Paris San Diego San Francisco São Paulo Sydney Tokyo Toronto

ACADEMIC PRESS, INC.
111 Fifth Avenue, New York, New York 10003

United Kingdom Edition published by
ACADEMIC PRESS, INC. (LONDON) LTD.
24/28 Oval Road, London NW1 7DX

LIBRARY OF CONGRESS CATALOG CARD NUMBER: 64-14220

ISBN 0-12-564124-9

PRINTED IN THE UNITED STATES OF AMERICA

82 83 84 85 9 8 7 6 5 4 3 2 1

CONTENTS

4. *A Brain Microtubule Protein Preparation Depleted of Mitochondrial and Synaptosomal Components*

Timothy L. Karr, Hillary D. White, Beth A. Coughlin, and Daniel L. Purich

5. *Purification and Reassembly of Tubulin from Outer Doublet Microtubules*

Kevin W. Farrell

6. *Production of Antisera and Radioimmunoassays for Tubulin*

Livingston Van De Water III, Susan D. Guttman, Martin A. Gorovsky, and J. B. Olmsted

7. *Immunofluorescence and Immunocytochemical Procedures with Affinity Purified Antibodies: Tubulin-Containing Structures*

Mary Osborn and Klaus Weber

12. *Dynein Decoration of Microtubules—Determination of Polarity*

Leah T. Haimo

13. *Microtubule Polarity Determination Based on Conditions for Tubulin Assembly* in Vitro

Steven R. Heidemann and Ursula Euteneuer

14. *The Use of Tannic Acid in Microtubule Research*

Keigi Fujiwara and Richard W. Linck

15. *The Cyclic Tyrosination/Detyrosination of Alpha Tubulin*

William C. Thompson

CONTRIBUTORS

Numbers in parentheses indicate the pages on which the authors' contributions begin.

MARTHA AYNARDI-WHITMAN,[1] Department of Biological Sciences, Carnegie-Mellon University, Pittsburgh, Pennsylvania 15213 (399)

CHRISTOPHER W. BELL, Pacific Biomedical Research Center, University of Hawaii, Honolulu, Hawaii 96822 (373)

LAWRENCE G. BERGEN, Laboratory of Molecular Biology and Department of Zoology, University of Wisconsin, Madison, Wisconsin 53703 (171)

GARY G. BORISY, Laboratory of Molecular Biology and Department of Zoology, University of Wisconsin, Madison, Wisconsin 53703 (171)

B. R. BRINKLEY, Department of Cell Biology, Baylor College of Medicine, Houston, Texas 77030 (1)

SUSAN S. BROWN, Department of Anatomy, University of Michigan Medical School, Ann Arbor, Michigan 48109 (291)

BETH A. COUGHLIN, Department of Chemistry, University of California, Santa Barbara, Santa Barbara, California 93106 (51)

URSULA EUTENEUER,[2] Department of Molecular, Cellular, and Developmental Biology, University of Colorado, Boulder, Colorado 80309 (207)

KEVIN W. FARRELL, Department of Biological Sciences, University of California, Santa Barbara, Santa Barbara, California 93106 (61)

MARTIN FLAVIN, Laboratory of Cell Biology, National Heart, Lung, and Blood Institute, National Institutes of Health, Bethesda, Maryland 20205 (257)

CLARENCE FRASER, Pacific Biomedical Research Center, University of Hawaii, Honolulu, Hawaii 96822 (373)

KEIGI FUJIWARA, Department of Anatomy, Harvard Medical School, Boston, Massachusetts 02115 (217)

I. R. GIBBONS, Pacific Biomedical Research Center, University of Hawaii, Honolulu, Hawaii 96822 (373)

ROBERT D. GOLDMAN,[1] Department of Biological Sciences, Carnegie-Mellon University, Pittsburgh, Pennsylvania 15213 (399)

MARTIN A. GOROVSKY, Department of Biology, University of Rochester, Rochester, New York 14627 (79)

SUSAN D. GUTTMAN, Department of Pharmacology, Stanford University School of Medicine, Stanford, California 94305 (79)

LEAH T. HAIMO, Department of Biology, University of California, Riverside, Riverside, California 92507 (189)

STEVEN R. HEIDEMANN, Department of Physiology, Michigan State University, East Lansing, Michigan 48824 (207)

TIMOTHY L. KARR, Department of Chemistry, University of California, Santa Barbara, Santa Barbara, California 93106 (51, 133)

TAKAAKI KOBAYASHI, Laboratory of Cell Biology, National Heart, Lung, and Blood Insti-

[1] *Present address:* Department of Cell Biology and Anatomy, Northwestern University Medical School, Chicago, Illinois 60611.

[2] *Present address:* Laboratory of Molecular Biology, University of Wisconsin, Madison, Wisconsin 53706.

tute, National Institutes of Health, Bethesda, Maryland 20205 (257)

DAVID KRISTOFFERSON, Department of Chemistry, University of California, Santa Barbara, Santa Barbara, California 93106 (133)

ELIAS LAZARIDES, Division of Biology, California Institute of Technology, Pasadena, California 91125 (313)

JAMES C. LEE, E. A. Doisy Department of Biochemistry, St. Louis University School of Medicine, St. Louis, Missouri 63104 (9)

RICHARD W. LINCK, Department of Anatomy, Harvard Medical School, Boston, Massachusetts 02115 (217)

THOMAS R. MALEFYT, Department of Chemistry, University of California, Santa Barbara, Santa Barbara, California 93106 (133)

ROBERT L. MARGOLIS, The Fred Hutchinson Cancer Research Center, Seattle, Washington 98104 (145, 159)

TODD M. MARTENSEN, Laboratory of Biochemistry, National Heart, Lung, and Blood Institute, National Institutes of Health, Bethesda, Maryland 20205 (257, 265)

DOUGLAS B. MURPHY, Department of Cell Biology and Anatomy, Johns Hopkins Medical School, Baltimore, Maryland 21205 (31)

J. B. OLMSTED, Department of Biology, University of Rochester, Rochester, New York 14627 (79)

MARY OSBORN, Max Planck Institute for Biophysical Chemistry, Karl Friedrich Bonhoeffer Institute, Göttingen, Federal Republic of Germany (97)

JOEL D. PARDEE, Department of Structural Biology, Stanford University School of Medicine, Stanford, California 94305 (271)

THOMAS D. POLLARD, Department of Cell Biology and Anatomy, Johns Hopkins University School of Medicine, Baltimore, Maryland 21205 (301, 333)

DANIEL L. PURICH, Department of Chemistry, University of California, Santa Barbara, Santa Barbara, California 93106 (51, 133)

WINFIELD S. SALE, Pacific Biomedical Research Center, University of Hawaii, Honolulu, Hawaii 96822 (373)

K. BRADFORD SNYDER, Department of Biological Sciences, University of California, Santa Barbara, Santa Barbara, California 93106 (159)

JAMES A. SPUDICH, Department of Structural Biology, Stanford University School of Medicine, Stanford, California 94305 (271)

PETER STEINERT, Dermatology Branch, National Cancer Institute, National Institutes of Health, Bethesda, Maryland 20205 (399)

WEN-JING Y. TANG, Pacific Biomedical Research Center, University of Hawaii, Honolulu, Hawaii 96822 (373)

WILLIAM C. THOMPSON, Departments of Biological Sciences and Chemistry, University of California, Santa Barbara, Santa Barbara, California 93106 (159, 235)

LIVINGSTON VAN DE WATER III, Center for Cancer Research, Massachusetts Institute of Technology, Cambridge, Massachusetts 02139 (79)

KLAUS WEBER, Max Planck Institute for Biophysical Chemistry, Karl Friedrich Bonhoeffer Institute, Göttingen, Federal Republic of Germany (97)

HILLARY D. WHITE, Department of Pharmacology, University of Washington School of Medicine, Seattle, Washington 98195 (51)

LESLIE WILSON, Department of Biological Sciences, University of California, Santa Barbara, Santa Barbara, California 93106 (159)

ROBERT ZACKROFF,[1] Department of Biological Sciences, Carnegie-Mellon University, Pittsburgh, Pennsylvania 15213 (399)

PREFACE

Knowledge of the organization of the cytoplasm of eukaryotic cells has expanded greatly during the past decade. The early view of the cytoplasm as a structureless "soup" has given way to our present understanding of it as being exquisitely ordered. This order is conferred by a three-dimensional network of fibrous structures, which include microtubules, microfilaments, and intermediate filaments, and the fiber-associated molecules that mediate the interactions and functions of the fibrous elements with each other and with cytoplasmic organelles. The three-dimensional network of fibrous structures has come to be known as the *cytoskeleton,* though the dynamic nature of the network is not captured by the term.

Early research on the cytoskeleton focused on identification of the major constituents of the cytoskeleton and their characterization, in terms of both their organization in cells and tissues and their biochemical properties. More recently, our attention has been turning strongly toward understanding the functions of cytoskeletal elements in living cells, and has focused on investigation of "less visible" cytoskeletal components (which are considered to interact functionally with the surfaces of the major types of filaments) and on the development of methods and model systems for investigating the functional interactions of cytoskeletal components with one another and with other cell components.

This volume consists of 24 chapters concerned with the isolation and characterization of cytoskeletal components, and with the development of research tools to enable the study of cytoskeletal components in living cells and *in vitro* cell models. The methods involved have been described in considerable detail and are often accompanied by a section of overview perspectives that should aid investigators new to this research. The next volume of this publication (Volume 25, The Cytoskeleton, Part B) consists of 18 chapters concerned with cell systems and *in vitro* model systems that are presently or potentially valuable for the elucidation of the functions of the cytoskeleton or its components in living cells.

I wish to thank all of the authors who have so generously contributed to this volume, and I apologize to readers who are searching for a technique that has not been included. I also wish to thank Susan Overton for her substantial help in preparing the Index.

LESLIE WILSON

Chapter 1

The Cytoskeleton: A Perspective

B. R. BRINKLEY

Department of Cell Biology
Baylor College of Medicine
Houston, Texas

This timely two-part volume on the cytoskeleton appears at the end of a productive decade in cell biology, accentuated by enormous progress in understanding the structural order and functional complexities of cytoplasm. The old concept of cytoplasm as an amorphous colloidal mass containing a suspension of organelles and inclusions must now be abandoned in favor of a more dynamic view in which organized arrays of fibrous elements interact to form a highly integrated structural network called the cytoskeleton. The capacity of cytoplasmic proteins such as tubulin, actin, and intermediate filament proteins to form polymers, to depolymerize, to intertwine and interact, and to elongate and contract in conjunction with subtle cell movements and shape changes gives cytoplasm its unique properties of life. Although the term *cytoskeleton* has become widely accepted, it is not totally descriptive. The expression applies to both muscle and nonmuscle cells and encompasses all of the fibrous elements of cytoplasm, including filamentous actin, myosin, intermediate filaments, microtubules, and a myriad of other proteins that anchor, cross-link, bind, or otherwise regulate the fibrous network in the cytoplasm. As the chapters in this volume will attest, however, the cytoskeleton provides a much more extensive function in cells than merely maintaining skeletal support and cytoplasmic consistency. Indeed, it appears to be actively involved in both force production and force transduction in various forms of cell motility, including cytoplasmic

ISBN 0-12-564124-9

streaming, organelle movement, cytokinesis, phagocytosis, secretion, axonal transport, and cell surface modulation. The suggested linkage of cytoskeletal components to transmembrane protein receptors implies a role in peptide hormone action. Components of the cytoskeleton may also be involved in mitogenesis and control of cell proliferation, but here the evidence is too new and incomplete to be fully convincing.

I. Microtubules

Like other developments in the field of cell biology, advances in knowledge of the cytoskeleton were largely made possible by a series of technical achievements dating back over two decades. Although a fibrous network was detected in silver-stained neurons by Cajal in the nineteenth century, we generally credit early electron microscopists with the discovery of microtubules and microfilaments. The vast improvements in fixation afforded by glutaraldehyde (Sabatini *et al.*, 1963) led to the widespread recognition of microtubules in most all eukaryotic cells. Prior to the glutaraldehyde era, discrete "filaments" were recognized in the mitotic spindles of amoebas (Roth and Daniels, 1962), and even earlier, fibrous elements were faintly seen to form a 9 + 2 pattern in dismembered and sectioned cilia (Manton and Clark, 1952; Fawcett and Porter, 1954).

A significant contribution to our knowledge of microtubule substructures came from EM studies in meristematic cells of juniper. As a result of the natural electron opacity afforded by the cell wall material of this plant, Ledbetter and Porter (1964) identified 13 globular subunits in the walls of microtubules, a finding that in recent years has been widely confirmed in many diverse species using tannic acid as a stain (Tilney *et al.*, 1973). Much of our current understanding of how subunits are arranged in the microtubule surface lattice has come from optical diffraction studies of negatively stained flagellar microtubules using optical filtering techniques and computer methods of image analysis (Amos and Klug, 1974; Erickson, 1974; Chasey, 1972). In addition, X-ray diffraction studies of unfixed, hydrated material has largely confirmed early electron microscopic studies (Mandelkow *et al.*, 1977).

In retrospect and considering all of the elegant electron microscopic studies, it is significant to point out that evidence for microtubules in living cells came a full decade before the EM era. By observing dividing marine oocytes through a polarizing microscope equipped with rectified, strain-free optics, Inoué detected weak form birefringence in the mitotic apparatus that was later shown to be due to highly oriented microtubules. Through a series of experimental manipulations Inoué and co-workers (1967, 1975) proposed that soluble subunits were in a dynamic equilibrium with their polymeric forms contained in spindle fibers. Moreover, Inoué proposed that the assembly was entropy driven and that the subunits were maintained in the polymer by weak hydrophobic bonds. This important study has now been confirmed by three decades of research, including *in vitro* assembly experiments using purified microtubule protein. Currently, renewed interest is being placed on studies of the cytoskeleton in living cells using microinjection techniques and videomicroscopy. Thus Inoué's pioneering experiments were well in advance of their time.

Although it is inappropriate to dwell excessively on microtubules in an overview of the cytoskeleton, I would be remiss in ignoring the exciting era of microtubule biochemistry that followed closely, and to some extent paralleled, the morphological studies of the 1960s and 1970s and continues unabated today. I need not elaborate the details here, since much of it will be covered in the chapters that follow. However, it should be pointed out that through the use of [^{3}H]colchicine as a probe and the specificity of binding of this drug to microtubule protein, E. W. Taylor and his students (Shelanski and Taylor, 1967; Weisenberg *et al.*, 1968; Borisy and Taylor, 1967a,b) and Wilson and Friedkin (1967) were able to isolate and characterize a single protein with a sedimentation coefficient of 6 S and a molecular weight of 110,000–120,000. Later it was shown that denaturation of the protein with guanidine hydrochloride produced two similar 55,000 M_r α and β subunits. It was concluded that in aqueous solutions the larger colchicine-binding molecules existed as a dimer that came to be known as tubulin (Mohri, 1968). It was also discovered that α- and β-tubulin subunits existed in a constant 1:1 molar ratio in microtubules forming an $\alpha\beta$ heterodimer (Bryan and Wilson, 1971).

The 1970s saw the rise of tubulin biochemistry as a bona fide discipline with worldwide participation. One of the more significant developments came when Weisenberg (1972) succeeded in attaining microtubule assembly *in vitro*. He demonstrated that microtubules would form spontaneously from supernatants of brain homogenates when the solution was warmed to 37°C in the presence of GTP and magnesium. The key to his success was the addition of the calcium chelating agent EGTA to the reassembly mixture. He concluded that free calcium concentrations as low as 6 μM could inhibit the *in vitro* assembly of microtubules, an observation that may have physiological relevance concerning cellular control of microtubule assembly. For assembly to occur, a "critical concentration" of tubulin (0.2 mg/ml) was essential. This concentration is the lowest level at which cooperative association of subunits occurs forming structures that nucleate the assembly of microtubules. Thus assembly appears to be a two-step condensation-polymerization process in which nuclei (seeds) form and then elongate. At steady state polymerized tubulin is in equilibrium with a critical concentration of soluble tubulin. The *in vitro* assembly procedure of Weisenberg as utilized and modified by many other laboratories has contributed significantly toward an understanding of the biology and biochemistry of microtubules in eukaryotic cells. It has also provided a convenient means of purifying tubulin from a variety of tissues.

The availability of highly purified cytoskeletal proteins has enabled numerous investigators to produce monospecific antibodies to these proteins and to use them as immunofluorescent and immunoelectron microscopic probes in cells. The procedure of indirect immunofluorescence originally introduced by Coons and co-workers (1941) has had a popular rebirth in cytoskeleton research. As will be discussed in the chapters by Lazarides and by Osborn and Weber, this method has had a major impact on our understanding of the holistic organization of cytoskeletal elements in cells. For example, through the use of tubulin antibodies and indirect immunofluorescence, two microtubule arrays have been identified in proliferating cells: the mitotic spindle and a delicate lacework of tubules in interphase cells called the cytoplasmic microtubule complex or CMTC (Brinkley *et al.*, 1975). The microtubules of the spindle and CMTC are morphologically and immunocytochemically similar and are both organized around discrete microtubule organizing centers. The two arrays differ, however, in overall morphology, numbers and length of microtubules, response to drugs and physical agents, time of appearance in the cell cycle, and, of course, function. These facts raise a number of new questions concerning regulation. Are there multiple populations of tubulin and microtubules in cells? How is the length, number, and distribution of microtubules maintained in the cytoplasm? How are the temporal events of microtubule assembly–disassembly regulated? How is the polarity of microtubules determined? What is the function(s) of the CMTC and how does it relate to the organization of microfilaments and other cytoskeletal components? These are questions for investigators to ponder during the next decade, but already significant progress has been made. *In vitro* assays for the analysis of MTOCs have been developed and progress has been made in the biochemical characterization of these components. Newer techniques have enabled investigators to determine the polarity and directionality of assembly of microtubules in the spindle and CMTC. Calcium, along with the calcium binding protein calmodulin, has been implicated in the control of microtubule assembly and disassembly (Marcum *et al.*, 1978). Recently, several laboratories have reported the cloning of tubulin genes, and the amino acid sequence of at least one α and β tubulin has been determined (Valenzuela *et al.*, 1980).

After three decades, microtubule research is still booming, but surprisingly, much remains to be learned, not the least of which is how microtubules actually function to achieve any one of their numerous roles in cells.

II. Microfilaments

One of the most significant episodes in modern cell biology began with the recognition of major musclelike proteins in the cytoplasm of nonmuscle cells—

even plant cells! Clearly, this area of cytoskeletal research is still enjoying exponential growth and much progress has been due to the availability of familiar techniques derived from an earlier decade of muscle biochemistry and physiology. Initially, electron microscopists recognized microfilaments as a somewhat heterogenous class of thick (>10 nm) and thin (4–6 nm) fibrous elements displaying various levels of organization and association in the cytoplasm. Sorting out the microfibrils into functional groups became an impossibility because of an identity problem until Ishikawa *et al.* (1969) adapted the heavy meromyosin labeling procedure of Huxley for "decorating" actin microfibrils in glycerol-extracted cells. Through this procedure actin filaments formed "arrowheads" displaying a pointed end and a barbed end. In every case, the thin 4–6 nm filaments were decorated with the heavy meromyosin or S_1. This procedure not only has aided in the identification of a specific class of microfibrils, but has provided valuable information on the polarity of F-actin. For example, actin filaments always decorate with their barbed ends facing the plasma membrane.

Through the use of a combination of techniques, including biochemistry, electron microscopy, and immunocytochemistry, much progress has been made in understanding the organization and function of actin in nonmuscle cells. Muscle and nonmuscle actin are very similar, as indicated by the fact that they differ in only 6% of their amino acid residues. Moreover, antibodies raised against vertebrate skeletal muscle actin cross-react with most forms of nonmuscle actin. On closer inspection, however, the nonmuscle (β- and α-isoactins) proteins are seen to differ significantly from muscle (α-actin) proteins in their isoelectric points. Mammalian α-actin displays 25 sequence differences from *Physarum* actin. Actins from vertebrate brain, platelet, and *Acanthamoeba* and *Physarum* have threonine at position 129, whereas heart and skeletal muscle actins substitute the hydrophobic residue valine (Geisow, 1979). Recent studies also suggest that the eukaryotic genome may contain multiple genes for actin—a surprising finding that may also apply to the tubulins. Obviously, recombinant DNA technology will provide much-needed information about the molecular structure, genetics, and evolution of cytoskeletal proteins.

Although much has been learned about the structure, biochemistry, and localization of actin filaments, little is yet known about how they are organized into functional units in the cytoplasm of nonmuscle cells. The functional analogy with the sarcomere and contraction in skeletal muscle may be an oversimplification. A survey of motility in nonmuscle cells, including amoeboid movement, chromosome movement, endocytosis, exocytosis, cleavage, and cell surface receptor mobility, suggests that a variety of functional associations have evolved. Although the characters of the play have been identified for the most part, the plot is yet to be revealed. For example, force production ranges from the explosive polymerization of G-actin in the acrosome reaction of echinoderms (Tilney *et al.*, 1973) to an apparent sliding filament mechanism in microfibril bundles such as that seen in microvilli, the contractile ring, and stress fibers. The latter have received much attention and through immunofluorescence studies appear to contain F-actin, myosin, α-actinin, tropomyosin, filamin, and calmodulin. Unlike the sarcomere, microfilament bundles are frequently labile and may be maintained by a steady-state polymerization of G-actin. Although ionic conditions in the cytoplasm favor the assembly of actin, much of the actin exists in nonfilamentous (G-actin) form. In nonmuscle cells, a protein called profilin binds to G-actin and prevents its nucleation into F-actin. Interesting differences also exist in the mode of calcium-activated contraction. In the sarcomere actin binds to troponin C causing a conformation change in the troponin–tropomyosin complex that enables actin to bind to myosin, thereby inducing myosin ATPase activation and contraction. In nonmuscle actin, activation of myosin ATPase apparently requires the phosphorylation of myosin light chain by a calmodulin-activated light chain kinase. Thus in muscle, calcium binding releases an inhibition; in nonmuscle cells calcium binding to calmodulin activates myosin light chain kinase that activates myosin.

Through studies of actin gels *in vitro,* progress has been made in identifying a variety of proteins that bind to actin and probably regulate the various structural and functional levels of organization of microfibrils seen in the cytoskeleton (Bryan, in Volume 25). Such *in vivo* models are invaluable in determining the function of actin microfibrils *in vitro*.

As yet, surprisingly little is known about the interaction of actin microfibrils with other cytoskeletal components and with various cell components such as membrane proteins. The viable proposal that actin may bind to or interact with transmembrane proteins and control their movements on the cell surface seems highly probable but is yet to be convincingly proved. The finding that both actin and tubulin co-cap with immunoglobulin in membranes of mouse B lymphocytes (Gabbiani *et al.*, 1977) and the strong association of actin with the histocompatability H-2 antigen in spontaneously shed plasma membrane (Koch and Smith, 1978) are examples of experiments that support the notion that the cytoskeleton interacts with membrane proteins to regulate their movement and perhaps controls transmembrane regulation of events in the cytoplasm and nucleus.

III. Intermediate Filaments

Intermediate filaments constitute another ubiquitous population of microfibrils in the cytoplasm. These 10-nm filaments were originally thought to be disaggregated forms of either microtubules or myosin, but more recent biochemical and immunocytochemical studies indicate that they are a distinct class of heterogenous proteins. Intermediate filaments may be classified into five major groups: (1)

desmin filaments, found predominantly in skeletal, cardiac, and smooth muscle; (2) keratin filaments (tonofilaments), found largely in epithelial cells; (3) vimentin filaments, found in mesenchymal cells; (4) neurofilaments, found in neurons; and (5) glial filaments, found exclusively in glial cells. Such classification is useful in terms of identifying specific proteins associated with these filaments. It is known, however, that more than one class of intermediate filaments may be found in a single cell, and more than one protein may constitute a filament. At this juncture, less is known about the subunit structure, assembly, and distribution of intermediate filaments than about any other of the microfibrils of the cytoskeleton. They exist in both muscle and nonmuscle cells and seem to have a major function as mechanical integrators of the other cytoskeletal elements (Lazarides, 1980). Other roles are likely to be found with additional research. All intermediate filament proteins identified so far can be phosphorylated *in vitro* and may be regulated *in vivo* by this mechanism. Little is known about their assembly *in vivo* and as yet no specific intermediate filament initiation sites or organizing center has been found. Their involvement with other cytoskeletal proteins is not well understood. Their distribution in cells as shown by indirect immunofluorescence is greatly altered by agents that disrupt cytoplasmic microtubules. This fact, along with the finding by Goldman and co-workers (1980) that intermediate filaments associate with centrioles, suggests a very close association of intermediate filaments with microtubules and tubulin initiation sites. Perhaps one of the most important tasks in cytoskeletal research will be to determine how intermediate filaments, microtubules, actin microfibrils, myosin, and various regulatory components interact and are modulated as an integrated unit in cells. Much emphasis has been placed on the components of the cytoskeleton as individual units and little is known about how overall coordination is carried out.

In conclusion, much has been learned about the filaments and tubules in the cytoskeleton. Have we now identified all of the major structural components of the cytoskeleton? Analysis of whole-mount and sectioned cells by high-voltage electron microscopy indicates a structural entity in cytoplasmic ground substance that Porter and co-workers (Wolosewick and Porter, 1979) call the microtrabecular lattice. Utilizing several fixation schedules and following vigorous protocols that minimize structural artifact, these investigators have described an ordered lattice composed of slender strands or microtrabeculae that interconnect membranes, polysomes, cytoskeletal elements, and other cell components. Whether or not the microtrabecular lattice constitutes a separate set of cytoplasmic proteins and structural components, is an artifact, or merely represents extensions of the familiar cytoskeletal elements remains to be determined. At the present it is not unreasonable to assume that new, uncharted dimensions of cytoplasm exist and that even more complicated levels of molecular organization will be found in cells. The rapid development of new technology for cell research and its application to studies of the cytoskeletal structure should make the next decade a decisive interval in defining the structure of cytoplasm.

References

Amos, L. A., and Klug, A. (1974). *J. Cell Sci.* **14,** 523–549.

Borisy, G. G., and Taylor, E. W. (1967a). *J. Cell Biol.* **34,** 525–533.

Borisy, G. G., and Taylor, E. W. (1967b). *J. Cell Biol.* **34,** 535–548.

Brinkley, B. R., Fuller, G. M., and Highfield, D. P. (1975). *Proc. Natl. Acad. Sci. U.S.A.* **72,** 4981–4985.

Bryan, J., and Wilson, L. (1971). *Proc. Natl. Acad. Sci. U.S.A.* **68,** 1762–1766.

Chasey, D. (1972). *Exp. Cell Res.* **74,** 140–146.

Coons, A. H., Creech, H. J., and Jones, R. N. (1941). *Proc. Soc. Exp. Biol. Med.* **47,** 200.

Erickson, H. P. (1974). *J. Cell Biol.* **60,** 153–167.

Fawcett, D. W., and Porter, K. R. (1954). *J. Morphol* **94,** 221–281.

Gabbiani, G. (1977). *Nature (London)* **269,** 697.

Geisow, N. (1979). *Nature (London)* **278,** 507–508.

Goldman, R. D., Hill, B. F., Steinert, P., Whitman, M. A., and Zackroff, R. B. (1980). *In* "Microtubules and Microtubule Inhibitors" (M. De Brabander and J. De Mey, eds.), pp. 91–102. Elsevier/North-Holland, Amsterdam.

Inoué, S., and Ritter, H. (1975). *In* "Molecules and Cell Movements" (S. Inoué and R. Stephens, eds.), pp. 3–30. Raven, New York.

Inoué, S., and Sato, H. (1967). *J. Gen. Physiol.* **50,** 259–292.

Ishikawa, H., Bischoff, R., and Holtzer, H. (1969). *J. Cell Biol.* **43,** 312–328.

Koch, G. L. E., and Smith, M. J. (1978). *Nature (London)* **273,** 274.

Lazarides, E. (1980). *Nature (London)* **283,** 249–256.

Ledbetter, M. C., and Porter, K. R. (1964). *Science* **144,** 872–874.

Mandelkow, E., Thomas, E., and Bensch, K. G. (1977). *Proc. Natl. Acad. Sci. U.S.A.* **74,** 3370–3374.

Manton, I., and Clarke, R. (1952). *J. Exp. Bot.* **3,** 265–275.

Marcum, J. M., Dedman, J. R., Brinkley, B. R., and Means, A. R. (1978). *Proc. Natl. Acad. Sci. U.S.A.* **75,** 3771–3775.

Mohri, H. (1968). *Nature (London)* **217,** 1053–1054.

Roth, L. E., and Daniels, E. W. (1962). *J. Cell Biol.* **12,** 57–78.

Sabatini, D. D., Bensch, K., and Barrnett, R. J. (1963). *J. Cell Biol.* **17,** 19–58.

Shelanski, M. L., and Taylor, E. W. (1967). *J. Cell Biol.* **38,** 304–315.

Tilney, L. G., Bryan, J., Bush, D. J., Fujiwara, K., and Mooseker, M. (1973). *J. Cell Biol.* **59,** 109–126.

Valenzuela, P., Zaldivar, J., Quiroga, M., Rutter, W., Cleveland, D., and Kirschner, M. (1980). *Eur. J. Cell Biol.* **22,** 14. (Abstr.)

Weisenberg, R. C. (1972). *Science* **177,** 1104–1105.

Weisenberg, R. C., Borisy, G. G., and Taylor, E. W. (1968). *Biochemistry* **7,** 4466–4467.

Wilson, L., and Friedkin, M. (1967). *Biochemistry* **6,** 3126–3135.

Wolosewick, J. J., and Porter, K. R. (1979). *J. Cell Biol.* **82,** 114–139.

Chapter 2

Purification and Chemical Properties of Brain Tubulin

JAMES C. LEE

E. A. Doisy Department of Biochemistry
St. Louis University School of Medicine
St. Louis, Missouri

I. Introduction

Microtubules are found in a wide variety of animal and plant cells. In dividing cells they are localized in the mitotic apparatus, and they are also found in axonal

ISBN 0-12-564124-9

and dendritic processions of neurons. They have been implicated in a number of different cellular functions. Structurally, microtubules are composed of protofilaments of the subunit protein, tubulin. Because of the wide distribution of this organelle and the large number of different cellular processes in which it seems to participate, microtubule has been the target of intensive investigation. The initial purification of tubulin from porcine and calf brains by Weisenberg and coworkers (Weisenberg *et al.*, 1968; Weisenberg and Timasheff, 1970) has stimulated many biochemical studies on the physical and chemical properties of the protein, hoping to establish a structure–function relationship.

The purpose of this chapter is to present the modified procedure involved in purifying brain tubulin developed originally by Weisenberg and to discuss the present information on the structure of tubulin. The techniques described will frequently be those adopted in this laboratory and with which the author is most familiar. Many excellent textbooks and articles on these methods are available (e.g., "Methods in Enzymology"), so the experimental details of these techniques will not be discussed in this chapter, the emphasis of which is a survey of a variety of techniques successfully employed in elucidating various characteristics of tubulin.

II. Purification of Brain Tubulin

The isolation procedure is primarily developed for brain tubulin, specifically calf brains, although the same procedure has been successfully applied to lamb, dog, and porcine brains. All operations are performed in a 4°C cold room. Approximately 4 kg of brain tissues is employed as the starting material.

A. Homogenization

Fresh brains are obtained within 1 hr of slaughter and placed in ice during transportation. The superficial blood vessels and meninges are removed with forceps and the tissue is minced and washed three times by suspension in 2–3 vol of 0.24 *M* sucrose in 10^{-2} *M* sodium phosphate, 5×10^{-4} *M* $MgCl_2$ buffer at pH 7.0. The wash is decanted by straining through cheesecloth. The mince is resuspended in 1.5–2 vol of the same buffer and homogenized in a domestic blender at the maximum setting for 30 sec. It is important not to denature the protein by excessive blending, as might occur with use of some of the more powerful laboratory blenders.

The cell debris and particulate components are removed by centrifuging the homogenate in a Sorvall GSA rotor at an average of 13,000 *g* for 30 min. The pellet is discarded.

B. Ammonium Sulfate Fractionation

The supernatant is brought to 32% saturation by the addition of 177 g/liter of solid ultrapure ammonium sulfate within an interval of approximately 5 min. An additional 10–15 min is allowed for total dissolution of the ammonium sulfate and precipitation of proteins. The precipitate is removed by centrifugation in a Sorvall GSA rotor at an average of 13,000 *g* for 30 min. The pellet is discarded.

The supernatant is brought to 43% saturation by dissolving an additional 71 g/liter of $(NH_4)_2SO_4$. The precipitate is again removed as mentioned. The supernatant is discarded.

C. DEAE-Sephadex Batch Elution

The 43% $(NH_4)_2SO_4$ precipitate is suspended in 150 ml of buffer consisting of 10^{-2} *M* sodium phosphate, 10^{-4} *M* GTP, 5×10^{-4} *M* $MgCl_2$ (PMG) at pH 7.0. Resuspension is facilitated by employing a 55 ml Potter-Elvehjem tissue grinder with Teflon pestle. One should avoid using a ground glass homogenizer and should exercise caution so as not to denature the protein as indicated by foaming.

The solution is mixed with 400 ml of gravity-packed DEAE-Sephadex (A-50) that has been washed thoroughly and equilibrated in PMG buffer. Allowing 10 min for protein adsorption, the slurry is then distributed into 50-ml centrifuge tubes. Usually it will require 16 tubes. The Sephadex beads are then pelleted by centrifugation using a Sorvall SS-34 rotor and by bringing the rotor up to 5000 rpm before stopping it. The supernatant is discarded by pouring. Enough 0.4 *M* KCl in PMG buffer is added to fill each tube and the Sephadex pellet is resuspended. The tubes are again centrifuged momentarily at 5000 rpm and the supernatant is discarded by pouring. The washing of Sephadex pellets with 0.4 *M* KCl is repeated.

Tubulin is then eluted from the resin by repeating the washing procedure twice with 0.8 *M* KCl in PMG buffer. In order to keep the elution volume to the minimum, the amount of 0.8 *M* KCl employed should be no more than twice that of the Sephadex pellet.

No precaution is necessary in preventing some Sephadex beads from being poured into the 0.8 *M* KCl eluant. They can be removed by passing the eluant over some glass wool. It is essential that at least 90% of the Sephadex beads are removed at this stage; otherwise they will clog the Sephadex G-25 column.

The eluant should be slightly turbid and the proteins are precipitated by adding 24.8 g/100 ml of $(NH_4)_2SO_4$. Again, approximately 10 min is allowed for equilibration after complete dissolution of the salt. The protein precipitate is collected by centrifugation at an average of 27,000 *g* using a Sorvall SS-34 rotor for 30 min.

The protein pellet is redissolved in 15 ml of PMG buffer with the aid of a

55-ml Potter-Elvehjem tissue grinder with Teflon pestle. The same precautions as mentioned should be exercised. The protein solution should be opaque and the total volume is approximately 50 ml.

D. Sephadex G-25 Chromatography

The protein solution is applied to either two columns (2 × 35 cm) or one column of equivalent-size of Sephadex G-25 medium. The column is eluted with PMG buffer at a rate of 1.5 to 2.0 ml/min to remove the ammonium sulfate completely. Two-milliliter fractions are collected. The elution of tubulin can easily be monitored by visual inspection of these fractions. Tubulin solutions are turbid.

E. $MgCl_2$ Precipitation

To each fraction that is turbid 2 drops of 1.0 *M* $MgCl_2$ in H_2O are added. If the Sephadex column is successful in removing the $(NH_4)_2SO_4$, the addition of $MgCl_2$ should precipitate tubulin in the fraction, as indicated by the solution becoming extremely turbid with a milky appearance immediately after the addition of $MgCl_2$. A slow precipitation usually indicates presence of $(NH_4)_2SO_4$. Fractions that show precipitation with $MgCl_2$ can then be pooled and the precipitate can be collected by centrifugation in a Sorvall centrifuge at 9000 *g* for 5 min. Subjecting the precipitate to excessive centrifugation is to be avoided.

The pellet is dissolved in approximately 5 ml of 1 *M* sucrose in PMG buffer with the aid of a 10-ml Potter-Elvehjem tissue grinder with Teflon pestle, preferably with grooves. Gentle handling of the pellet is absolutely essential at this stage. Short and slow strokes of the pestle are recommended so that the tubulin pellet will not form a paste between the bottom of the grinder and the pestle. Having resuspended the pellet, the purified tubulin solution is dialyzed overnight against 250 ml of 1 *M* sucrose in PMG buffer.

F. Storage

The protein concentration of the solution is determined and adjusted to 50–60 mg/ml and rapidly frozen in small aliquots and stored in liquid nitrogen. Under these conditions the protein can be stored for at least 14 days without observable loss of colchicine binding activity or its ability to undergo *in vitro* microtubule assembly. Generally about 500 mg of highly purified tubulin can be obtained by this procedure.

Figure 1 presents the results of polyacrylamide gel electrophoresis experiment in the presence of 0.1% sodium dodecyl sulfate–6 *M* urea showing the purity of tubulin at various stages of the purification procedure. It is apparent that the

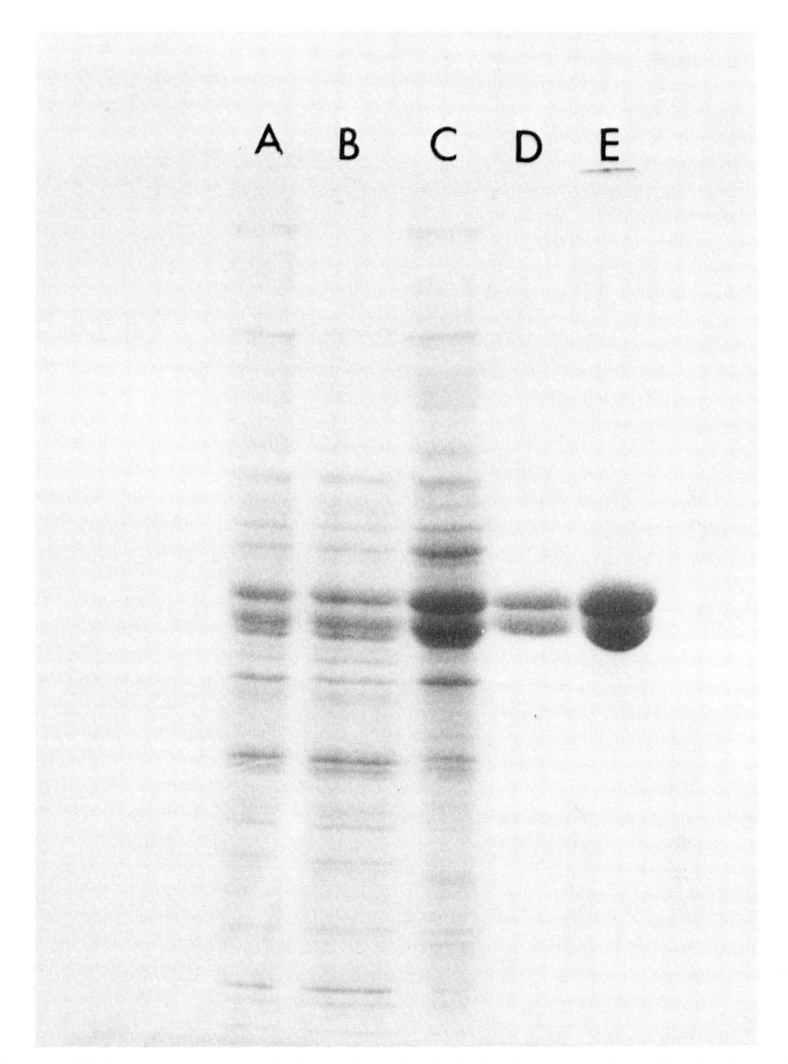

FIG. 1. SDS–polyacrylamide gel electrophoresis of tubulin fractions collected during the purification procedure. The experiment was performed according to Laemmli (1970) on a slab gel with a linear gradient of 4–15% acrylamide. Proteins were stained with 0.1% (w/v) Coomassie Brilliant Blue R. The identity of the sample and amount of protein loaded are (A) brain homogenate, 95 μg; (B) 32% $(NH_4)_2SO_4$ precipitate, 90 μg; (C) 43% $(NH_4)_2SO_4$ precipitate, 110 μg; (D) 0.8 *M* KCl eluent of DEAE-Sephadex, 36 μg; (E) $MgCl_2$ precipitate, 21 μg.

present procedure would yield protein samples of ~95% homogeneous tubulin from brain tissues.

III. Chemical Properties of Brain Tubulin

A. Subunit Molecular Weight

The subunit molecular weight of brain tubulin has been determined by a variety of techniques. Weisenberg and co-workers (Weisenberg *et al.*, 1968) reported a molecular weight of 57,000 in 6 *M* guanidine hydrochloride and 0.12 *M* mercaptoethanol by employing sedimentation equilibrium using a value of 0.73 for partial specific volume $\bar{v}$. An average value of 54,000 ± 1000 was reported for calf brain tubulin as a result of measurements in the presence of denaturants by various methods including sedimentation equilibrium and light scattering in 6 *M* guanidine hydrochloride, sodium dodecyl sulfate gel electrophoresis, chromatography on an agarose column in the presence of 6 *M* guanidine hydrochloride, and viscometry of *S*-carboxymethylated, reduced protein in 6 *M* guanidine hydrochloride (Lee *et al.*, 1973). The difference in molecular weight reported for porcine and calf brain tubulin leads to the following questions: Is the difference in the reported molecular weight significant and does it reflect a difference in the protein from different sources? The reported difference, however, may also be due to not taking into account the preferential solvent interaction between protein and guanidine hydrochloride in the calculation of molecular weight employing the technique of sedimentation equilibrium. The importance of preferential solvent interaction, specifically between protein and guanidine hydrochloride, has been discussed by Tanford (1968, 1970) and Lee and Timasheff (1974). Such interaction is reflected in the value of $\bar{v}$. Let us examine the effect of $\bar{v}$ on the estimation of subunit molecular weight of tubulin in guanidine hydrochloride. If the experimental value of 0.725 (Lee and Timasheff, 1974) is employed for $\bar{v}$ instead of the assumed value of 0.73 (Weisenberg *et al.*, 1968), the corrected value of 54,700 is obtained for tubulin from porcine brain. Such a value is in good agreement with that reported for calf brain (Lee *et al.*, 1973). Therefore, the reported difference most likely is due to the omission in considering protein–solvent interaction. Rat brain tubulin has also been reported to have a molecular weight of 56,000 ± 2000 based on sodium dodecyl sulfate–polyacrylamide gel electrophoresis (Eipper, 1974). It may, therefore, be concluded that within experimental uncertainties tubulin samples from calf, porcine, and rat brains have identical or very similar size.

B. Heterogeneity of Subunits

Although the results from SDS–gel electrophoresis using high ionic strength, neutral pH conditions (Weber and Osborn, 1969) and other chromatographic or hydrodynamic techniques indicate the presence of a single component for tubulin subunits, incorporation of 8 *M* urea into the gel system enables a resolution of a single protein into two of approximately equal intensity (Feit *et al.*, 1971; Fine, 1971; Lee *et al.*, 1973). It was concluded that tubulin consists of two nonidentical subunits that differ in charge, molecular weight, or ability to interact with urea; it may also reflect protein–urea interactions of the nature described by Cann (1970). The resolution of the bands and their relative intensities were found to be independent of loading concentration, indicating that an effect of the kind described by Cann is not likely. The major difference is apparently not in molecular weight, since only a single, sharp band is present in SDS–gel electrophoresis. More recently, however, the use of different SDS–polyacrylamide gel electrophoresis systems has resolved tubulin into two bands in the presence of SDS (Forgue and Dahl, 1979; Nelles and Bamburg, 1979; Dahl and Weibel, 1979), suggesting that the subunits may also differ in molecular size. Bryan and co-workers (Wilson and Bryan, 1974) found that the mobility of the α subunit exhibits both pH and ionic-strength-dependent changes, whereas the β subunit and some standard protein markers do not exhibit such property. The behavior of the α subunit may be the consequence of differential SDS binding or residual electrostatic interactions within the SDS–protein complex. At present, however, there is no definitive report on the molecular weight of the α subunit.

In connection with the question of heterogeneity of tubulin one final comment seems desirable. Molecular-weight measurements of tubulin by gel electrophoresis have been reported using 8 *M* urea–0.1% SDS as the denaturing agent (Feit *et al.*, 1971; Fine, 1971). The basic principle of molecular-weight measurements by SDS-polyacrylamide gel electrophoresis or gel chromatography is that the proteins must all assume an essentially hydrodynamically identical structure. Tanford and co-workers (Reynolds and Tanford, 1970a,b; Fish *et al.*, 1969, 1970) have shown this to be true for SDS–gel electrophoresis, a conclusion recently confirmed and subjected to further quantitative analysis by Mattice *et al.* (1976). A mixture of 8 *M* urea and 0.1% SDS consists of two denaturants that favor different conformations. It has been shown recently that calf brain tubulin in such a medium assumes an intermediate structure determined by circular dichroism (Lee *et al.*, 1978b). It seems unlikely that in such a medium all proteins would assume the same intermediate structure. Hence the rate of migration of proteins in urea–SDS gels must not be taken as a measure of molecular weight without confirmation by an independent rigorous technique, such as sedimentation equilibrium or osmotic pressure.

The α and β subunits of brain tubulin may be further resolved into more components by isoelectric focusing (Feit *et al.*, 1971), as shown in Fig. 2. More recently, Gozes and Littauer (1978) reported that at least nine bands can be resolved from purified cytoplasmic tubulin from rat brain. Furthermore, the reported microheterogeneity apparently increases with brain maturation. The generation of these bands may be a consequence of protein–ampholyte interaction described by Cann and co-workers (Hare *et al.*, 1978; Cann and Stimpson, 1977; Stimpson and Cann, 1977; Cann *et al.* 1978). Cognizant of these potential pitfalls, George *et al.* (1981) recently subjected calf brain tubulin to an extensive series of tests of isoelectric focusing and tryptic peptide map analysis. Results from isoelectric focusing experiments showed a total number of 17 well-resolved protein peaks. The number of peaks and the mass distribution under each peak remained the same when the concentration of protein or ampholyte was altered. When the protein was subjected to two-dimensional isoelectric focusing, a diagonal pattern was observed, indicating that the multiple peaks observed are

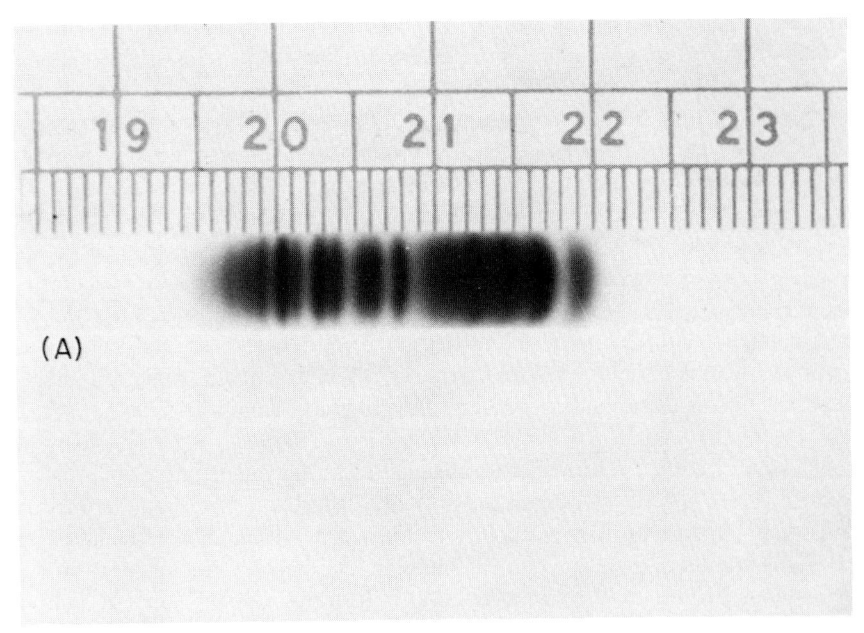

FIG. 2. Isoelectric focusing of purified tubulin. The experiment was performed according to O'Farrel (1975) in 9.2 *M* urea, 5% (v/v) mercaptoethanol, 2% (v/v) Non-Idet P-40, 2% ampholines (LKB) that consisted of 1.2% and 0.8% of pH range 4–6 and 5–7, respectively. Seventy-five μg of carboxymethylated tubulin at 3 mg/ml was applied. On termination of the run, the protein was precipi-

not a manifestation of tubulin–ampholyte interaction. Further investigation by isolating these individual subspecies and subjecting them to isoelectric focusing yielded single peaks corresponding to the original ones without generating the initial pattern of multiple peaks. Tryptic peptide maps showed that among the subspecies of the α subunit there are 26 spots that are common in each subspecies. There are, however, 7 ± 1 spots that are unique in each subspecies. Similar observations were obtained for the subspecies of the β subunit, although there are only 2 ± 1 unique spots in each subspecies. These results suggest that tubulin subunits probably consist of polypeptides with both constant and variable regions in their sequences.

Identical results were obtained for canine and rabbit brain tubulin, indicating that tubulin polymorphism is common among brain tissues. Tubulin isolated by either the polymerization–depolymerization or the modified Weisenberg procedures yielded identical results. These results show that the same subspecies of tubulin are extracted by both isolation procedures. These conclusions are consistent with the recent reports of resolution of a second β chain in pig brain tubulin

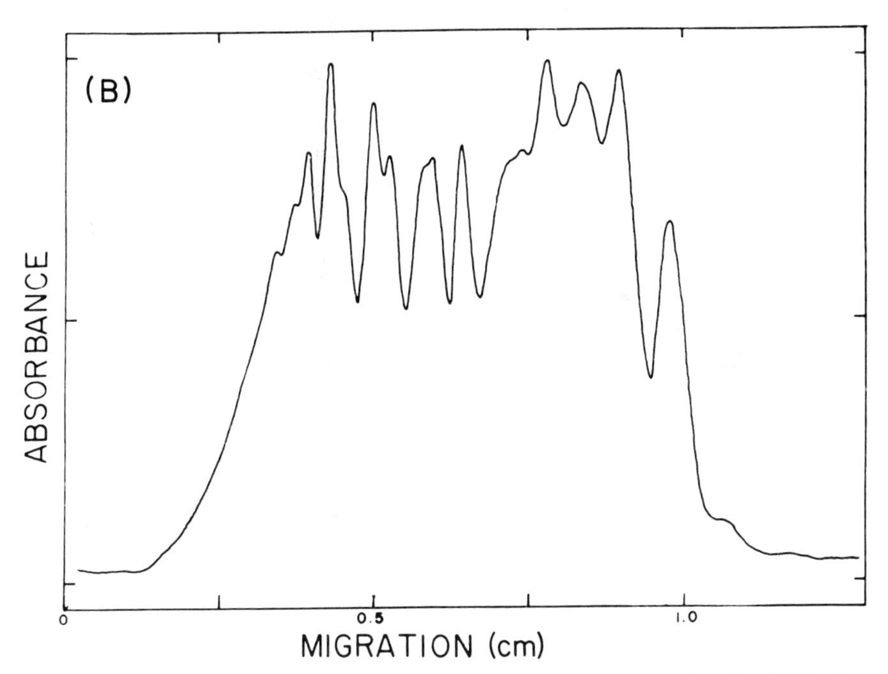

tated with 5% (w/v) trichloroacetic acid, stained with 0.1% (w/v) Coomassie Brilliant Blue R (A). The densitometric trace of the destained gel was obtained by scanning at 560 nm with a Gilford 250 spectrophotometer (B).

by hydroxylapatite column chromatography (Little, 1979), *in vitro* translation of rat brain mRNA (Gozes *et al.*, 1980), and the results on amino acid sequence analysis from porcine brain by Ponstingl *et al.* (1981). However, the identity and significance of these bands remain to be determined.

C. Primary Structure

The existence of α and β subunits is definitely established by the separation of these subunits, subjecting these polypeptide chains to peptide mapping and sequence analysis. Having separated α and β chains by preparative polyacrylamide gel electrophoresis, Luduena and Woodward (1973) characterized the separated subunits by cyanogen bromide cleavage. The cyanogen bromide peptide maps are distinctly different between α and β chains isolated from chick embryo brains.

The carboxy terminal sequence of calf brain chain has been determined by Lu and Elzinga (1977). The segment of 25 amino acids determined is highly acidic, containing 11 glutamic residues. Such a sequence is identical to that of porcine brain (Ponstingl *et al.*, 1979). The entire segment is devoid of proline and glutamine. It is one of the most acidic sequences known.

Ponstingl and co-workers have recently determined the complete sequence of the α subunit of porcine brain tubulin (Ponstingl *et al.*, 1981) as shown in Fig. 3. These data indicate heterogeneity in the sequence. There are apparently positions that are variable with more than one amino acid occupying the same position.

The α and β tubulin subunits of chick brain have sequences very similar to that of the porcine (Valenzuela *et al.*, 1981). This observation is consistent with the report by George *et al.* (1981) that tubulin subspecies with the same isoelectric points are present in brain tubulin isolated from calf, canine, and rabbit.

Lee *et al.* (1973) reported the presence of one intrachain disulfide bond in calf brain tubulin. Such a conclusion is based on the difference between the total number of half cystine determined by amino acid composition and the number of free sulfhydryl residues determined by titration with sulfhydryl specific reagents. It was further substantiated by intrinsic viscosity measurements. The presence of intrachain disulfide linkage is also inferred from the study of changes in electrophoretic mobility in SDS–polyacrylamide gel with and without reducing the disulfide bond (Lee *et al.*, 1975a). Furthermore, the results from circular dichroic studies also indicate the presence of such a linkage (see Section II,D). Eipper (1974), however, reported that rat brain protein contains no disulfide bridges at all. This difference in chemical properties, although surprising, may be a reflection of species specificity of tubulin. It may, nevertheless, be due also to a consequence of the different conditions employed in purifying the proteins (Ikeda and Steiner, 1978). Work is required to resolve the differences and also to locate the disulfide bond in these tubulin subunits.

```
                                                                                                   25
MET-ARG-GLU-CYS-ILE-SER-ILE-HIS-VAL-GLY-GLN-ALA-GLY-VAL-GLN-ILE-GLY-ASN-ALA-CYS-TRP-GLU-LEU-TYR-CYS-
                                                                                                   50
LEU-GLU-HIS-GLY-ILE-GLN-PRO-ASP-GLY-GLN-MET-PRO-SER-ASP-LYS-THR-ILE-GLY-GLY-GLY-ASP-ASP-SER-PHE-ASN-
                                                                                                   75
THR-PHE-PHE-SER-GLU-THR-GLY-ALA-GLY-LYS-HIS-VAL-PRO-ARG-ALA-VAL-PHE-VAL-ASP-LEU-GLU-PRO-THR-VAL-ILE-
                                                                                                  100
ASP-GLU-VAL-ARG-THR-GLY-THR-TYR-ARG-GLN-LEU-PHE-HIS-PRO-GLU-GLN-LEU-ILE-THR-GLY-LYS-GLU-ASP-ALA-ALA-
                                                                                                  125
ASN-ASN-TYR-ALA-ARG-GLY-HIS-TYR-THR-ILE-GLY-LYS-GLU-ILE-ILE-ASP-LEU-VAL-LEU-ASP-ARG-ILE-ARG-LYS-LEU-
                                                                                                  150
ALA-ASP-GLN-CYS-THR-GLY-LEU-GLN-GLY-PHE-SER-VAL-PHE-HIS-SER-PHE-GLY-GLY-GLY-THR-GLY-SER-GLY-PHE-THR-
                                                                                                  175
SER-LEU-LEU-MET-GLU-ARG-LEU-SER-VAL-ASP-TYR-GLY-LYS-LYS-SER-LYS-LEU-GLU-PHE-SER-ILE-TYR-PRO-ALA-PRO-
                                                                                                  200
GLN-VAL-SER-THR-ALA-VAL-VAL-GLU-PRO-TYR-ASN-SER-ILE-LEU-THR-THR-HIS-THR-THR-LEU-GLU-HIS-SER-ASP-CYS-
                                                                                                  225
ALA-PHE-MET-VAL-ASP-ASN-GLU-ALA-ILE-TYR-ASP-ILE-CYS-ARG-ARG-ASN-LEU-ASP-ILE-GLU-ARG-PRO-THR-TYR-THR-
                                                                                                  250
ASN-LEU-ASN-ARG-LEU-ILE-GLY-GLN-ILE-VAL-SER-SER-ILE-THR-ALA-SER-LEU-ARG-PHE-ASP-GLY-ALA-LEU-ASN-VAL-
                                                        ILE                                       275
ASP-LEU-THR-GLU-PHE-GLN-THR-ASN-LEU-VAL-PRO-TYR-PRO-ARG-ALA-HIS-PHE-PRO-LEU-ALA-THR-TYR-ALA-PRO-VAL-
                                                        GLY ILE                 ARG PHE ASX
                                                                                                  300
ILE-SER-ALA-GLU-LYS-ALA-TYR-HIS-GLU-GLN-LEU-SER-VAL-ALA-GLU-ILE-THR-ASN-ALA-CYS-PHE-GLU-PRO-ALA-ASN-
                                                                                                  325
GLN-MET-VAL-LYS-CYS-ASP-PRO-ARG-HIS-GLY-LYS-TYR-MET-ALA-CYS-CYS-LEU-LEU-TYR-ARG-GLY-ASP-VAL-VAL-PRO-
                                                                                                  350
LYS-ASP-VAL-ASN-ALA-ALA-ILE-ALA-THR-ILE-LYS-THR-LYS-ARG-THR-ILE-GLN-PHE-VAL-ASP-TRP-CYS-PRO-THR-GLY-
                                                        SER
                                                                                                  375
PHE-LYS-VAL-GLY-ILE-ASN-TYR-GLU-PRO-PRO-THR-VAL-VAL-PRO-GLY-GLY-ASP-LEU-ALA-LYS-VAL-GLN-ARG-ALA-VAL-
                                                                                                  400
CYS-MET-LEU-SER-ASN-THR-THR-ALA-ILE-ALA-GLU-ALA-TRP-ALA-ARG-LEU-ASP-HIS-LYS-PHE-ASP-LEU-MET-TYR-ALA-
                                                                                                  425
LYS-ARG-ALA-PHE-VAL-HIS-TRP-TYR-VAL-GLY-GLU-GLY-MET-GLU-GLU-GLY-GLU-PHE-SER-GLU-ALA-ARG-GLU-ASP-MET-
                                                                                                  450
ALA-ALA-LEU-GLU-LYS-ASP-TYR-GLU-GLU-VAL-GLY-VAL-ASP-SER-VAL-GLU-GLY-GLU-GLY-GLU-GLU-GLU-GLY-GLU-GLU-(TYR)
```

FIG. 3. Amino acid sequence of porcine brain α-tubulin. (From Postingl *et al.*, 1981.)

D. Secondary and Tertiary Structures

Tubulin is known to be the site of interaction with such antimitotic drugs as daunomycin (Na and Timasheff, 1977), colchicine, vinblastine, and podophyllotoxin (Wilson *et al.*, 1974) and to possess two sites for binding GTP and possible additional sites for binding divalent cations (Weisenberg *et al.*, 1968; Frigon and Timasheff, 1975a). *In vitro* it has been shown to polymerize to microtubules, 42 S double-ring structures (Frigon and Timasheff, 1975a), and a variety of other associated systems (Lee *et al.*, 1975b). A possible explanation of the great versatility of this protein might reside in the ability of brain tubulin to assume a variety of conformations imposed by physical and chemical interactions.

The conformation of porcine tubulin was studied by circular dichroism spectroscopy (Ventilla *et al.*, 1972). The secondary structure of the native protein at pH 6.5 is reported to vary as a function of temperature. This is in contrast to the report (Lee *et al.*, 1978b) that, at pH 7.0, the circular dichroic spectrum of calf brain tubulin, as shown in Fig. 4, does not undergo change in both the far- and near-UV region. Furthermore, the addition of daunomycin, vinblastine, and $MgCl_2$ does not seem to perturb the circular dichroism spectra below 250 nm, indicating that binding of these ligands is not accompanied by any major net changes in the secondary structure of calf brain tubulin.

The effect of pH on the structure of brain tubulin was monitored by circular dichroism spectroscopy for both porcine and calf proteins. In both reports (Ventilla *et al.*, 1972; Lee *et al.*, 1978b) the far-UV circular dichroic spectra indicate changes in the secondary structure of the protein. In the case of calf brain tubulin it can be interpreted as a gradual loosening of the structure by increasing the pH from 7.0 to 9.0 with drastic qualitative changes occurring at pH 10.0.

The high pH conformational transition of calf brain tubulin was further probed by spectrophotometric titration of the tyrosine residues as shown in Fig. 5. By analyzing these data in terms of the Linderstrøm-Lang equation it can be shown that between pH 10 and 12 the contribution of the electrostatic free energy term is small and that there is little electrostatic interaction between charged groups on the protein. This suggests that the protein is expanded above pH 10 and that its charged groups are considerably more separated from each other than would be expected of a tightly folded globular protein (Timasheff, 1970; Tanford, 1961). The fact that at least 14 of the tyrosines ionize only after unfolding of the protein, and with identical pK_{app}, further suggests that these residues are not fully accessible to protons in the native protein.

Circular dichroism has also been useful in probing the tertiary structure of tubulin. The near-UV spectrum of calf brain tubulin, as shown in Fig. 4, is dominated by a strong positive absorption band centered above 275 nm, which is due to tryptophan or disulfide residues. At pH values at which the protein is tightly

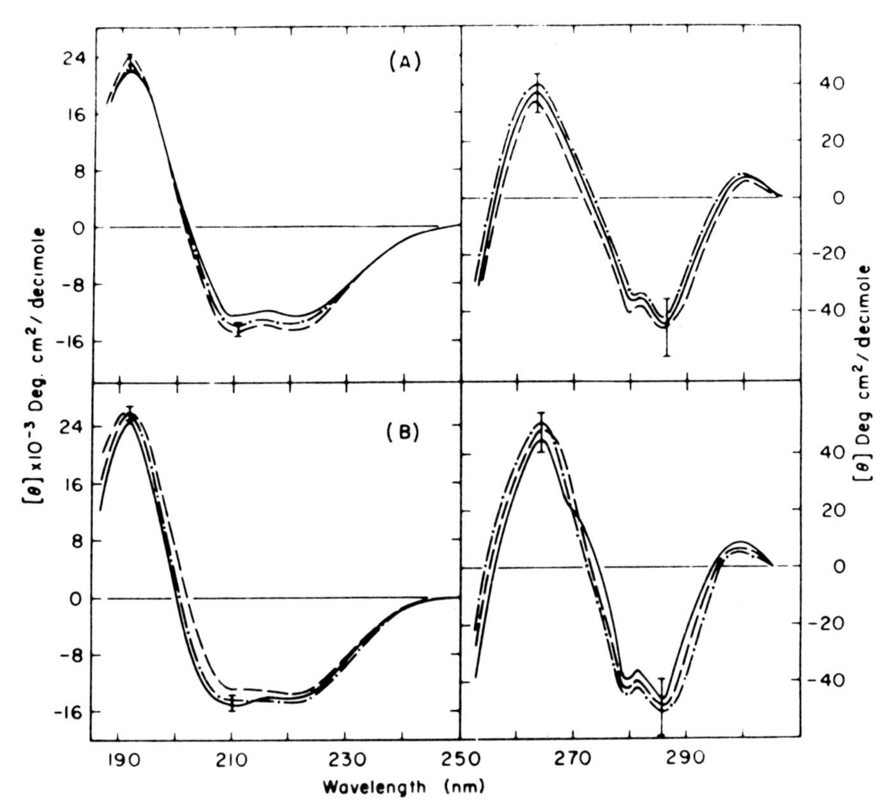

FIG. 4. (A) Circular dichroism spectra of calf brain tubulin in 10^{-2} *M* sodium phosphate – 10^{-4} *M* GTP, pH 7.0, at 22–23°C as a function of divalent cation concentrations. The protein concentration ranges from 1.2 to 1.4 mg/ml. The vertical lines represent the maximum deviations observed. -·-, no Mg^{2+}; –, 1.6×10^{-2} *M* Mg^{2+}, --, 10^{-4} *M* Ca^{2+}. (B) Circular dichroism spectra of calf brain tubulin in 10^{-2} *M* sodium phosphate – 10^{-4} *M* GTP, pH 7.0, as a function of temperature -·-, no Mg^{2+} at 5°C; –, no Mg^{2+} at 37°C; --, 1.6×10^{-2} *M* Mg^{2+} at 37°C. (From Lee *et al.*, 1978b.)

folded, this band is canceled to a major extent by negative bands at 280 and 286 nm, which most probably correspond to optically active buried tyrosines. The positive extreme at 265 and 300 nm would then be simply the two edges of a strong single positive band. In an attempt to arrive at better assignments of these bands, circular dichroic studies were carried out in the presence of denaturants, with and without cleavage of the disulfide bridge. In the pH range between 5 and 7, the near-UV circular dichroic spectrum of calf brain tubulin was found to be identical within experimental error in the presence of a variety of denaturants, as shown in Fig. 6. This spectrum is characterized by a positive band with well-reproducible twin maxima at 267 and 271 nm, a positive shoulder at 285 nm, and

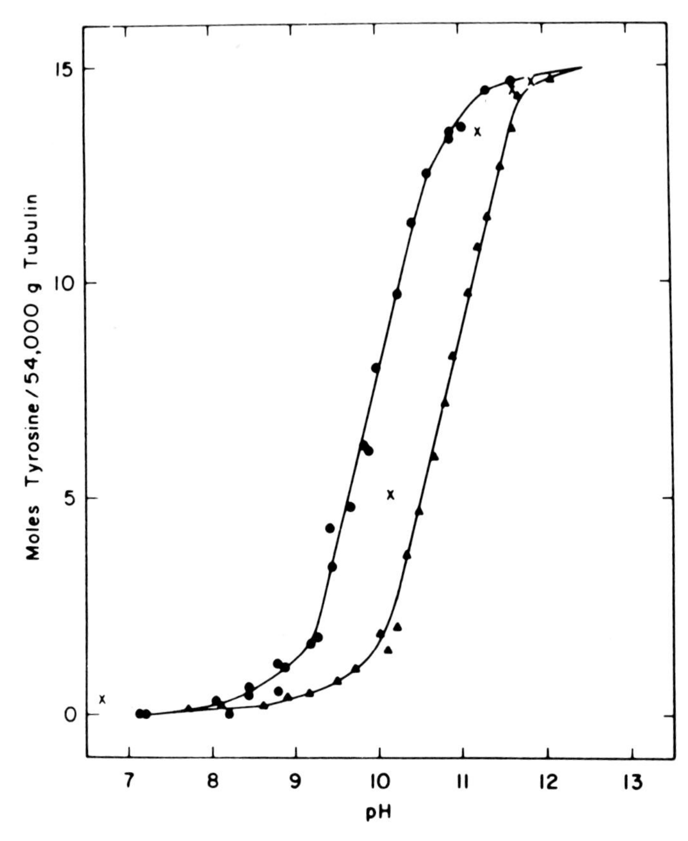

FIG. 5. Spectrophotometric titration of calf brain tubulin in 6 *M* GuHCl (●—●) and 10^{-2} *M* sodium phosphate or glycine (▲—▲) buffers at 22–23°C. The protein concentration in the latter two solvents was 1.3 mg/ml. x—x represents the back titration after exposure to pH 12. (From Lee *et al.*, 1978b.)

negative absorption between 295 and 310 nm. Reduction of the single disulfide bridge in tubulin, followed by carboxymethylation, eliminates totally the optical activity in the near-UV in all of the denaturants, as shown by the dashed lines in Fig. 6. When the pH of the reduced-carboxymethylated tubulin in 6 *M* GuHCl was raised to 12.9, optical activity reappeared in the form of a broad positive band maximal at 258 nm and extending up to 295 nm, as shown in Fig. 6 by the dot–dash line. The appearance of this band at high pH can be due only to tyrosine ionization (Beychok and Fasman, 1964), and its great similarity to the lower-wavelength band at pH 12.6 in the absence of denaturants confirms the previous assignments. The positive rotation seen between 253 and 295 nm at neutral pH contains too much fine structure for assignments to disulfide transition dipoles and is most likely due to tryptophan transitions (Sears and Beychok, 1973), as is the weak negative band at 300 nm (Ananthanarayan and Bigelow, 1969). Its elimination by cleavage of the disulfide bond suggests that one or more of the six tubulin tryptophan residues are constrained by the disulfide bridge, perhaps in a highly stable structural domain.

The tertiary structure of tubulin may also be probed by the reactivity of sulfhydryl residues toward cysteine specific reagents. *p*-Mercuric benzoate is not a suitable reagent, since it is too reactive with the tubulin sulfhydryl residues. DTNB [5,5′-Dithiobis(2-nitrobenzoic acid)], however, can be employed for monitoring the pseudo-first-order rate of its reaction with tubulin sulfhydryl

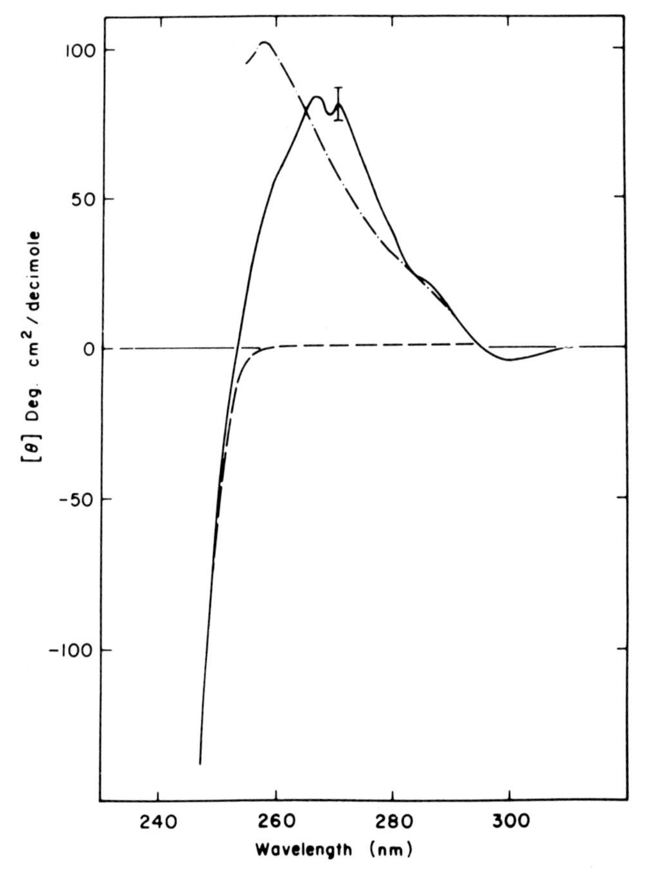

FIG. 6. Near-UV circular dichroism spectra of calf brain tubulin in denaturants. —, pH 5–7 with disulfide bridge intact (average of 10 experiments in 6 *M* GuHCl, 8 *M* urea, 0.1% SDS, and 8 *M* urea–0.1% SDS); --, same denaturants at pH 5–7 after reduction and *S*-carboxymethylation (average of six experiments in all denaturants); -·-, 6 *M* GuHCl at pH 12.9 with the disulfide bridge reduced and S-carboxymethylated. (From Lee *et al.*, 1978b.)

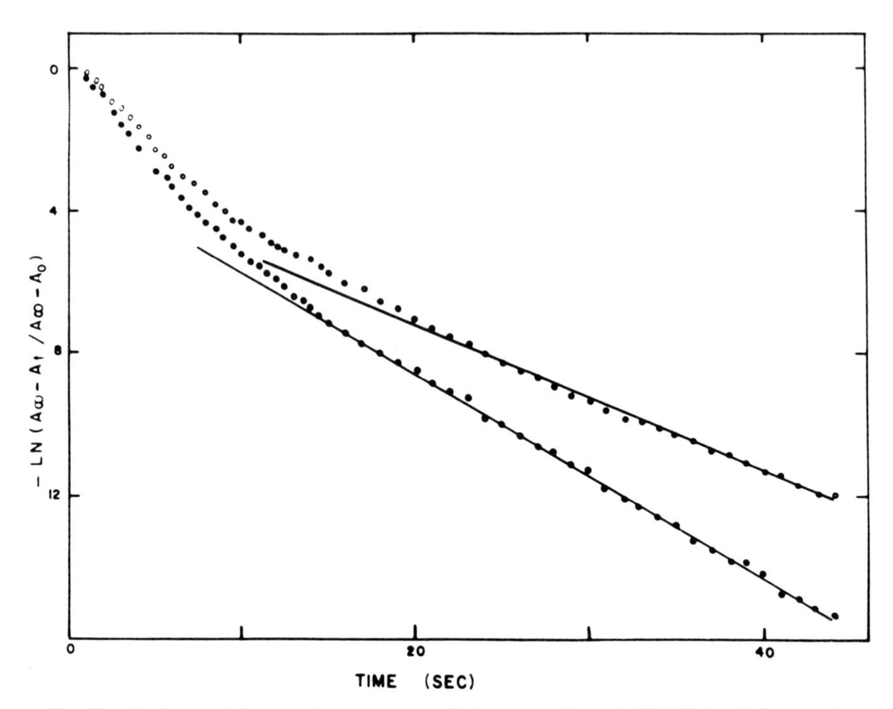

FIG. 7. Analysis of the reaction between calf brain tubulin and DTNB in 10^{-2} *M* sodium phosphate-10^{-4} *M* GTP, pH 7.0 and 22–23°C. Data were analyzed in terms of pseudo-first-order kinetics and plotted as $\ln(A_\infty - A_t/A_\infty - A_0)$ versus time. A_0, A_t, and A_∞ are the absorbance at zero time, at various time *t*, and at completion of the reaction, respectively. The slope of the plot yields a negative value for *k*, the pseudo-first-order rate constant. The tubulin concentration employed was approximately 0.2 mg/ml. The experimental conditions and symbols are (○) no Nocodazole; and (●) 10^{-5} *M* Nocodazole.

groups. Lee and Lee (1979) have monitored the change in reactivity of —SH groups in calf brain tubulin in the presence and absence of Nocodazole, a new synthetic antitumor drug, as shown in Fig. 7. It is evident that the formation of a tubulin-Nocodazole complex alters the accessibility of the sulfhydryl residues. They are probably more accessible for reacting with DTNB.

E. Quaternary Structure

Having established the presence of α and β subunits for tubulin, we are led to ask, "How do these nonidentical subunits interact with each other?" "What kind of complexes are formed?" Williams and co-workers (Detrich and Williams, 1978) studied the self-association of tubulin at low concentration (e.g., 0.02 mg/ml by molecular sieve chromatography and sedimentation equilibrium). It was reported that at those concentrations tubulin can dissociate into α,β subunits with a dissociation constant of 8×10^{-7} *M*. Hence at protein concentrations usually employed, the native and stable state of isolated tubulin is a dimer of 110,000 molecular weight and a sedimentation coefficient $S_{20,w}$ of 5.8 S (Frigon and Timasheff, 1975a). It is generally believed that the predominant dimer is α-β (Luduena *et al.*, 1977).

The self-association of tubulin in the presence of magnesium ions and vinblastine has been most thoroughly studied by Timasheff and co-workers (Frigon and Timasheff, 1975a,b; Na and Timasheff, 1980). A typical progression of the sedimentation patterns obtained as a function of tubulin concentration in the presence of magnesium ions is shown in Fig. 8. An increase of protein concentration first results in an increase in the area under the slow peak, with concomitant gradual increase in its sedimentation velocity, as shown by the displacement of this peak from the vertical line bisecting that obtained at the lowest protein concentration. Once a definite concentration has been reached, bimodality sets in, and all further added protein leads only to an increase in the area under the

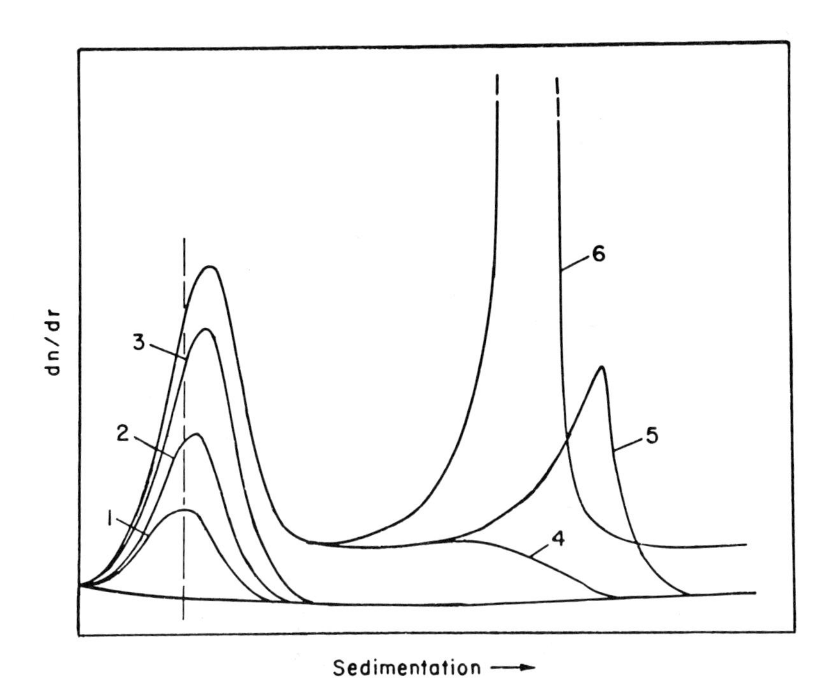

FIG. 8. Sedimentation velocity profiles of calf brain tubulin in 10^{-2} *M* sodium phosphate–10^{-4} *M* GTP–10^{-2} *M* $MgCl_2$ at pH 7.0 and 20°C. Protein concentrations (in mg/ml): 1, 0.65; 2, 1.35; 3, 2.7; 4, 7.9; 5, 12.7. (From Frigon and Timasheff, 1975a. Reprinted with permission from *Biochemistry*. Copyright 1975, American Chemical Society.)

rapid peak. By expressing the data as a dependence of the weight-average sedimentation coefficient $\bar{s}$ on protein concentration and by comparing the experimental results with theoretical curve fitting, it was established that in the presence of magnesium ions tubulin self-association proceeds first by the stepwise addition of subunits to a growing chain, the association constants of all the steps except the last one being identical. The addition of the last subunit occurs with a higher association constant and brings the polymerization process to an end with the formation of a thermodynamically favored polymer. The final polymer sediments with a $S_{20,w} = 42$ S. The best value of the degree of polymerization of the final product of self-association was found to be 26 ± 2 of tubulin dimers. The structures of the final product as deduced from hydrodynamic parameters is determined to be a ring structure, a conclusion substantiated by electron microscopy as shown in Fig. 9.

Interaction between vinblastine and brain tubulin leads to an isodesmic self-association of tubulin. Weisenberg and Timasheff (1970) initially reported *in vitro* self-association of calf brain tubulin induced by vinblastine and recognized the presence of bimodality in the sedimentation profile of tubulin at low vinblastine concentration as ligand mediated dimerization of tubulin in accordance to the theory derived by Cann and Goad (1972; Cann, 1970). Recently, Na and Timasheff (1980) convincingly demonstrated by sedimentation velocity study and computer data fitting that the self-association is one-ligand-molecule-

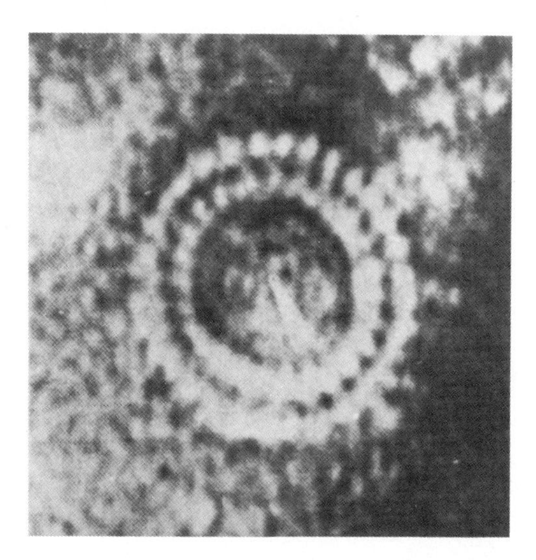

FIG. 9. Electron micrograph of calf brain tubulin in 10^{-2} *M* sodium phosphate–10^{-4} *M* GTP–1.6×10^{-2} *M* $MgCl_2$, pH 7.0, negatively stained with uranyl acetate. (From Frigon and Timasheff, 1975a. Reprinted with permission from *Biochemistry*. Copyright 1975, American Chemical Society.)

mediated (i.e., the binding of one vinblastine molecule is coupled to the formation of each intertubulin bond). The intrinsic association equilibrium constant for dimerization of the vinblastine-liganded tubulin was found to be $1.8 \times 10^5 M^{-1}$. Apparent binding isotherms of vinblastine to tubulin were calculated and were found to be consistent with the vinblastine binding results reported previously (Lee *et al.*, 1975b). Comparison of apparent binding curves calculated with different values of the self-association constants suggested that cooperativity between ligand-binding and self-association induced by magnesium ions may account for the disparity of the reported vinblastine-tubulin binding constants that range from $2.2 \times 10^4 M^{-1}$ to $6.2 \times 10^6 M^{-1}$ (Lee *et al.*, 1975b; Wilson *et al.*, 1975; Bhattacharyya and Wolff, 1976).

IV. Posttranslational Modification

Brain tubulin has been reported to be the substrate for glycosylation, phosphorylation, and tyrosylation. At present the extent of these modifications on tubulin and the biological significance of them are still undetermined.

A. Glycosylation

There are two opposing viewpoints on the possible glycosylation of brain tubulin. One maintains that tubulin is a glycoprotein containing approximately 1% carbohydrate, i.e., about 8 monosaccharide residues per mole of tubulin dimer (Falxa and Gill, 1969; Goodman *et al.*, 1970; Margolis *et al.*, 1972; Prus and Mattisson, 1979). On the other hand, Eipper (1972) demonstrated that in purifying rat brain tubulin employing pyrophosphate buffer instead of phosphate buffer, a highly purified sample of tubulin can be obtained. It contains little or no covalently linked carbohydrates. Microtubule protein purified in phosphate buffers, however, contains up to 10 moles of neutral sugar per mole of dimer, 4 moles each of galactosamine and glucosamine per mole of dimer. Material with spectral and chemical properties similar to nucleic acids can be resolved from tubulin by gel filtration on Sepharose 4B in 0.2% SDS. Although the presence and absence of carbohydrates may be a reflection of species specificity, the question on glycosylation of tubulin is still open and more work is required to clarify this point.

B. Phosphorylation

There is growing evidence indicating that brain tubulin can be phosphorylated and the site of phosphorylation is the β subunit (Goodman *et al.*, 1970; Lagnado *et al.*, 1972; Eipper, 1972). The incorporation of phosphate as detected by the presence of ^{32}P varies from 20 mmole to 0.4 mole per mole of monomer. The role of phosphorylation, however, remains to be identified.

C. Tyrosylation

Posttranslational incorporation of aromatic amino acids (tyrosine, phenylalanine or dihydroxyphenylalanine) into the C-terminus of the α subunit of tubulin has been reported (Arce *et al.*, 1975; Raybin and Flavin, 1975; Argarana *et al.*, 1977). The biological role of the raction remains unknown. It has been reported that tubulin can assemble *in vitro* with or without C-terminal tyrosine (Raybin and Flavin, 1977). In addition, Rodriquez and Borisy (1979) reported that covalent incorporation of either phenylalanine or tyrosine to the C-terminus of the α subunit does not prevent *in vitro* reconstitution of microtubules. These results suggest that the removal or addition of an aromatic amino acid does not affect the polymerizability of tubulin. It may, however, affect the equilibrium constant of the propagation step, a quantitative measurement that has yet to be carried out.

V. Conclusion

Significant progress has been made toward the understanding of the structures of brain tubulin since the initial isolation of the protein by Weisenberg *et al.*, (1968). The purification procedure enables one to obtain tubulin that is greater than 95% homogeneous. By employing highly purified samples it is much easier to establish a meaningful structure–function relationship with minimal complication from other components. Employing such protein has helped to establish the minimum conditions in which the protein can undergo *in vitro* reconstitution (Lee and Timasheff, 1975, 1977; Lee *et al.*, 1978a) and ligand-induced polymerization (Frigon and Timasheff, 1975a,b; Na and Timasheff, 1980). Having established these conditions and knowing the basic physicochemical parameters that govern these reactions, it may then be possible to explore the effect of other ligands, be they small molecules or macromolecules, on the behavior of tubulin. These macromolecules may include other components of the cytoskeletal system so as to elucidate the interplay of these protein components in the normal functioning of cells.

Much is yet to be learned about the structure–function relationship of tubulin. The role of the highly acidic carboxyl terminal sequence is still not established. The topographical relationship between binding sites of various ligands that include antimitotic drugs and nucleotides has yet to be investigated. The physiological significance of the posttranslational modifications is still unknown. They may not play any role in microtubule assembly, but they may be important for the association of microtubule with other cellular organelles such as vesicles or other protein components in the cytoskeletal system. It is clear that much investigation is required to provide answers to these questions.

ACKNOWLEDGMENT

This work is supported by NIH grants NS-14269 and AM-21489 and the Council for Tobacco Research—U.S.A., Inc.

REFERENCES

Ananthanarayan, V. S., and Bigelow, C. C. (1969). *Biochemistry* **8,** 3717–3722, 3723–3728.
Arce, C. A., Rodriguez, J. A., Barra, H. S., and Caputto, R. (1975). *Eur. J. Biochem.* **59,** 145–149
Argaraña, E. E., Arce, C. A., Barra, H. S., and Caputto, R. (1977). *Arch. Biochem. Biophys.* **180,** 264–268.
Beychok, S., and Fasman, G. D. (1964). *Biochemistry* **3,** 1675–1678.
Bhattacharyya, B., and Wolff, J. (1976). *Proc. Natl. Acad. Sci. U.S.A.* **73,** 2375–2378.
Cann, J. R. (1970). "Interacting Macromolecules." Academic Press, New York.
Cann, J. R., and Goad, W. B. (1972). *Arch. Biochem. Biophys.* **153,** 603–609.
Cann, J. R., and Stimpson, D. I. (1977). *Biophys. Chem.* **7,** 103–114.
Cann, J. R., Stimpson, D. I., and Cox, D. J. (1978). *Anal. Biochem.* **86,** 34–49.
Dahl, J. L., and Weibel, V. J. (1979). *Biochem. Biophys. Res. Commun.* **86,** 822–828.
Detrich, H. W., III, and Williams, R. C., Jr. (1978). *Biochemistry* **17,** 3900–3907.
Eipper, B. (1972). *Proc. Natl. Acad. Sci. U.S.A.* **69,** 2283–2287.
Eipper, B. (1974). *J. Biol. Chem.* **249,** 1407–1416.
Falxa, M. L., and Gill, T. J., III (1969). *Arch. Biochem. Biophys.* **135,** 194–200.
Feit, H., Slusarek, L., and Shelanski, M. (1971). *Proc. Natl. Acad. Sci. U.S.A.* **68,** 2028–2031.
Fine, R. E. (1971). *Nature (London), New Biol.* **233,** 283–284.
Fish, W., Mann, K., and Tanford, C. (1969). *J. Biol. Chem.* **244,** 4989–4994.
Fish, W., Reynolds, J., and Tanford, C. (1970). *J. Biol. Chem.* **245,** 5166–5168.
Forgue, S. T., and Dahl, J. L. (1979). *J. Neurochem.* **32,** 1015–1025.
Frigon, R. P., and Timasheff, S. N. (1975a). *Biochemistry* **14,** 4559–4566.
Frigon, R. P., and Timasheff, S. N. (1975b). *Biochemistry* **14,** 4567–4573.
George, H. J., Misra, L., Field, D. J., and Lee, J. C. (1981). *Biochemistry* **20,** 2402–2409.
Goodman, D. B. P., Rasmussen, H., DiBella, F., and Guthrow, C. E. (1970). *Proc. Natl. Acad. Sci. U.S.A.* **67,** 652–659.
Gozes, J., and Littauer, U. Z. (1978). *Nature (London)* **276,** 412–413.
Gozes, J., Baetselier, A., and Littauer, U. Z. (1980). *Eur. J. Biochem.* **103,** 13–20.
Hare, D. L., Stimpson, D. I., and Cann, J. R. (1978). *Arch. Biochem. Biophys.* **187,** 274–275.
Ikeda, Y., and Steiner, M. (1978). *Biochemistry* **17,** 3454–3459.
Laemmli, U. K. (1970). *Nature (London)* **227,** 680–685.
Lagnado, J. R., Lyons, C., Weller, M., and Phillipson, O. (1972). *Biochem. J.* **128,** 95P.
Lee, J. C., and Lee, L. L. Y. (1979). *Fed. Proc., Fed. Am. Soc. Exp. Biol.* **38,** 796.
Lee, J. C., and Timasheff, S. N. (1974). *Biochemistry* **13,** 257–265.

Lee, J. C., and Timasheff, S. N. (1975). *Biochemistry* **14,** 5183–5187.
Lee, J. C., and Timasheff, S. N. (1977). *Biochemistry* **16,** 1754–1764.
Lee, J. C., Frigon, R. P., and Timasheff, S. N. (1973). *J. Biol. Chem.* **248,** 7253–7262.
Lee, J. C., Frigon, R. P., and Timasheff, S. N. (1975a). *Ann. N.Y. Acad. Sci.* **253,** 284–291.
Lee, J. C., Harrison, D., and Timasheff, S. N. (1975b). *J. Biol. Chem.* **250,** 9276–9282.
Lee, J. C., Tweedy, N., and Timasheff, S. N. (1978a). *Biochemistry* **17,** 2783–2790.
Lee, J. C., Corfman, D., Frigon, R. P., and Timasheff, S. N. (1978b). *Arch. Biochem. Biophys.* **185,** 4–14.
Little, M. (1979). *FEBS Lett.* **108,** 283–286.
Lu, R. C., and Elzinga, M. (1977). *Anal. Biochem.* **77,** 243–250.
Luduena, R. F., and Woodward, D. O. (1973). *Proc. Natl. Acad. Sci. U.S.A.* **70,** 3594–3598.
Luduena, R. F., Shooter, E. M., and Wilson, L. (1977). *J. Biol. Chem.* **252,** 7006–7014.
Margolis, R. K., Margolis, R. U., and Shelanski, M. L. (1972). *Biochem. Biophys. Res. Commun.* **47,** 432–437.
Mattice, W. L., Riser, J. M., and Clark, D. S. (1976). *Biochemistry* **15,** 4264–4272.
Na, G. C., and Timasheff, S. N. (1977). *Arch. Biochem. Biophys.* **182,** 147–154.
Na, G. C., and Timasheff, S. N. (1980). *Biochemistry* **19,** 1347–1354.
Nelles, L. P., and Bamburg, J. R. (1979). *J. Neurochem.* **32,** 477–489.
O'Farrel, P. H. (1975). *J. Biol. Chem.* **250,** 4007–4021.
Ponstingl, H., Little, M., Krauhs, E., and Kempf, T. (1979). *Nature (London)* **282,** 423–424.
Ponstingl, H., Krauhs, E., Little, M., and Kempf, T. (1981). *Proc. Natl. Acad. Sci. U.S.A.* **78,** 2757–2761.
Prus, K., and Mattisson, A. (1979). *Histochemistry* **61,** 281–289.
Raybin, D., and Flavin, M. (1975). *Biochem. Biophys. Res. Commun.* **65,** 1088–1095.
Raybin, D., and Flavin, M. (1977). *J. Cell Biol.* **73,** 492–504.
Reynolds, J. A., and Tanford, C. (1970a). *Proc. Natl. Acad. Sci. U.S.A.* **66,** 1002–1007.
Reynolds, J. A., and Tanford, C. (1970b). *J. Biol. Chem.* **245,** 5161–5165.
Rodriguez, J. A., and Borisy, G. G. (1979). *Science* **206,** 463–465.
Sears, D. W., and Beychok, S. (1973). *In* "Physical Principles and Techniques of Protein Chemistry" (S. J. Leach, ed.), Part C, pp. 445–593. Academic Press, New York.
Stimpson, D. I., and Cann, J. R. (1977). *Biophys. Chem.* **7,** 115–119.
Tanford, C. (1961). "Physical Chemistry of Macromolecules," pp. 457–525. Wiley, New York.
Tanford, C. (1968). *Adv. Protein Chem.* **23,** 122–282.
Tanford, C. (1970). *Adv. Protein Chem.* **24,** 2–95.
Timasheff, S. N. (1970). *In* "Biological Polyelectrolytes" (A. Veis, ed.), pp. 1–64. Dekker, New York.
Valenzuela, P., Quiroga, M., Zaldivar, J., Rutter, W. J., Kirschner, M. W., and Cleveland, D. W. (1981). *Nature (London)* **289,** 650–655.
Ventilla, M., Cantor, C. R., and Shelanski, M. (1972). *Biochemistry* **11,** 1554–1561.
Weber, K., and Osborn, M. (1969). *J. Biol. Chem.* **244,** 4406–4412.
Weisenberg, R. C., and Timasheff, S. N. (1970). *Biochemistry* **9,** 4110–4116.
Weisenberg, R. C., Borisy, G. G., and Taylor, E. W. (1968). *Biochemistry* **7,** 4466–4479.
Wilson, L., and Bryan, J. (1974). *Adv. Cell Mol. Biol.* **3,** 21–72.
Wilson, L., Bamburg, J. R., Mizel, S. B., Grisham, L. M., and Creswell, K. M. (1974). *Fed. Proc., Fed. Am. Soc. Exp. Biol.* **33,** 158–166.
Wilson, L., Brewsell, K. M., and Chin, D. (1975). *Biochemistry* **14,** 5586–5592.

Chapter 3

Assembly–Disassembly Purification and Characterization of Microtubule Protein without Glycerol

DOUGLAS B. MURPHY

Department of Cell Biology and Anatomy
Johns Hopkins Medical School
Baltimore, Maryland

I. Introduction

Microtubules are prominent components of the cytoplasmic matrix and perform important functions as cytoskeletal elements for the determination of cell

ISBN 0-12-564124-9

shape and as key elements in intracellular motility such as mitosis and the translocation of cell organelles. These functions are thought to depend on the controlled assembly and disassembly of microtubules in the cytoplasm and on the interaction of microtubules with each other and with other cytoplasmic components. Weisenberg (1973) was the first to demonstrate that tubulin[1] would polymerize in brain extracts to form microtubules when prepared in the presence of Ca^{2+}-chelating buffers and GTP. Since this report several methods have been developed for the purification of microtubule protein[1] from mammalian brain tissue and from several other neuronal and nonneuronal cell types. The goal of this paper is to describe the procedures for isolating microtubule protein from brain tissue and for fractionating the tubulin and microtubule-associated proteins (MAPs) by chromatographic methods.

The purification methods are based on the temperature dependence of microtubule polymerization (Borisy *et al.*, 1975; Shelanski *et al.*, 1973). Microtubule assembly is an equilibrium process involving free tubulin subunits and microtubule polymers. Warm temperature (37°C) shifts the equilibrium to favor polymer, whereas cold temperature (0°C) shifts the equilibrium to favor free dissociated subunits. As shown in the scheme in Fig. 1, microtubules can be purified by differential centrifugation under conditions that favor microtubule assembly. In the first cycle of *in vitro* purification microtubules are polymerized at 37°C and pelleted by centrifugation at 37°C. The polymer is resuspended in ice-cold buffer to depolymerize the tubules and then centrifuged at 5°C to remove large protein aggregates and microsomal contaminants. The protein is then passed through one or two additional cycles of assembly and disassembly to remove nearly all soluble and particulate cytoplasmic contaminants. An SDS–polyacrylamide gel of hog brain microtubule protein purified by two cycles of *in vitro* assembly–disassembly in buffer without glycerol is shown in Fig. 2. These preparations contain 75% tubulin and 25% microtubule-associated proteins (MAPs)[2] of which 60% is composed of two classes of high-molecular-weight polypeptides: MAP-1, which is observed on gels as several trace components of 345,000 MW, and MAP-2, a doublet band of 286,000 and 271,000 MW. The remaining 40% consists of the tau proteins (approximately 70,000 MW) and other trace components (Borisy *et al.*, 1975). In general, two assembly–disassembly methods have been commonly employed to isolate microtubules from brain tissue. One involves the use of a buffer that contains glycerol, and the other uses a buffer in which glycerol is not included.

The procedure in which a buffer containing glycerol is used has been applied to isolate microtubules from hog, calf, beef, and rat brain (Shelanski *et al.*,

[1]In this paper, microtubule protein refers to preparations of microtubules containing tubulin and microtubule-associated proteins as obtained by the *in vitro* assembly procedure. Tubulin is used to indicate the MAP-free 6 S tubulin fraction obtained by ion-exchange chromatography.

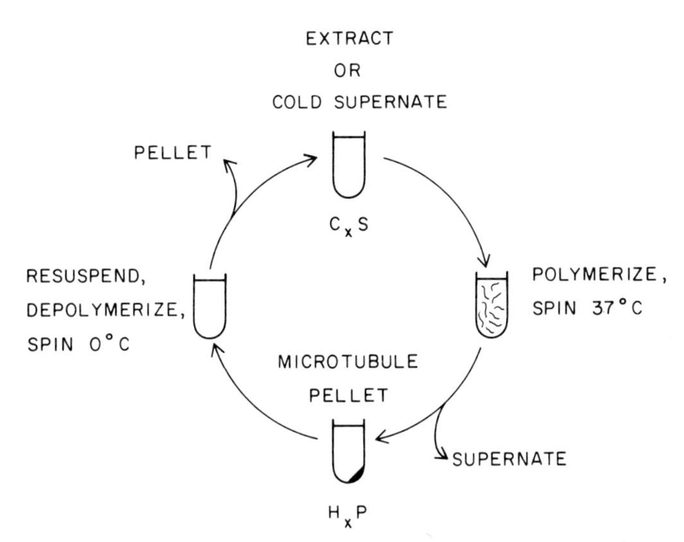

FIG. 1. Scheme for the purification of microtubule protein by cycles of temperature-dependent polymerization and depolymerization. One cycle of purification (assembly and centrifugation at 37°C followed by disassembly and centrifugation at 5°C) is shown. C_xS and H_xP refer to "cold" supernate and "hot" pellet obtained during a cycle of purification, x. The corresponding C_xP and H_xS fractions are discarded. For details of the purification procedure see text.

1973). The buffer composition is 0.1 *M* Mes[2] [2-(*N*-morpholino)ethanesulfonic acid]at pH 6.4 supplemented with 0.1 m*M* Mg^{2+} and 1.0 m*M* GTP plus 3.4 *M* glycerol. The advantages of this method are that it is adaptable to brain tissue from a number of different species, and it consistently gives large yields of microtubule polymer. Its disadvantages are that the glycerol may in some ways modify the native properties of tubulin and reduce the amount of MAPs present in the purified polymer. Glycerol has been demonstrated by Detrich *et al.* (1976) to bind to tubulin with high affinity, with 2 moles of glycerol binding per mole of 6 S tubulin dimer (however, see also Zabrecky and Cole, 1979), and to alter the number of free sulfhydryls on the tubulin subunit (Mellon and Rebhun, 1976). Glycerol may also reduce the affinity of tubulin for MAPs, since the stoichiometric ratio of the high-molecular-weight MAPs to tubulin becomes reduced during cycles of *in vitro* purification in the presence of glycerol (Scheele and Borisy, 1976). Whereas microtubules prepared in the absence of glycerol contain 25% microtubule-associated proteins, microtubules isolated in the presence of glycerol have been reported to contain only 5–10% nontubulin material after several cycles of assembly–disassembly (Cleveland *et al.*, 1977, 1979). The reason for the depletion of MAPs during cycles of *in vitro* purification in high concentrations of glycerol is unclear. Glycerol has been described as a thermodynamic booster that is capable of reducing the critical concentration required

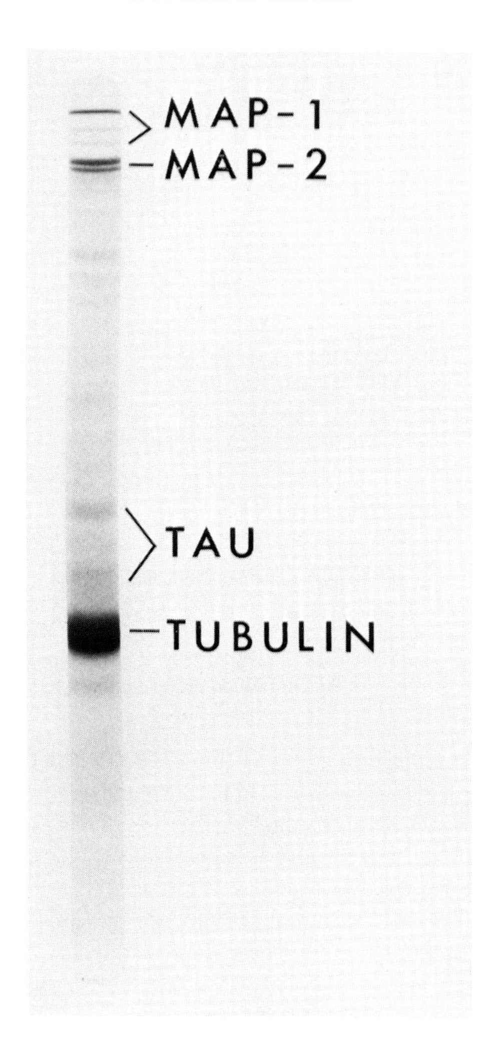

FIG. 2. SDS polyacrylamide gel (5%) of unfractionated hog brain microtubule protein after two cycles of *in vitro* purification by assembly–disassembly. Microtubule-associated proteins (MAP-1, MAP-2, tau factors) and tubulin are indicated.

for the polymerization of pure tubulin in the presence of Mg–GTP (Lee and Timasheff, 1975). Thus polymerization in the presence of glycerol may be supported by two factors, MAPs and glycerol. The gradual depletion of MAPs during purification suggests that glycerol may displace the MAPs and serve as a substitute in stabilizing microtubule polymer. Alternatively, the kinetics of glycerol-supported tubulin assembly may be sufficiently rapid to reduce the amount of tubulin available for a competing but slower MAP-dependent tubulin assembly process.

An assembly–disassembly procedure that does not employ glycerol was originally used to purify microtubules from extracts of hog brain (Borisy *et al.*, 1975). The buffer composition is 0.1 *M* PIPES[2] [piperazine-*N*,*N*′-bis(2-ethanesulfonic acid)] at pH 6.9 supplemented with 0.1 m*M* Mg^{2+} and 1.0 m*M* GTP. We have selected this procedure as the method of choice for obtaining MAPs in large quantity for biochemical studies and for examining the assembly properties and composition of microtubule protein from other cell systems. This method has two major advantages. First, MAPs bind to tubulin in large quantity; second, they copurify in constant stoichiometry to tubulin through several successive cycles of *in vitro* assembly–disassembly (Borisy *et al.*, 1975). The disadvantage of this method is that it has been necessary to alter certain ionic conditions such as pH and Mg^{2+} concentration in order to apply the method to tissues and cell types other than hog brain tissue (Murphy and Hiebsch, 1979; Bulinski and Borisy, 1979). The methods for isolating microtubule protein from hog and beef brain are presented in this paper along with the procedures for fractionating microtubule protein into separate tubulin and MAP fractions.

II. General Remarks on the Isolation of Microtubule Protein

A. Timing

There are two important time constraints that greatly affect the yield of microtubule protein: the freshness of the brain tissue used for tubule purification and the time required for the purification procedure itself.

Brain tissue should be obtained as soon as possible after the time of slaughter (within 20 min) and packed in ice. For reasons that are not yet understood, the assembly competency of microtubule protein declines very rapidly after the time of slaughter, but this decay can be substantially reduced by bringing the tissue to 0°C. For beef brain we have been able to show that the polymerization activity decays with a half-time of only 19 min (Fig. 3). This phenomenon is so consistent that the yield of purified microtubule polymer from beef brain tissue can be accurately predicted if the precise time of slaughter is known. Thus, when obtaining brain tissue from a slaughterhouse, it is important to ascertain the time at which slaughter took place so that only material that is less than 20–30 minutes old will be used. For hog brain the timing has not been investigated in detail, but

[2]Abbreviations used: PIPES, piperazine-*N*,*N*′-bis(2-ethanesulfonic acid); EGTA, ethyleneglycol-bis(β-aminoethyl ether)-*N*,*N*′-tetraacetic acid; Mes, 2-(*N*-morpholino)ethanesulfonic acid; MAP, microtubule-associated protein.

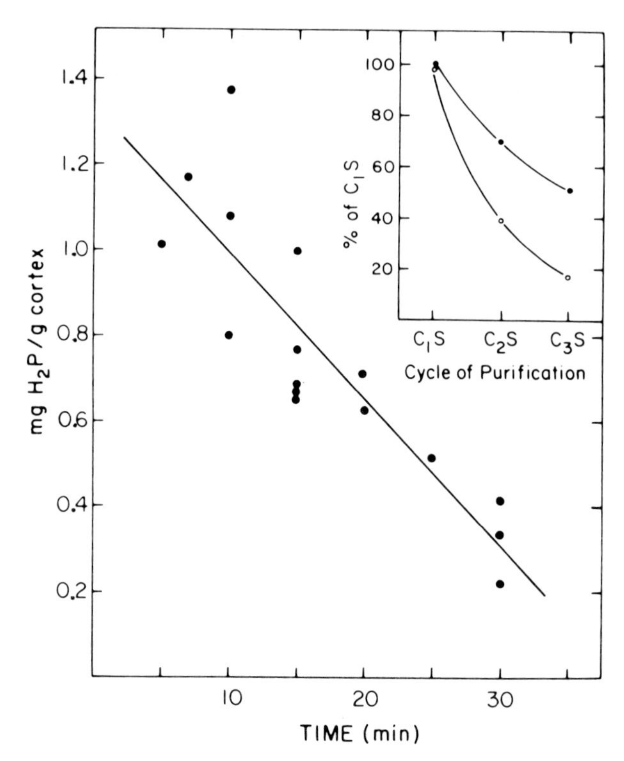

FIG. 3. Decrease in the assembly activity of microtubule polymerization after slaughter. The time indicated is the interval between the time of slaughter and the time the brain tissue was packed in ice. The yield of microtubule protein is expressed as mg H_2P per g brain cortex. Data are shown for 16 separate isolations of microtubule protein. Inset: Differences in the efficiency of microtubule polymerization during cycles of *in vitro* assembly for two different preparations of microtubule protein. Data are shown for preparations whose H_2P/extract ratio was 0.39 (○) and 1.38 (●). The amount of protein present in samples of C_1S is designated 100%. (From Murphy and Hiebsch, 1979.)

it appears to be less critical. Ideally one should use brain tissue obtained within 20 min of slaughter, but reasonable yields may be obtained from brains obtained from 20 to 60 min after slaughter.

A second important factor involving time is the schedule employed for *in vitro* purification. Olmsted and Borisy (1975) determined that the assembly activity of microtubule protein in hog brain extracts decays with a half-life of 3.5 hr and that the half-time for purified microtubule protein at 3–6 mg/ml is 19 hr. This decay in assembly activity is due to the lability of the cold-dissociated tubulin subunits. Thus after the brain tissue is homogenized, one should proceed without delay in the purification procedure.

B. Viscometric Assay for Monitoring Microtubule Assembly

It is useful to monitor microtubule assembly by viscometry during the purification procedure. The viscometric data are useful for two reasons. They indicate how polymerization for a given preparation is proceeding so that it is possible to anticipate the polymer yield for any given cycle of purification. Second, the viscosity data indicate the assembly activity of the preparation. As seen in the inset to Fig. 3, not all preparations of purified microtubule protein are alike in assembly activity. Samples of protein from a low-yield isolation (○) continue to

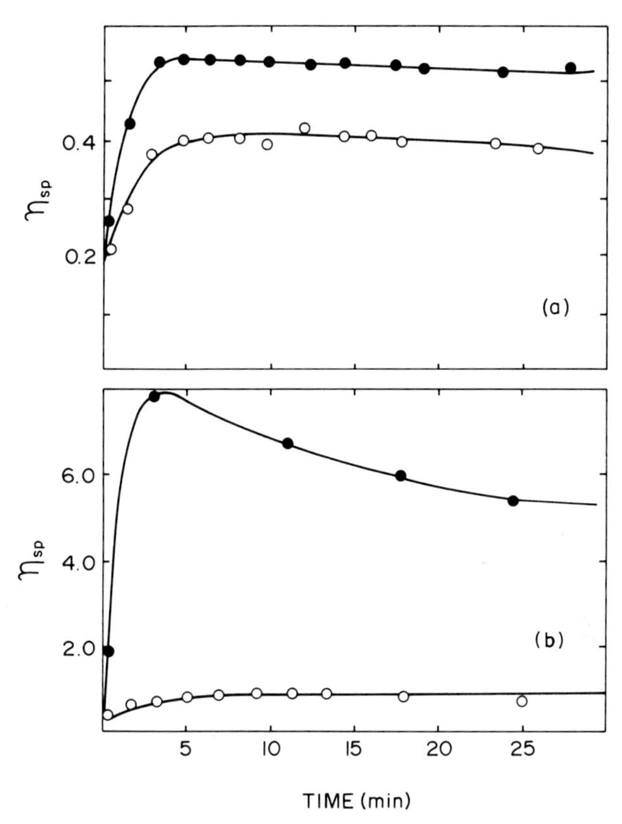

FIG. 4. Comparison of the polymerization of microtubule protein from beef brain at pH 6.62 for preparations giving high (●) and low (○) yields of microtubule polymer. (a) Brain tissue extract; (b) microtubule protein (C_1S). Microtubules were polymerized in medium containing 0.1 *M* PIPES, pH 6.62, supplemented with 0.1 m*M* Mg^{2+} and 1.0 m*M* GTP. Microtubule assembly was monitored by viscometry at 37°C using Ostwald capillary viscometers as described in the text. Protein concentrations in the brain extract and purified material were 10.3 and 11.2 mg/ml, respectively, for the high-yield preparation and 9.2 and 7.0 mg/ml for the low-yield preparation.

assemble with reduced activity during subsequent cycles of *in vitro* purification compared to protein obtained from more successful high-yield isolations (●). Thus the viscosity data obtained on the day of protein purification are also useful at a later date for predicting the polymerization activity of a given sample.

Specific viscosity is easily measured using Ostwald capillary viscometers, as described by Olmsted and Borisy (1973). Viscometers (Cannon-Manning semimicro viscometer, Type 100; Cannon Instrument Company; State College, Pa.) are immersed in an aquarium maintained at 37° ± 0.05°C. Samples (0.6 ml) are placed in the viscometer and the outflow time (OT) is determined with a stopwatch to the nearest 0.2 sec. The outflow time of the buffer (OT_b) and of the sample (OT_s) are used to calculate the specific viscosity: $\eta_{sp} = (OT_s - OT_b)/(OT_b)$. For reference the viscosity data from two different isolations of microtubule protein from beef brain, one excellent and the other fair, are shown in Fig. 4. In the case of the good isolation, the viscosity was high and stable in the extract and very high in the second polymerization (C_1S).[3] In the poor isolation the viscosity is not as high in the extract and is very low (and often unstable) in the second assembly step. If the viscosity data for each isolation are tabulated, one can, by comparison with previous data, evaluate the assembly activity of a preparation on the day of purification.

III. Purification of Microtubule Protein by Cycles of Assembly and Disassembly

A. Isolation of Microtubule Protein from Hog Brain

The following procedure is based on the method described by Borisy *et al.* (1975). Hog brains should be obtained within 20 min of slaughter and cooled down as rapidly as possible by packing in ice. Brains that are obtained after 60 min should not be used. Superficial blood clots and vessels are removed and the cortex dissected away from underlying white matter and the brain stem. The brain cortex is homogenized (5–10 sec) in a prechilled Waring blender with 1.5 vol (ml) of ice-cold 0.1 *M* PIPES, adjusted to pH 6.94 at 23°C with NaOH, containing 1.0 m*M* EGTA[2] and 0.1 m*M* GTP. The disrupted tissue is then homogenized (2–3 passes) in a Potter-Elvehjem tissue grinder with a motor-driven Teflon pestle operated at 3000 rpm. The method of homogenization does not appear to be critical, as reasonable yields of polymer may be obtained using the blender alone. The advantage of the glass–Teflon homogenizer is that it

[3]HS, HP, CS, and CP refer to the "hot" and "cold" supernates and pellets obtained by the *in vitro* assembly procedure for purifying microtubule protein. The numerical subscript refers to the cycle of purification.

disrupts cells without shearing apart the nuclei and mitochondria. The homogenate is brought to 5°C by swirling on ice and centrifuged at 32,000 $g_{(max)}$ for 60 min at 4°C (14,000 rpm in a Sorvall GSA rotor or a Beckman JA-14 rotor) to produce a supernate that is called the brain extract.

In the procedure originally described by Borisy *et al.* brain extract is prepared by two clarifying spins at 5°C: an initial centrifugation at 15,000 $g_{(max)}$ for 15 min (12,000 rpm in a Sorvall GSA rotor) to remove large cellular debris followed by a second ultracentrifugation step at 55,000 $g_{(max)}$ for 60 min (20,000 rpm in a Beckman-type 21 rotor) to remove microsomes and protein aggregates. However, microtubule protein with essentially the same composition may be obtained using a single centrifugation step.

Microtubule protein is purified from the brain extract by successive cycles of temperature-dependent assembly and disassembly as follows:

1. Extract is made 1.0 m*M* in GTP, brought to 37°C, and incubated at 37°C for an additional 30 min. To save time, the extract may be brought to 37°C by swirling in a flask in a 50°C bath. During this time it is useful to monitor microtubule assembly by viscometry.
2. Microtubule polymer is pelleted by centrifugation at 39,000 $g_{(max)}$ for 30 min at 37°C (18,000 rpm in a Sorvall SS-34 or Beckman JA-20 rotor).
3. The volume of the warm supernate (H_1S) is recorded and the supernate is then discarded. Resuspend (homogenize) the microtubule pellets (H_1P) in ice-cold polymerization buffer (0.1 *M* PIPES, pH 6.94, containing 0.1 m*M* Mg^{2+} and 1.0 m*M* GTP) using one-eighth the H_1S volume and keep at 0°C for a total of 30 min to depolymerize the microtubules.
4. Centrifuge the depolymerized protein to remove contaminating microsomes and protein aggregates at 39,000 $g_{(max)}$ for 30 min at 5°C (18,000 rpm in a Sorvall SS-34 or Beckman JA-20 rotor). The resulting cold supernate contains the tubulin monomers and MAPs and is processed further for microtubule purification. The cold pellet (C_1P) is discarded.

These four steps constitute one cycle of *in vitro* purification. For additional cycles, resuspend the microtubule pellets (H_xP) in 0.4 the H_xS volume using the same polymerization buffer. After two cycles of purification, hog brain microtubule protein (H_2P) may be frozen in liquid nitrogen and stored at −80°C with full retention of polymerization activity.

The expected yield of protein using this method is from 0.3 to 0.7 mg H_2P per 1 g brain cortex homogenized.

B. Large-Scale Isolation of Microtubule Protein from Hog Brain

If the aim of the microtubule isolation is to obtain tubulin or MAPs from brain for biochemical studies, the following modification of the preceding method may be used for large-scale preparations (600–1500 g cortex) in which glycerol is included in the first step of purification. The advantages of this modification are twofold: (1) the yield of polymer is high and very consistent from preparation to preparation; (2) the percentage of MAPs in the purified sample is approximately the same as that for the glycerol-free method described earlier (15% of H_2P protein in MAPs 1 and 2), since 25% glycerol is used only for the first cycle of purification.

1. Homogenize brain cortex 1:1 with homogenization buffer (0.1 *M* PIPES, pH 6.94, containing 4 m*M* β-mercaptoethanol) and centrifuge as described earlier to obtain brain extract.
2. Measure the extract volume. Supplement the extract with the following additives prior to polymerization to give the following final concentrations, being sure to include the glycerol volume as part of the extract volume: 3.4 *M* glycerol (one-third of the original extract volume), 1.0 m*M* EGTA, 0.1 m*M* Mg^{2+}, 2.4 m*M* ATP, and 0.1 m*M* GTP. The nucleotides may be added as a concentrated stock solution (0.24 *M* ATP, 0.1 *M* GTP made up fresh). If the nucleotides are added in powdered form, it is essential to add them before adding the glycerol; 2.4 m*M* ATP and 0.1 m*M* GTP are used in place of 1.0 m*M* GTP, for reasons of economy. In this case it is important to add the ATP to the extract just before polymerization. As noted by Olmsted and Borisy (1973), ATP may be substituted for GTP, but only when it is added to the extract. If brain tissue is homogenized in ATP-containing buffer, microtubule assembly in the extract is minimal.
3. Pellet the microtubule polymer by ultracentrifugation at 100,000 $g_{(max)}$ for 45 min at 25°C (40,000 rpm in a Beckman 45 Ti rotor, or by some equivalent procedure).
4. Resuspend the H_1P pellets in ice-cold polymerization buffer (0.1 M PIPES, pH 6.94, 1.0 mM EGTA, 0.1 mM Mg^{2+}, 1.0 m*M* GTP) using ⅛ the H_1S volume and keep the protein at 0°C for a total of 30 min to depolymerize the microtubules. Remove protein aggregates from the tubulin subunits by centrifugation at 39,000 $g_{(max)}$ for 30 min at 5°C (18,000 rpm in a Sorvall SS-34 or Beckman JA-20 rotor). No glycerol is added to the C_1S prior to polymerization, nor is it added prior to subsequent polymerizations. Purified protein may be frozen and stored after two cycles (H_2P) with maintenance of full polymerization activity.

The expected yield of protein using this method is from 1.0 to 1.3 mg H_2P per 1 g cortex homogenized.

C. Isolation of Microtubule Protein from Beef Brain

Microtubule protein may be isolated from beef brain using the procedures described earlier for isolating microtubules from hog brain with the following modifications: (1) It is absolutely necessary to use fresh brain tissue (obtained within 20 min of slaughter). If brain tissue obtained more than 30 min after slaughter is used, microtubule assembly is minimal (see Fig. 3). (2) The microtubules must be purified through three cycles (H_3P) prior to freezing and storing at −80°C in order to maintain full assembly activity. (3) The homogenization and assembly buffers contain 0.1 *M* PIPES adjusted to pH 6.62 at 23°C with NaOH. If isolation of microtubules is attempted using Pipes buffer at pH 6.94, one observes that whereas microtubule assembly in the extract appears normal, microtubule assembly during purified stages (C_1S) is minimal. The difference in assembly conditions is due to the fact that purified beef tubulin requires a lower pH for optimal assembly (Murphy and Hiebsch, 1979). Asnes and Wilson (1979) have found that microtubule protein may also be purified from beef brain in the absence of glycerol using phosphate–glutamate buffer at pH 6.74. The purification of tubulin obtained by this method for a particular preparation is presented in Table I. Approximately 29% of the tubulin present in the brain extract is isolated during the first cycle of purification. In the second cycle 89% of the tubulin is retained, and in the third cycle 86% of the tubulin is recovered. The final yield of tubulin after three cycles of purification is 22.1%.

The composition of microtubule protein purified by two cycles of purification from beef brain by this method (Murphy and Hiebsch, 1979) and from hog brain by the method of Borisy *et al.* (1975) is shown in Table II. Whereas the hog protein contains 75% tubulin and 25% nontubulin components, the material from beef contains 50% tubulin and 50% nontubulin proteins. High-molecular-weight MAPs comprise 20% of the total protein in beef and 15% in hog microtubule protein, and the ratio of the high-molecular-weight proteins to tubulin is 0.40 for beef and 0.20 for hog. Thus although proteins with similar electrophoretic mobilities are present in both preparations, the samples differ in the relative amounts of the nontubulin proteins contained.

TABLE I

PURIFICATION OF TUBULIN[a]

Cycle	Protein (mg)	Tubulin (%)	Tubulin (mg)	Tubulin yield[b] (%)
Extract	3296.0	15.1	497.7	100.0
C_1S	427.8	33.6	143.7	28.9
C_2S	294.7	43.4	127.9	25.7
C_3S	230.8	47.7	110.1	22.1

[a] From Murphy and Hiebsch (1979).

[b] The yield of tubulin from 228 g of beef brain cortex through three cycles of purification in 0.1 *M* Pipes, pH 6.62. Percentage tubulin was determined by quantitative gel electrophoresis.

TABLE II

COMPOSITION OF MICROTUBULE PROTEIN ISOLATED FROM BRAIN TISSUE FROM BEEF AND HOG[a,b]

Brain tissue	Tubulin (%)	MAP 1,2 (%)	Other (%)	MAP 1,2: Nontubulin ratio	MAP 1,2: Tubulin ratio
Beef	50	20	30	0.40	0.40
Hog	75	15	10	0.60	0.20

[a] From Murphy and Hiebsch (1979).

[b] The composition of microtubule protein was determined by quantitative SDS-gel electrophoresis. Microtubule protein (C_2S) was isolated from beef brain in 0.1 *M* PIPES at pH 6.62 and from hog brain in 0.1 *M* PIPES at pH 6.94. "Nontubulin components" refer to all the components detected on the gel minus the component due to tubulin. "MAP 1,2" refers to the high-molecular-weight proteins. "Other" refers to all the nontubulin proteins minus the high-molecular-weight MAPs.

IV. Purification of Tubulin and MAPs from Microtubule Protein

Microtubule protein may be fractionated by a variety of chromatographic methods, including gel filtration chromatography (Haga and Kurokawa, 1975) and ion exchange chromatography on DEAE-Sephadex (Murphy *et al.,* 1977b) and on phosphocellulose (Weingarten *et al.,* 1975). The following methods are recommended for rapidly fractionating tubulin, tau proteins, and high-molecular MAPs into highly purified fractions.

A. Single-Column Fractionation on DEAE-Sephadex

Single-column fractionation on DEAE-Sephadex is recommended for the rapid fractionation of microtubule protein into highly purified, but dilute fractions. The protein fractions are removed from the column in a stepwise elution procedure (Murphy *et al.,* 1977b). The purified tubulin obtained by this method retains its assembly activity and has an estimated critical concentration of 2 mg/ml. Since the tubulin and MAP fractions are eluted in high salt, they must be desalted before they are used for studies involving microtubule polymerization.

Column: 30 ml DEAE-Sephadex A-50 equilibrated with 5 bed vol of 0.1 *M* PIPES, pH 6.94. To assure a rapid flow rate, prepare the column so that the bed height is only two times the bed diameter.

Sample: 100 mg C_2S in 100–150 ml 0.1 *M* PIPES, pH 6.94. If samples more concentrated than 1 mg/ml are applied to the column, microtubule oligomers form and the unbound fraction contains significant amounts of MAPs and tubulin. If tubulin is to be used in polymerization studies, all resuspension and chromatography buffers should contain 0.1 m*M* Mg^{2+} and 0.1 m*M* GTP.

Elution: (a) The unbound protein fraction contains the tau polypeptides. It includes two components of approximately 70,000 MW and two components of 80,000 MW as well as several other components but contains only a small amount of tubulin (7%) and a trace amount of the high-molecular-weight MAP proteins (see Fig. 5b). Since this fraction is dilute (<0.1 mg/ml), it may be necessary to concentrate the protein by pressure or vacuum dialysis before use. (b) The high-molecular-weight MAP-1 and MAP-2 polypeptides are eluted using 2 bed vol of 0.1 *M* PIPES containing 0.3 *M* KCl. The protein peak eluted with this pulse is 2 mg/ml and contains MAP-1 and MAP-2 (71%), some tubulin

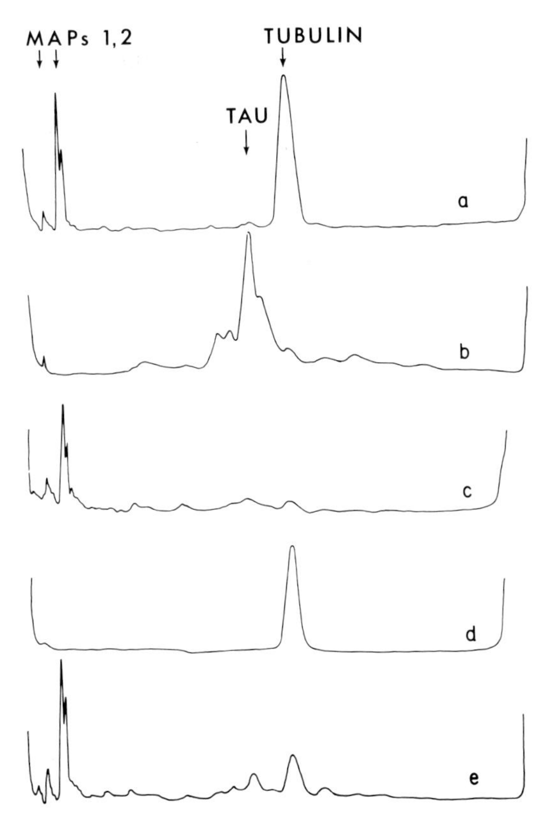

FIG. 5. Gel electrophoretic analysis of column fractions from chromatography on DEAE-Sephadex. Densitometer tracings of 5% polyacrylamide gels containing 4 μg loads of unfractionated microtubule protein (a), unbound protein (b), high-molecular-weight MAPs (c), tubulin (d), and total nontubulin proteins (e). (From Murphy *et al.*, 1977b.)

(4%), and other components (25%) (Fig. 5c). The column must be thoroughly washed (10 bed vol, 0.3 *M* KCl solution) prior to the elution of tubulin to remove any possible MAP contaminants. (c) Tubulin is eluted with 2 bed vol of 0.1 *M* PIPES containing 0.8 *M* KCl. The peak tubes are approximately 4 mg/ml and contain greater than 99% tubulin (Fig. 5d). When 100 μg are loaded on a gel, no high-molecular-weight MAP polypeptides and only traces of other components can be identified. The MAP and tubulin fractions should be immediately desalted into microtubule assembly buffer (dialysis or gel filtration chromatography on Sephadex G-25 coarse). The desalted fractions may be frozen in liquid nitrogen and stored at −80°C for several weeks with full retention of polymerization activity. The total recovery of protein from this column is 60–70%. Densitometer tracings of SDS–polyacrylamide gels of these protein fractions are shown in Fig. 5.

B. One-Step Fractionation of Tubulin from Nontubulin Proteins on DEAE-Sephadex

A modification of the preceding procedure can be used to obtain all of the nontubulin protein together in a single fraction plus a second fraction containing pure tubulin (Vallee and Borisy, 1976; Murphy *et al.*, 1977b).

Column: 30 ml DEAE-Sephadex A-50 of the same design as described earlier, but preequilibrated with 5 bed vol of PIPES containing 0.3 *M* KCl.

Sample: 100 mg C_2S in 10 ml 0.1 *M* PIPES, pH 6.94 containing 0.3 *M* KCl.

Elution: (a) The unbound protein fraction contains both the tau and high-molecular-weight MAPs. The peak tubes are approximately 4 mg/ml and contain 51% MAP-1 and MAP-2, 12% tubulin, and 37% tau and other components (Fig. 5e). As earlier, the column is washed extensively (10 bed vol, PIPES containing 0.3 *M* KCl) prior to the elution of tubulin. (b) Tubulin is eluted with 2 bed vol of PIPES containing 0.8 *M* KCl. Densitometer tracings of the nontubulin fraction are shown in Fig. 5e.

C. Purification of Tubulin on Phosphocellulose

Microtubule protein may be fractionated on phosphocellulose (Whatman phosphocellulose P-11, precycled prior to use) to obtain tubulin very rapidly and in highly purified form (Weingarten *et al.*, 1975). In order to maintain full assembly activity of the protein, however, it is necessary to include Mg-GTP in the chromatograph buffer (Williams and Detrich, 1979). These authors recommend that one equilibrate the column with 10 bed vol of buffer containing 10 m*M* Mg^{2+} (to saturate the Mg^{2+}-binding sites of the column). Therefore, one should either supplement the eluted tubulin fractions with Mg^{2+}-GTP or preequilibrate the column with buffer containing 10 m*M* Mg^{2+} in order to saturate the Mg^{2+}-binding sites of the column. The tubulin fractions can then be equilibrated with the buffer of interest by gel filtration chromatography on Sephadex G-25 and frozen in liquid nitrogen and stored at −80°C until use. The critical concentration of the purified tubulin fraction is approximately 2 mg/ml.

Column: 10 ml phosphocellulose slurry preequilibrated with at least 10 bed vol of 0.1 *M* Mes, pH 6.4 with or without 3.4 *M* glycerol.

Sample: 100 mg C_2S resuspended to 6 ml (final volume) with 0.1 *M* Mes buffer. The sample should be 10–15 mg/ml in order to obtain the tubulin fraction at a high concentration.

Elution: The unbound protein fraction is pure tubulin at approximately 6 mg/ml. The purity of the fraction sometimes depends on the source of the microtubule protein. If there is a problem with contamination by MAP proteins, elution with a buffer containing 3.4 *M* glycerol or 0.1% Triton X-100 can be tried.

V. Perspectives on the Role of MAPs in Microtubule Assembly and Function

Two classes of MAPs have been isolated and studied with respect to their ability to bind to microtubules *in vitro:* a group of high-molecular-weight components known as MAP-1 and MAP-2 and a group of lower-molecular-weight polypeptides known as tau. Since both classes of proteins bind to tubules with high affinity and have been demonstrated to be localized on microtubules *in situ,* it is very likely that these proteins are key structural components of cytoplasmic microtubules. MAPs are presently being examined for their role in regulating microtubule assembly and for their possible role as microtubule crossbridges.

A. The Role of MAPs in Microtubule Assembly

A consequence of the high binding affinity of MAPs for microtubules is their ability to stimulate the polymerization of tubulin *in vitro.* But so far it is not known if this activity is a function normally performed in the cell. Purified tubulin does not require MAPs in order to polymerize *in vitro* and will do so in the presence of aqueous buffers without solvents and at physiological pH and ionic strength (Murphy *et al.,* 1977b; Herzog and Weber, 1977). However, the critical protein concentration for the polymerization of column-purified tubulin is more than 10 times higher than that found for microtubule protein containing MAPs. The addition of MAPs (either the high-molecular-weight proteins or tau) to tubulin has been demonstrated to have a striking effect on microtubule self-assembly (Murphy *et al.,* 1977a; Sloboda and Rosenbaum, 1979; Weingarten *et al.*, 1975). In a concentration-dependent manner, MAPs lower the critical tubulin concentration for self-assembly and thereby increase the amount of microtubule polymer formed at steady state. The mechanism of the stimulatory activity of the high-molecular-weight proteins has been investigated by kinetic experiments involving the titration of tubulin with MAPs and measurements of the rates of polymer growth off of microtubule seeds (Murphy *et al.*, 1977a). It was demonstrated that MAPs reduce the rate of tubule depolymerization in the monomer–polymer equilibrium, thereby shifting the equilibrium to favor increased amounts of polymer. A consequence of this interaction is that microtubules are stabilized by MAPs. Thus one might account for the differential stability of certain cytoplasmic microtubules, such as those present in spindles and in axons, on the basis of varying affinities of MAPs for microtubules in these systems. As striking as this ability of MAPs to stimulate tubule assembly might seem, the effect is not specific for these proteins alone, since several basic proteins and polycationic substances (Erickson and Voter, 1976; Murphy *et al.*, 1977b) as well as organic solvents such as glycerol (Lee and Timasheff, 1975; Murphy *et al.*, 1977b) and dimethylsulfoxide (Himes *et al.*, 1976) and high concentrations of magnesium (Herzog and Weber, 1977) are also capable of stimulating tubulin polymerization. This has raised the question of the specificity and physiological relevance of the observed MAP-induced stimulation.

One way in which this is presently being explored is to examine the biochemical properties of the MAPs to see if they are subject to some chemical modification that would determine their ability to stimulate tubulin assembly *in vivo*. Sloboda *et al.* (1975) found that MAP-2 can be phosphorylated *in vitro* by a cAMP-dependent protein kinase. If this process also occurred *in vivo*, it might mean that microtubule assembly is regulated in the cell by the binding of MAP-2, which is in turn activated by a specific phosphorylation event. Another possibility being examined is the association of MAPs with other regulatory enzymes such as ATPase (White *et al.*, 1980) and transphosphorylase (Jacobs and Huitorel, 1979).

B. MAPs as Microtubule Crossbridges

Another important finding is that at least one of the high-molecular-weight MAPs (MAP-2) has been demonstrated to correspond to hairlike projections on microtubule surfaces (Dentler *et al.*, 1975; Murphy and Borisy, 1975; Vallee and Borisy, 1977). The projections appear to be uniform in length and bind in an ordered helical fashion (Amos, 1977; Kim *et al.*, 1979). The images of tubules decorated with MAP-2 are similar to those of tubules in intact neurons seen in thin sections by electron microscopy. These observations further support the idea that they may correspond to sidearms or crossbridges observed on microtubules *in situ* that may link tubules to each other or to other cytoplasmic components such as actin or membrane vesicles.

Griffith and Pollard (1979) recently demonstrated that microtubules containing MAPs can interact with actin. Assuming that such interactions are maintained by weak bonding forces, they used a low-shear viscometric assay (falling ball viscometry) to demonstrate that microtubules could not interact with actin unless MAPs were present. Another line of analysis for the possible cross-linking function of MAPs has been to examine MAP preparations for associated ATPase activity. Several workers have described a dyneinlike ATPase activity that binds to neuronal microtubules (Burns and Pollard, 1974; Gaskin *et al.*, 1974; Gelfand *et al.*, 1978). We have found that at least one ATPase activity of approximately 50,000 MW may be associated with the high-molecular-weight proteins (Hiebsch *et al.*, 1979). However, despite the numerous *in vitro* studies, there still has been no demonstration of the function of these proteins *in vivo*. However, given the distinct hairlike morphology of the high-molecular-weight proteins, it is possible that they are the crossbridges that have been postulated to determine some of the binding and motile functions of microtubules in neuronal cells.

C. Identification of MAPs in Nonneuronal Cells

When microtubule assembly methods are applied to cultured neuronal cells such as glial C6 cells, similar MAP proteins are identified as are seen in preparation of microtubule protein from brain tissue (Wiche and Cole, 1976). Microtubule protein isolated from blood platelets and from the spindles of clam oocytes contain high-molecular-weight polypeptides, but it is not known if they are homologous to the brain MAPs. However, Connolly *et al.* (1977), Sherline and Schiavone (1977), and Sheterline (1978) using antibodies to the high-molecular-weight proteins and the tau factors and indirect immunofluorescence microscopy observed fibrous patterns in nonneuronal cells for both of these proteins, and these patterns looked similar to those expected for the intracellular distribution of microtubules. In addition, Cleveland *et al.* (1979) copolymerized labeled SV-3T3 fibroblast extracts with unlabeled brain microtubule protein and identified several fibroblast MAPs, one of which had the same peptide composition as MAP-2 from brain.

However, using an antibody directed specifically against MAP-2, Peloquin and Borisy (1979) failed to observe MAP-2 in any cell type but neuronal cells. Furthermore, altogether different polypeptides have been identified by SDS–gel electrophoresis in preparations of microtubules purified from Ehrlich ascites tumor cells (Doenges *et al.*, 1977) and HeLa cells (Bulinski and Borisy, 1979; Weatherby *et al.*, 1978). This has raised the question of whether the brain MAPs are universal components of cytoplasmic microtubules or whether they may be unique to nerve cells. Thus in the future it remains to be determined to what extent MAPs and the tau proteins are represented in nonneuronal cells. Since the work to date indicates that several MAPs may be involved, it will also be of primary importance to determine whether these different MAPs are engaged in different intracellular functions.

References

Amos, L. A. (1977). *J. Cell Biol.* **72,** 642–654.

Asnes, C. F., and Wilson, L. (1979). *Anal. Biochem.* **98,** 64–73.

Borisy, G. G., Marcum, J. M., Olmsted, J. B., Murphy, D. B., and Johnson, K. A. (1975). *Ann. N.Y. Acad. Sci.* **253,** 107–132.

Bulinski, J. C., and Borisy, G. G. (1979). *Proc. Natl. Acad. Sci. U.S.A.* **76,** 293–297.

Burns, R. G., and Pollard, T. D. (1974). *FEBS Lett.* **40,** 274–280.

Cleveland, D. W., Hwo, S., and Kirschner, M. W. (1977). *J. Mol. Biol.* **116,** 207–225.

Cleveland, D. W., Spiegelman, B. M., and Kirschner, M. W. (1979). *J. Biol. Chem.* **254,** 12670–12678.

Connolly, J. A., Kalnins, V. I., Cleveland, D. W., and Kirschner, M. W. (1977). *Proc. Natl. Acad. Sci. U.S.A.* **74,** 2437–2440.

Dentler, W. L., Granett, S., and Rosenbaum, J. L. (1975). *J. Cell Biol.* **65,** 237–241.

Detrich, H. W., Berkowitz, S. A., III, Kim, H., and Williams, R. C., Jr. (1976). *Biochem. Biophys. Res. Commun.* **68,** 961–968.

Doenges, K., Nagle, B. W., Uhlmann, A., and Bryan, J. (1977). *Biochemistry* **16,** 3455–3459.

Erickson, H. P., and Voter, W. A. (1976). *Proc. Natl. Acad. Sci. U.S.A.* **73,** 2813–2817.

Gaskin, F., Kramer, S. B., Cantor, C. R., Adelstein, R., and Shelanski, M. L. (1974). *FEBS Lett.* **40,** 281–286.

Gelfand, V. I., Gyoeva, F. K., Rosenblat, V. A., and Shanina, N. A. (1978). *FEBS Lett.* **88,** 197–200.

Griffith, L. M., and Pollard, T. D. (1978). *J. Cell Biol.* **78,** 958–965.

Haga, T., and Kurokawa, M. (1975). *Biochim. Biophys. Acta* **392,** 335–345.

Herzog, W., and Weber, K. (1977). *Proc. Natl. Acad. Sci. U.S.A.* **74,** 1860–1864.

Hiebsch, R. R., Hales, D. D., and Murphy, D. B. (1979). *J. Cell Biol.* **83,** 345a.

Himes, R. H., Burton, P. R., Kersey, R. N., and Pierson, G. B. (1976). *Proc. Natl. Acad. Sci. U.S.A.* **73,** 4397–4399.

Jacobs, M., and Huitorel, P. (1979). *Eur. J. Biochem.* **99,** 613–622.

Kim, H., Binder, L. I., and Rosenbaum, J. L. (1979). *J. Cell Biol.* **80,** 266–276.

Lee, J. C., and Timasheff, S. N. (1975). *Biochemistry* **14,** 5183–5187.

Mellon, M. G., and Rebhun, L. I. (1976). *J. Cell Biol.* **70,** 226–238.

Murphy, D. B., and Borisy, G. G. (1975). *Proc. Natl. Acad. Sci. U.S.A.* **72,** 2696–2700.

Murphy, D. B., and Hiebsch, R. R. (1979). *Anal. Biochem.* **96,** 225–235.

Murphy, D. B., Johnson, K. A., and Borisy, G. G. (1977a). *J. Mol. Biol.* **117,** 33–52.

Murphy, D. B., Vallee, R. B., and Borisy, G. G. (1977b). *Biochemistry* **16,** 2598–2605.

Olmsted, J. B., and Borisy, G. G. (1973). *Biochemistry* **12,** 4282–4289.

Olmsted, J. B., and Borisy, G. G. (1975). *Biochemistry* **14,** 2996–3005.

Peloquin, J. G., and Borisy, G. G. (1979). *J. Cell Biol.* **83,** 338a.

Scheele, R. B., and Borisy, G. G. (1976). *Biochem. Biophys. Res. Commun.* **70,** 1–7.

Shelanski, M. L., Gaskin, F., and Cantor, C. R. (1973). *Proc. Natl. Acad. Sci. U.S.A.* **70,** 765–768.
Sherline, P., and Schiavone, K. (1977). *Science* **198,** 1038.
Sheterline, P. (1978). *Exp. Cell Res.* **115,** 460–464.
Sloboda, R. D., and Rosenbaum, J. L. (1979). *Biochemistry* **18,** 48–55.
Sloboda, R. D., Rudolph, S. A., Rosenbaum, J. L., and Greengard, P. (1975). *Proc. Natl. Acad. Sci. U.S.A.* **72,** 177–181.
Vallee, R. B., and Borisy, G. G. (1976). *Biophys. J.* **16,** 1772a.
Vallee, R. B., and Borisy, G. G. (1977). *J. Biol. Chem.* **252,** 377–382.
Weatherby, J. A., Luftig, R. B., and Weihing, R. R. (1978). *J. Cell Biol.* **78,** 47–57.
Weingarten, M. D., Lockwood, A. H., Hwo, S., and Kirschner, M. W. (1975). *Proc. Natl. Acad. Sci. U.S.A.* **72,** 1858–1862.
Weisenberg, R. C. (1973). *Science* **177,** 1104–1105.
White, H. D., Coughlin, B. A., and Purich, D. L. (1980). *J. Biol. Chem.* **255,** 486–491.
Wiche, G., and Cole, R. D. (1976). *Proc. Natl. Acad. Sci. U.S.A.* **73,** 1227–1231.
Williams, R. C., and Detrich, H. W. (1979). *Biochemistry* **18,** 2499–2503.
Zabrecky, J. R., and Cole, R. D. (1979). *Biochem. Biophys. Res. Commun.* **91,** 755–760.

Chapter 4

A Brain Microtubule Protein Preparation Depleted of Mitochondrial and Synaptosomal Components

TIMOTHY L. KARR, HILLARY D. WHITE,[1] BETH A. COUGHLIN, AND DANIEL L. PURICH

Department of Chemistry
University of California, Santa Barbara
Santa Barbara, California

I. Introduction

Microtubule protein may be isolated by two principal approaches: ion-exchange chromatography on DEAE-Sephadex and a recycling procedure involving warm-induced assembly and cold disassembly with intervening ultracentrifugation steps (Borisy *et al.*, 1975; Eipper, 1975; Shelanski *et al.*, 1973; Weisenberg *et al.*, 1968). The former yields tubulin, the principal component of microtubules as well as the pharmacologically active receptor of such well-known drugs as colchicine, vinblastine, podophyllotoxin, and griseofulvin. The second

[1]*Current affiliation:* Department of Pharmacology, School of Medicine, University of Washington, Seattle, Washington

ISBN 0-12-564124-9

method provides a microtubule protein preparation that contains a number of less abundant proteins that characteristically co-purify with tubulin, and these proteins are commonly called microtubule-associated proteins (MAPs). Of these MAPs there are mainly two high-molecular-weight components (HMW_1 and HMW_2), a heterogeneous 70,000 MW fraction [referred to by Weingarten *et al.* (1975) as tau protein] and a number of minor proteins (Borisy *et al.*, 1974; Sloboda *et al.*, 1976). The total MAPs fraction comprises 15–25% of the total protein of microtubules derived from brain tissue by recycling methods.

During experiments designed to assay microtubule protein for contaminating dehydrogenase activities, we observed the presence of glutamate dehydrogenase even after multiple warm–cold recycling (Karr *et al.*, 1979). The early studies of Christie and Judah (1953) had demonstrated, from differential contrifugation experiments, that glutamate dehydrogenase is entirely of mitochondrial origin. Likewise, Hogeboom and Schneider (1953) found the enzyme distributed in three fractions: 29% in the nuclear, 73% in the mitochondrial, and 4% in the microsomal fraction. They showed, nonetheless, that the nuclear activity could be attributed to contaminating mitochondria and became negligible when these were removed. Indeed, glutamate dehydrogenase may be considered a marker enzyme for the mitochondrial matrix (Grenville, 1969), and our observation of this dehydrogenase revealed that considerable mitochondrial disruption may attend the early steps in tissue extraction. This hypothesis was subsequently verified by measuring cytochrome c oxidase, another enzymatic activity characteristically of mitochondrial origin. We have found that many microtubule purification methods apply one or both of the following methods that are capable of disrupting osmotically fragile organelles: hypotonic extraction of whole brain tissue and/or electromechanical blenders (Borisy *et al*, 1975; Eipper, 1975; Shelanski *et al.*, 1973; Weisenberg *et al.*, 1968). These extraction methods contribute to destruction of brain mitochondria and synaptosomes, particularly emulsifying effects by the lipid-rich myelinated white matter in whole brains. This report outlines a new protocol minimizing such disruption and contamination while providing excellent yields. Additional information about the molecular characterization of sucrose purified microtubule protein is also presented.

II. Experimental Procedures

A. Materials and Special Equipment

It is convenient to prepare the following buffer solutions in the quantities indicated on the night prior to isolation. Storage at 4°C helps to stabilize buffers and provides ice-cold buffer as required in the specified steps.

"Transport buffer" 0.4 *M* sucrose–0.001 *M* EDTA (adjusted to pH 7.0 with potassium hydroxide). Make 100 ml per brain.

"Extraction buffer" 0.52 *M* sucrose–0.001 *M* EGTA–0.001 *M* ATP (adjusted to pH 7.0 with potassium hydroxide). Prepare 1 ml per gram of brain tissue.

"10X MEM buffer" 1.0 *M* MES[1]–0.01 *M* EGTA–0.01 *M* magnesium sulfate (adjusted to pH 6.8 with potassium hydroxide). Make 1 ml for each gram of initial brain tissue scrapings. This buffer may be diluted 10-fold to prepare "MEM buffer" required in later purification steps. GTP is added immediately before polymerization to minimize nonenzymatic hydrolysis commonly observed in divalent metal ion buffers.

For large-scale purification, a 500-ml Teflon and glass homogenizer is essential. An ideal version was fabricated from an 18-in. true bore glass tubing (1.55-in. inside diameter closed to a rounded bottom and fitted with a matching 2-in.-long Teflon plunger machined to provide 0.0005 in. radial clearance. The two-piece 0.5-in. diameter stainless steel drive shaft was joined near the top with J-12211-3 flexible rubber coupling (Lord, Inc., Erie, Pa.) to facilitate entry and removal from the homogenizer. The most important feature of this large-scale homogenizer is the tubular aluminum shroud that protects the glass homogenizer and anchors the homogenizer during use. The considerable torque exerted by a 0.5 hp drill press required that the homogenizer be immobilized with respect to rotation. This is achieved by cementing the glass tubular base to the aluminum shroud with an elastic silicone glue, which also served as a shock absorber as the drill press is actuated. We have found that this homogenizer is an indispensable part of large-scale preparation methods, and we are grateful to Professor Stanley M. Parsons for these essential design details.

B. Tissue Handling and Extraction

Undamaged bovine brains are obtained from exsanguinated animals at the slaughterhouse within 10–20 min of death. The brains are removed from the mechanically cleaved cranium and freed of any obvious bone chips or grit. They are promptly immersed in "transport buffer" and placed in a portable ice chest. Although it is generally advantageous to minimize the elapsed time between death and extraction, a period of up to 2.5 hr does not result in significant deterioration. Experience has shown, however, that yields are lower from tissue that has become darker gray in appearance, and it is advisable to secure several additional brains to eliminate the occasionally visibly darker tissue.

[1]Abbreviations used are MES; 2-(N-morpholino) ethane sulfonic acid. EGTA; ethyleneglycol bis(-aminoethylether)-N,N,tetraacetic acid. MEM buffer; 0.1 *M* MES, 0.001 *M* $MgS0_4$, 0.001 *M* EGTA, pH to 6.8 with KOH.

In the laboratory, the cerebral hemispheres are separated from the other brain parts, and the meninges is removed with forceps. Frequent immersion in ''transport buffer'' helps to maintain the required ice-water temperatures. The pink cortical tissue is cut or scraped from the cerebral hemispheres, with care taken to minimize gross contamination with the underlying white matter. A pair of scissors or a dull weighing spatula is adequate for this purpose. A group of three workers can prepare these cortical tissue scrapings from six to eight brains within 40 min, and it is apparently unnecessary to remove the white matter completely. Rapid handling and temporary storage of the scrapings in ''transport buffer'' should be encouraged. The well-drained scrapings are weighed in a tared beaker; for every gram of scrapings, it is necessary to add 1 ml of ''extraction buffer.'' In the cold room the tissue is homogenized by at least five passes in a Teflon and glass tissue homogenizer (0.005-in. radical clearance). The thick homogenate is then centrifuged at 50,000–70,000 *g* for 1 hr at 4°C using large-capacity ultracentrifuge rotors (e.g., Beckman 21 or 35 fixed-angle rotors). The supernatant fluid is carefully decanted into a large Erlenmeyer flask, and the pelleted material is discarded.

C. Microtubule Isolation and Storage

The crude extract fluid is buffered by the addition of ''10X MEM buffer'' in a proportion of 0.11 ml for every 1.0 ml of the extract. GTP is also added to a final concentration of 0.1 m*M*[2]. Addition of 35 ml reagent-grade glycerol for every 100 ml of the preceding fluid brings the glycerol to 3.5 *M* concentration, and warming to 37°C induces microtubule assembly. Here it is useful to employ a large vessel to effect adequate heat exchange. After 45 min the solution may be centrifuged at 75,000 *g* for 90 min at 30–37°C.[3] The pellets (referred to as the H_1P fraction) are resuspended to a final protein concentration of 10–20 mg/ml in MEM buffer containing 1 m*M* GTP at 4°C, homogenized gently with one pass in a 50-ml Teflon–glass homogenizer, depolymerized on ice for 30 min, and subsequently centrifuged for 60 min at 75,000 *g*. The supernatant fluid (referred to as the C_1S) is rich in microtubule protein, and it is again brought to 3.5 *M* in glycerol and then raised to 37°C. Microtubules are harvested as the pelleted material (H_2P fraction) after centrifugation at 75,000 *g* for 60 min. This protein

[2]To conserve GTP, one may utilize an acetate kinase regenerating system to replenish the GTP and minimize the accumulation of GDP. The optimal conditions are 50–100 *M* GTP (or GDP), 5–10 m*M* acetyl phosphate, and 0.01–0.05 units/ml of acetate kinase (freed of most ammonium sulfate by centrifugation in a conical plastic tube and subsequent removal of the supernatant with a Kimwipe).

[3]For large-scale work the centrifugation conditions must be suited to the rotor's k factor. With the Beckman 21 rotor, the relative centrifugal force is 45,000 *g;* the spin times are 1 and 2.5 hr for pelleting the brain homogenate and polymerized crude extract, respectively.

may be further purified by another cycle of cold-induced depolymerization and warm-induced assembly as described earlier (giving the C_2S and H_3P fractions, respectively). A flow chart with typical yields is given in Fig. 1.

Two methods of storage are suitable. The H_3P pellets may be frozen after the supernatant fluid is decanted. Alternatively, one may cold-depolymerize the H_3P fraction, add glycerol to 3.5 *M*, and freeze aliquots of the microtubule protein. In either case, rapid freezing in liquid nitrogen is necessary, and the protein should be stored at or below −80°C.

Finally, it is noteworthy that sucrose-purified microtubule pellets are larger

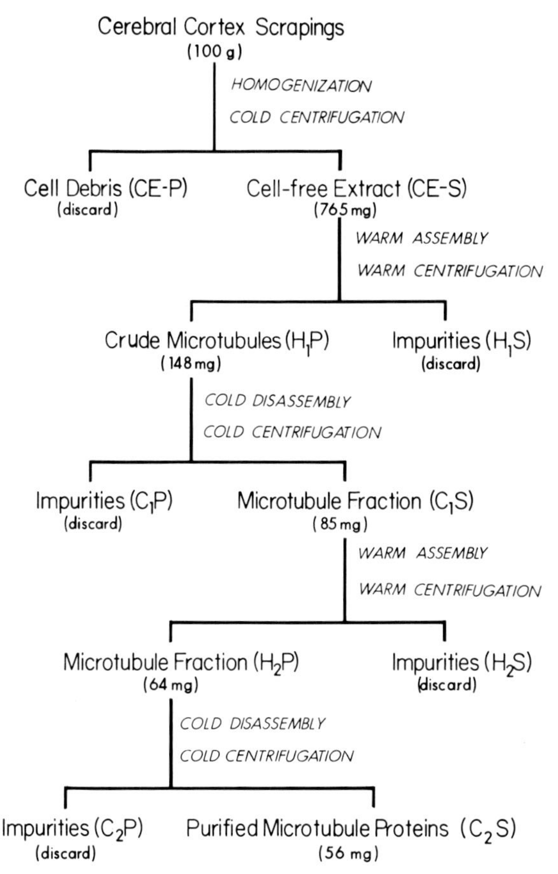

FIG. 1. Flow chart for the purification of microtubule protein by the sucrose extraction technique. Abbreviations used are CE-P and CE-S: cell extract pellet and cell extract supernatant, respectively; C_xS and C_xP; the 4°C supernatant and pellet, respectively, obtained following centrifugation at 4°C (x = 1, 2, 3, . . . and designates the cycle number); H_xS and H_xP: the warm supernatants and pellets, respectively, obtained following centrifugation at 37°C.

and softer in consistency than corresponding pellets from hypotonic preparations. Even at the H_1P stage, they are essentially colorless and demonstrate a bluish opalescence. We have repeatedly found that such pellets are also more readily resuspended in cold "MEM buffer."

III. Molecular Characteristics of Purified Microtubule Protein

It has been possible to gain a better understanding of the molecular properties of microtubule proteins prepared by the new sucrose extraction method and the Shelanski method. This is best approached by examining the yields, electrophoretic properties, associated enzymatic activities, and the assembly–disassembly behavior.

A. Electrophoretic Properties

Figure 2 depicts the results of our SDS–gel electrophoretic analysis of microtubule protein prepared by the sucrose extraction procedure and the hypotonic extraction method of Shelanski. At the C_1S stage of purification (i.e., after one complete purification cycle) there are notable differences revealed on the electropherograms. The yield of HMW_1 and HMW_2 is considerably higher with the sucrose-purified tubules, but this material is depleted in at least five and as many as eight other proteins. At the H_3P stage there are still many more proteins associated with microtubule protein from the hypotonic method. This indicates that many proteins that become associated with the microtubules in crude extracts tend to recycle with the microtubule fraction. A densitometer tracing for each H_2P fraction is shown in Fig. 3 as a further indication of the greatly improved purity possible with the sucrose extraction method. It is noteworthy that Berkowitz *et al.* (1977) identified proteins from 100 Å neurofilaments in typical microtubule preparations. Further work by Schlaepfer and Freeman (1978) and Keats and Hall (1975) suggests that neurofilaments are comprised of three proteins having subunit molecular weights of 68,000, 160,000, and 210,000. Of the proteins marked with asterisks in Fig. 2 there are protein bands corresponding to 68,000 and 220,000, but none in the 160,000 range. The fact that these are not present in the sucrose method at the C_2S stage of purification suggests that the slightly hypertonic conditions (approximately 400 milliosmoles) do not effectively extract neurofilaments.

B. Enzymatic Activities

To evaluate the extent of mitochondrial and synaptosomal breakage in both extraction methods, it was possible to measure the relative activities of two

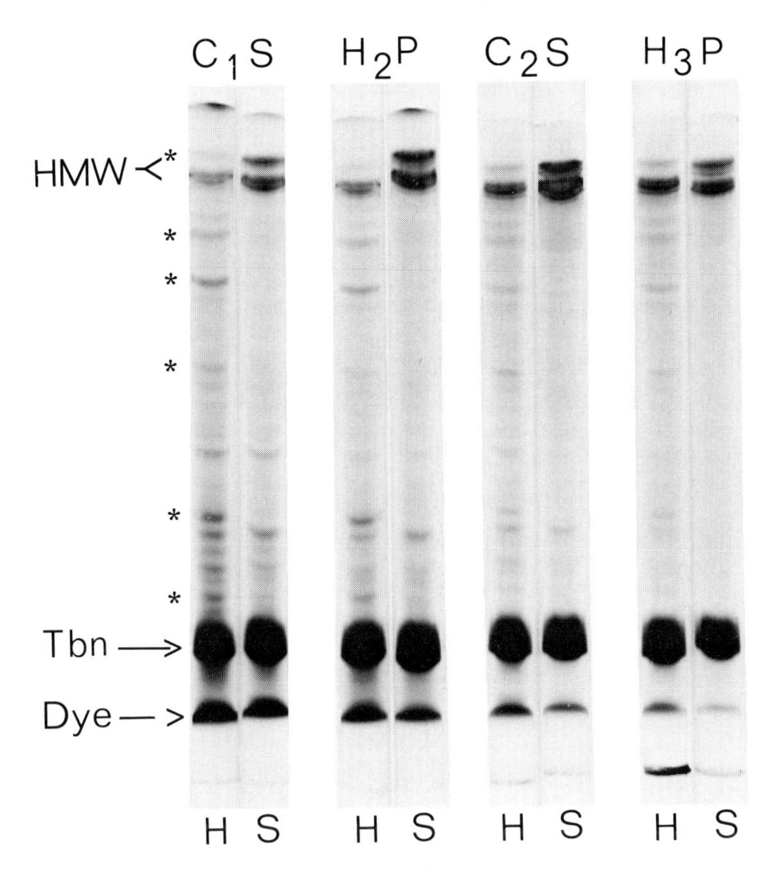

FIG. 2. Comparison of microtubule protein prepared by the sucrose extraction procedure and the standard Shelanski method; 7% polyacrylamide gels were loaded with 80 μg of microtubule protein from either the hypotonic protocol (H) or the sucrose procedure (S) from various stages of the isolation as indicated at the top of each gel. (For additional experimental details, see Karr *et al.*, 1979. Reprinted with permission from the *Journal of Biological Chemistry*.)

mitochondrial marker enzymes: glutamate dehydrogenase and cytochrome c oxidase. The activities of each were greatly diminished in cell extracts obtained by the sucrose extraction protocol: the former by at least a factor of 20 and the latter by a factor of 7. There is no evidence of the mitochondrial F_1 ATPase in purified microtubules from either extraction method, but the F_1 ATPase is known to be cold labile (Lien and Racker, 1971). Finally, there is no GDPase activity in either bovine brain preparation, and this contrasts with porcine brain microtubule protein (Carlier and Pantaloni, 1978; Zeeberg *et al.*, 1977).

The levels of other enzymatic activities typically associated with microtubules has also been determined. Nucleoside diphosphate kinase is relatively comparable in both sucrose and hypotonic purification methods. If GDP and GTP are not rigorously removed by charcoal treatment (Penningroth and Kirschner, 1978) or alkaline phosphatase action (Purich and MacNeal, 1978), then CTP, UTP, and ATP will support assembly by an indirect phosphorylation and exchange mechanism (Terry and Purich, 1979). Adenosine triphosphatase activity is associated with microtubule protein obtained by either method (White *et al.*, 1979). Recently, Couglin *et al.*, 1980) demonstrated that the endogenous protein kinase activity of microtubule protein from the sucrose protocol is considerably higher. This may reflect the increased level of HMW_2, the principal component that undergoes "autophosphorylation." On the other hand, these investigators found increased dependence on cyclic AMP, suggesting that either the protein kinase is more stable in the new preparation or the contaminating adenylate cyclase is lower.

C. Assembly–Disassembly Behavior

Karr *et al.* (1979) demonstrated that there were no significant differences in the rates of microtubule assembly or podophyllotoxin-induced disassembly of microtubule protein obtained by either purification method. Likewise, cold- and calcium ion-induced disassembly were about same. The only significant change in polymerization behavior identified to date regards the critical concentration of microtubule protein required for assembly. With the hypotonic method a value of around 0.2 mg/ml is typical (Borisy *et al.*, 1975), but this is reduced to only

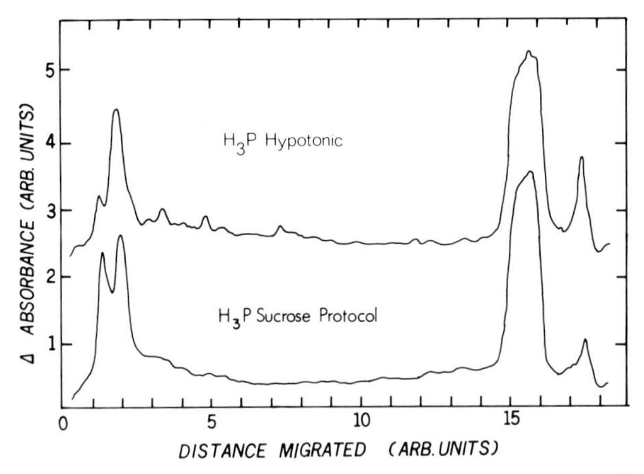

FIG. 3. Densitometric tracings of sucrose and hypotonic extraction procedures. Gels from the H_3P purification step shown in Fig. 2 were scanned in a Gelman DCD-16 densitometer at 600 nm. (Reprinted with permission from the *Journal of Biological Chemistry*.)

0.075 mg/ml with the method described in this report. This is most probably related to the enriched levels of HMW_1 and HMW_2 in these samples.

IV. Concluding Remarks

The recycling method for purifying microtubule protein has become a routine method, but the relationship of co-purified associated proteins to microtubule physiology is still unclear. The ability of a protein to recycle with tubulin may be completely unrelated to microtubule function (Lee *et al.*, 1978), and the goal of identifying logical connections to *in vivo* behavior is made more obscure by osmotic shock damage to various cellular organelles. The object of the sucrose method for microtubule extraction and isolation is to minimize such contamination. On the other hand, it will be necessary to investigate other cell components to learn more about the cellular distribution of microtubule proteins. For example, recent work in our laboratory with the synaptosomol fraction reveals low levels of HMW_1 and HMW_2 in isolated microtubules (B. A. Coughlin and D. L. Purich, unpublished observations). The significance of such compositional changes in microtubules from various cell regions obviously will be the subject of future chapters on this cytoskeletal structure. In any event, the microtubule protein preparation is a logical start toward such goals.

References

Berkowitz, S. A., Katagiri, J., Binder, H. K., and Williams, R. C., Jr. (1977). *Biochemistry* **16,** 5610–5617.

Borisy, G. G., Olmsted, J. B., Marcum, J. M., and Allen, C. (1974). *Fed. Proc., Fed. Am. Soc. Exp. Biol.* **33,** 167–174.

Borisy, G. G., Marcum, J. M., Olmsted, J. B., Murphy, D. B., and Johnson, K. A. (1975). *Ann. N.Y. Acad. Sci.* **253,** 107–132.

Carlier, M., and Pantaloni, D. (1978). *Biochemistry* **17,** 1908–1915.

Christie, G. S., and Judah, J. D. (1953). *Proc. R. Soc. London Ser. B.* **141,** 420.

Coughlin, B. A., White, H. D., and Purich, D. L. (1980). *Biochem. Biophys. Res. Commun.* **92,** 89.

Eipper, B. A. (1975). *Ann. N.Y. Acad. Sci.* **253,** 239–246.

Greville, G. D. (1969). *In* "Citric Acid Cycle" (J. M. Lowenstein, ed.), pp. 33–36. Dekker, New York.

Hogeboom, G. H., and Schneider, W. C. (1953). *J. Biol. Chem.* **204,** 233.

Karr, T. L., White, H. D., and Purich, D. L. (1979). *J. Biol. Chem.* **254,** 6107–6111.

Keats, R. A. B., and Hall, R. H. (1975). *Nature (London)* **257,** 418.

Lee, J. C., Tweedy, N., and Timasheff, S. N. (1978). *Biochemistry* **17,** 2783.

Lien, S., and Racker, E. (1971). *J. Biol. Chem.* **246,** 4298.

Penningroth, S. M., and Kirschner, M. W. (1978). *Biochemistry* **17,** 734.

Purich, D. L., and MacNeal, R. K. (1978). *FEBS Lett.* **96,** 83.

Schlaepfer, W. W., and Freeman, L. A. (1978). *J. Cell Biol.* **78,** 653.

Shelanski, M. L., Gaskin, F., and Cantor, C. R. (1973). *Proc. Natl. Acad. Sci. U.S.A.* **70,** 765–768.

Sloboda, R. D., Dentler, W. L., and Rosenbaum, J. L. (1976). *Biochemistry* **15,** 4497–4505.

Terry, B. J., and Purich, D. L. (1979). *J. Biol. Chem.* **254,** 9469.

Weingarten, M. D., Lockwood, A. H., Hwo, S. Y., and Kirschner, M. W. (1975). *Proc. Natl. Acad. Sci. U.S.A.* **72,** 1858–1862.

Weisenberg, R. C., Borisy, G. G., and Taylor, E. W. (1968). *Biochemistry* **7,** 4466.

White, H. D., Coughlin, B. A., and Purich, D. L. (1979). *J. Biol. Chem.* **255,** 486.

Zeeberg, B., Hassid, A., and Caplow, M. (1977). *J. Biol. Chem.* **252,** 2101.

Chapter 5

Purification and Reassembly of Tubulin from Outer Doublet Microtubules

KEVIN W. FARRELL

Department of Biological Sciences
University of California, Santa Barbara
Santa Barbara, California

I. Introduction

The discovery of conditions that promoted the *in vitro* polymerization of microtubules from vertebrate brain tubulin solutions (Weisenberg, 1972; Borisy and Olmsted, 1972) rapidly led to an extensive description of the *in vitro* assembly properties of microtubules (e.g., Olmsted and Borisy, 1973, 1975; Shelanski *et al.*, 1973; Gaskin *et al.*, 1974; Kuriyama and Sakai, 1974; Kuriyama, 1975; Murphy and Borisy, 1975; Weingarten *et al.*, 1975; Sloboda *et al.*, 1976; Lee and Timasheff, 1977). In contrast, the way in which microtubule properties and functions are determined remains largely obscure. In principle, two factors could

ISBN 0-12-564124-9

exert major but not necessarily exclusive influences: the tubulin composition of the microtubules and the nontubulin proteins associated with them. Although a large body of experimental data indicates a role for these factors, much of the evidence is essentially inferential.

For example, microtubules have been observed in association with a wide range of supramolecular structures, although direct physical linkage to the microtubules has not always been demonstrated. These include the numerous flagellar axonemal proteins (e.g., Gibbons, 1965; Stephens, 1969; Warner, 1972), intermicrotubule linkages of axopodia, axostyles and mitotic and oral apparatuses (Wilson, 1969; Roth *et al.*, 1970; Tucker, 1968, 1970; Tilney, 1971; Mooseker and Tilney, 1973), various biomembranes (Burgess and Northcote, 1969; Smith *et al.*, 1970; Franke, 1971a,b; Gray and Westrum, 1976), and chromosome kinetochores (Forer, 1969; Bajer and Mole-Bajer, 1972, 1975; McIntosh *et al.*, 1975).

Frequently, microtubules associated with such structures exhibit a greater stability to various experimental treatments than nonassociated microtubules in the same cytoplasm. Burgess and Northcote (1969) noted a greater resistance to colchicine for microtubules lying close to the endoplasmic reticulum, and chromosomal microtubules (associated with the chromosome kinetochore region) were refractory to low temperature and colcemid treatments that disrupted other nonassociated spindle microtubules (Roth, 1967; Brinkley and Cartwright, 1971, 1975; Brinkley and Stubblefield, 1970; Brinkley *et al.*, 1967). Several workers have also drawn a correlation between microtubule stability and the number and type of intermicrotubule linkage proteins bound to heliozoan axopodial microtubules (Tilney and Gibbins, 1968; Roth *et al.*, 1970).

In most cases, however, the role of associated structures in microtubule stability is equivocal, since the possibility cannot be excluded that the microtubules, by virtue of their intrinsic stability, preferentially associate with these structures.

In other aspects of microtubule behavior, associated proteins are clearly required. The most notable example is the role of axonemal proteins in flagellar and ciliary operation. The dynein arms associated with A-subfiber microtubules are essential for generating a shear force between adjacent outer doublet microtubules, and the radial spoke proteins appear to be involved in translating this shear force into a bending motion (e.g., Gibbons, 1975; Warner and Satir, 1974; Warner, 1976).

However, in a sense these observations do not clarify the question of the control of microtubule properties and functions, but merely redefine it at another level. That is, if microtubule properties and functions are contingent on a precisely defined spatial organization of associated protein structures, then what directs the ordered binding of these proteins?

The most logical explanation is that this information is encoded in the surface chemistry of the microtubule lattice. Localized changes in the lattice conformation could conceivably define regions of high probability for accessory protein binding. The wide variation in tubulin chemistry, not only within a single cell but also within a single organelle, suggests a potential basis for generating functional specificity in the microtubule lattice. For example, amino acid analysis and gel electrophoretic techniques have demonstrated differences in the chemistry of tubulins isolated from sea urchin flagellar, ciliary, and mitotic apparatus microtubules (Bibring *et al.*, 1976; Stephens, 1978), as well as in flagellar tubulins from *Chlamydomonas* (Whitman *et al.*, 1972; Piperno and Luck, 1976) and *Asterias* (Kobayashi and Mohri, 1977). Similar heterogeneity has also been reported for vertebrate brain tubulins (Feit *et al.*, 1977; Marotta *et al.*, 1978; Sullivan *et al.*, 1979).

In addition to a possible role in specifying accessory protein binding, the tubulin composition of a microtubule may in itself determine certain microtubule properties. Current evidence for this suggestion is entirely circumstantial, however, and derives from observations on the differential stability of flagellar microtubules.

On the basis of their stability to a variety of chemical, physical, or enzymatic treatments flagellar microtubules fall into three distinct classes: A-subfiber, B-subfiber, and central pair and accessory microtubules (Behnke and Forer, 1967). In certain cases, specific regions of the outer doublet microtubules solubilize in a characteristic sequence (Behnke and Forer, 1967; Kiefer, 1970; Witman *et al.*, 1972). It is these cases that form the most compelling arguments for intrinsic differences in the microtubules, since it is otherwise difficult to explain why regions of a B-subfiber microtubule, for example, should solubilize in a precise sequence when it is apparently devoid of associated proteins.

Other examples of differential microtubule solubilization have been documented; however, in these cases the presence of associated proteins obscured the influence of tubulin composition *per se* on microtubule stability (e.g., Tilney and Gibbins, 1968; Roth *et al.*, 1970; Stephens, 1970a, 1978).

To investigate these questions a microtubule system is required that possesses certain desirable features. First, it must be possible to isolate and purify large amounts of native, assembly-competent tubulin. It must also be possible to isolate tubulin populations of defined chemical composition that derived from microtubules of known properties. Finally, it is desirable to be able to isolate and purify proteins or protein assemblies associated with the microtubules.

Currently, the most commonly used microtubule system is that of vertebrate brain. This system has the advantage of yielding large quantities of native tubulin; however, the tubulin derives from the total brain microtubule population, which shows considerable heterogeneity (e.g., Feit *et al.*, 1977; Marotta *et al.*, 1978). If microtubule functional specificity is encoded in the tubulin composition, these directives may become "scrambled" during tubulin isolation. Microtubules reassembled *in vitro* from such tubulin populations may therefore represent ''consensus'' microtubules that reflect tubulin directives common to all classes of brain microtubules but which fail to reflect functional information unique to microtubule subclasses.

In contrast, the outer doublet subfiber A and B microtubules of flagellar axonemes are morphologically distinct (e.g., Warner, 1972) and may be selectively solubilized by a variety of techniques (Behnke and Forer, 1967; Gibbons, 1965; Stephens, 1970a; Witman *et al.*, 1972). Furthermore, each subfiber microtubule has a chemically unique tubulin composition (Stephens, 1978; Bibring *et al.*, 1976). Selected tubulin populations may therefore be obtained with the outer doublet system and offer the potential for examining the influence of tubulin chemistry on microtubule properties.

This system has the additional advantage of being the only microtubule system about which we have some understanding of its mechanical operation (Gibbons, 1975; Warner and Satir, 1974; Warner, 1976). The ultrastructural location of several of the proteins involved in flagellar motility are known and in some cases have been isolated and partially purified (Gibbons and Rowe, 1965; Stephens, 1970b; Gibbons and Fronk, 1972; Linck, 1976). The ability of these proteins to rebind to *in vitro* reassembled microtubules in a morphologically and functionally correct manner may provide an assay for the fidelity with which *in vitro* reassembled microtubules duplicate *in vivo* specifications.

II. Methodologies

A. Isolation and Storage of Sperm

Most commonly the starting material for *in vitro* reassembly studies using outer doublet tubulin has been sea urchin sperm flagella, owing to the high yield of material. Outer doublet tubulin has been successfully reassembled *in vitro* from the following species: *Strongylocentrotus drobachiensis, S. purpuratus, Pseudocentrotus depressus, Anthrocidaris crassispina* and *Hemicentrotus pulcherrimus* (Kuriyama, 1976; Farrell and Wilson, 1978; Binder and Rosenbaum, 1978). Cilia from *Tetrahymena pyriformis* have also been used (Kuriyama, 1976).

Sperm can be isolated from sea urchins by one of two methods. Either the animals are induced to spawn by injecting 0.52 *M* KCl into the body cavity or the whole gonads are excised, minced in calcium-free seawater, and filtered through cheesecloth to remove pieces of gonad (Kuriyama, 1976; Farrell and Wilson, 1978; Binder and Rosenbaum, 1978). Although the first method gives a more homogeneous preparation, consisting mainly of mature sperm, the second method is preferred, since the yields are considerably higher and the technique less laborious. Furthermore, the second method does not rely on the sea urchins being gravid, although decreased sperm yields will obviously be obtained under these conditions.

Isolated sperm may be stored either as a packed pellet at −80°C or as a 50% glycerol suspension at −20°C for several months. In this laboratory, we have not found significant differences in terms of subsequent tubulin assembly competence between the various sperm isolation and storage procedures.

B. Preparation and Storage of Outer Doublet Microtubules

To prepare demembranated axonemes from whole sperm the sperm are physically sheared to separate heads and tails. The tails are purified away from the heads by differential centrifugation and finally demembranated in a 1% Triton X-100 buffer (Stephens, 1970a).

To prepare outer doublet microtubules, methods based on the procedures of Gibbons (Gibbons, 1965; Gibbons and Fronk, 1972) have been generally used. Axonemes either are dialyzed against a low-ionic-strength buffer (1 m*M* Tris-HCl, pH 8.0, 0.1 m*M* EDTA, 0.1 m*M* dithiothreitol (DTT) for up to 36 hr at 4°C (e.g., Stephens, 1970a; Kuriyama, 1975, 1976; Binder and Rosenbaum, 1978) or are resuspended in a high-ionic-strength buffer (0.6 *M* KCl, 5 m*M* EDTA, 1 m*M* DTT, 1 m*M* ATP, 10 m*M* Tris-phosphate, pH8.0) for up to 1.5 hr at 4°C (Farrell and Wilson, 1978). This latter method may not be suitable in every case, however, since in certain species the outer doublet microtubules are solubilized by high salt treatment. Both treatments solubilize the central pair microtubules and dynein arms of subfiber A but leave the outer doublet microtubules largely intact. Tubulin forms approximately 70–80% of these preparations, with radial spoke proteins and nexin links constituting the major contaminants visible by electron microscopy.

The outer doublets may be stored as a pellet in liquid nitrogen or at −80°C for up to three months (longest period examined).

C. Solubilization of Outer Doublet Microtubules

Two methods have been reported that yield assembly-competent tubulin from outer doublet microtubules. Pfeffer *et al.* (1978) solubilized *S. purpuratus* outer doublets by shearing through a French Press. However, subsequent microtubule reassembly was very poor and was detectable only by electron microscopy.

In contrast, sonication routinely solubilizes 20–40% of the outer doublet microtubules and the tubulin is largely native as assayed by colchicine-binding activity (Kuriyama, 1976; Binder and Rosenbaum, 1978; Farrell and Wilson, 1978). Outer doublet solubilization is optimized if the sonication step is carried out in a low-ionic-strength buffer [5 m*M* MES, 1 m*M* EGTA, 0.5 m*M* $MgSO_4$,

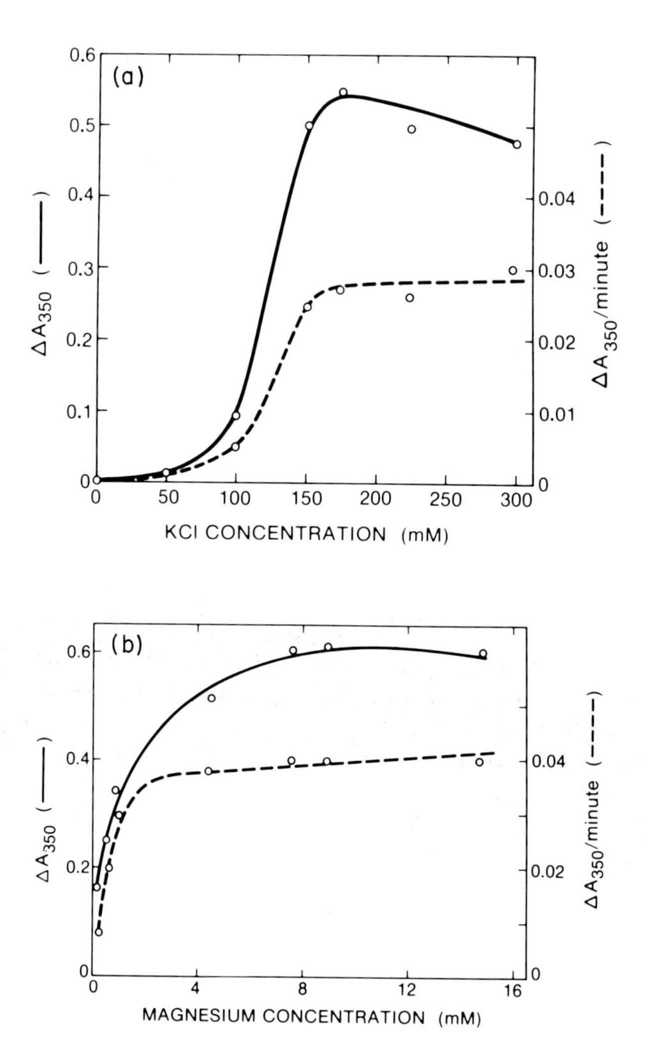

FIG. 1. (a) Effect of KCl concentration on the reassembly of outer doublet tubulin. Suspensions of outer doublets were sonicated in sonication buffer. Aliquots were adjusted to varying concentrations of KCl and centrifuged at 200,000 *g* for 1 hr at 4°C. Polymerization in the resulting supernatants was initiated by adding 2 m*M* GTP and warming to 37°C. Maximum extents of polymerization (—); initial rates (---). (b) Effect of magnesium concentration on the reassembly of outer doublet tubulin. Suspensions of outer doublets were sonicated in 5 m*M* Mes, 1 m*M* EGTA, 150 m*M* KCl, pH 6.7, adjusted to the requisite magnesium concentration and centrifuged at 200,000 *g*. Reproduced from Farrell *et al.* (1979a), with permission.

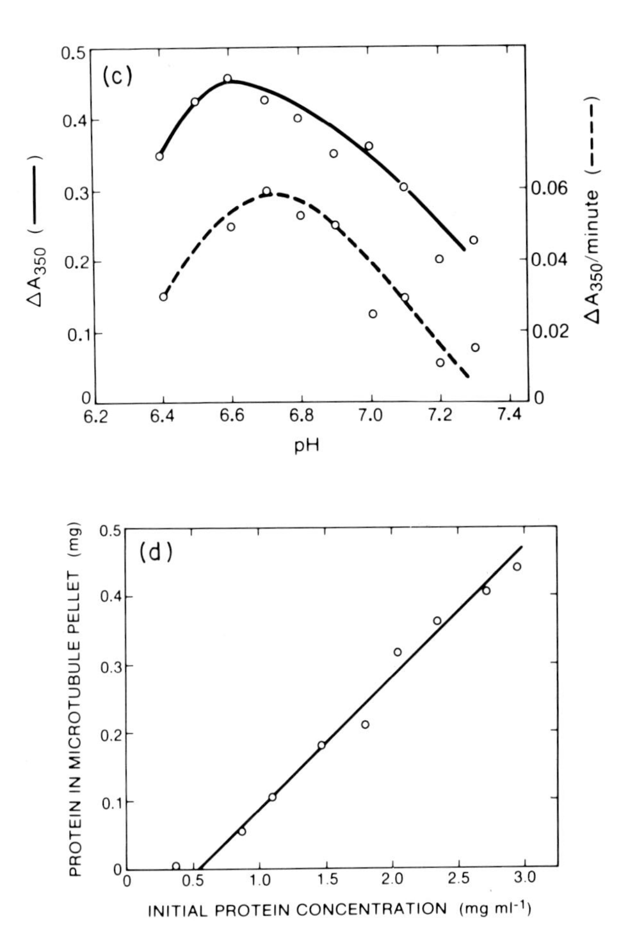

for 1 hr at 4°C. Polymerization was initiated by adding 2 m*M* GTP to the supernatants and warming to 37°C. Maximum extents of polymerization (—); initial rates (---). (c) Effect of pH on the maximum extents (—) and initial rates (---) of polymerization of outer doublet tubulin. Outer doublets were sonicated in 50 m*M* Mes, 1 m*M* EGTA, 1 m*M* $MgSO_4$, pH 6.7, and 150 m*M* KCl was added immediately following sonication. Aliquots of 200,000 *g* supernatants were then adjusted to varying pH values, and polymerization initiated by adding 2 m*M* GTP and warming to 37°C. The pH values given on the ordinate are correct for 37°C. (d) Determination of the critical protein concentration for reassembly of outer doublet tubulin. Varying concentrations of outer doublet 200,000 *g* supernatant in sonication buffer, 150 m*M* KCl, pH 6.7, were incubated for 1 hr at 37°C with 2 m*M* GTP to promote assembly. Assembled microtubules were pelleted by centrifugation at 40,000 *g* for 45 min at 30°C, and the protein in the pellets was determined. Reproduced from Farrell *et al.* (1979a), with permission.

pH 6.7, (MEM)]. Immediately following sonication the outer doublet suspension may be adjusted to 150 m*M* KCl to stabilize the solubilized tubulin and then centrifuged (40,000–200,000 *g*, 30–60 min, 0–4°C).

The resulting supernatant constitutes the starting material for all subsequent reassembly studies and is routinely 75–85% tubulin. Pelleted outer doublet microtubules not solubilized by the first sonication can be resuspended, washed in MEM to remove KCl, and resonicated. We have repeated this procedure up to four times and have obtained assembly-competent tubulin at each repetition. However, the yield of tubulin from subsequent steps is significantly lower than that from the initial sonication.

III. Reassembly Properties of Outer Doublet Tubulin

The solution conditions optimal for outer doublet tubulin reassembly are remarkably similar to those of vertebrate brain, especially when one considers the diverse properties and functions of the microtubule sources.

Outer doublet tubulin reassembly requires a high ionic strength (150 m*M* KCl), magnesium ions (1 mm), and pH 6.7–6.8 (Fig. 1) (Kuriyama, 1976; Binder and Rosenbaum, 1978). GTP is also required and neither GDP nor GMP will promote assembly at tubulin concentrations up to 2–3 mg/ml. The ability of

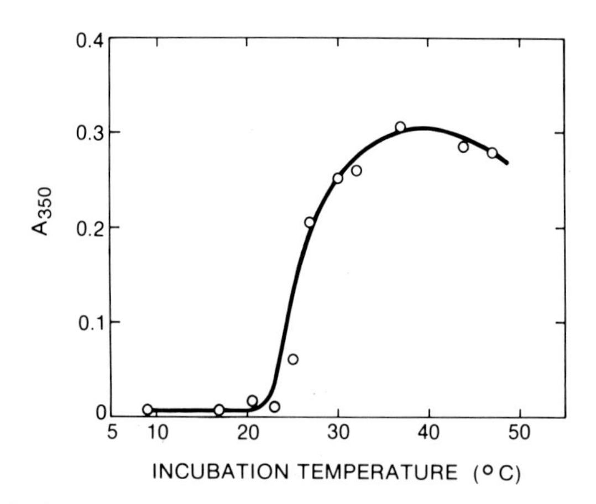

FIG. 2. Determination of the optimal temperature for reassembly of outer doublet tubulin. Aliquots of 200,000 *g* supernatants in sonication buffer, 150 m*M* KCl, 2 m*M* GTP, pH 6.7, were incubated at varying temperatures for 1 hr. Following incubation, the absorbance at 350 nm of each aliquot was determined and then corrected for aggregate formation by subtracting from this initial value the absorbance of each aliquot remaining after the aliquots were chilled to 4°C.

other nucleoside triphosphates to promote microtubule assembly is most likely the result of a nucleoside diphosphokinase activity in the tubulin preparations (Farrell *et al.*, 1979a).

Curiously, the temperature optimum for reassembly is 37°C (Fig. 2). By contrast, the physiological growth temperature for *S. purpuratus* is 11–15°C; in fact, 37°C would be lethal. Microtubule assembly at physiological temperatures was not observed, unless high tubulin concentrations (4.55 mg/ml) were used. Even under these conditions, assembly was only detectable by electron microscopy.

Under the preceding optimal conditions, microtubule formation is not observed below a tubulin concentration of 0.55–0.72 mg/ml (Fig. 1) (Binder and Rosenbaum, 1978). Above this concentration the mass of assembled microtubules is directly proportional to the initial tubulin concentration. An identical mechanism has been observed for vertebrate brain microtubules (Olmsted and Borisy, 1973; Olmsted *et al.*, 1974; Gaskin *et al.*, 1974; Weisenberg and Deery, 1976) and is indicative of a condensation polymerization reaction (Oosawa and Kasai, 1962).

IV. Properties of Microtubules Reassembled from Outer Doublet Tubulin

A. Morphology

Microtubules reassembled *in vitro* from outer doublet tubulin are closed singlet microtubules. This is in sharp contrast to the doublet microtubules from which the tubulin was derived. A further dissimilarity is that the reassembled microtubules are composed of 14 or 15 protofilaments, whereas the A-subfiber and central pair microtubules in flagellar axonemes are composed of only 13 (Fig. 3), the most common number for microtubules *in vivo* (Tilney *et al.*, 1973). The reason for this is unclear but may be related to buffer conditions (Pierson *et al.*, 1978).

In contrast to *in vitro* reassembled vertebrate brain microtubules, microtubules reconstituted from outer doublet tubulin lack the high-molecular-weight MAPs (HMW MAPs) and exhibit smooth profiles by thin-section electron microscopy (Fig. 4A, B). Nevertheless, vertebrate brain HMW MAPs will stimulate outer doublet tubulin reassembly and the reassembled microtubules now show the "hairy" profiles characteristic of brain microtubules (Fig. 4C, D). The vertebrate- and invertebrate-derived microtubules are not exactly equivalent, however, since the HMW MAPs constitute a lower proportion of the total microtubule mass in the invertebrate-derived microtubules than in the vertebrate-derived microtubules

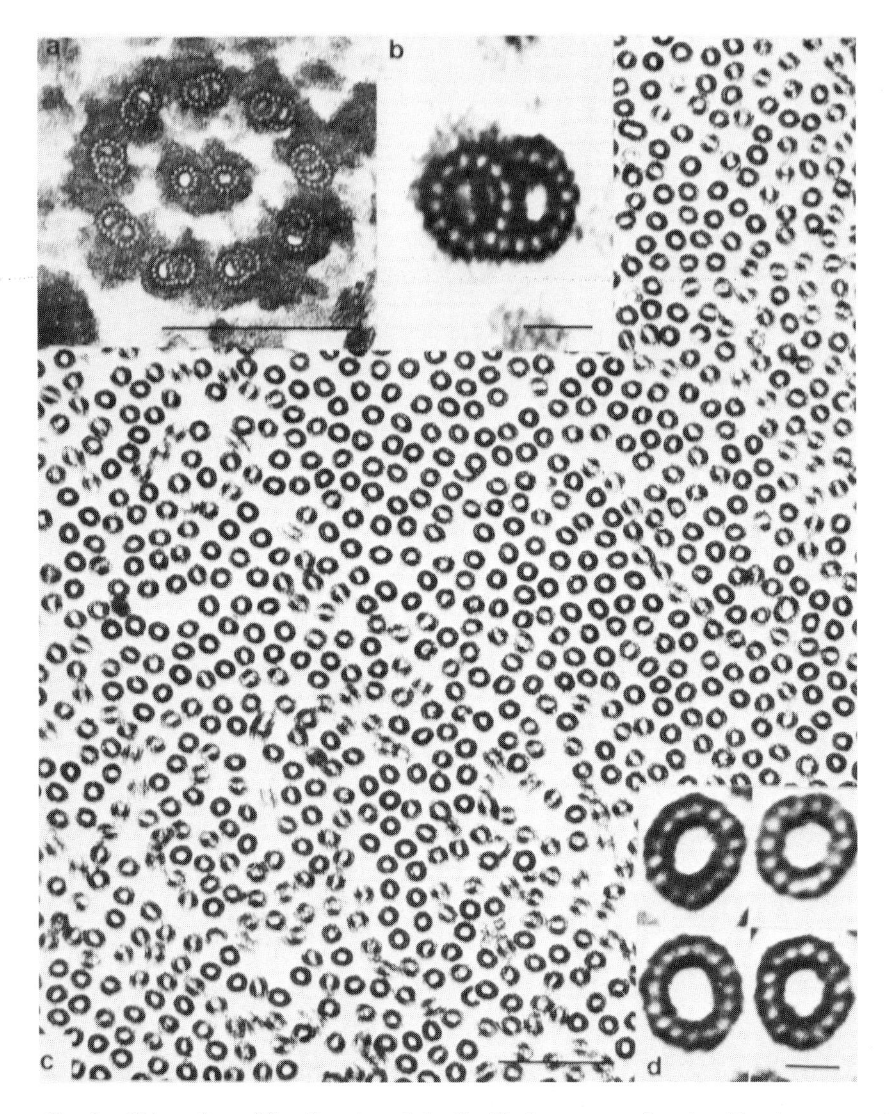

FIG. 3. Thin sections of flagellar microtubules fixed in the presence of tannic acid as they appear *in situ* and after assembly of S-1 outer doublet tubulin *in vitro*. (a) Intact axoneme showing nine outer doublets surrounding two central pair microtubules; bar, 0.2 μm; (b) outer doublet, demonstrating 13 protofilaments in the intact A-subfiber and 10 protofilaments in the C-shaped B-subfiber; bar, 20 nm; (c) single microtubules assembled from outer doublet, S-1 tubulin *in vitro;* bar, 0.2 μm; (d) higher magnification of the *in vitro*-assembled microtubules in (c), showing 14 and 15 protofilaments in the *in vitro*-assembled microtubules; bar = 20 nm. Reproduced from Binder and Rosenbaum (1978), with permission.

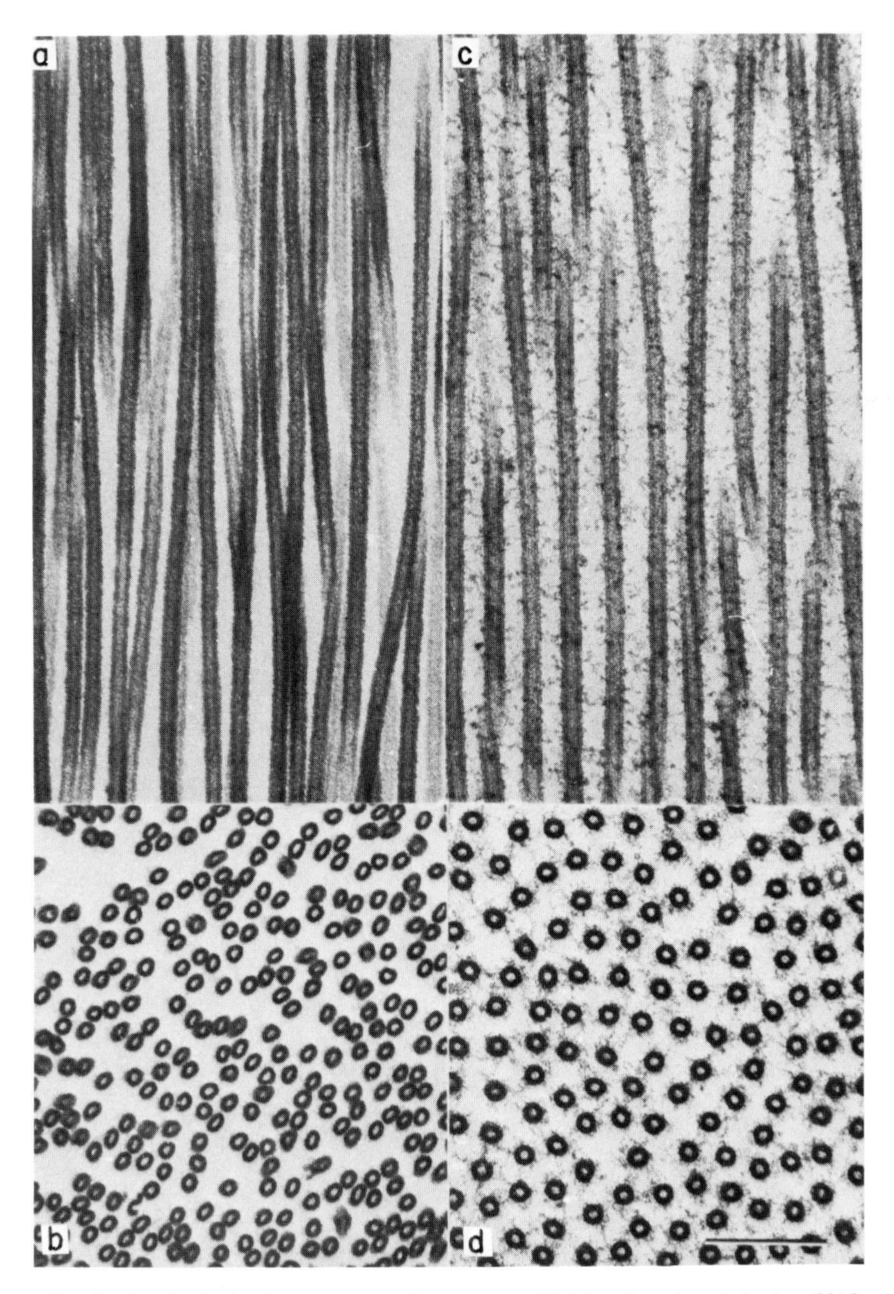

FIG. 4. Longitudinal and cross sections of *in vitro*-assembled flagellar microtubules (a and b) in the absence of brain MAPs and (c and d) in the presence of brain MAPs; bar = 0.2 μm. Reproduced from Binder and Rosenbaum (1978), with permission.

(i.e., 10% versus 25%). Furthermore, MAP 2 is preferentially bound to the invertebrate-derived microtubules (Farrell *et al.*, 1979b).

Microtubules reassembled *in vitro* from outer doublet tubulin in the absence of brain HMW MAPs tend to be very long, the average microtubule length of a population routinely lying in the 9–12 μm range. These microtubules also appear to be very flexible structures by negative stain electron microscopy, and even after repeatedly shearing them by passage through a 25 gauge needle the population average length is still routinely 4–8 μm. Similar treatment of vertebrate brain microtubules yields a population average length of 1–2 μm, suggesting that HMW MAPs may confer some mechanical rigidity on the microtubules.

A further peculiarity of microtubules reassembled from outer doublet tubulin is that they tend to stick together to form "sheets" of singlet microtubules when observed by negative stain electron microscopy (Kuriyama, 1976; Farrell and Wilson, 1978). The reason for this is unclear but may be related to the absence of associated proteins.

B. Stability

Microtubules reconstituted from outer doublet tubulin differ from the doublet microtubules not only in morphology but also with respect to their stability. Outer doublet microtubules are resistant to a variety of experimental treatments, including low temperature, calcium, and antimitotic drugs, and are classified as "stable" microtubules (Behnke and Forer, 1967). In contrast, the reconstituted singlet microtubules are depolymerized by each of the preceding treatments (Kuriyama, 1976; Binder and Rosenbaum, 1978; Farrell and Wilson, 1978; Farrell *et al.*, 1979b), a behavior more characteristic of "labile" cytoplasmic microtubules and reconstituted brain microtubules.

The cold lability of the reconstituted singlet microtubules allowed their purification by the assembly–disassembly method of Borisy *et al.* (1974). Twice recycled microtubules are at least 95% tubulin (Farrell and Wilson, 1978; Binder and Rosenbaum, 1978). Although HMW MAPs are not present in these preparations, traces of lower-molecular-weight proteins have been detected. The significance of these proteins for microtubule assembly is currently unclear.

C. Mechanism of Microtubule Elongation

Outer doublet tubulin assembly onto flagellar axonemes or basal bodies *in vitro* exhibits a biased polarity. Kinetically, growth is favored at the distal ends of these seeds (Binder and Rosenbaum, 1978). The heterologous assembly of vertebrate brain tubulin onto flagellar axonemes and basal bodies also shows the same phenomenon (Allen and Borisy, 1974; Snell *et al.*, 1974; Binder *et al.*, 1975; Rosenbaum *et al.*, 1975; Bergen and Borisy, 1980).

Biased directional growth is not a peculiarity of "seeded" assembly. Dimer additional to and loss from microtubules reassembled to apparent equilibrium from outer doublet and vertebrate brain tubulins occur at opposite ends of the microtubules. Under these steady-state conditions, dimer addition and loss occur at both microtubule ends, but net addition occurs at one of the microtubule ends, and net loss at the opposite end, resulting in an apparent unidirectional flux of tubulin dimers through the microtubules. Consequently, although net dimer addition can occur at the microtubule disassembly ends, growth is favored at the assembly ends (Margolis and Wilson, 1978; Farrell *et al.*, 1979b; Farrell and Wilson, 1980).

This microtubule behavior provides a basis for explaining the differential sensitivity of microtubules to antimitotic drugs, such as colchicine. Colchicine itself does not detectably inhibit dimer addition to microtubules, but must first complex with a tubulin dimer (CD complex) (Margolis and Wilson, 1977). CD complexes add to the growing ends of microtubules and decrease the apparent rate of subsequent tubulin dimer addition (Sternlicht and Ringel, 1979; Farrell and Wilson, 1980). Microtubule depolymerization occurs since dimer loss continues from the opposite ends of the microtubules. The marked difference in colchicine sensitivity for outer doublet microtubules and singlet microtubules reconstituted from outer doublet tubulin could therefore be attributed to the fact that the singlet microtubules are in a state of flux, whereas the outer doublets are "static" microtubules (Wilson and Margolis, 1978).

V. Conclusions

The optimal *in vitro* assembly requirements of outer doublet and vertebrate brain tubulins are very similar, as are the properties of the assembled microtubules. This appears to be generally true for all tubulins so far examined, irrespective of the properties, functions, cellular location, or phylogeny of the microtubules from which the tubulin derives. The reconstituted microtubules are "labile" (Behnke and Forer, 1967) singlet microtubules, archetypical of which are reconstituted vertebrate brain microtubules (Barnes *et al.*, 1975; Green *et al.*, 1975; Weisenberg and Rosenfeld, 1975; Kuriyama, 1976; Wiche and Cole, 1976; Doenges *et al.*, 1977; Nagle *et al.*, 1977; Wiche *et al.*, 1977; Binder and Rosenbaum, 1978; Maekawa and Sakai, 1978; Wetherbee *et al.*, 1978; Farrell *et al.*, 1979a; Stearns and Brown, 1979). These data argue that both the assembly properties of tubulins and the mechanism of microtubule formation have been highly conserved at a gross level of observation.

Structurally *bona fide* singlet microtubules can be formed *in vitro* from outer doublet and dogfish brain tubulins in the absence of associated proteins (Langford, 1977; Farrell and Wilson, 1978). This is also true of microtubules normally found in association with nontubulin proteins. For example, purified vertebrate brain tubulin can be induced to form microtubules under conditions of high magnesium or in the presence of dimethyl sulfoxide (Himes *et al.,* 1976; Lee and Timasheff, 1977; Herzog and Weber, 1977). There appears to be, therefore, sufficient information intrinsic to the tubulin dimer to specify construction of a labile singlet microtubule.

In contrast, under *in vitro* conditions tubulin dimers alone contain insufficient information to specify the diversity in microtubule properties observed *in vivo* (e.g., stability, doublet structure). There would appear to be a requirement for additional factors to generate this diversity, of which nontubulin associated proteins may be of paramount importance. Conceptually, microtubules may therefore be imagined as the structural framework to which nontubulin proteins, specifying precise properties or functions, may bind.

For example, vertebrate brain HMW MAPs stimulate both the rate and extent of tubulin assembly and the assembled microtubules are more stable to cold or drug treatments than microtubules reassembled in their absence (Sloboda *et al.,* 1976; Sloboda and Rosenbaum, 1979; Herzog and Weber, 1977). The HMW MAPs slow the flux rate of tubulin dimers in microtubules reassembled from outer doublet tubulin (Farrell *et al.,* 1979b) and may account for their assembly-promoting and stabilizing influences. Although the microtubule-associated proteins examined to date do not generate *in vitro* the observed *in vivo* diversity in microtubule properties, the studies clearly demonstrate the potential of nontubulin proteins for modifying microtubule behavior.

Failure to bind the correct nontubulin protein(s) may account for the formation of labile, singlet microtubules from outer doublet tubulin, instead of stable doublets. Conceivably, this class of proteins could be responsible for directing B-subfiber microtubule formation onto the A-subfiber wall; a potential candidate for this role has been reported in ciliary doublets (Amos *et al.,* 1976; Linck, 1976). Similarly, tubulin assembly *in vitro* at low physiological temperatures may occur efficiently in the presence of the correct regulator protein.

The influence of microtubule surface chemistry on associated protein binding is unclear, but circumstantial evidence suggests an important role. Flagellar and ciliary axonemal proteins repeat with periodicities based on the tubulin repeat in the microtubule lattice. Differences in the tubulin lattices of A- and B-subfiber microtubules (e.g., Linck and Amos, 1974; Amos and Klug, 1974; Amos *et al.,* 1976) may therefore account for selectivity in axonemal protein binding to these microtubules.

Posttranslation chemical modifications of tubulin (Eipper, 1972; Kuriyama and Sakai, 1974; Arce *et al.,* 1975; Raybin and Flavin, 1975; Piperno and Luck,

1976) suggest a potential basis for changing dimer conformation and microtubule lattice structure. This is clearly illustrated from the work on tubulin tyrosylation. From the predicted secondary structure of tubulin, Ponstingl *et al.* (1979) have suggested that tubulin tyrosylation could lead to structural rearrangement at the α-chain C-terminus. Other workers have shown that tubulin detyrosylation occurs more rapidly from tubulin in microtubules than from tubulin in solution (Thompson *et al.*, 1979) and indicates that tyrosylated tubulin undergoes a conformational change on incorporation into MTs.

Direct evidence that tubulin conformational changes alter the microtubule lattice is lacking. However, studies on the mechanism of colchicine poisoning of microtubule assembly are consistent with this hypothesis. Colchicine poisons microtubule assembly by first complexing with tubulin dimers, which then copolymerize with uncomplexed tubulin into the microtubules and decrease the apparent rate of subsequent drug-free dimer addition (Sternlicht and Ringel, 1979; Farrell and Wilson, 1980). Since colchicine binding induces a conformational change in the tubulin dimer (Garland, 1978), the altered drug-bound tubulin may disrupt the normal microtubule lattice, causing the observed decreased rate of tubulin addition. The action of colchicine on the tubulin dimer may be mimicking physiologically relevant conformational changes in tubulin, and endogenous colchicine-like molecules have been reported in vertebrate brain (Lockwood, 1979). This suggests the possibility that endogenous regulators may control microtubule properties via the formation of copolymers with altered structures.

If the composition and arrangement of structurally distinct tubulin dimers within microtubules are crucial to microtubule properties or functions, this information may become ''scrambled'' on microtubule reconstitution *in vitro*. The consequences of this scrambling could be the formation of microtubules that display features common to all tubulin populations (i.e., labile singlets) rather than features unique to ordered tubulin arrangements. By exploiting the advantages offered by the flagellar system of microtubules, it should be possible to examine the roles of tubulins and associated nontubulin proteins on microtubule properties and functions.

REFERENCES

Allen, C., and Borisy, G. G. (1974). *J. Mol. Biol.* **90,** 381–402.

Amos, L. A., and Klug, A. (1974). *J. Cell Sci.* **14,** 523–550.

Amos, L. A., Linck, R. W., and Klug, A. (1976). *Cold Spring Harbor Conf. Cell Proliferation* **3,** 847–867.

Arce, C. A., Rodriguez, J. A., Barra, H. S., and Caputto, R. (1975). *Eur. J. Biochem.* **59,** 145–149.

Bajer, A. S., and Mole-Bajer, J. (1972). *Int. Rev. Cytol., Suppl.* No. 3.

Bajer, A. S., and Mole-Bajer, J. (1975). *In* "Molecules and Cell Movement" (S. Inoué and R. Stephens, eds.), pp. 77–96. Raven, New York.
Barnes, L. D., Engel, A. G., and Dousa, T. P. (1975). *Biochim. Biophys. Acta* **405,** 422–433.
Behnke, O., and Forer, A. (1967). *J. Cell Sci.* **2,** 169–192.
Bergen, L., and Borisy, G. G. (1980). *J. Cell Biol.* **84,** 141–150.
Bibring, T., Baxandall, J., Denslow, S., and Walker, B. (1976). *J. Cell Biol.* **69,** 301–312.
Binder, L. I., and Rosenbaum, J. L. (1978). *J. Cell Biol.* **79,** 500–515.
Binder, L. I., Dentler, W. L., and Rosenbaum, J. L. (1975). *Proc. Natl. Acad. Sci. U.S.A.* **72,** 1122–1126.
Borisy, G. G., and Olmsted, J. B. (1972). *Science* **177,** 1196–1197.
Borisy, G. G., Olmsted, J. B., Marcum, J. M., and Allen, C. (1974). *Fed. Proc., Fed. Am. Soc. Exp. Biol.* **33,** 167–174.
Brinkley, B. R., and Cartwright, J., Jr. (1971). *J. Cell Biol.* **50,** 416–431.
Brinkley, B. R., and Cartwright, J., Jr. (1975). *Ann. N.Y. Acad. Sci.* **253,** 428–439.
Brinkley, B. R., and Stubblefield, E. (1970). *Adv. Cell Biol.* **1,** 119–185.
Brinkley, B. R., Stubblefield, E., and Hsu, T. C. (1967). *J. Ultrastruct. Res.* **19,** 1–18.
Burgess, J., and Northcote, D. H. (1969). *J. Cell Sci.* **5,** 433–451.
Doenges, K. H., Nagle, B. W., Uhlmann, A., and Bryan, J. (1977). *Biochemistry* **16,** 3449–3455.
Eipper, B. A. (1972). *Proc. Natl. Acad. Sci. U.S.A.* **69,** 2283–2287.
Farrell, K. W., and Wilson, L. (1978). *J. Mol. Biol.* **121,** 393–410.
Farrell, K. W., and Wilson, L. (1980). *Biochemistry* **19,** 3048–3054.
Farrell, K. W., Morse, A., and Wilson, L. (1979a). *Biochemistry* **18,** 905–911.
Farrell, K. W., Kassis, J. A., and Wilson, L. (1979b). *Biochemistry* **18,** 2642–2647.
Feit, H., Neudeck, U., and Gaskin, F. (1977). *J. Neurochem.* **28,** 697–706.
Forer, A. (1969). *In* "Handbook of Molecular Cytochemistry" (A. Lima-de-Faria, ed.), pp. 553–604. North-Holland Publ., Amsterdam and Wiley (Interscience), New York.
Franke, W. W. (1971a). *Exp. Cell Res.* **66,** 486–489.
Franke, W. W. (1971b). *Protoplasma* **73,** 263–292.
Garland, D. (1978). *Biochemistry* **17,** 4266–4272.
Gaskin, F., Cantor, C. R., and Shelanski, M. L. (1974). *J. Mol. Biol.* **89,** 737–755.
Gibbons, I. R. (1965). *Arch. Biol.* **76,** 317–374.
Gibbons, I. R. (1975). *In* "Molecules and Cell Movement," (S. Inoué and R. Stephens, eds.), pp. 207–232. Raven, New York.
Gibbons, I. R., and Fronk, E. (1972). *J. Cell Biol.* **54,** 365–381.
Gibbons, I. R., and Rowe, A. J. (1965). *Science* **149,** 424–426.
Gray, E. G., and Westrum, L. E. (1976). *Cell Tissue Res.* **168,** 445–454.
Green, L. H., Brandis, J. W., Turner, F. R., and Raff, R. A. (1975). *Biochemistry* **14,** 4487–4491.
Herzog, W., and Weber, K. (1977). *Proc. Natl. Acad. Sci. U.S.A.* **74,** 1860–1864.
Himes, R. H., Burton, P. R., Kersey, R. N., and Pierson, G. B. (1976). *Proc. Natl. Acad. Sci. U.S.A.* **73,** 4397–4399.
Kiefer, B. I. (1970). *J. Cell Sci.* **6,** 177–193.
Kobayashi, Y., and Mohri, H. (1977). *J. Mol. Biol.* **116,** 613–617.
Kuriyama, R. (1975). *J. Biochem. (Tokyo)* **77,** 23–31.
Kuriyama, R. (1976). *J. Biochem. (Tokyo)* **80,** 153–165.
Kuriyama, R., and Sakai, H. (1974). *J. Biochem. (Tokyo)* **76,** 651–654.
Langford, G. M. (1977). *Fed. Proc., Fed. Am. Soc. Exp. Biol.* **36,** 899.
Langford, G. M. (1978). *Exp. Cell Res.* **111,** 139–151.
Lee, J. C., and Timasheff, S. N. (1977). *Biochemistry* **16,** 1754–1764.
Linck, R. W. (1976). *J. Cell Sci.* **20,** 405–440.

Linck, R. W., and Amos, L. (1974). *J. Cell Sci.* **14,** 551–559.
Lockwood, A. H. (1979). *Proc. Natl. Acad. Sci. U.S.A.* **76,** 1184–1188.
McIntosh, J. R., Cande, Z., Snyder, J., and Vanderslice, K. (1975). *Ann. N.Y. Acad. Sci.* **253,** 407–427.
Maekawa, S. and Sakai, H. (1978). *J. Biochem. (Tokyo)* **83,** 1065–1075.
Margolis, R. L., and Wilson, L. (1977). *Proc. Natl. Acad. Sci. U.S.A.* **74,** 3466–3470.
Margolis, R. L., and Wilson, L. (1978). *Cell* **13,** 1–8.
Marotta, C. A., Harris, J. L., and Gilbert, J. M. (1978). *J. Neurochem.* **30,** 1431–1440.
Mooseker, M. S., and Tilney, L. G. (1973). *J. Cell Biol.* **56,** 13–26.
Murphy, D. B., and Borisy, G. G. (1975). *Proc. Natl. Acad. Sci. U.S.A.* **72,** 2696–2700.
Nagle, B. W., Doenges, K. H., and Bryan, J. (1977). *Cell* **12,** 573–586.
Olmsted, J. B., and Borisy, G. G. (1973). *Biochemistry* **12,** 4284–4289.
Olmsted, J. B., and Borisy, G. G. (1975). *Biochemistry* **14,** 2996–3005.
Olmsted, J. B., Marcum, J. M., Johnson, K. A., Allen, C., and Borisy, G. G. (1974). *J. Supramol. Struct.* **2,** 429–450.
Oosawa, F., and Kasai, M. (1962). *J. Mol. Biol.* **4,** 10–21.
Pfeffer, T., Asnes, C., and Wilson, C. (1978). *Cytobiologie* **16,** 367–372.
Pierson, G. B., Burton, P. R., and Himes, R. H. (1978). *J. Cell Biol.* **76,** 223–228.
Piperno, G., and Luck, D. (1976). *J. Biol. Chem.* **251,** 2161–2167.
Ponstingl, H., Little, M., Krauhs, E., and Kempf, T. (1979). *Nature (London)* **282,** 423–424.
Raybin, D., and Flavin, M. (1975). *Biochem. Biophys. Res. Commun.* **65,** 1088–1095.
Rosenbaum, J. L., Binder, L. I., Granett, S., Dentler, W. L., Snell, W., Sloboda, R., and Haimo, L. T. (1975). *Ann. N.Y. Acad. Sci.* **253,** 147–177.
Roth, L. E. (1967). *J. Cell Biol.* **34,** 47–59.
Roth, L. E., Pihlaja, D. J., and Shigenaka, Y. (1970). *J. Ultrastruct. Res.* **30,** 7–37.
Shelanski, M. C., Gaskin, F., and Cantor, C. R. (1973). *Proc. Natl. Acad. Sci. U.S.A.* **70,** 765–768.
Sloboda, R. D., and Rosenbaum, J. L. (1979). *Biochemistry* **18,** 48–55.
Sloboda, R. D., Dentler, W. L., and Rosenbaum, J. L. (1976). *Biochemistry* **15,** 4497–4505.
Smith, D. S., Jarlfors, U., and Beraneck, R. (1970). *J. Cell Biol.* **46,** 199–219.
Snell, W. J., Dentler, W. L., Haimo, L. T., Binder, L. I., and Rosenbaum, J. L. (1974). *Science* **185,** 357–360.
Stearns, M. E., and Brown, D. L. (1979). *Proc. Natl. Acad. Sci. U.S.A.* **76,** 5745–5749.
Stephens, R. E. (1969). *Q. Rev. Biophys.* **1,** 377–390.
Stephens, R. E. (1970a). *J. Mol. Biol.* **47,** 355–363.
Stephens, R. E. (1970b). *Biol. Bull. (Woods Hole, Mass.)* **139,** 438.
Stephens, R. E. (1978). *Biochemistry* **17,** 2882–2891.
Sternlicht, H., and Ringel, I. (1979). *J. Biol. Chem.* **254,** 10504–10550.
Sullivan, K. F., Farrell, K. W., and Wilson, L. (1979). *J. Cell Biol.* **83,** 351a.
Thompson, W. C., Deanin, G. G., and Gordon, M. W. (1979). *Proc. Natl. Acad. Sci. U.S.A.* **76,** 1318–1322.
Tilney, L. G. (1971). *J. Cell Biol.* **51,** 837–854.
Tilney, L. G., and Gibbins, J. R. (1968). *Protoplasma* **65,** 167–179.
Tilney, L. G., Bryan, J., Bush, D. J., Fujiwara, K., Mooseker, M. S., Murphy, D. B., and Snyder, D. H. (1973). *J. Cell Biol.* **59,** 267–275.
Tucker, J. B. (1968). *J. Cell Sci.* **3,** 493–514.
Tucker, J. B. (1970). *J. Cell Sci.* **6,** 385–429.
Warner, F. D. (1972). *Adv. Cell Mol. Biol.* **2,** 193–235.
Warner, F. D. (1976). *Cold Spring Harbor Conf. Cell Proliferation* **3,** 891–914.

Warner, F. D., and Satir, P. (1974). *J. Cell Biol.* **63,** 35–63.
Weingarten, M. A., Lockwood, A. H., Hwo, S. Y., and Kirschner, M. C. (1975). *Proc. Natl. Acad. Sci. U.S.A.* **72,** 1858–1862.
Weisenberg, R. C. (1972). *Science* **177,** 1104–1105.
Weisenberg, R. C., and Deery, W. J. (1976). *Nature (London)* **263,** 792–793.
Weisenberg, R. C., and Rosenfeld, A. C. (1975). *J. Cell Biol.* **64,** 146–158.
Wetherbee, J. A., Luftig, R. B., and Weihing, R. R. (1978). *J. Cell Biol.* **78,** 47–57.
Wiche, G., and Cole, R. D. (1976). *Proc. Natl. Acad. Sci. U.S.A.* **73,** 1227–1231.
Wiche, G., Lunblad, V. J., and Cole, R. D. (1977). *J. Biol. Chem.* **252,** 794–796.
Wilson, H. (1969). *J. Cell Biol.* **40,** 854–859.
Wilson, L., and Margolis, R. L. (1978). *In* "Cell Reproduction: Essays in Honor of Daniel Mazia" (E. R. Dirksen, D. M. Prescott, and C. F. Fox, eds.), pp. 241–257. Academic Press, New York.
Witman, G. B., Carlson, K., and Rosenbaum, J. L. (1972). *J. Cell Biol.* **54,** 540–555.

Chapter 6

Production of Antisera and Radioimmunoassays for Tubulin

LIVINGSTON VAN DE WATER III,[1] SUSAN D. GUTTMAN,[2] MARTIN A. GOROVSKY, AND J. B. OLMSTED

Department of Biology
University of Rochester
Rochester, New York

I. Introduction

Microtubules are widely distributed organelles that have a variety of functions (for reviews, see Dustin, 1979; Goldman *et al.*, 1976). The amount of mi-

[1]*Current affiliation:* Department of Pathology, Beth Israel Hospital, Boston, Massachusetts
[2]*Current affiliation:* Department of Pharmacology, Stanford University School of Medicine, Stanford, California

ISBN 0-12-564124-9

crotubule protein has been measured in a number of cell types using quantitative gel analyses and drug-binding assays. Recently, antisera to tubulin have been generated for use in both cytochemical and quantitative studies. This chapter discusses the production of antisera raised against electrophoretically purified and SDS-treated *Tetrahymena* tubulin; these antisera react with tubulin from a number of species. Reports on the reactivity of these antisera with *Tetrahymena* tubulin (Guttman, 1978; Guttman and Gorovsky, 1979) and mammalian neuronal tubulins (Van De Water, 1979; Van De Water and Olmsted, 1978, 1980) have appeared. This chapter will emphasize the methodology used for raising antisera of this type and the parameters that were found most useful in developing reliable radioimmunoassays.

II. Methods

A. Production of Tubulin Antisera

Cilia were isolated from *Tetrahymena* using the ethanol–sucrose procedure of Gibbons (1965). Isolated cilia were pelleted at 16,000 *g* for 20 min, and axonemes were prepared by dialysis against Tris-EDTA buffer according to procedure 3 of Renaud *et al.* (1968). Isolated axonemes were precipitated with 6 vol of acetone and dried under vacuum. Precipitates were dissolved in sample buffer (Laemmli, 1970), boiled for 3 min, and run on SDS-containing polyacrylamide gels (see Section II,F). Tubulin was localized by scanning gels or gel slices at 280 nm, marking the tubulin band with ink, and excising the strip corresponding to tubulin from the unstained gel.

Protein was eluted from the unstained gel slices using a modification of the method of Lazarides (1976). Gel slices were placed in a glass tube to which a dialysis bag was attached; the slices were held in place with absorbent tissue and the tube was filled with Laemmli sample buffer. Electrophoresis was carried out at 240 V for 18–24 hr. Eluted tubulin was dialyzed for 18 hr at room temperature against three 1-liter changes of 0.2 *M* NH_4HCO_3 containing 0.05% SDS. The sample was concentrated in the dialysis bag to a volume of about 1 ml using dry Sephadex G-25 (300 mesh) and was then precipitated overnight with a sixfold volume of acetone. The acetone precipitate was collected by centrifugation at 10,000 rpm for 10 min, and the resulting pellet was dried under vacuum and then redissolved in 0.5–1.0 ml of complete Laemmli sample buffer by boiling for 3 min.

The SDS-treated samples were rerun on SDS gels to determine the purity and concentration of the tubulin. Gels run with known amounts of tubulin were stained with Fast Green (Gorovsky *et al.*, 1970) and quantitative densitometry was used to establish a standard curve from which the tubulin concentration of the immunogen was determined. In a typical preparation, acetone-precipitated axonemes from 1 liter of *Tetrahymena* (2–3 $\times$ 10^5 cells/ml) were dissolved in 1.1 ml of SDS sample buffer, and 0.8 ml was run on a 4-mm-thick 7.5% acrylamide slab gel for elution. The final yield of purified eluted tubulin was between 500 and 800 μg. Preparations that showed lower-molecular-weight contaminants that might be indicative of breakdown products were not used for immunization.

For injection, 350 μg of tubulin in 0.7 ml of SDS sample buffer was mixed with 0.7 ml of complete Freund's adjuvant. This suspension was emulsified by 20–30 vigorous passes through a glass syringe, although a typical emulsion was not formed because of the presence of SDS. Rabbits (New Zealand White) were injected with a total of 1.4 ml of the emulsion (350 μg/rabbit) placed subcutaneously in a total of two sites on either side of the backbone (0.7 ml/site). Rabbits were boosted one week later with the same amount of protein in complete Freund's adjuvant prepared as described earlier. Animals were first bled two weeks after the second injection and were subsequently bled every 5–7 days. Preimmune serum was obtained from animals one week prior to the first injection. The data presented in this paper were obtained using antisera from a total of four rabbits injected according to this procedure.

An alternate injection schedule was used in earlier experiments (Guttman, 1978; Guttman and Gorovsky, 1979; Van De Water, 1979; Van De Water and Olmsted, 1980). Animals were injected subcutaneously on days 1 and 8 as above, followed by intravenous injections in the upper marginal ear vein with 150 μg tubulin in 0.2 ml SDS sample buffer on days 22, 29, 36, 38, and 40. They were also boosted biweekly with intravenous injections. The animals were first bled 45 days after the initial injection and 5–7 days after each booster injection. These antisera had lower titers than the antisera produced by the preceding procedure.

B. Preparation of Protein A Adsorbent (PAA)

Quantitation of immune complexes was carried out using protein-A-bearing strains of *Staphylococcus aureus* as an immunoadsorbent. The Cowan I strain of *S. aureus* was grown, harvested, fixed, and heat-treated after 16–17 hr of exponential growth as described by Kessler (1976). Stock solutions of the protein A adsorbent (PAA) were stored at 4°C in phosphate-buffered saline (PBS: 0.14 *M* NaCl, 3 m*M* KCl, 1.5 m*M* KH_2PO_4, 6.7 m*M* $Na_2HPO_4 \cdot 7\,H_2O$, pH 7.2) containing 0.2% sodium azide. As described by Kessler (1976), PAA was washed and incubated for 15 min at room temperature in NET buffer (0.15 *M* NaCl, 5 m*M* EDTA, 0.05 *M* Tris, 0.02% sodium azide, pH 7.4) containing 0.5% Nonidet P-40, followed by washing in NET containing 0.05% NP-40. Samples were resuspended to 10% v/v in NET-0.05% NP-40, 5 mg/ml BSA for use in the

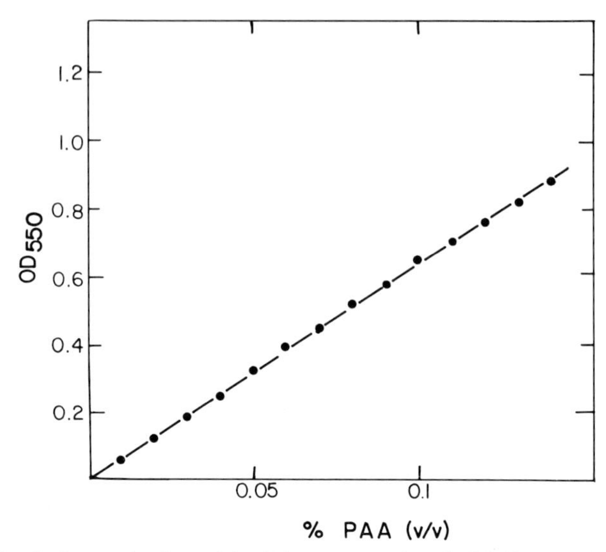

FIG. 1. Standard curve for determining PAA concentration. Serial dilutions of PAA were prepared in NET–BSA containing 0.05% NP-40 and the absorbance at 550 nm was determined. Dilutions were then centrifuged in hematocrit tubes to obtain a v/v measure of PAA content.

assay. The concentration of this solution was adjusted by determining the OD_{550} of diluted samples (Fig. 1).

C. Purification and Iodination of Antigens

Microtubule protein was purified from hog brain by repetitive cycles of assembly–disassembly (Borisy *et al.*, 1975). Tubulin was purified from microtubule-associated proteins by DEAE-Sephadex (Murphy *et al.*, 1977) or phosphocellulose (Sloboda *et al.*, 1976) chromatography. Purity was assessed by quantitative densitometry of gels stained with Fast Green (Gorovsky *et al.*, 1970) or Coomassie Blue (Fairbanks *et al.*, 1971).

Competitor tubulin and iodinated tubulin were prepared from samples determined by quantitative gel analyses to be greater than 95% pure. Iodinations were carried out using the method of Hunter and Greenwood (1962). Typically, 10 μg of protein in 10 μl or less was mixed with 25 μl of 0.5 *M* sodium phosphate buffer, pH 7.2. To the sample was added 250 μCi of $Na^{125}I$ and 10 μl of 5 mg/ml chloramine T in 0.05 *M* sodium phosphate buffer. After 60 sec the reaction was stopped by the addition of 25 μl of 2.5 mg/ml sodium metabisulfite in 0.05 *M* sodium phosphate buffer, pH 7.2. Free iodine was separated by chromatography of the sample on a 0.7 cm I.D. × 13 cm G-25 (150 mesh) Sephadex column equilibrated in 0.05 *M* sodium phosphate buffer and prerun with 1 ml of 2%

BSA. Fractions (0.5 ml) were collected into tubes containing 50 μl of 2% BSA in 0.05 *M* phosphate buffer, and those containing bound radioactivity were aliquoted, frozen in liquid nitrogen, and stored at −80°C. The specific activity of the labeled tubulin was routinely 10,000 cpm/ng, corresponding to approximately 0.5–1.0 iodine molecules/110,000 dalton tubulin dimer.

D. Binding Assays

1. Serum Dilution Assays

The extent of binding of antisera to labeled tubulin and the relative binding capacity of various antisera were tested by carrying out serum dilution experiments. Dilutions of sera were obtained by serial passage of a fixed volume of serum (usually 10, 25, or 50 μl) into reaction test tubes (12 × 75 mm) that contained 25 μl of SBA–BSA (0.15 *M* NaCl, 0.05 *M* sodium borate, 0.02% sodium azide, pH 7.4 containing 5 mg/ml BSA). Iodinated tubulin diluted in SBA–BSA to contain 10,000 cpm/25 μl was added, and incubation carried out for 2 hr at 0°C. A volume of 10% PAA sufficient to absorb the highest concentration of serum was then added (usually 50 or 100 μl), and incubation at 0°C continued for 10 min. The samples were mixed with 2 ml SBA–BSA and centrifuged at 3000 rpm for 10 min. The supernatant was decanted, and the pellet resuspended in 2 ml SBA–BSA. Following centrifugation, the pellet was resuspended in 2 ml NET–BSA. The resuspended pellets were poured onto 0.2-μm cellulose acetate filters that had been precoated with 1 ml NET–BSA. The tubes were rinsed twice with 1 ml of NET–BSA, and suction was then applied. Filters were subsequently washed twice with 1 ml of NET–BSA under suction, dried, and counted in toluene-based fluor in a liquid scintillation counter. Controls included incubations of labeled tubulin with serial dilutions of preimmune serum and with buffer alone (background). The number of input counts/tube was determined by TCA precipitation of an aliquot of the diluted tubulin. The percent bound was calculated as the fraction of the TCA precipitable counts bound at a given serum concentration, after subtraction of background counts from both samples.

2. Competition Assays

For competition assays, serial dilutions of a known amount of tubulin were prepared in the reaction tubes as described earlier. Typically, unlabeled tubulin at concentrations from 1 ng to 1 μg was used to establish a standard curve. To 25 μl of the diluted protein was added 25 μl of iodinated tubulin (10,000 cpm, 1 ng), followed by 50 μl of antiserum at a dilution that bound 50% of the input counts. Incubation of the reaction mixture, addition of 10% PAA (50 μl), and processing of the filters were carried out as previously described. Controls included samples in which unlabeled tubulin was deleted (uncompeted samples) and in which preimmune serum or buffer and labeled tubulin were incubated (background samples).

To determine tubulin content in cellular extracts, serial dilutions of the experimental samples were made and incubated with labeled tubulin and antiserum as described earlier. Values for the experimental samples were computed from the standard curves using the logit transformation of Rodbard *et al.* (1969).

E. Immunostaining Procedures

1. Gels

The distribution of antigenic species fractionated on SDS gels was determined using the immunostaining procedure of Burridge (1978). Typically, gels were fixed overnight in 10% acetic acid, 50% methanol, agitated for 4 hr in several changes of the same solution, and then equilibrated in washing buffer (WB: 0.15 *M* NaCl, 10 m*M* Tris, 0.1% azide, pH 7.4) until a pH of 7.4 was attained. Strips of gel were overlayered with antiserum (1/10 dilution in WB containing 10 mg/ml BSA) and left at room temperature for 24 hr. The gel strips were then washed for four days with several changes of WB before overlayering with iodinated protein A (Amersham; 30 mCi/mg; final of 2–5 × 10^6 cpm/ml diluted in WB containing 10 mg/ml BSA). Washing continued for four additional days before gels were stained with Coomassie Blue (Fairbanks *et al.*, 1971) and autoradiographed with X-Omat XR-5 film.

2. Cells

Cells plated on coverslips were fixed for 15–30 min in 3.7% formalin in PBS, washed with several changes of PBS, and then extracted with −20°C acetone for 30–60 sec. Coverslips were incubated sequentially with immune serum (1/30 dilution in PBS) and fluoroisothiocyanate-labeled goat antirabbit IgG antibody (Miles; 1/30 in PBS) for 30 min at 37°C with intervening washes in PBS. Coverslips were mounted in water on pieces of coverslip and sealed to slides with nail polish. Cells were photographed using a Zeiss narrow-band FITC filter and a 40 or 63 × planapochromat objective, and using Tri-X film developed with Diafine at an ASA of 1600.

F. Gel Electrophoresis

Electrophoresis was carried out according to the method of Laemmli (1970), using a 7.5% acrylamide running gel and a 3% stacking gel. Analytical slab gels were 1.3 mm thick, and preparative slab gels were 4 mm thick. Electrophoresis was normally carried out at 30 mA for 2–6 hr, and gels were stained either with Coomassie Blue (Fairbanks *et al.*, 1971) or with Fast Green (Gorovsky *et al.*, 1970).

III. Results

A. Assay Method

Because of our interest in developing antisera to use for measuring tubulin content in cell extracts, assay methods were sought that would give accurate quantitation of antibody–antigen binding. Double immunodiffusion and precipitin curve analyses were used for characterization of the initial antiserum and its reaction with *Tetrahymena* tubulin (Guttman, 1978; Guttman and Gorovsky, 1979). However, these methods required large amounts of serum, and precipitation of immune complexes did not occur as readily with mammalian cell extracts. Antibody–antigen binding was therefore measured by an indirect precipitation method in which protein-A-bearing strains of *S. aureus* were used as the precipitating agent. This precipitation method was as efficient in binding immune complexes as conventional second antibody techniques, but resulted in lower backgrounds and was more economical. The following discusses the general conditions for the optimization of this assay.

1. Protein-A-adsorbent (PAA) Concentration

For a given concentration of serum to be used in serum dilution curves or the radioimmunoassay, it was necessary to establish the optimum amounts of PAA needed to quantitatively absorb the immune complexes. Figure 2 shows a typical PAA saturation curve in which fixed amounts of labeled tubulin and antibody were incubated with a fixed volume of PAA diluted to various concentrations. For each preparation of PAA, this type of curve was established. Typically, a volume of 10% PAA was used that would give binding in the middle of the plateau region of the curve.

2. Preparation of Labeled Tubulin

Iodination by the chloramine T method was carried out using $Na^{125}I$ that had been commercially prepared within two weeks. If older iodine was used, less efficient incorporation of label and lower binding to tubulin was obtained. Iodination using the Bolton-Hunter reagent (Bolton and Hunter, 1973) was not satis-

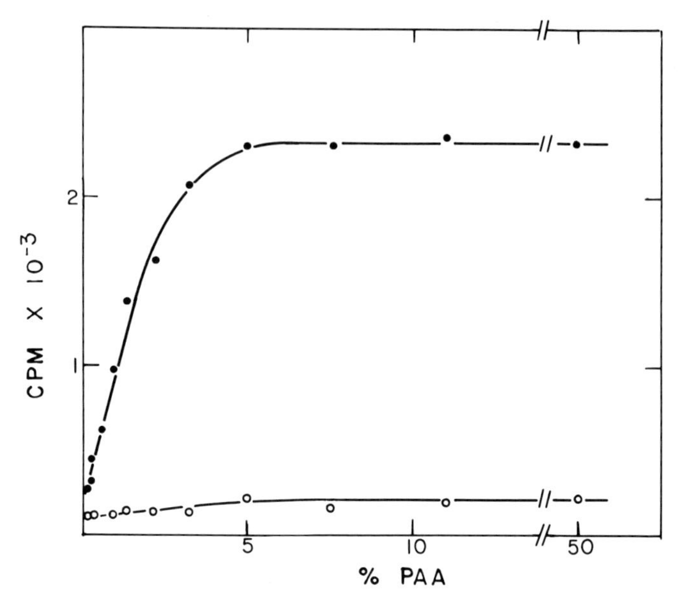

FIG. 2. PAA saturation curve. Twenty-five μl of immune serum (1/100 dilution) (●) or SBA-BSA (○) was incubated with 25 μ1 (2500 cpm) of iodinated tubulin for 2 hr at 0°C; 50 μ1 of PAA, serially diluted between 0.1 and 50% v/v, was added to each sample, and the samples were then processed as described in Section II,D.

factory, presumably because of reaction with sites with which the antibody complexed. Iodinated proteins were stable for up to three months at −80°C, with less than 10–15% loss of binding to antibody.

3. Assay Procedures

Procedures were optimized such that greater than 90% binding of antibody to antigen occurred under conditions of antibody excess, and background levels were routinely less than 10% of the input counts. Inclusion of BSA in all incubation buffers was essential to minimize nonspecific absorbtion. Background levels were dependent on the number of washes and the age of the PAA preparation used. A minimum of one 2-ml wash by centrifugation and two rinses of the filter were essential to obtain quantitative recovery and low backgrounds. Precoating the filters with 1 ml of NET–BSA also reduced nonspecific trapping. Backgrounds increased as the PAA preparation increased in age, although binding capacity did not appear to change. Routinely, PAA was prepared immediately before use by incubation with the solutions indicated in Section II,B; if less than 24 hr elapsed before the next assay, PAA was only rewashed with NET containing 0.05% NP 40 and BSA.

The times necessary for maximum binding were also investigated. With these sera, maximum binding of 90% of the input iodinated antigen to saturating amounts of antibody required a minimum of 2 hr at 0°C. In contrast, the interaction of PAA with immune complexes occurred rapidly and with high affinity; no change was observed in the amount of immune complex precipitated for incubation times of 10 min to 2 hr at 0°C.

The reproducibility from assay to assay was very high. However, in order to minimize variation within each assay, glass capillary pipets were used for all dilutions and additions of reagents; use of automatic pipeting devices significantly decreased the accuracy of the assay.

B. Characterization of Antisera Binding

Bleeds of the four immunized rabbits were initially assayed using serum dilution curves. A previously tested antiserum was assayed simultaneously to allow normalization of the data from the various bleeds. Figure 3 shows a serum dilution curve typical of the animals immunized by the methods outlined, in which greater than 90% of the labeled tubulin is bound under conditions of antibody excess. Sera obtained from three of the animals two weeks after the

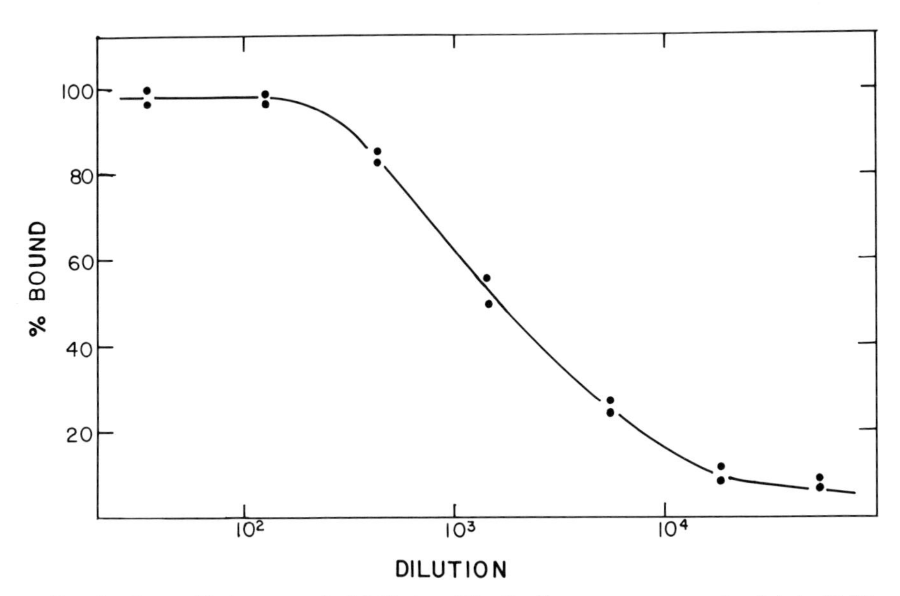

FIG. 3. Serum dilution curve. Serial dilutions (25 μl) of immune serum were incubated with 25 μl of purified iodinated hog brain tubulin (10,000 cpm, 1 ng) as described in Section II,D,1. Background counts for preimmune controls were 220 cpm (2.2% of input counts) and those for buffer controls were 150 cpm (1.5%).

final injection all showed maximum binding to tubulin, and the fourth serum bound approximately 70% of the labeled tubulin. After an additional week, and for the succeeding two months, all four sera bound greater than 90% of the labeled antigen, and all showed 50% binding at serum dilutions of 1/500 to 1/1000. At the time of sacrifice (four months post injection), tubulin binding capacities of the sera had started to decrease, but 50% binding still ranged between dilutions of 1/200 and 1/1000, depending on the animal; all animals still bound greater than 90% of the labeled antigen. Because the binding to tubulin remained high over a long period, no additional booster injections were given. In fact, antibody titers decreased rapidly when the earlier protocol utilizing frequent booster injections was employed.

C. Characterization of Binding Specificity

The specificity of the antisera for binding to tubulin was characterized in a number of ways. Although direct precipitation of immune complexes with similar antisera had been reasonably efficient for *Tetrahymena* tubulin (Guttman, 1978; Guttman and Gorovsky, 1979), mammalian tubulins did not form precipitating complexes. It was therefore felt that Ouchterlony analyses and rocket immunoelectrophoresis could not be routinely used as indicators of antibody specificity and titer. Instead, the binding of the antisera to labeled tubulin under conditions of antibody excess was routinely monitored. All of the sera maximally bound purified labeled tubulin at levels of 90% or greater. This was a qualitative indication not only that the sera contained tubulin antibodies, but also the sera did not react with a subset of tubulin, as had previously been found for another tubulin antiserum (Van De Water and Olmsted, 1978). As described below, the specificity of antiserum binding to tubulin was also demonstrated by experiments in which tubulin from *Tetrahymena,* hog, mouse, and neuroblastoma cells competed the binding of iodinated hog brain tubulin to background levels.

The use of antibodies for immunofluorescent staining has demonstrated the network of microtubules that exists in various cell types. As shown in Fig. 4, the antisera raised against *Tetrahymena* tubulin stained the microtubules of neuroblastoma cells. The background levels of staining with these sera were very low, indicating little nonspecific absorption, and staining was abolished if preimmune serum or immune serum absorbed with tubulin was used in the primary incubation. These data also demonstrated that the antisera reacted well with aldehyde-fixed microtubules.

To determine the protein species to which the antisera bound, cell extracts to be assayed (usually brain or neuroblastoma cells) were fractionated on SDS gels, and "stained" with the antibody and iodinated protein A (Burridge, 1978). As shown in Fig. 5, only the tubulin doublet in a complex extract of mouse neuroblastoma cells reacted with the antiserum; no bands were observed with preim-

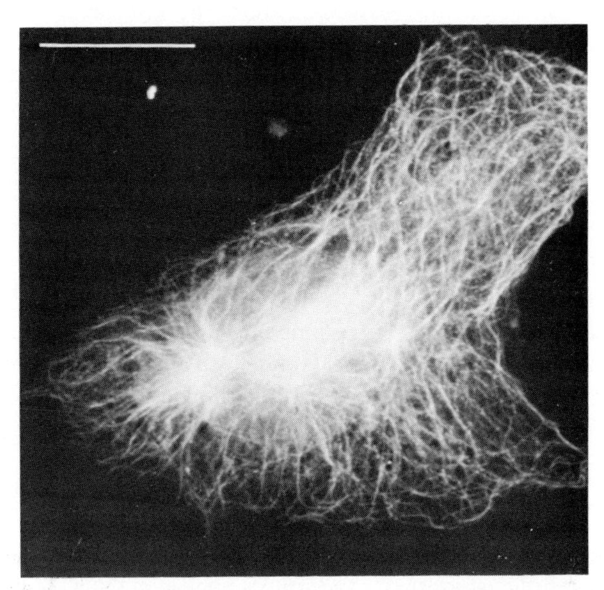

FIG. 4. Immunofluorescent staining of microtubules in neuroblastoma cells. Mouse neuroblastoma cells (Nb2a-AB-1) were cultured on coverslips for 24 hr before being processed for immunofluorescent staining as described in Section II,E,2. Magnification: ×1120. Bar = 20 μm.

mune serum. However, it has been proposed that some antigens may lose reactivity after fractionation on SDS gels (Burridge, 1978). Therefore, in separate experiments, extracts from neuroblastoma cells labeled *in vivo* were incubated with saturating amounts of antiserum, treated with PAA, and the number of counts bound determined. It was found that the bound fraction was equivalent to the percentage of counts in the alpha and beta tubulin bands isolated from gels. In addition, these bound counts could be displaced by incubation with purified tubulin. These experiments taken together demonstrate that the antisera are spe-

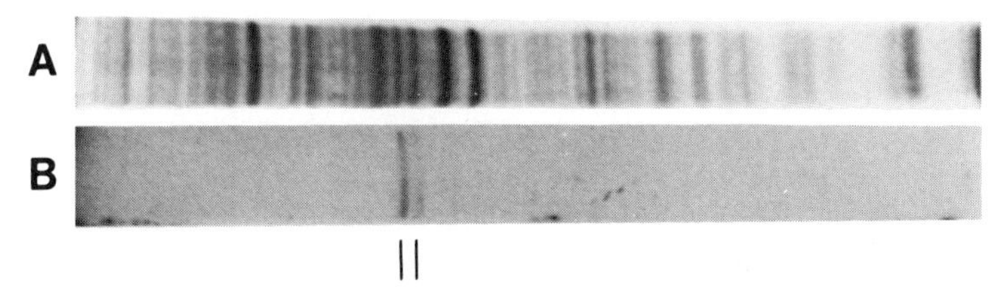

FIG. 5. Gel electrophoresis of neuroblastoma extract. A postmitochondrial supernatant from neuroblastoma cells was electrophoresed on an SDS-polyacrylamide gel, and incubated with immune serum and protein A as described in Section II,E,1. (A) Gel stained with Coomassie Blue. (B) Autoradiogram of same gel. Lines denote tubulin bands.

cific for tubulin in mammalian extracts and that the antisera react with both alpha and beta tubulins.

Cross reactivity of the antisera with mammalian tubulin was also demonstrated by the adsorption of the immune IgG fraction to an affinity column prepared with purified tubulin isolated from hog brain. This affinity purified material bound efficiently to hog and neuroblastoma tubulins, and also to *Tetrahymena* tubulin (F. Calzone, unpublished observations).

D. Radioimmunoassay

In order to develop a reliable radioimmunoassay (RIA) for quantitation of tubulin, a number of parameters were explored. In all the experiments described, purified hog brain tubulin was used as the iodinated antigen (tracer). For the RIA, dilutions of serum were chosen at which 50% binding of input tracer occurred; this value was determined from a serum dilution curve, such as that shown in Fig. 3, and usually corresponded to a 1/500 to 1/1000 dilution of serum. It has been reported that the sensitivity of a RIA may be increased by prolonged incubations with lower serum concentrations (Haber and Poulson, 1974). However, the majority of experiments for which the RIA was developed involved incubation of cell extracts, and although proteolysis did not appear to be a problem (Van De Water and Olmsted, 1980), attempts were made to keep incubation times to a minimum.

Figure 6 shows a typical standard curve in which purified hog brain tubulin was used as unlabeled competitor. The variable Y (or % bound) was calculated for samples with standard and unknown amounts of tubulin by the equation

$$Y = [(B - N)/(B_0 - N)] \times 100$$

where the amount of radioactivity bound to the antibody in the presence (B) or absence (B_0) of competitor is normalized for background counts (N).

Figure 6 also demonstrates the same data recalculated using the logit function of Rodbard *et al.* (1969), which converts the sigmoidal binding curve to a linear function. The logit function was calculated as

$$\text{logit } Y = \ln(Y/100 - Y)$$

and results in a straight line with the equation:

$$\text{logit } Y = a + b(\log x)$$

where x is amount of competitor tubulin and a and b are derived empirically. The line derived from the logit function was fitted using a linear least squares computer program and was used for calculating the amounts of tubulin in the experimental samples. Typically, experimental samples containing between 0.5 and 50 μg/ml of tubulin could be assayed reliably.

In order to demonstrate that tubulin from various sources would compete

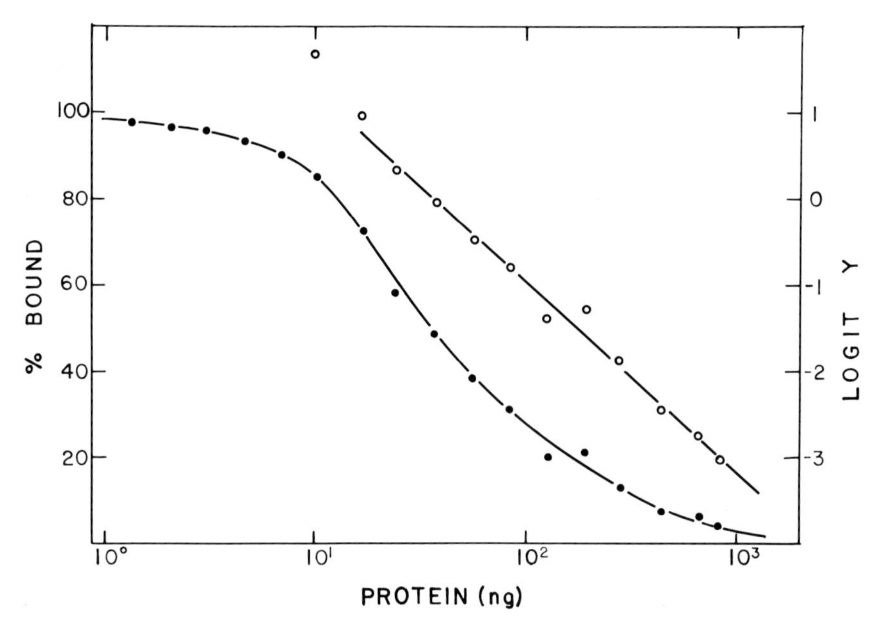

FIG. 6. Standard competition curve. Purified hog brain tubulin (25 μl) was diluted (1/1.5) over a range of 1 ng to 1 μg (*x* axis). Iodinated tubulin (25 μl, 10,000 cpm, 1 ng) was added to each dilution, followed by 50 μl of antiserum (1/1000 dilution). Samples were processed as described in Section II,D,2. Left *y* axis and closed circles: % bound. Right *y* axis and open circles: logit *Y*.

effectively in the RIA, studies were carried out to examine the characteristics of the antisera for binding to tubulin from various species and in different polymeric states. As previously shown for the original antiserum raised with *Tetrahymena* tubulin (Van De Water and Olmsted, 1980), the sera described in this report showed broad species reactivity. Tubulin derived from *Tetrahymena,* hog and mouse brain, and neuroblastoma cells competed iodinated hog brain tracer to background and had similar competition curves. Fixed microtubules and monomeric tubulin were also essentially equivalent in competition of labeled tubulin. These properties make these antisera generally useful for quantitating tubulin in a variety of cell types.

Although soluble materials can be readily assayed in most buffers, it is often problematical to assay cell extracts because of the presence of particulate or insoluble material that does not dilute reproducibly enough for quantitation in the RIA. Under conditions where buffers other than those usually employed in the RIA were used, it was necessary to determine whether they affected the RIA by preparing standard curves in the same solutions in which the cell extracts were prepared. A number of buffers have now been tested, the majority in the pH range of 7.0–7.4. The presence of up to 10% glycerol, 0.1% spermidine, or high magnesium (25 m*M*) had no effect on the slope of the competition curve; 2%

Triton X-100 depressed the slope slightly, but concentrations of 1% or less had little effect. In addition, as previously described for immunoprecipitation (Lingappa *et al.*, 1978), samples can be treated with SDS sample buffer lacking mercaptoethanol, and then diluted into Triton X-100 containing buffer such that the final ratio (%) of SDS to Triton X-100 is 1:10. This procedure works reliably in the RIA and allows solubilization of most cellular fractions such that dilutions can be accurately made.

IV. Discussion

Tubulin is apparently a highly conserved protein (Luduena and Woodward, 1975), and production of antisera, even with moderate titers, has proved problematical. A number of different immunogen preparations have been used; these include (1) native purified tubulins (Bhattacharyya and Wolff, 1975; Crawford *et al.*, 1976; Frankel, 1976; Gozes *et al.*, 1977; Hiller and Weber, 1978; Ikeda and Steiner, 1976; Joniau *et al.*, 1977; Meier and Jorgensen, 1977; Van De Water and Olmsted, 1978); (2) glutaraldehyde cross-linked tubulin (Fuller *et al.*, 1975; Gordon *et al.*, 1977; Karsenti *et al.*, 1978; Morgan *et al.*, 1977, 1978); (3) SDS-denatured species, either injected directly (Zenner and Pfeuffer, 1976) or purified from gels (Aubin *et al.*, 1976; Cande *et al.*, 1977; Connolly *et al.*, 1977; Eckert and Snyder, 1978; Guttman, 1978; Koehn and Olsen, 1978; Piperno and Luck, 1977; Wiche and Cole, 1976); and (4) other tubulin-containing preparations [e.g., *Arabacia* axonemes (Fulton *et al.*, 1971); *Naegleria* axonemes (Kowit and Fulton, 1974); and vinblastine–tubulin crystals from cultured cells (Dales, 1972) and sea urchin eggs (Fujiwara and Pollard, 1978)]. The majority of these preparations have been used for immunofluorescent localization of tubulin; a few have been used for radioimmunoassays (Bhattacharyya and Wolff, 1975; Gozes *et al.*, 1977; Hiller and Weber, 1978; Joniau *et al.*, 1977; Koehn and Olson, 1978; Kowit and Fulton, 1974; Morgan *et al.*, 1977; Van De Water, 1979; Van De Water and Olmsted, 1978, 1980).

The majority of antisera have been tested for specificity by double immunodiffusion tests, immunoelectrophoresis, or staining of microtubule networks by indirect immunofluorescence. These assays have given variable results. Some immune sera have given high backgrounds in immunofluorescence, and further purification by affinity chromatography has been necessary to obtain satisfactory images. In addition, the results that depend on precipitation of immune complexes may be difficult to analyze because of the apparent weak precipitation properties of most tubulin antisera; this property may make assessment of the presence of minor contaminants in either the immunogen or test antigens difficult.

For those sera that precipitate antigens efficiently, the properties of immune complexes can be tested directly. These sera have been useful in demonstrating different species reactivities (Fulton *et al.*, 1971; Gordon *et al.*, 1977; Joniau *et al.*, 1977) and in quantitating tubulin synthesis (Guttman, 1978; Guttman and Gorovsky, 1979; Kowit and Fulton, 1974; Piperno and Luck, 1977). However, the titers of the tubulin antisera produced usually have been low, necessitating the use of large amounts of serum even in those cases where precipitation reactions could be used. In addition, animals have not generally appeared to respond well to booster injections, further limiting the amount of serum that might be produced. It therefore seemed desirable to develop techniques for the production of higher titer antisera.

The basic technique of purifying the immunogen off of SDS gels was originally used to produce actin antibodies (Lazarides, 1976) and, as noted previously, has also been successful for tubulin. In the case described here, the excellent response of all four animals to immunogen prepared in this manner suggests that the technique may be generally useful for antibody production. The strong immune reaction may also be due to the use of tubulin isolated from an organism distantly related to the injected animal; this is consistent with other reports that demonstrated that tubulin derived from lower eukaryotes (e.g., sea urchin, Fujiwara and Pollard, 1978; *Naegleria,* Kowit and Fulton, 1974; *Chlamydomonas,* Piperno and Luck, 1977) appear to produce more highly reactive sera. In addition, isolation of gel-purified *Tetrahymena* tubulin reduces the probability of eliciting antibodies to contaminating protein species that would then cross-react with proteins present in mammalian cell extracts.

In those cases where antiserum binding has been measured quantitatively (by binding to labeled antigen), the efficiency of binding has varied widely, with 50% binding of labeled antigen (usually 0.5–1.0 ng) occurring over serum dilutions from 1/25 (Zenner and Pfeuffer, 1976) to 1/30,000 (Kowit and Fulton, 1974). The majority of antisera appear to have 50% binding to nanogram amounts of tubulin in the dilution range of 1/50–1/100. The antisera described here show 50% binding at dilutions of 1/500 to 1/1000. However, normalization of these data is difficult, since not all reports demonstrate serum dilution curves with a typical sigmoidal shape, or binding that reached a maximum of 100% under conditions of antibody excess.

Tubulin antisera have been produced that show generalized binding to a variety of tubulins or that are more reactive with particular species. Under the conditions used for immunofluorescence, most tubulin antisera appear to react with a number of cell types and species. With some antisera, more limited reactivities have been shown by Ouchterlony tests (Fulton *et al.*, 1971), precipitin tests (Gordon *et al.*, 1977), and radioimmunoassays (Hiller and Weber, 1978; Morgan *et al.*, 1978). For example, antiserum raised to lamb brain tubulin has been shown to discriminate between various mammalian brain tubulins (Morgan *et al.*, 1978), and a chick brain tubulin antiserum reacted differently with tubulins from various tissues and cultured cells (Hiller and Weber, 1978). In contrast, the antisera described here, as well as others (Dales, 1972; Joniau *et al.*, 1977), react equally well with tubulin from diverse species. It is not possible to generalize about which preparation of immunogen results in either narrow or broad antigenic reactivity. However, in the cases reported, the use of SDS-denatured, electrophoretically purified tubulins seems to result in antisera that recognize determinants common to several tubulins.

The types of variable properties with which tubulin antiserum reacts with various antigens indicates the care with which these reactions must be characterized in order to develop a reliable RIA. The sensitivity of the RIA has allowed discrimination between antigenic determinants that are distinct from species to species (Morgan *et al.*, 1978) and from one cell type to the next (Hiller and Weber, 1978). However, these examples make clear the necessity for characterizing the binding properties of the antisera carefully in order to assay various tissue types. For example, if the reactivity of a cellular tubulin is markedly different from the tracer or competitor used for obtaining the standard curve, uninterpretable measurements could be obtained. This is especially true in those cases where heterologous tubulin does not compete the tracer to background levels, indicating that the tubulins may have different antigenic determinants. The resulting measurements may reflect only a subset of the tubulin in the extract. In addition, it is particularly difficult to normalize these data if the competition curves with homologous competitor and heterologous experimental material have different slopes over either part or all of the curves.

It is also important to establish the reactivity of the antisera with different oligomeric forms of tubulin. Previous work (Meier and Jorgensen, 1977; Van De Water and Olmsted, 1978) suggested that some tubulin antisera react preferentially with aggregated or polymerized forms of tubulin. Under solution conditions where tubulin might be largely in subunit form, RIA utilizing these sera would underestimate actual tubulin content. The antisera described here react with both SDS-denatured tubulin and glutaraldehyde-fixed microtubules equally efficiently. It therefore appears that these antibodies react with determinants exposed on the subunit regardless of whether it is in monomeric or oligomeric form.

In summary, in order to determine tubulin content accurately in heterogeneous solutions by RIA, a number of parameters must be investigated. These include (1) demonstrating the efficient reaction of the labeled tracer with the antibody, (2) measuring the competability of the tracer with homologous antigen for a standard curve, and (3) assessing the cross-species and oligomeric reactivity of the antibody. With these factors taken into consideration, the use of tubulin antibodies should greatly increase the sensitivity with which tubulin distribution and metabolism in a number of cell types can be measured.

ACKNOWLEDGMENTS

Special thanks go to F. Calzone and C. Kenny, who participated in obtaining the antisera described in this chapter. This work has been supported by grants from the NIH and ACS to J.B.O. and M.A.G.

REFERENCES

Aubin, J. E., Subrahmayan, L., Kalnins, V., and Ling, V. (1976). *Proc. Natl. Acad. Sci. U.S.A.* **73,** 1246–1249.

Bhattacharyya, B., and Wolff, J. (1975). *J. Biol. Chem.* **250,** 7693–7746.

Bolton, A. E., and Hunter, W. M. (1973). *Biochem. J.* **133,** 529–539.

Borisy, G. G., Marcum, J. M., Olmsted, J. B., Murphy, D. B., and Johnson, K. A. (1975). *Ann. N.Y. Acad. Sci.* **253,** 107–132.

Burridge, K. (1978). *In* "Complex Carbohydrates," Part C (V. Ginsberg, ed.), Methods in Enzymology, Vol. 50, pp. 54–64. Academic Press, New York.

Cande, W. A., Lazarides, E., and McIntosh, J. R. (1977). *J. Cell Biol.* **72,** 552–567.

Connolly, J. A., Kalnins, V. I., Cleveland, D. W., and Kirschner, M. W. (1977). *Proc. Natl. Acad. Sci. U.S.A.* **74,** 2437–2440.

Crawford, N., Trenchev, P., Castle, A. J., and Hoborow, E. J. (1976). *Cytobios* **14,** 121–130.

Dales, S. (1972). *J. Cell Biol.* **52,** 748–754.

Dustin, P. (1979). "Microtubules." Springer-Verlag, Berlin and New York.

Eckert, B. S., and Snyder, J. A. (1978). *Proc. Natl. Acad. Sci. U.S.A.* **75,** 334–338.

Fairbanks, G., Steck, T., and Wallach, D. (1971). *Biochemistry* **10,** 2606–2617.

Frankel, F. R. (1976). *Proc. Natl. Acad. Sci. U.S.A.* **73,** 2798–2802.

Fujiwara, K., and Pollard, T. D. (1978). *J. Cell Biol.* **77,** 182–195.

Fuller, G. M., Brinkley, B. R., and Boughter, J. M. (1975). *Science* **187,** 948–950.

Fulton, C., Kane, R., and Stephens, R. (1971). *J. Cell Biol.* **50,** 762–773.

Gibbons, I. R. (1965). *Arch. Biol.* **76,** 317–352.

Goldman, R. D., Pollard, T. D., and Rosenbaum, J. L., eds. (1976). "Cell Motility," Vols. 1-3. Cold Spring Harbor Lab., Cold Spring Harbor, New York.

Gordon, R. E., Lane, B. P., and Miller, F. (1977). *J. Cell Biol.* **75,** 586–592.

Gorovsky, M. A., Carlson, K., and Rosenbaum, J. L. (1970). *Anal. Biochem.* **35,** 359.

Gozes, I., Geiger, B., Fuchs, S., and Littauer, U. Z. (1977). *FEBS Lett.* **73,** 109–114.

Guttman, S. D. (1978). Ph.D. Thesis. Dep. Biol., Univ. of Rochester, Rochester, New York.

Guttman, S. D., and Gorovsky, M. A. (1979). *Cell* **17,** 307–318.

Haber, E., and Poulsen, K. (1974). *In* "The Antigens" (M. Sela, ed.), Vol. 2, pp. 249–275. Academic Press, New York.

Hiller, G., and Weber, K. (1978). *Cell* **14,** 795–804.

Hunter, W. M., and Greenwood, F. C. (1962). *Nature (London)* **194,** 495–496.

Ikeda, Y., and Steiner, M. (1976). *J. Biol. Chem.* **251,** 6135–6141.

Joniau, M., de Brabander, M., de Mey, J., and Hoebeke, J. (1977). *FEBS Lett.* **78,** 307–312.

Karsenti, E., Guilbert, B., Bornens, M., Avrameas, S., Whalen, R., and Pantaloni, D. (1978). *J. Histochem. Cytochem.* **26,** 934–947.

Kessler, S. W. (1976). *J. Immunol.* **117,** 1482–1490.

Koehn, J. A., and Olsen, R. W. (1978). *Biochem. Biophys. Res. Commun.* **80,** 391–397.

Kowit, J. D., and Fulton, C. (1974). *J. Biol. Chem.* **249,** 3638–3646.

Laemmli, U. K. (1970). *Nature (London)* **227,** 680.

Lazarides, E. (1976). *J. Supramol. Struct.* **5,** 531.

Lingappa, V. R., Lingappa, J. R., Prasad, R., Ebner, K., and Blobel, G. (1978). *Proc. Natl. Acad. Sci. U.S.A.* **75,** 2338–2342.
Luduena, R. F., and Woodward, D. O. (1975). *Ann. N.Y. Acad. Sci.* **253,** 272–283.
Meier, E., and Jorgensen, O. S. (1977). *Biochim. Biophys. Acta* **494,** 354–366.
Morgan, J. L., Rodkey, L. S., and Spooner, B. S. (1977). *Science* **197,** 578–580.
Morgan, J. L., Holladay, C. R., and Spooner, B. S. (1978). *Proc. Natl. Acad. Sci. U.S.A.* **75,** 1414–1417.
Murphy, D. B., Vallee, R. B., and Borisy, G. G. (1977). *Biochemistry* **16,** 2598–2606.
Piperno, G., and Luck, D. J. (1977). *J. Biol. Chem.* **252,** 383–391.
Renaud, F. L., Rowe, A. J., and Gibbons, I. R. (1968). *J. Cell Biol.* **36,** 79–90.
Rodbard, D., Bridson, W., and Rayford, P. L. (1969). *J. Lab. Clin. Med.* **74,** 770–781.
Sloboda, R. D., Dentler, W. L., and Rosenbaum, J. L. (1976). *Biochemistry* **15,** 4497–4505.
Van De Water, L. (1979). Ph.D. Thesis. Dep. Biol., Univ. of Rochester, Rochester, New York.
Van De Water, L., and Olmsted, J. B. (1978). *J. Biol. Chem.* **253,** 5980–5984.
Van De Water, L., and Olmsted, J. B. (1980). *J. Biol. Chem.* **255,** 10744–10751.
Wiche, G., and Cole, R. D. (1976). *Exp. Cell Res.* **99,** 15–22.
Zenner, H. P., and Pfeuffer, T. (1976). *Eur. J. Biochem.* **71,** 177–184.

Chapter 7

Immunofluorescence and Immunocytochemical Procedures with Affinity Purified Antibodies: Tubulin-Containing Structures

MARY OSBORN AND KLAUS WEBER

Max Planck Institute for Biophysical Chemistry
Goettingen, Federal Republic of Germany

ISBN 0-12-564124-9

I. Introduction

Histochemical and immunological procedures can be valuable aids in the identification of structural proteins in supramolecular structures. Although Coons and Kaplan (1950) described the indirect immunofluorescence procedure in 1950, and it subsequently became a method of choice for studying the spatial location of viral antigens, appreciation of its potential use in studying cytoskeletal elements in cells in culture took more than 20 years. In retrospect this was probably for two reasons. First, it was not generally realized that the fluorescence microscope possesses sufficient resolution to yield information on the distribution of cytoskeletal elements such as microtubules, microfilaments, and intermediate filaments. Second, there was the widely held belief that it would be difficult if not impossible to raise antibodies to proteins that were not only highly conserved in sequence but that were also common constituents of most, if not all, eucaryotic cells. However, in the last six years immunofluorescence microscopy demonstrated the arrangements of actin-containing structures (Lazarides and Weber, 1974) and subsequently the arrangements of microtubules (Weber *et al.*, 1975a; Fuller *et al.*, 1975) and of the different classes of intermediate filaments (see, e.g., Hynes and Destree, 1978a; Bennett *et al.*, 1978; Franke *et al.*, 1978; Lazarides and Balzer, 1978). This approach has made a major impact on cell biology and has helped in understanding the organization and possible functions of the different cytoskeletal elements.

Immunofluorescence microscopy is a very sensitive method theoretically requiring only one antigenic site on the structural protein. It offers several unique advantages for the study of cytoskeletal structures. First, only the protein of choice will be visualized by use of its specific antibody. Second, if the protein forms part of a supramolecular structure, the organization of that structure throughout the cell can be demonstrated. Third, numerous cells can be observed simultaneously, and thus the unique features of the organization under different experimental conditions can be observed. Thus immunofluorescence microscopy can bridge the gap between light microscopy and electron microscopy. These advantages make immunofluorescent studies a useful screen to decide which objects or conditions to study later at the higher resolution of the electron microscope. Comparative immunofluorescent and electron microscopy either on samples processed in parallel or on the same object show a convincing one-to-one relationship. Immunoelectron microscopical studies have the further advantage that the high-resolution studies allow possible interactions between the immunochemically labeled structures and other unlabeled structures to be assessed.

The purpose of this review is to describe in detail some procedures to study the intracellular location of proteins by immunocytochemical procedures. Although the methods given are generally applicable, they are documented for tubulin-containing structures and antibodies to tubulin. Further examples of the usefulness of these methods in studying cell structure can be found in several recent reviews (see, e.g., Brinkley *et al.*, 1980; Lazarides, 1980; Weber and Osborn, 1979, 1981a,b; Osborn, 1981), as well as in the other references quoted here.

II. Antibody Production, Purification, and Characterization

A. Antibodies

Although an antibody raised against vinblastine-induced tubulin paracrystals isolated from drug-treated cultured cells had already been described in 1972 (Dales, 1972), direct visualization of cytoplasmic microtubules by fluorescence microscopy was not obtained. Using an improved technology and an antibody prepared against sea urchin sperm outer doublet tubulin, direct visualization was achieved in 1975 (Weber *et al.*, 1975a). Today the most commonly used tubulin is isolated from mammalian brain tissue (Fuller *et al.*, 1975; Weber *et al.*, 1976; Frankel, 1976; Aubin *et al.*, 1976). Microtubular protein rich in tubulin is prepared by standard procedures (see, e.g., Shelanski *et al.*, 1973) and homogeneous 6 S tubulin free of associated proteins is isolated from this source by chromatographic procedures (see, e.g., Weingarten *et al.*, 1975). Various tubulin derivatives, including SDS-denatured tubulin obtained from preparative gels, urea-denatured *S*-carboxymethylated tubulin, as well as glutaraldehyde cross-linked tubulin, have been used in addition to tubulin. The different antigens and the various immunization schemes are covered in another chapter. It has been our experience over the last five years that homogeneous pig brain tubulin used either directly or after carboxymethylation in 8 *M* guanidine HCl and subsequent dialysis against water gives rise to a reproducible immune response (see legend Fig. 1). Highly suitable antibodies against homogeneous tubulin have been elicited in various animals. Although rabbits are most commonly used, excellent results have also been obtained with goats, sheep, and guinea pigs.

B. Characterization of Presera

It is very important to establish that immune IgGs have been specifically raised in response to immunization by the antigen. Thus it is useful to test preimmunization sera in order to identify those animals whose sera even at low dilution show no positive decoration in immunofluorescence microscopy on cultured cells using various fixation methods. It has been our experience that a large percentage of the rabbits available to us have preimmune sera, which decorate the cytokeratin filaments of epithelial cells at dilutions of 1:6 or 1:10 (Osborn *et al.*, 1977). Autoantibodies to the vimentin type of intermediate filaments (Gordon *et al.*, 1978) and antibodies decorating centrioles (Connolly and Kalnins, 1978) have also been occasionally recognized. Although appreciable levels of IgGs reacting in immunodiffusion with tubulin have been reported for pooled sera from rabbits, and from several other mammals (Karsenti *et al.*, 1977), the identity of these IgGs has remained unclear, since immunoperoxidase staining of cultured cells revealed vinblastine-induced tubulin crystals but not the normal cytoplasmic microtubular complex. Our own experience in the last five years shows that although the serum from an occasional rabbit or guinea pig decorates microtubules, or bundles of microfilaments, when used at a dilution of 1:10, this phenomenon is rare and found in less than 1% of the animals tested. Obviously certain autoantibodies can be very useful (see preceding examples) but in cases where an induced antibody is the aim, animals with positive preimmune sera should be avoided.

C. Testing for Induced Antibodies

Methods to determine the presence of specific antibodies in the sera of immunized animals include: immunodiffusion, immunoelectrophoresis, complement fixation, immune replicas of proteins separated by polyacrylamide gel electrophoresis, characterization of immune complexes isolated from cell extracts and separated by gel electrophoresis, as well as various forms of radioimmune assays. Many of these methods relying on radioactively labeled antigens are very much more sensitive than immunodiffusion analysis and easily reveal low titer antibodies. They are general immunological procedures that can be found in various textbooks and are also covered in Chapter 6 in this volume.

In addition a variety of cell biological controls using standard tissue culture cells in indirect immunofluorescence microscopy can be very useful. It may be possible, for example, to show that the antibody gives a novel pattern of staining on cells, or a pattern that has been previously reported and that therefore identifies the structures stained by the antibody under study. For instance in the case of tubulin antibodies the display of cytoplasmic microtubules in interphase cells (Weber *et al.*, 1975a; Brinkley *et al.*, 1975; Frankel, 1976; Osborn and Weber, 1977b) as well as in enucleates (Weber *et al.*, 1975a), and the different mitotic microtubular arrangements (Fuller *et al.*, 1975; Weber *et al.*, 1975b) as well as the decoration of cilia and flagella (Aubin *et al.*, 1976; Osborn and Weber, 1976;

Weber *et al.*, 1977) are well documented. Cytoplasmic microtubules are also sensitive to antimitotic drugs such as colcemid and colchicine although cilia, flagella, and centrioles are more stable (earlier references). Typical patterns of vinblastine-induced paracrystals have also been given (Weber *et al.*, 1975b). Indeed if a new tubulin antibody is elicited, it is wise to characterize it by indirect immunofluorescence microscopy on standard tissue culture cells before proceeding to a new type of cell or organism. One popular control, which has led to dozens of uninformative black pictures, is to adsorb an aliquot of the antiserum with the purified antigen and to demonstrate that after this treatment the immunofluorescence reaction has vanished. Although this control may in certain instances be useful, it rarely deserves extensive documentation in print. In addition it should only be done with a truly homogeneous antigen used at the correct concentration, since otherwise absorption on a contaminant can occur.

D. Antigen-Affinity-Purified IgGs

Tubulin is rarely a good antigen and titers may vary quite strongly. Although some strong sera after appropriate dilution may give very good results in immunofluorescence microscopy, it is advisable to concentrate the antibodies by preparing a purified IgG fraction. Even better results can generally be obtained if the tubulin antibodies are isolated by antigen-affinity chromatography (Fuller *et al.*, 1975; Weber *et al.*, 1976). For this purpose, homogeneous tubulin free of any associated proteins is covalently coupled to Sepharose-4B, or a corresponding carrier, and washed until only covalently bound antigen is retained on the support. When the IgG fraction is passed through the column specific antitubulin IgGs are absorbed, whereas the bulk IgGs are removed. Antigen-affinity-purified IgGs can be released by elution with solvents which decrease the stability of the antigen–antibody complexes. Suitable solvents include 0.2 *M* glycine–HCl pH 2.6, 5 *M* $MgCl_2$, and 3 *M* KSCN. The procedures are given elsewhere and are also discussed in Chapter 6. Although some very tightly binding antibodies are not recovered, the population of antitubulin IgGs is very pure. We and others have taken advantage of such antigen-affinity-purified antibodies. They can be used at very low concentrations (10–50 μg/ml) resulting in extremely low background staining in indirect immunofluorescence microscopy, giving results much superior to similar micrographs obtained with diluted antiserum or total IgGs. Two points may be of interest. We routinely use pH 2.6 elution at 4°C and have found that immediate neutralization of the fractions containing the specific IgGs is very important to avoid loss in titer. Specific IgGs when present at high enough concentrations (in general 0.1 mg/ml or higher) can be stored at −70°C. When diluted and stored at 4°C, addition of a neutral carrier protein (such as bovine serum albumin at 0.5 mg/ml) is advantageous.

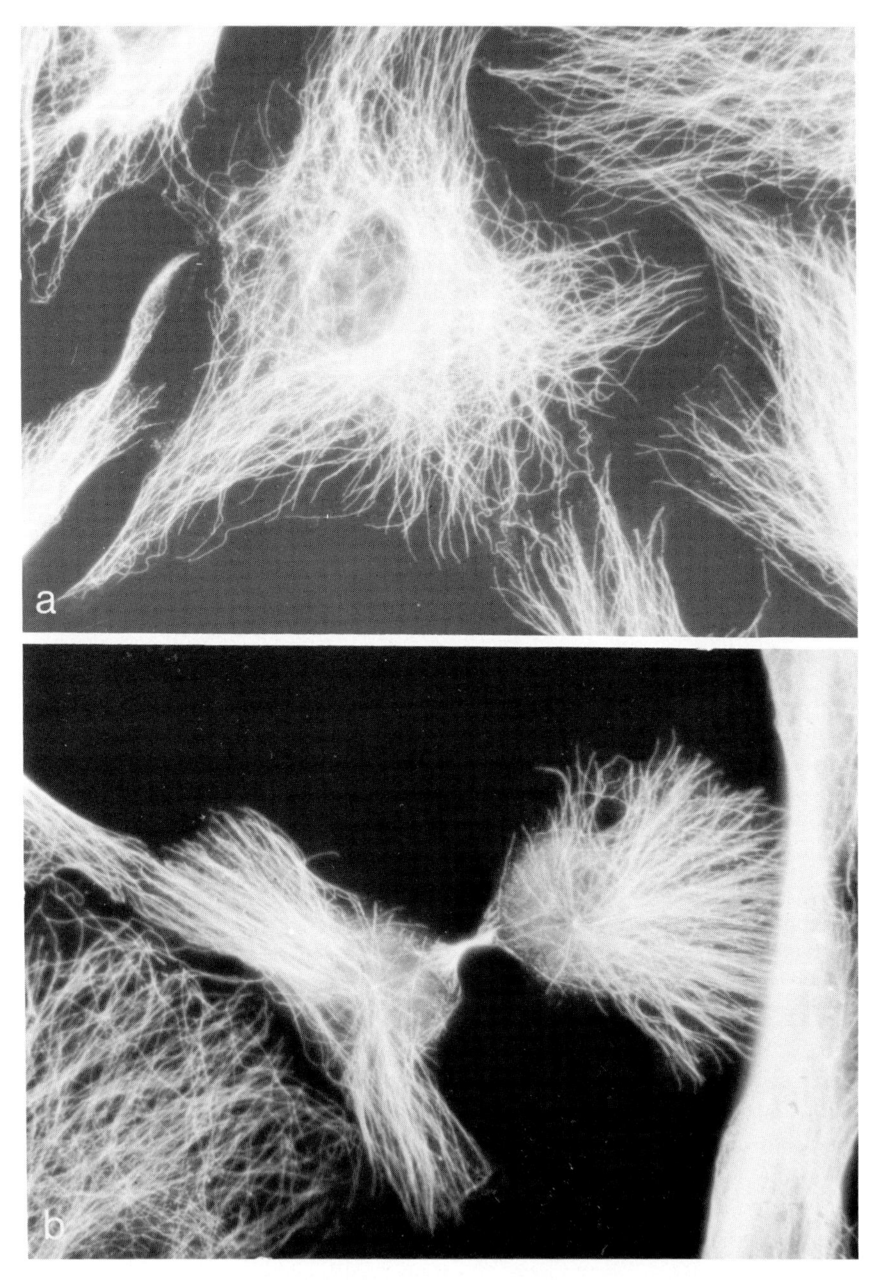

FIG. 1. Rabbit kidney cells stained with antibody to tubulin purified from porcine cerebrum. Microtubular protein was prepared from porcine cerebrum by three cycles of temperature-dependent assembly and disassembly. Highly purified tubulin, free of microtubular-associated proteins, was

E. Cross-Species Reactivity of Tubulin Antibodies

Tubulin is generally considered to be highly conserved during eucaryotic evolution (see, e.g., Luduena and Woodward, 1973). Thus it is not surprising that antibodies to mammalian brain tubulin show a broad cross-species reactivity covering species as diverse as man and plants (Franke *et al.*, 1977; Weber *et al.*, 1977; Lloyd *et al.*, 1979). Therefore it seems advisable to use in general antibodies elicited against mammalian brain tubulin, because this antigen is easily purified to homogeneity. However, there are reports of antibodies that distinguish different tubulin antigens (see, e.g., Fulton and Simpson, 1979), thus indicating the potential usefulness of other antigens. In the latter case precautions are necessary to assure the highest possible purity of the antigen used for immunization to avoid undesirable mixtures of immune IgGs. If this criterium cannot be met, subsequent experiments perhaps involving tubulin purified by preparative gel electrophoresis may be helpful if such material is used to isolate antigen-affinity-purified antibodies.

In view of the strong preservation of the tubulin structure, and the fact that tubulin is an intracellular protein, it is not surprising that experimentally raised antibodies recognize tubulin structures present in the cells of the immunized animal. Figure 1 shows the cytoplasmic microtubular network present in primary rabbit kidney cells. These cells were derived from the immunized rabbit after it was bled, and were decorated in indirect immunofluorescence by tubulin antibodies raised in this rabbit. Thus the production of antibodies to tubulin resem-

---

derived by phosphocellulose chromatography. Homogeneity of the tubulin has been documented for an equivalent fraction by polyacrylamide gel electrophoresis (Weber *et al.*, 1976). Homogeneous tubulin was denatured with 8 *M* urea, reduced with 2-mercaptoethanol, and carboxymethylated with iodoacetic acid using standard procedures (Crestfield *et al.*, 1963). The protein solution was freed of urea by extensive dialysis against water and then dialyzed against phosphate-buffered saline. The partially precipitated protein was resuspended in the same solvent at approximately 1 mg/ml. The injection schedule was as follows: Day 0 total 1 mg given intradermally, intramuscularly, and in lymph nodes in the hind leg, in the presence of an equal volume of Freund's complete adjuvant. After 6, 8, 11, 14, and 17 weeks, a total of 1 mg was given at each injection intradermally and intramuscularly in the presence of an equal volume of Freund's incomplete adjuvant. The rabbit was bled 12 days after the final injection. An IgG fraction was prepared and processed by affinity chromatography on nondenatured 6 S porcine cerebrum tubulin.

After exsanguination of the rabbit the kidneys were excised and primary cultures established from minced fragments of the tissue using conventional techniques (Gallagher, 1973, p. 102). Growth medium was Dulbecco's Modified Eagles Medium plus 10% fetal calf serum. Cells were grown until confluence and passaged 2–4 times before use in immunofluorescence microscopy.

The figure shows kidney cells (a) in interphase and (b) in early G_1 stained in indirect immunofluorescence microscopy using tubulin antibody elicited in the rabbit from which the kidney cells were isolated. Thus this figure emphasizes not only the microtubular profiles characteristic of many types of cells in tissue culture, but also shows that rabbit antibodies to tubulin stain the cells of the rabbit in which the antibody was raised.

bles the production of autoimmune antibodies and proves that antibodies can be raised that recognize a cellular antigen also present in the animal's own cells. This observation has also been made with actin antibodies (M. Osborn and K. Weber, unpublished observations) where full amino acid sequence analyses (for review, see Vandekerckhove and Weber, 1979) prove that the original antigen—β,γ-nonmuscle actin—is identical with the nonmuscle actin present in the corresponding cells of the experimental animal.

F. Monoclonal Antibodies

Monoclonal antibodies secreted by hybridoma cells (Köhler and Milstein, 1975) present an alternative source of immunological reagents with which to study the cytoskeleton. They can be generated by injecting an animal—a mouse or rat—with a purified antigen and after antibodies are produced fusing spleen cells from the animal with a myeloma line derived from the same species. Hybridoma cells which secrete antibody molecules directed against the injected antigen can then be isolated and have the advantage that they often continue to secrete antibody molecules of defined specificity into the medium. Thus often the culture medium can be used as a source of antibody, although because of the presence of various serum components in the culture medium, as well as the possible presence of cellular substances and their remnants or of proteases it may be important to concentrate and purify the secreted monoclonal IgG before use. Stringent characterization of monoclonal antibody specificity is necessary since the isolation of individual hybridoma cells can result in antibodies unrelated to the injected antigen, and also because, since monoclonal antibodies recognize only a single limited region of the corresponding antigen, some monoclonals may recognize similar sequences in functionally unrelated proteins. However, it is already clear that in some instances monoclonal antibodies can be extremely useful, and in certain cases, as when the antigen cannot be purified to homogeneity, use of hybridoma technology may be the only way to generate an antibody with the required specificity.

G. Fluorescently Labeled Second Antibodies

These are usually purchased from commercial sources. Not all commercial products are of equivalent quality and indeed some preparations are rather poor. They either show a high general background or reveal strong fluorescent staining of particular cellular structures (i.e., nuclei, nucleoli, Golgi, mitochondria, or membranous structures) when assayed on cells without the use of a first antibody. If such reactions are encountered, preabsorption of the second antibody on fixed cells is indicated. A convenient protocol is as follows: Two almost confluent monolayers of the cell type under study grown on plastic dishes of 9-cm diameter are fixed using the formaldehyde, methanol, or the methanol fixation procedure (Section IV,B). After rehydration in PBS, the PBS is removed and 0.4 ml of a 1:6 dilution of the commercial antibody is incubated with occasional agitation on the first petri dish at 37°C in the dark. After 45 min the antibody solution is transferred to the second dish and the process is repeated. After a further 45 min incubation the antibody solution is removed and appropriately diluted, so that the final dilution is in the range 1:20 to 1:50 (approximately 0.5 mg/ml).

Given the cost of second antibodies we have found it often quite useful to raise these antibodies ourselves. Antigen-affinity purification yields an ample supply of unlabeled IgGs. These can be labeled by standard protocols using fluoresceinisothiocyanate, tetramethylrhodamineisothiocyanate, or dichlorotriazinylaminofluorescein (see e.g., Cebra and Goldstein, 1965; Blakeslee and Baines, 1976). The latter procedure has a specific advantage. Simple salt fractionation selects those IgGs to which the most suitable number of fluorescent groups are linked, thus avoiding the need for elaborate ion exchange fractionation procedures (for details, see Blakeslee and Baines, 1976).

Alternatively, commercially available fluorescently labeled second antibodies can be fractionated by antigen-affinity chromatography (see above). In our hands the second antibody can be used after the antigen-affinity purification in indirect immunofluorescence microscopy at standard dilutions of 0.05 mg/ml or less. The resulting micrographs have particularly low backgrounds.

III. Direct versus Indirect Immunofluorescence Microscopy

Should the direct or the indirect immunofluorescence technique be used? In the direct technique the fluorochrome is directly coupled to the immune IgG molecule. In the indirect technique cells are first reacted with unlabeled immune IgGs, which are in turn visualized by a fluorescently labeled second antibody directed against IgGs from the animal species in which the first antibody was raised (for instance, fluorescently labeled sheep-anti-rabbit IgGs to detect rabbit-antitubulin IgGs). There are three arguments that favor the use of the indirect method: First, the indirect technique, because of its inherent amplification effect, is more sensitive than the direct technique. Estimates vary between 4- and 10-fold. Second, substitution of IgGs by fluorescent components generally gives rise to a population of molecules with varying number of fluorochromes. Subsequent fractionation is aimed at isolating the particular IgG molecules with an intermediate number of fluorescent groups. Too lightly labeled IgGs are not very useful and too strongly labeled IgGs give rise to "stickiness," i.e., unspecific binding. Thus direct labeling is necessarily accompanied by a loss of potentially valuable immune IgGs. Third, time is saved, because the fluorochrome-labeled second antibody can be obtained commercially and used for several different antigens. In this case most controls on antigen specificity of the second antibody only have to be performed once.

Several groups have used rabbit or sheep antitubulin IgGs and corresponding commercial fluorescent second antibodies in the indirect technique to document the display of cytoplasmic microtubules in various cells (see preceding references). The micrographs obtained are clearly of superior quality to those provided in a study by Fujiwara and Pollard (1978) using the direct technique. This is most likely due to the different light intensities provided (see above). There are, however, situations, such as for instance in double immunofluorescence microscopy, where the direct technique is faster and avoids certain blocking steps (see Section V). If the amount of immune IgGs available is not limiting, and if the antigen forms a heavily decorated structure, most shortcomings of the direct technique are readily overcome.

IV. Indirect Immunofluorescence Technique

A. Sample Preparation

Round glass coverslips (12 mm in diameter, 0.16–0.19 mm thick) are washed in 100% ethanol, wiped dry with a Kimwipe, and placed one layer thick on successive layers of aluminum foil in a glass petri dish prior to sterilization at 160°C for 2 hr.

1. Cells in Culture That Attach to a Substratum

Cells are seeded after appropriate dilution in medium on glass coverslips 1–3 days before use. A 3.5-cm diameter tissue culture dish accommodates 4 coverslips, a 5.5-cm diameter culture dish up to 10 coverslips. Therefore it is possible to obtain multiple samples from a single culture.

2. Cells in Suspension and Samples That Do Not Attach to Glass

These objects present more of a challenge. Although it is possible to process them for immunofluorescence microscopy in suspension, using repeated centrifugation steps, this is time consuming and requires comparatively large amounts of antibody. The following tricks may help to attach the specimens to coverslips:

a. Coat the coverslips with polylysine (Mazia *et al.*, 1975). We use 100 μg/ml polylysine (P.1886, Sigma Chemical Co., St. Louis, MO) in water, wash well with water, and air dry; for some specimens it seems helpful to coat the coverslips just before use. The specimen is then pipetted onto the coverslip and allowed to settle.

b. Spin the cells gently onto a coverslip or a microscope slide, using a Cytospin Centrifuge (Shandon, Southern Products, U.K.).

c. Pipette the sample onto the coverslip and allow it to dry partially either in an air stream or under a 60-W bulb of a desk lamp.

3. Cytoskeletons

Visualization of cytoskeletal elements can sometimes be improved, if instead of using the one- or two-step fixation methods given later, the cells are first converted into cytoskeletons (for example, by treatment with a nonionic detergent) and are then fixed. When procedures for making cytoskeletons are discussed, it is important to note that "cytoskeleton" is a subjective term and that the type and possibly the amount of cytoskeletal elements preserved depends in large part on (1) the detergent and (2) the buffer used. Thus depending on the aim of the experiment it can be worthwhile to consider a change in the detergent or in the buffer.

a. Methods That Destroy Microtubules But Preserve Actin and Intermediate Filaments. Suitable procedures include the following:

(i) Rinse monolayer or coverslips twice with 5 ml Tris-glucose to which 0.5 m*M* $MgCl_2$ and 0.025 m*M* $CaCl_2$ have been added (TGMC buffer). For a 9-cm diameter dish add 1 ml TGMC-buffer containing 0.5% Triton X-100 (J. T. Baker Co., Phillipsburg, NJ) and incubate 10 min at room temperature. Wash with 5 ml TGMC buffer (Brown *et al.*, 1976).

(ii) Rinse monolayer with buffer A:0.01 *M* Tris-HCl pH 7.8, 0.14 *M* NaCl, and 5 m*M* $MgCl_2$. Treat with buffer A containing 0.1% NP40 (Shell Chemical Co., London, U.K.) or 0.5% Triton X-100 for 4 min at 37°C. Wash with two changes buffer A (Osborn and Weber, 1977b).

(iii) Wash with buffer B: 100 m*M* KCl, 3 m*M* $MgCl_2$, 10 m*M* HEPES buffer (pH 6.8), 1 m*M* $CaCl_2$, and 200 m*M* sucrose. Treat with buffer B containing 0.5% Triton X-100 and 0.15 mg/ml of phenylmethylsulfonylfluoride for 2 min. Wash with buffer B. This method results in the additional retention of polyribosomes (Lenk *et al.*, 1977; Fulton *et al.*, 1980).

b. Methods That Preserve Microtubules. These methods are based on buffers that are known from *in vitro* studies (Weisenberg, 1972) to preserve microtubules. Thus cytoskeletons are usually made at a pH of approximately 6.9, and the buffers usually contain EGTA, in addition to a stabilizing agent such as polyethylene glycol.

A suitable protocol in which all steps are performed at room temperature is as follows (Osborn *et al.*, 1978a; Webster *et al.*, 1978a):

(i) Wash cells with stabilization buffer: 0.1 *M* piperazine-*N*,*N*′-bis(2-ethanesulfonic acid)sodium salt adjusted to pH 6.9 with KOH, 1 m*M* ethyleneglycolbis(2-aminoethyl ether)-*N*,*N*′-tetraacetic acid (EGTA), 2.5 m*M* GTP, and 4% polyethylene glycol 6000 (Serva, Heidelberg, F.R.G.), for 30 secs.

(ii) Incubate for 4 min in the same buffer containing 0.2% Triton X-100.

(iii) Wash twice for 30 sec each with stabilization buffer.

(iv) Fix the preparation (e.g., by procedure 6, Section IV,B) and combine with the normal indirect immunofluorescence procedure (Section IV,C). Addition of 5 m*M* Mg^{2+} to the preceding stabilization buffer results in the additional retention of ribosomes. Other protocols that also result in microtubular preservation, and that it may be possible to adapt for fluorescence microscopy, include those given by Small and Celis (1978) and by Heuser and Kirschner (1980).

In some cases the micrographs of cytoskeletons are of a higher quality than those of the cells from which the cytoskeletons were derived. This is presumably either because the detergent treatment can remove some proteins, which especially in the case of a non-affinity-purified antiserum can result in nonspecific staining, or because the nonpolymerized subunits (e.g., of tubulin) are extracted from the cell, and therefore in the cytoskeleton the contrast between the microtubules and the rest of the cell is increased. Thus, for example, use of cytoskeletons helped in the demonstration of cytoplasmic microtubules in transformed cells (Osborn and Weber, 1977b) and has been of use in some laboratories to demonstrate microtubular organizing centers (Spiegelman *et al.*, 1979) or to provide starting materials for freeze-dried cytoskeletons (Heuser and Kirschner, 1980).

B. Fixation

1. Normal Cells

The fixation procedures used for immunofluorescence microscopy of whole cells must preserve the structure and, in addition, either fully dissolve the lipid bilayer or extract sufficient material so that antibodies can penetrate the fixed cell. This can be achieved either by a two-step fixation or by a one-step fixation. The coverslips with the specimen attached are placed either with the cell side toward the observer in a useful ceramic rack that can accommodate up to 12 coverslips (staining rack, Catalogue No. 8542-E40, A. H. Thomas, Philadelphia, PA) or cell side up in a suitable glass or plastic container (note, however, that normal tissue culture plastic is attacked by 100% acetone). All reagents used in the following fixation procedures are of reagent grade, and unless otherwise stated all steps are performed at room temperature. Examples of microtubules stained by each of these techniques can be found in the indicated references.

a. Recipe for PBS. PBS composition per liter: 8 g NaCl, 0.2 g KCl, 0.2 g KH_2PO_4, 1.15 g Na_2HPO_4 adjusted to pH 7.3 with NaOH.

b. Two-step Fixation Procedures

(1) *Formaldehyde, methanol, acetone* (Lazarides and Weber, 1974; Weber *et al.*, 1975a): 3.7% formaldehyde in phosphate-buffered saline (PBS) for 10 min, methanol at −10°C for 6 min, acetone at −10°C for 30 sec, then PBS for 30 sec.

(2) *Formaldehyde, methanol* (Weber *et al.*, 1975b): As (1), but omitting the acetone step.

(3) *Formaldehyde, Triton* (Heggeness *et al.*, 1977): 3.7% formaldehyde in PBS for 10 min, then 0.2% Triton X-100 in PBS for 2 min followed by PBS wash.

(4) *Acetone, formaldehyde* (Sato *et al.*, 1976, p. 419): acetone at −5°C for 8–10 min, rehydrate in PBS, fix with 1–1.5% formaldehyde 6–8 min.

(5) *Glutaraldehyde* (Weber *et al.*, 1978): (for details see Section VI).

Note that since commercially available formaldehyde solutions often contain substantial amounts of methanol (up to 11%), the formaldehyde solutions used for fixation (methods 1–4) are usually made from solid paraformaldehyde. Thus 18.5 g paraformaldehyde is dissolved in 500 ml PBS, heated on a warm plate to approximately 60°C, and then filtered through a 0.45 μm Millipore filter. The solution is stored at room temperature and appears stable.

c. One-step Fixation Procedures

(6) *Methanol* (Osborn and Weber, 1977b): Methanol at −10°C for 6 min, then PBS for 30 sec (Fig. 2).

(7) *Dimethylsulfoxide* (Osborn and Weber, 1980): Cells are fixed with 50% dimethylsulfoxide.

d. Use of EGTA in Fixation. In certain cases, e.g., the sea urchin egg, it may be essential to add EGTA to chelate excess calcium ions present at the cellular surfaces or released from cellular calcium stores. A 0.5 *M* EGTA solution is made and the pH adjusted to approximately 7.0 with NaOH.

(8) *Formaldehyde, methanol, EGTA:* As (2), except that EGTA (5–50 m*M*) is added to both the formaldehyde and methanol solutions.

(9) *Methanol, EGTA* (Harris *et al.*, 1980a): As (5), except that EGTA (usually at 5 m*M* but up to 50 m*M* can be used) is added to the methanol.

The fixation methods given all reveal cytoplasmic microtubular profiles of reasonable quality in tissue culture cells (see Figs.). In general, at least with the antibodies we have available, an air-drying step instead of rehydration in PBS from the organic solvent results in poorer profiles, and the substitution of acetone for methanol, or the use of a slow dehydration–rehydration series using acetone, gives micrographs of similar or somewhat poorer quality than those obtained with methanol. Because of the known *in vitro* and *in vivo* sensitivity of mi-

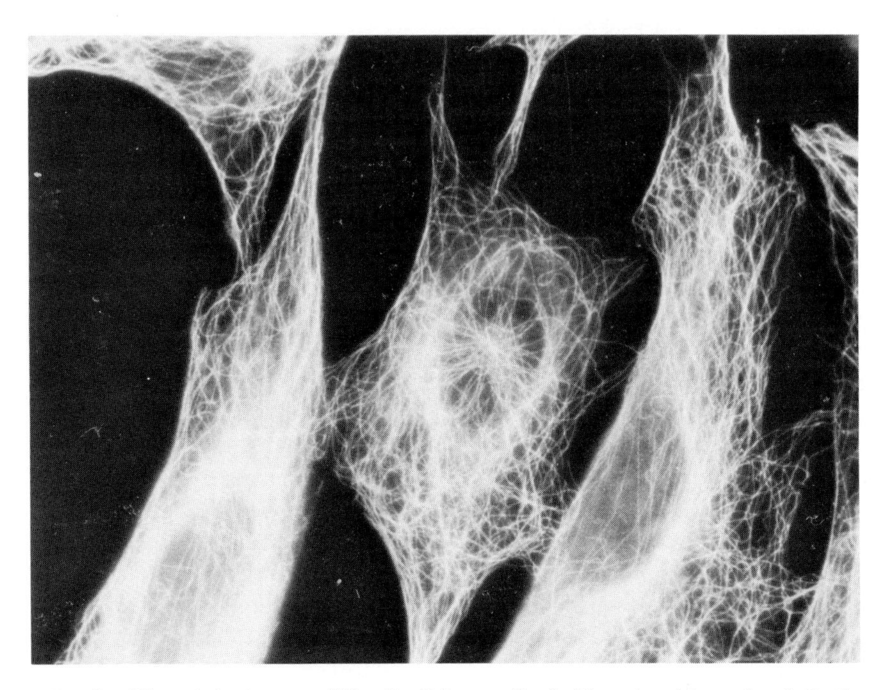

FIG. 2. Microtubules in mouse 3T3 cells. Cells were fixed with methanol (procedure 6, Section IV,B) before being treated with tubulin-specific antibody. Note the uniform thickness of the fluorescent fibers, most of which probably correspond to individual microtubules and the predominantly radial nature of the display. These cells were fixed and processed for immunofluorescence microscopy on 02.11.1977. The cell in this figure was photographed almost 3 years later on 01.08.1980 (exposure time approximately 6 sec). Note the good preservation of the specimen even after this length of time (see Section IV,D). Magnification ×900.

crotubules to cold and to excess calcium, care should be taken not to expose the cells to low temperatures prior to fixation, and if poor cytoplasmic microtubular profiles are seen with methods 1-6, the addition of EGTA to the formaldehyde solution or to the methanol, or both, should be considered. Procedures 1 and 2, although very popular in many laboratories, sometimes lead in our hands to some fragmentation of microtubules. Procedure 5 shows very good three-dimensional preservation, as well as fluorescent fibers of uniform thickness, and in addition allows a direct comparison of the same or parallely processed samples by electron microscopy (see below). However, it may not be applicable for all tubulin antibodies and is time consuming. Therefore we routinely use procedure 6 for cells in culture. Other specimens—for example, the sea urchin egg or plant cells—can present special problems. In the case of sea urchin eggs neither glutaraldehyde nor formaldehyde treatments were satisfactory and good preservation was obtained only with methanol and 50 m*M* EGTA (procedure 9) (Harris *et al.*, 1980a,b). In the case of plant cells certain highly specialized preparations that lack cell walls can be processed for immunofluorescence microscopy by procedures similar to those used for animal cells. Thus it has been possible to document microtubular arrangements in mitotic cells of liquid endosperm (e.g., Franke *et al.*, 1977), and interphase arrays in protoplasts (Lloyd *et al.*, 1979). Recently it has proved possible to extend such studies to meristematic plant cells, and procedures have been described involving enzymatic digestion of the plant cell wall and EGTA treatment that preserve to a large extent the three-dimensional arrangements of microtubules seen within interphase and mitotic cells in organized plant tissues (for further details and pictures see Wick *et al.*, 1981).

In general, and especially if a new antibody is assayed, it is important to note that treatment with formaldehyde or glutaraldehyde can render certain antigens unreactive. In addition there are examples of antibodies in which positive staining is only observed after formaldehyde fixation (e.g., the tropomyosin antibody described in Webster *et al.*, 1978b). Thus for these reasons new antibodies should be tested on both formaldehyde-fixed cells and on cells treated only with organic solvents (e.g., by procedure 6). However, the dozen different tubulin antibodies we have elicited work on both formaldehyde-fixed cells and on cells fixed only with organic solvents.

The fixation methods given allow excellent penetration of the IgGs into the cytoplasm and, as judged by control experiments with nuclear antigens, also into the nucleus.

2. Cytoskeletons

Since cytoskeletons prepared as in Section IV,A,3 have already been treated with nonionic detergents, they are permeable to antibodies without subsequent organic solvent treatment. It is therefore possible to add the first antibody immediately after preparing the cytoskeleton and to proceed as in Section IV,C,3. In this case, when dealing with labile elements such as microtubules, the buffer used to produce the cytoskeleton should be retained for all the subsequent immunocytochemical steps including the washes; otherwise microtubular breakdown can and does occur while the samples are being processed (e.g., cf. Figs. 11 and 12 in Osborn and Weber, 1977b). Alternatively the cytoskeletal preparations can be fixed by one of the fixation procedures given earlier, and indirect immunofluorescence microscopy can be performed exactly as described in Section IV,C. Specimens are not stored after the fixation steps but are processed immediately as described in the next section.

C. Protocol for Indirect Immunofluorescence Microscopy

1. A humid chamber is prepared by placing three to four circles of filter paper in a plastic petri dish (9- or 13-cm diameter). Numbers that identify the

coverslips are written on the top sheet of filter paper with pencil or ballpoint pen. Sufficient water is added so that the filter paper layer is moist but not wet.

2. After fixation the coverslips are briefly washed in PBS and the excess PBS is removed by touching the edge of each coverslip to dry filter paper. The coverslip is then placed cell side up over the appropriate number on the filter paper.

3. Five to 10 μl of the appropriate dilution of the first antibody in PBS is pipetted on to each coverslip and gently spread over the surface with the plastic tip of an Eppendorf pipette. Care should be taken not to allow the antibody solution to contact the filter paper, as it may then drain on the paper. The petri dish with its top in place is transferred to a 37°C incubator with high humidity.

4. After 30–60 min the coverslips are washed to remove excess antibody. This is usually done by handling each coverslip individually with pointed tweezers (size 5, Dumont & Fils, Switzerland) and dipping it three times each into three 30-ml beakers containing PBS. Excess PBS is drained before moving to a new beaker by touching the side of the coverslip to filter paper. A useful alternative procedure, particularly if a large number of coverslips are to be processed and all are being reacted with the same antibody, is to reinsert the coverslips into the rack used for fixation and transfer them through several changes of PBS. If, as sometimes happens during these washing procedures, a coverslip is mishandled, it can be difficult to decide which is the "cell side" of the coverslip. A useful trick is to compare the coverslip in question with a coverslip on which the cell side is known, under the 10× lens of a normal cell microscope (e.g., Leitz Diavert). If the cells on the two coverslips are in the same focal plane, they are the same side up.

5. After the washing step the coverslips are drained of PBS but are not allowed to dry completely. They are replaced on their appropriate numbers in the humid chamber (see above) and 5–10 μl of a suitable dilution of the second antibody carrying the fluorescent label is added. The coverslips are incubated at 37°C.

6. After 30–45 min the coverslips are washed as in step 4.

7. Microscope slides are identified by using small adhesive labels containing details of each specimen (e.g., date, specimen number, antibody etc.). Processing of a number of specimens at the same time is facilitated by the use of aluminum trays (Microslide Tray, Catalogue No. 6709-C6, A. H. Thomas, Philadelphia, PA) constructed so as to hold 20 slides horizontally. These trays can be stacked one above the other or inserted into appropriate storage cabinets. A small drop of mounting medium (see next section) is then placed in the center of each slide. The drained coverslip is inverted and mounted cell side down. Excess mounting medium is removed by placing filter paper over the coverslip and pressing very gently. If specimens are to be examined directly, they are secured with nail polish at three or four positions around the circumference.

Otherwise they are held in the dark for 1–2 hr for the mounting medium to harden and then examined.

D. Mounting Media

Although it is well documented that the fluorescence emission of FITC-labeled antibody preparations is maximal at pH 8.5 (Nairn, 1976, p. 153; Jongsma *et al.*, 1971), many workers still use a mounting medium of glycerol and phosphate-buffered saline at neutral pH. Use of polyvinyl alcohol-based mounting media such as Elvanol 51-05 (DuPont Co. Ltd., London, U.K.) (Rodriguez and Deinhardt, 1960; Heimer and Taylor, 1974), or more recently, because Elvanol has proved difficult to obtain, Mowiol 4-88 (Hoechst, Frankfurt, F.R.G.) at pH 8.5 has two further advantages. First, although the mounting medium is liquid when the sample is mounted, it solidifies after several hours, thus forming a permanent bond between the coverslip and the slide; second, the fluorescence is stable if the sample is held in the dark and at 4°C. Thus slides can be reexamined and if necessary rephotographed after several months or even years (e.g., Fig. 2), an important point particularly if fluorescence of pathological specimens is considered. Redistribution of fluorescent label has not been observed after mounting samples in Elvanol or Mowiol.

Our standard recipe is that of Heimer and Taylor (1974):

Six grams of analytical-grade glycerol is placed in a 50-ml disposable plastic conical centrifuge tube. Mowiol 4-88 (or Elvanol 51-05), 2.4 g, is added and stirred thoroughly to mix the Mowiol with the glycerol. Six milliliters of distilled water is added and the solution is left for 2 hr at room temperature. Twelve milliliters of 0.2 *M* Tris buffer (2.42 g Tris/100 ml water, pH adjusted to 8.5 with HCl) is added and the solution is incubated in a water bath at 50°C for 10 min with occasional stirring to dissolve the Mowiol. The mixture is clarified by centrifugation at 5000 *g* for 15 min and placed in aliquots in glass vials.

If the vials are subsequently stored at −20°C, the solution is stable, and the pH is retained for at least 12 months (M. Osborn,unpublished observations). Vials are unfrozen as required, and the mounting medium is stable at room temperature for at least one month.

E. Photography

Because of the rapid bleaching of the fluorescein label, long exposure of the slide to the light during examination can be detrimental. The time for photographic exposure should be kept to a minimum. We routinely use films with a high ASA rating such as Kodak 35-mm Tri-X film (Kodak Co., Rochester, NY) (normal ASA rating of 400). If special developers such as Diafine are used, the ASA setting can be increased to 800–1600 (30-33 DIN). The Diafine is obtain-

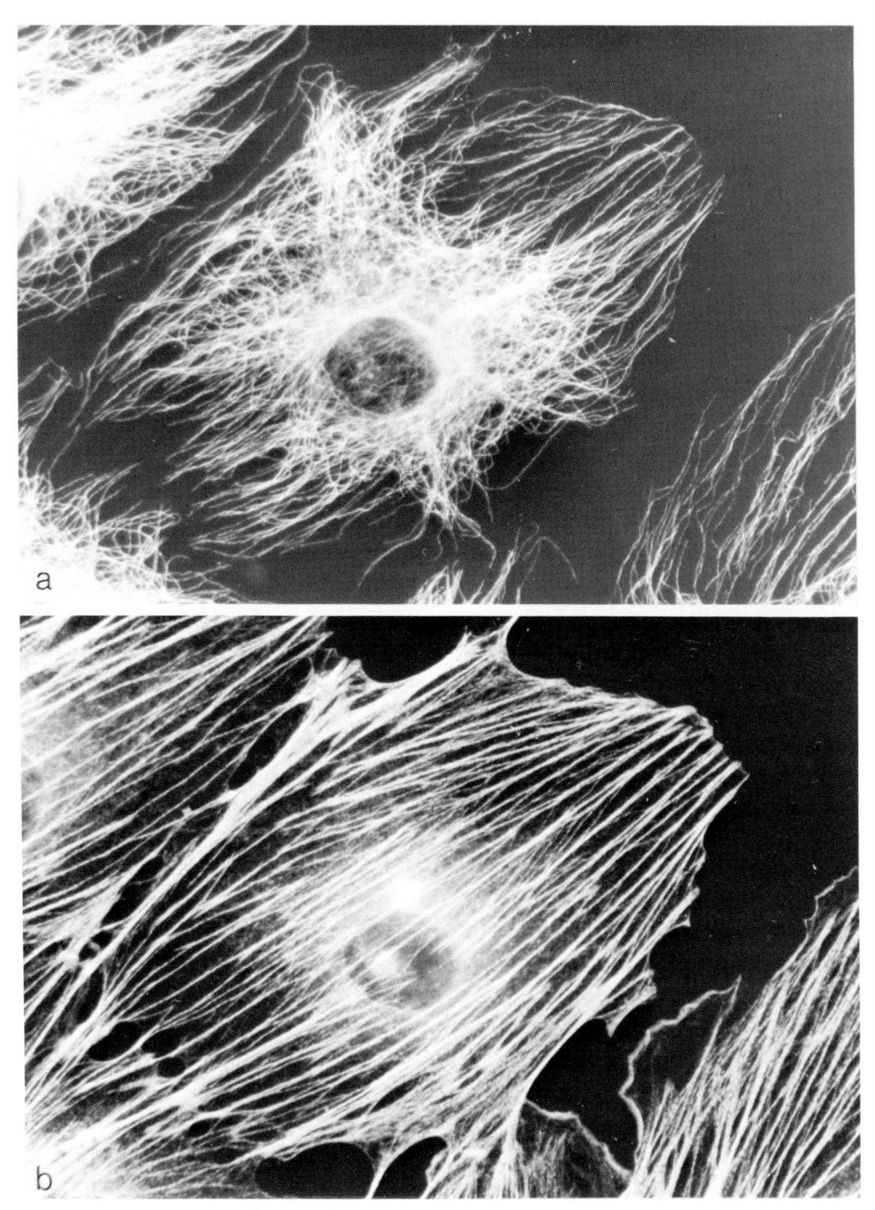

FIG. 3. The same rat kangaroo PtK2 cell stained in double immunofluorescence microscopy (a) rabbit antibody to tubulin and (b) guinea pig antibody to filamin. The microtubules were photographed in the rhodamine channel and the filamin, which is found associated with microfilamentous structures including the stress fibers, was photographed in the FITC channel. Note the clearly different arrangements of microtubules and of stress fibers, but also the tendency of some of the microtubules to align with the stress fibers. Magnification ×850.

able from Acufine Inc., 439-447 E. Illinois Street, Chicago, IL 60611, or from Fa. Bremaphot, Biedebach & Co., Kurt-Schumacher-Str. 31, Kassel, F.R.G. and development follows the manufacturers directions for the increased ASA setting. Microtubules are very fine structures and much more difficult to document than actin-containing stress fibers. Thus it is important that the microscope is well adjusted and equipped with epifluorescence optics and high-quality objectives. The length of the light path and the percentage of light used for photographic recording can vary, depending on the instrument design. Under our conditions micrographs such as those provided in this text are usually obtained with exposure times of 2–10 sec.

V. Double Immunofluorescence Microscopy

Often it can be advantageous to identify two different cellular structures in the same specimen using different antibodies (e.g., Fig. 3). There are three general methods available.

A. Direct Procedure

Both antibodies are used in the direct technique, i.e., rabbit-antitubulin and rabbit- or sheep-antiactin, with one antibody carrying the fluorescein and the other the rhodamine chromophore.

B. Indirect Procedures

1. Both Antibodies Elicited in the Same Species

Here one structure is reacted by the indirect technique using one fluorochrome, e.g., FITC, and the other in the direct technique using a second fluorochrome, e.g., rhodamine.

The following procedure is adapted from Hynes and Destree (1978b).

(i) Stain with first antibody, e.g., rabbit-antitubulin.

(ii) Incubate with FITC-labeled goat-anti-rabbit IgGs.

(iii) Add a 1:10 dilution of normal rabbit serum (or an appropriate concentration of rabbit IgGs) and incubate.

(iv) Add rhodamine-conjugated antibody to the second structure to be visualized, e.g., rhodamine-conjugated actin antibody.

Step iii with normal rabbit serum or IgGs is essential to block free antigen binding sites on the goat-anti-rabbit IgGs used to stain microtubules. If this step is not performed, the antibody added in step iv will also bind to the structure stained by antitubulin in step i.

2. ANTIBODIES ELICITED IN DIFFERENT ANIMAL SPECIES

There are several strategies that can be used, depending on which FITC- and rhodamine-conjugated second antibodies are available (see, e.g., Osborn *et al.*, 1980).

a. The first and simpler possibility depends on being able to obtain the two second antibodies, e.g., in the preceding example FITC-labeled anti-rabbit IgGs and rhodamine-conjugated anti-guinea pig IgGs from the same species, e.g., goat. In this case a possible procedure is as follows:

(i) Stain with both first antibodies simultaneously, e.g., rabbit-antitubulin and guinea pig-antiactin.

(ii) Stain with both labeled antibodies simultaneously, e.g., fluorescein-labeled goat-anti-rabbit IgGs plus rhodamine-labeled goat-anti-guinea pig IgGs.

Control experiments are necessary to assure that the two second antibodies are indeed species specific—that, for instance, the fluorescein-labeled goat-anti-rabbit antibody recognizes only the rabbit IgGs and not the guinea pig IgGs. In case of a cross-species reactivity preabsorption of the second antibodies on the corresponding IgGs coupled to a support should remove the cross-reacting antibodies (see below).

b. An alternative procedure is to use FITC- and rhodamine-labeled second antibodies made in different species, e.g., FITC-labeled rabbit-anti-guinea pig IgGs and rhodamine-labeled goat-anti-rabbit IgGs. Also in this case it seems important to cross-absorb both the FITC- and the rhodamine-labeled antibodies to remove some cross-species-reacting IgGs. Thus FITC-labeled rabbit-anti-guinea pig IgGs can be passed through a Sepharose 4B column to which goat IgGs were covalently bound, and the rhodamine-labeled goat-anti-rabbit IgGs are absorbed on guinea pig IgGs bound to Sepharose 4B. In this case the procedure is as follows:

(i) Stain with both first antibodies simultaneously, e.g., rabbit-antitubulin and guinea pig-antiactin.

(ii) Rhodamine-labeled goat-anti-rabbit IgGs.

(iii) Blocking step with rabbit IgGs—see 1,(iii) above.

(iv) Fluorescein-labeled rabbit-anti-guinea pig IgGs.

In all cases the cells are usually incubated for 30–45 min at 37°C for each step and are well washed in PBS between steps. After the last wash they are mounted in mounting medium and viewed in a microscope equipped with appropriate filters to separate fluorescein and rhodamine fluorescence. Potential overlap between the two channels, as well as the quality of the staining, must be controlled using cells labeled only for the single structures. Obviously it is of utmost importance that the filter combinations provided in the microscope allow a total separation of the two chromophores.

Another alternative includes the use of a biotin–avidin label in fluorescence microscopy (for single label, see, e.g., Heggeness and Ash, 1977; for double label, Heggeness *et al.*, 1977; Gottlieb *et al.*, 1979).

It can sometimes also be helpful to view the DNA in relation to the microtubular profiles. In this case after completing step 6 of the immunofluorescence procedure described in Section IV,C, the samples are immersed for 2–5 min in a 10 μM Hoechst solution, and then mounted as in Section IV,C. A suitable recipe is to dissolve 0.5 mg bisbenzidine H33258 (Riedel de Haen, Seelze, F.R.G.) in 25-ml ethanol, and after it has dissolved add 75 ml PBS. The solution can be stored in the dark at 4°C. By an appropriate choice of filters Hoechst fluorescence can be discriminated from both fluorescein and from rhodamine fluorescence (see, e.g., Berlin *et al.*, 1979). A second procedure in which peroxidase staining is used to identify microtubules (cf. Section X) and the chromosomes are identified by toluidine blue staining has been described by De Brabander *et al.* (1979).

VI. Use of Glutaraldehyde in Immunofluorescence Microscopy

Fixation with glutaraldehyde was initially avoided in immunofluorescence microscopy because attempts to include it as the primary fixative resulted in samples with very high backgrounds (Cande *et al.*, 1977), presumably because of an excess of covalently bound aldehyde groups in the strongly cross-linked cellular matrix. However, the introduction of a sodium borohydride reducing step after the glutaraldehyde step not only results in immunofluorescence pictures of excellent quality but also allows a direct comparison of the same structure by both immunofluorescence and, after embedding and sectioning, by electron microscopy (Weber *et al.*, 1978).

Two procedures have been suggested for normal cells:

Procedure 1. (a) Fix in 2.5% glutaraldehyde (Serva, Heidelberg, F.R.G.), in 0.1 *M* Na cacodylate/10 m*M* KCl/5 m*M* $MgCl_2$, pH 7.2, for 10 min at room temperature, followed by 10 min on ice. (b) Rinse four times, 7 min each, with ice-cold 0.1 *M* Na cacodylate, pH 7.2. (c) Serially dehydrate through ice-cold 50%, 70%, 80%, 90%, and 96% ethanol for 15 min each. (d) Treat with $NaBH_4$ (Merck Co., F.R.G.), 0.5 mg/ml in 96% ethanol, three times for 6 min each at 4°C (under these conditions, $NaBH_4$ dissolves slowly; solutions are made up 3–4 min prior to the addition of the coverslips). Wash two times for 5 min each with 96% ethanol at 4°C (e). Serially rehydrate through ethanol at 50% (4°C), 20%, and 10% at room temperature and equilibrate with PBS at room temperature. (f) Add tubulin antibody (0.05 mg/ml in PBS) and incubate 45 min at 37°C; wash well with PBS. (g) Add fluorescein-labeled second antibody and incubate 45 min at 37°C; wash well with PBS. (h) For immunofluorescence microscopy, coverslips are mounted in Mowiol and viewed in the fluorescence microscope. (i) For electron microscopic analysis, steps a, b, and c are repeated. (j) Dehydrate the cells in water-free ethanol followed by water-free propylene oxide and embed as monolayers in Epon 812. Ultrathin sections are cut parallel to the original substratum and stained with uranyl acetate and lead citrate and examined with the electron microscope. The conditions for the $NaBH_4$ step have not yet been optimized. Preliminary experiments suggest that the $NaBH_4$ treatment can also be done either in step c in 50% ethanol, in which case the further serial dehydration in this step can be omitted, or in PBS after step e using steps b and c from procedure 2.

Procedure 2. (a) Fix in 1% glutaraldehyde in PBS for 15 min at room temperature; treat with methanol at −10°C for 15 min. (b) Treat with $NaBH_4$, 0.5 mg/ml in PBS, three times for 4 min each at room temperature. The $NaBH_4$ solution is made just before use. (c) Wash with PBS two times for 3 min each at room temperature. (d) Use steps f and g of procedure 1. For fluorescence microscopy, use step h above; for electron microscopy, use step i above.

In control experiments, other treatments—for instance, extensive washing with glycine or Tris buffers—were not nearly as successful as the $NaBH_4$ treatment in reducing the nonspecific background staining. Autofluorescence due to glutaraldehyde was not observed. Procedure 1 was designed to stay close to normal electron microscopical procedures. Procedure 2 is less time consuming and the results are equivalent, at least for our actin and tubulin antibodies, to those obtained by procedure 1. Both procedures give, as expected, excellent preservation of the three-dimensional structure (e.g., Fig. 6). Electron microscopical analysis reveals heavily antibody-decorated microtubules. Because of their circumferential decoration by two antibodies (indirect method), the diameter of the microtubules increases as expected from 22 nm to 55 nm. This decoration effect is specific. It is only seen when tubulin antibody is used and involves only microtubules and not microfilaments, intermediate filaments, or other recognizable structures. Although general ultrastructural preservation is reduced because of the excessive treatments, many cellular structures are clearly preserved and identifiable. Thus a parallel assessment of ultrastructure by immunofluorescence and electron microscopy is achieved.

VII. Comparison of the Same Specimen in Immunofluorescence Microscopy and in Electron Microscopy

This procedure allows the uniequivocal identification of the same structure in the light and in the electron microscope. Two general procedures have been used:

first, comparison of the fluorescence micrograph of a cytoskeletal preparation with that obtained by viewing the same specimen as whole mount in the electron microscope (see Fig. 4); second, comparison of the fluorescence micrograph with the images obtained after the same specimen is processed by embedding and serial sectioning for normal electron microscopical analysis.

In procedure 1 it is important to establish that the conditions necessary to make a "cytoskeleton" preserve the structure under study. Thus in the case of microtubules the procedure in Section IV,A,3,b is used. Viewing of the whole mount in the electron microscope is possible because a large fraction of the noncytoskeletal proteins are solubilized by the nonionic detergent treatment allowing resolution of structures by the electron beam. Procedure 1 is performed as follows: Gold grids—selected because gold is nontoxic for cells—are cleaned by boiling and sonication in 50% HNO_3, followed by boiling and sonication in three changes of distilled water. They are then attached to 12-mm glass coverslips by a 1.2% Formvar film, which is subsequently coated with polylysine (Section IV,A). Cells are allowed to attach to the grid for 24–36 hr, and then cytoskeletons are made using the procedure given in Section IV,A,3,b. After step d cells are fixed for 10 min at room temperature in stabilization buffer containing 1% glutaraldehyde. The resulting glutaraldehyde-fixed cytoskeletons are treated with 0.5 mg/ml sodium borohydride (see Section VI) in PBS twice 4 min at room temperature and washed with PBS twice 3 min at room temperature. They are processed for immunofluorescence microscopy as above (Section IV,C). After the final PBS wash each coverslip is mounted carefully over a small well made from several layers of sticky tape containing 0.04 *M* Tris-HCl, pH 8.5, 0.14 *M* NaCl and secured with nail polish around the circumference. Selected cells are then photographed using epifluorescent optics, and their position on the grid is recorded. The coverslip is carefully removed using acetone to dissolve the nail polish and washed twice with the Tris-HCl buffer, twice with 0.3 *M* KCl, and then treated with 1% aqueous uranyl acetate for 45 sec. The grid is then detached from the coverslip and screened in the electron microscope. Cells that have been photographed in the fluorescence microscope are located and photographed.

The results demonstrate clearly the specificity of the antibody preparations used. In the case of microtubules (Osborn *et al.*, 1978a) and also for other cytoskeletal elements including microfilament bundles and keratin-type intermediate filaments (Webster *et al.*, 1978b) a 1:1 correspondence between the structures seen in fluorescence and in electron microscopy is obtained. In addition the technique demonstrates that immunofluorescence microscopy under optimal conditions visualizes individual microtubules (see Section VIII).

Procedure 2 is useful when the structure in question cannot be readily identified in whole mounts and therefore thin sectioning is required. In this case cells are grown directly on glass coverslips, and after making the fluorescent picture the sample is flat embedded and then serially sectioned. Using this approach a

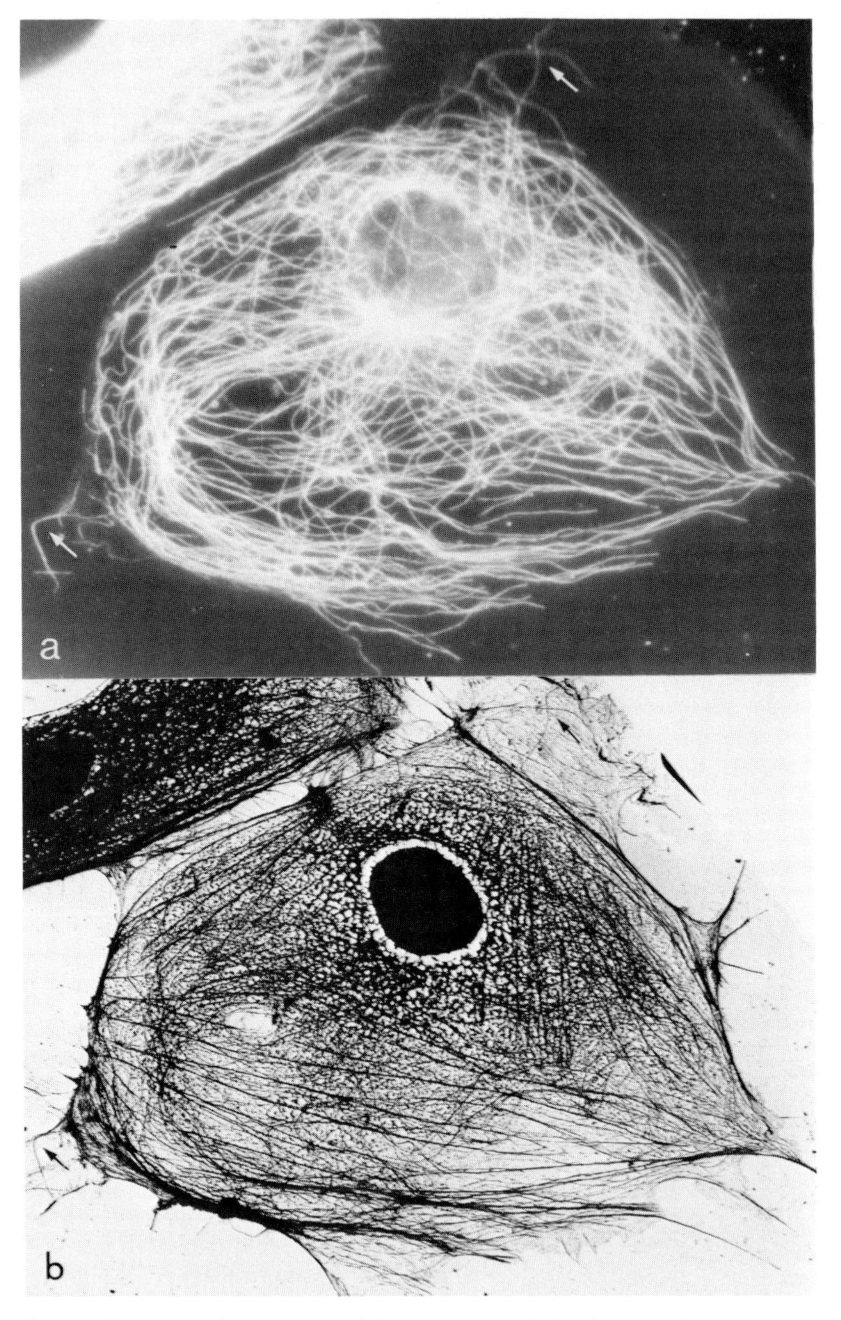

FIG. 4. Fluorescent micrograph (a) and electron micrograph (b) of the same PtK2 cytoskeleton. The cell was treated with detergent, fixed in glutaraldehyde, and treated with tubulin-specific anti-

1:1 correspondence in immunofluorescence and electron microscopy has been demonstrated for the tubulin-containing primary cilium of mouse 3T3 cells (Albrecht-Buehler and Bushnell, 1980). Similarly the multiple microtubular organizing centers characteristic of certain neuroblastoma cells (Spiegelman *et al.*, 1979; Marchisio *et al.*, 1979), as well as the structures decorated by autoantibodies assumed to recognize centrioles (Connolly and Kalnins, 1978), have been formally identified as centriolar bodies (Sharp *et al.*, 1981).

VIII. Limit of Resolution of Fluorescence Microscopy

The limit of resolution possible by fluorescence microscopy is not always appreciated. When the same cytoskeleton is first viewed by indirect immunofluorescence microscopy and then as whole mount by electron microscopy, microtubules are easily discerned by both procedures. Fluorescent fibers of an approximate diameter of 200–250 nm measured on the fluorescent micrographs correspond to fibrous structures, which in the electron microscope reveal a diameter of only approximately 55 nm. The latter structures are individual microtubules (diameter 20–25 nm) decorated in the indirect technique circumferentially by two layers of IgG molecules. Since an IgG molecule is approximately 9 nm in diameter, the indirect technique should provide an increase in diameter of 4×9 nm, i.e., 36 nm, giving rise to a final diameter of 56–61 nm in excellent agreement with the values found in the electron microscope (Osborn *et al.*, 1978a). Such a fiber is revealed in the fluorescent micrograph with a diameter of 200–250 nm, a value closer to the calculated theoretical limit of resolution when apertures of 1.4 and light of wavelength 515 nm are used in the fluorescence microscope. Thus individual microtubules that come closer than 200–250 nm will not be resolved in the light microscope. This explains why structures such as the mitotic spindle, the postmitotic intercellular bridge, or the microtubular complex of melanophores, which contain in electron micrographs very closely packed microtubules (distance perhaps 40 nm), appear as strongly fluorescent structures with disappointingly little substructure. Thus, in general, fluorescence microscopy is not a good method to study the distribution of structures that from

---

body followed by fluorescent second antibody before being photographed in the fluorescence microscope (a). The cytoskeleton was then stained with uranyl acetate and photographed in the electron microscope (b). The fluorescent fibers seen in (a) correspond to uninterrupted fibers of approximately 55-nm diameter visualized in (b). Each 55-nm fiber is assumed to be an individual antibody-decorated microtubule (see text). The arrows indicate areas where it is particularly easy to see a correspondence, but almost all the fluorescent fibers seen in (a) can also be traced in (b). Magnification approximately ×1400. (Reprinted from Weber and Osborn, 1979; with permission from "Microtubules." Copyright by Academic Press Inc. (London) Ltd. For further details and a comparison at higher magnification, see Osborn *et al.*, 1978a.)

ultrastructural studies are known to be closer than approximately 200 nm. Nevertheless in well-spread cultured cells in interphase microtubules are generally separated by a larger distance and in majority revealed as individual microtubules. When two microtubules come closer than the limit of resolution, this situation is usually appreciated in good micrographs because of the local relative increase in fluorescence intensity.

IX. Stereo Immunofluorescence Microscopy

This procedure, introduced by Osborn *et al.* (1978b), yields information as to the three-dimensional arrangement of structures within the cell. The fixation and processing of samples for stereo immunofluorescence microscopy are the same as for normal specimens. Glutaraldehyde fixation is advantageous because it preserves the three-dimensional nature of the specimens very well. It is important that the cells show strongly fluorescent patterns since two exposures are made from the same specimen. Control experiments have shown that it is unimportant whether the cell is photographed from below—as in normal epifluorescent optics—or whether a ''flip-flop'' technique is employed and the cell is photographed from above. In the case of a fibrous system such as microtubules, identical results are obtained.

The parts required for stereo microscopy, illustrated in Fig. 5, can be constructed from parts available commercially for other purposes. On the left is an intermediate ring containing a lens to compensate for the increase in tube length (Carl Zeiss, Oberkochen, F.R.G., part no. 474465) which can be inserted into a commercial microscope adapted for epifluorescent optics between the revolving nosepiece and the objective. On the right is the ''stereo insert'' constructed by welding a half-moon diaphragm into a differential interference contrast slider (Carl Zeiss, part no. 474571, but without the prism). The metal pins in the intermediate ring are removed so that the slider can be easily moved in and out of the ring. To take a stereo micrograph the stereo insert is slid into the intermediate

FIG. 5. Parts required for stereo fluorescence microscopy (see text). Approximately half real size. (Reprinted from Osborn and Weber, 1979; courtesy of the *European Journal of Cell Biology*.)

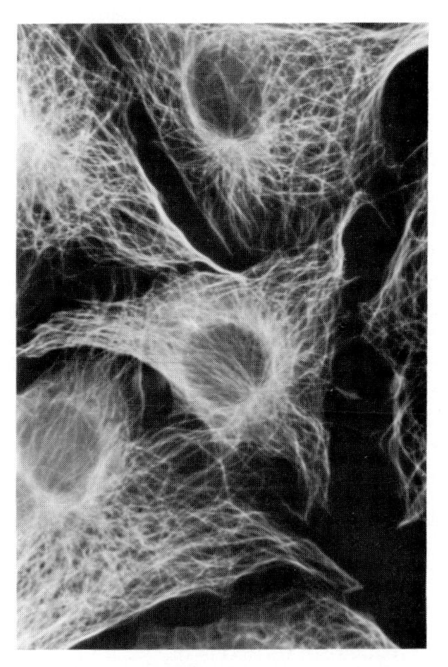
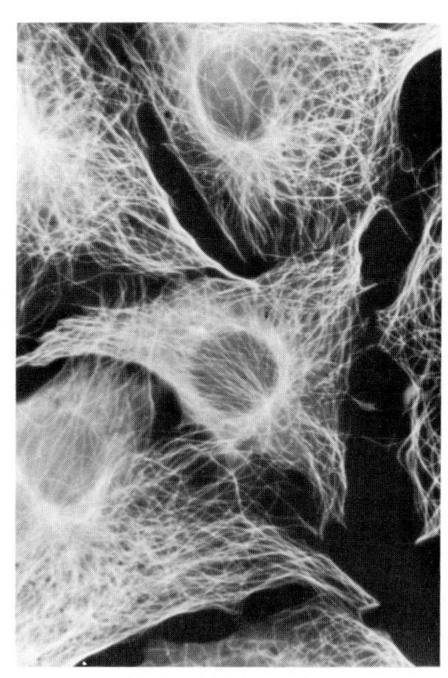

FIG. 6. Stereomicrograph of microtubules in mouse 3T3 cells after glutaraldehyde fixation. This figure should be viewed with a 2× stereo viewer placed above the stereo pair. Note that microtubules appear to be present at all levels in the cell and that, in some cells, cilia, which mark the microtubular organizing centers, can be detected above and below the nucleus. (Reprinted from Osborn *et al.*, 1978b; copyright MIT Press.)

ring at the arrow, and the first micrograph is taken while one-half of the field is obscured. The stereo insert is then removed, inverted, and replaced so that the other half of the field is blocked, and the second micrograph is taken.

An optimal three-dimensional impression is obtained by rotating the two micrographs through 90° before mounting them. In the example provided (Fig. 6) it appears as though the cell is viewed from above the nucleus. If the picture now on the left were placed to the right of the second micrograph, it would appear as though the cell was viewed from the side attached to the substratum. Thus before the two pictures are mounted, the filament system can be viewed either from on top or below (see also Fig. 11 in Osborn *et al.*, 1978b). The extent of the stereo impression can be varied by changing the percentage of the field blocked by the aperture. The 50% aperture gives a good stereo impression although clearly it exaggerates the vertical dimension in relation to the horizontal. Different effects can sometimes be obtained by focusing on different areas of the cell. If a high-power objective such as the 63 Planapo is used, it is important to realize that at times the whole depth of the cell may not be included in the micrographs.

Some observers can fuse the images of a stereo pair and obtain a three-dimensional impression without a viewer. The use of a 2× stereo viewer (Polaron, Watford, England, or Polysciences, Warrington, PA), adjusted to the eye distance of the observer and placed symmetrically over the stereo pair, is however recommended.

Stereomicrographs have been taken not only for microtubules but also for microfilament arrangements, the cytokeratin filaments in epithelia cells (Osborn *et al.*, 1978b, Osborn and Weber, 1979), and chromosomes in mitotic cells (Osborn, 1981). In addition, stereo fluorescence often can help to determine whether particular structures are present close to the upper or the lower cellular surfaces (see, e.g., Wehland *et al.*, 1979; Hiller *et al.*, 1980).

X. Peroxidase Immunoferritin and Gold Labeling

Given the success in visualizing microtubules by immunofluorescence microscopy several laboratories turned to peroxidase ferritin and gold labeling. Given the variety of different fixation and antibody permeabilization procedures used, as well as the different experimental material (i.e., whole cells, ultrathin frozen sections, or cytoskeletal preparations), the reader should consult the individual reports to get an opinion on the most suitable protocol for his own problem.

Immunoperoxidase with tubulin antibodies at the light microscopic level using phase contrast optics (De Brabander *et al.*, 1977a,b; De Mey *et al.*, 1978; Karsenti *et al.*, 1978; Henderson and Weber, 1979), performed using the techniques of Sternberger (1979), gives an overview of the cytoplasmic microtubular profiles but the image quality is poorer than that obtainable in immunofluorescence microscopy. This is not surprising because of the higher sensitivity of fluorescence microscopy and the further advantage that microtubules in fluorescence are visualized against a black background. Immunoperoxidase does offer, however, the possibility of using electron microscopy. Two approaches were taken. Cells were very lightly fixed with glutaraldehyde and then dehydrated with acetone before reaction with tubulin antibodies and subsequent processing for ultrastructural observation in thin sections (De Brabander *et al.*, 1977a,b; De Mey *et al.*, 1978). As expected, because only light fixation was used, the preservation of ultrastructure was clearly suboptimal but microtubule decoration was very convincing. An easier procedure is to use cytoskeletons prepared with Triton and then fixed with glutaraldehyde. Subsequent immunoperoxidase staining of these specimens revealed strongly decorated microtubules against well-preserved undecorated microfilaments and intermediate filaments. The preservation was suitable for examination of critically point dried whole mounts by stereo electron microscopy (Henderson and Weber, 1979) (Fig. 7).

Ferritin labeling requires the use of higher resolution electron microscopy. Successful use of glutaraldehyde-fixed cytoskeletal preparations in stereo-

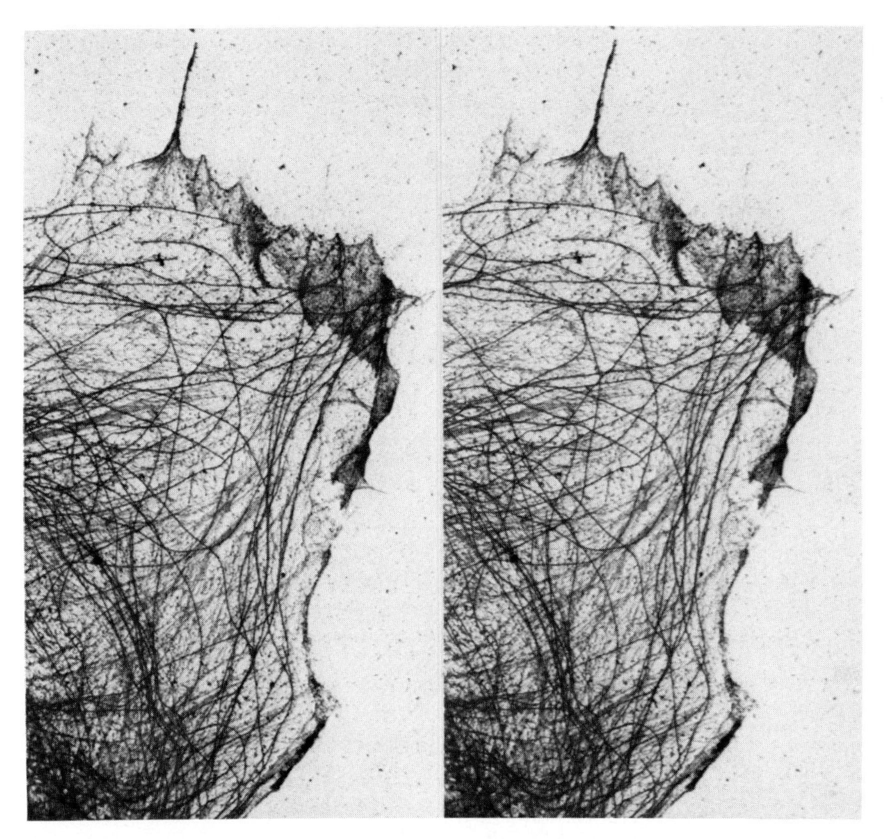

FIG. 7. Low-power stereo-electron micrograph of a part of a cytoskeleton of a rat mammary cell (RMCD) labeled with tubulin antibody by the peroxidase-antiperoxidase technique. Microtubules are clearly visible and labeled, and their profiles are similar to those seen in immunofluorescence. Microfilament bundles can also be distinguished and are unlabeled. Magnification ×3000. (Reprinted from Henderson and Weber, 1979; with permission from *Exp. Cell Res.*)

electron microscopy after tubulin antibody decoration has been reported (Webster *et al.*, 1978a) (Fig. 8). In addition, the EGS procedure (Willingham and Yamada, 1979) can be used to permeabilize the cells, the microtubules can be decorated using a ferritin bridge method with tubulin antibody (Willingham *et al.*, 1980).

Successful visualization of microtubules using gold labeled IgGs in the indirect technique has been reported recently. The microtubules can also be visualized in bright field microscopy as red stained strands (De Mey *et al.*, 1981).

All these methods clearly reveal cytoplasmic microtubules in interphase cells. The profiles obtained show the same features as those described above with fluorescence techniques.

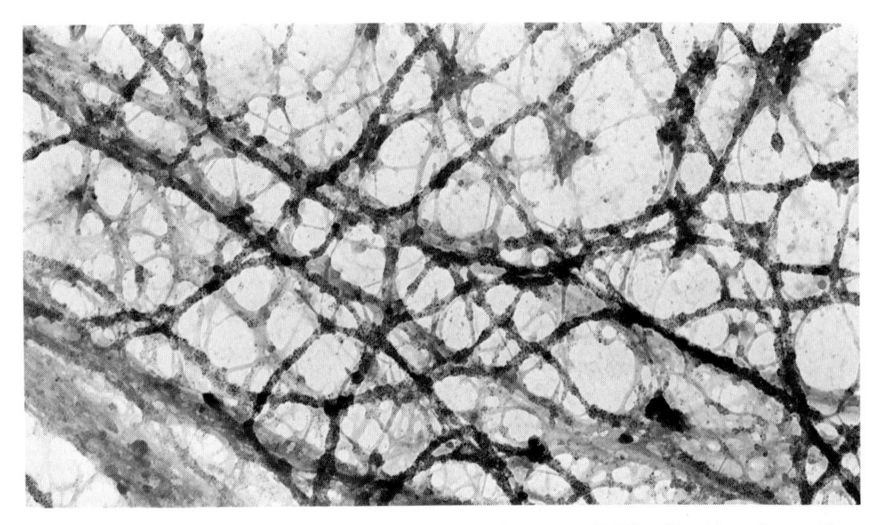

FIG. 8. Electron micrograph of a part of a cytoskeleton of a PtK2 cell subjected to indirect immunoferritin decoration with antibody to tubulin. Many microtubules are present. They are strongly decorated by the ferritin molecules and can often be traced over the whole field. Other cytoskeletal structures and stress fibers are not labeled. Magnification approximately ×40,000. Micrograph courtesy of Dr. R. Webster. (For further details, see Webster *et al.*, 1978a.)

XI. Protein-Chemical Anatomy

Immunofluorescence or immunoelectron microscopic methods allow the identification of a specific cellular structure if the antigen is a component of this organization. An early example of this approach was the demonstration by immunofluorescence microscopy that myosin was often associated with actin-containing microfilament bundles (stress fibers) (Weber and Groeschel-Stewart, 1974). Later experiments extended the protein-chemical anatomy of the microfilament organization by the identification of a variety of other associated proteins including α-actinin, filamin, tropomyosin, vinculin, and fimbrin (Lazarides and Burridge, 1975; Lazarides, 1975; Wang *et al.*, 1975; Geiger, 1979; Bretscher and Weber, 1980). In addition, different microfilament arrangements were recognized by different spectra of associated proteins for the ruffling edge, the stress fibers, microvilli of intestinal epithelial cells, vaccinia virus-induced microvilli, and nuclear actin paracrystals (for references, see Weber and Osborn, 1981b).

A similar approach has been started for tubulin-containing structures including cytoplasmic microtubules, the mitotic spindle, and cilia. Current results have concentrated on antigens related to the two high-molecular-weight proteins (HMW proteins), MAP_1 and MAP_2 (polypeptide molecular weights approximately 320,000) (Murphy and Borisy, 1975; Sloboda *et al.*, 1976), and the lower-molecular-weight tau proteins (polypeptide molecular weights 55,000–65,000) (Cleveland *et al.*, 1977). Results with antibodies specific for one or the other of these classes of microtubular-associated proteins indicate that cytoplasmic microtubules as well as mitotic microtubules contain HMW proteins and tau (Sherline and Schiavone, 1977; Connolly *et al.*, 1977, 1978). In contrast cilia, although containing tubulin and tau, may not contain the MAP proteins (Connolly and Kalnins, 1980). An additional microtubular component—Tap—has also been found in association with microtubules (Lockwood, 1978), and it seems likely that the number of associated proteins will increase in the future. It will be interesting to see whether different classes of microtubules can be identified within the same cell on grounds of different associated proteins. Ferritin labeling and electron microscopy may also help to elucidate details of the arrangement of associated proteins along microtubular structures (see Section X).

XII. Tissue Sections

The methods so far discussed have been concerned with the description of microtubular arrays in animal cells in culture, or in isolated cells of animal or plant origin. Although methods to study the location of antigens in tissue sections have proved to be generally useful for a variety of antigens (see, e.g., Hökfelt, 1980, Franke *et al.*, 1979; for detailed protocols used in the study of frozen sections, see Osborn, 1981), use of frozen sections has not resulted in good location *in situ* of microtubular networks. Presumably this is because as discussed earlier microtubules are sensitive to cold and calcium, and thus do not survive the freezing, cutting, and staining steps involved in the frozen section preparation. Recently, however, Matus *et al.* (1981) have been able to demonstrate *in situ* microtubular networks in rat brain by perfusing the brain with a mixture of 4% paraformaldehyde and 0.5% glutaraldehyde in 50m*M* cacodylate buffer. Vibratome sections were then processed with antisera to tubulin, or to high-molecular-weight protein by the usual techniques. Again these results emphasize the necessity of choosing appropriate fixation procedures before preceeding with the immunofluorescence microscopy.

XIII. Troubleshooting

Failure to decorate particular cellular structures even though they contain the antigen under study may arise for a variety of reasons.

a. Certain antibodies may have a very limited cross-species reactivity. This was clearly the case with some of the earlier antibodies against certain intermediate filament proteins. We have never observed this situation with tubulin antibodies (see Section II,E), but it cannot be excluded. Therefore controls on cells of the appropriate species should be performed.

b. Certain antigens may be sufficiently modified by the fixation procedure to render them immunologically unreactive. Thus aldehyde fixation seems to inactivate the immunological reactivity of intermediate filament proteins against certain antisera (see Weber and Osborn, 1981a). This situation is unfortunate because aldehyde fixatives have to be avoided and one is restricted to other fixation protocols or to very light fixation (see Section IV,B). Again we have not recognized this problem with some dozen different tubulin antibodies but again one cannot exclude the possibility.

c. The antigen may be only poorly fixed or even be partially extracted. This problem does not arise with tubulin organizations but can be encountered with other antigens. In this situation one has to try stronger fixation procedures, e.g., formaldehyde and glutaraldehyde fixation for prolonged times or at higher fixative concentrations. Alternatively one has to explore the possibility that interfering substances may be responsible for the result. For instance, in the case of microtubules one may consider the release of excessive amounts of free calcium ions from cellular stores during the fixation step itself. Since calcium can depolymerize microtubules, such a situation can give rise to serious problems with certain objects (Section IV,B).

d. The antibody may not be able to gain access to the antigen. This is a rather general problem which is often neglected. In the case of microtubules the midbody of the postmitotic intercellular bridge is an obvious example. Although known from electron microscopical studies to contain tightly packed microtubules (see, e.g., Mullins and Biesele, 1977), the midbody is usually revealed as a dark zone against the highly fluorescent structure of the bridge. The presence of highly osmophilic substances in the midbody probably prevents access of the tubulin antibodies to this structure (Weber *et al.*, 1975b, Fuller *et al.*, 1975).

e. Cellular morphology may have an influence. Severe problems may be experienced when rounded cells are studied by immunofluorescence microscopy because of the limited depth of field of the available objectives. Photographic recording may be especially cumbersome because of the superposition of several fluorescent structures in the same focal plane. This problem may be further aggravated by poor fixation procedures or because some antibodies give rise even in well-spread cells to high and unspecific backgrounds. In such cases extreme care is advisable in evaluating the results, and trivial explanations should first be considered. Examples for such situations are amply found in the literature of microtubules (for a detailed discussion, see Weber and Osborn, 1981a). Thus it was claimed that transformed cells are characterized by defects in the microtubular networks. Descriptions ranged from broken and fragmented microtubules, and less microtubules, to virtually no microtubules (see, e.g., Brinkley *et al.*, 1975; Edelman and Yahara, 1976; Miller *et al.*, 1977). In addition, it was suggested that cells derived from animals suffering from muscular distrophy (Shay and Fuseler, 1979) and Chediak-Higashi-Syndrome (Oliver, 1976) display defective cytoplasmic microtubular profiles. In no case have subsequent studies upheld these first conclusions (Osborn and Weber, 1977b; Tucker *et al.*, 1978; de Mey *et al.*, 1978; Connolly *et al.*, 1979; Frankel *et al.*, 1978), since more careful analyses clearly reveal a more or less full complement of microtubules. Obviously when the morphologies of two cells differ, also the general images of the cytoplasmic microtubules will be somewhat different. These critical remarks should by no means discourage the use of rounded cells since beautiful microtubular images can also be recorded in such cells, as for instance documented for lymphocytes by Yahara and Kakimoto-Sameshima (1978) or for platelets by Debus *et al.* (1981).

Acknowledgments

We wish to acknowledge the friendly hospitality experienced in Dr. R. Pollack's laboratory in Cold Spring Harbor in 1972–1974, where we first were exposed to immunofluorescence techniques used for the study of SV40 T antigen and also benefited from discussions with Rex Risser and Art Vogel. In the subsequent time at Göttingen, F.R.G., we were greatly helped by the technical expertise of H.-J. Koitzsch, W. Koch, S. Schiller, and T. Born, and also by discussions with our colleagues Dr. J. Aubin, Dr. A. Bretscher, Dr. D. Henderson, Dr. G. Hiller and Dr. J. Wehland. In addition, many improvements in technique were only achieved by the disturbing problems and nagging questions posed by the systems studied or proposed by sabbatical visitors. We thank Dr. P. Harris, Drs. J. and T. Steitz, and especially Dr. R. E. Webster.

References

Albrecht-Buehler, G., and Bushnell, A. (1980). *Exp. Cell Res.* **126,** 427–438.

Aubin, J. E., Subrahmanyan, L., Kalnins, V. I., and Ling, V. (1976). *Proc. Natl. Acad. Sci. U.S.A.* **73,** 1246–1249.

Bennett, G. S., Fellini, S. A., Croop, J. M., Otto, J. J., Bryan, J., and Holtzer, H. (1978). *Proc. Natl. Acad. Sci. U.S.A.* **75,** 4364–4368.

Berlin, R. D., Oliver, J. M., and Walter, R. J. (1979). *Cell* **15,** 327–341.

Blakeslee, D., and Baines, M. G. (1976). *J. Immunol. Methods* 13, 305.

Bretscher, A., and Weber, K. (1980). *J. Cell Biol.* **86,** 335–340.

Brinkley, B. R., Fuller, G. M., and Highfield, D. P. (1975). *Proc. Natl. Acad. Sci. U.S.A.* **72,** 4981–4985.

Brinkley, B. R., Fistel, S. H., Marcum, J. M., and Pardue, R. L. (1980). *Int. Rev. Cytol.* **63,** 59–95.

Brown, S., Levinson, W., and Spudich, J. A. (1976). *J. Supramol. Struct.* **5,** 119–130.

Cande, W. Z., Lazarides, E., and McIntosh, J. R. (1977). *J. Cell Biol.* **72,** 552–567.

Cebra, J. J., and Goldstein, G. (1965). *J. Immunol.* **95,** 230–245.

Cleveland, D. W., Hwo, S.-Y., and Kirschner, M. W. (1977). *J. Mol. Biol.* **116,** 207-225.
Connolly, J. A., and Kalnins, V. I. (1978). *J. Cell Biol.* **79,** 526-532.
Connolly, J. A., and Kalnins, V. I. (1980). *Exp. Cell Res.* **127,** 341-350.
Connolly, J. A., Kalnins, V. I., Cleveland, D. W., and Kirschner, M. W. (1977). *Proc. Natl. Acad. Sci. U.S.A.* **74,** 2437-2440.
Connolly, J. A., Kalnins, V. I., Cleveland, D. W., and Kirschner, M. W. (1978). *J. Cell Biol.* **76,** 781-786.
Connolly, J. A., Kalnins, V. I., and Barber, B. H. (1979). *Nature (London)* **282,** 511-513.
Coons, A. H., and Kaplan, M. H. (1950). *J. Exp. Med.* **91,** 1-13.
Crestfield, A. M., Moore, S., and Stein, W. H. (1963). *J. Biol. Chem.* **238,** 622-627.
Dales, S. (1972). *J. Cell Biol.* **52,** 748-753.
De Brabander, M., De Mey, J., Joniau, M., and Gueuens, G. (1977a). *Cell Biol. Int. Reports* **1,** 177-183.
De Brabander, M., De Mey, J., Joniau, M., and Gueuens, G. (1977b). *J. Cell Sci.* **28,** 283-301.
De Brabander, M., Gueuens, G., De Mey, J., and Joniau, M. (1979). *Biol. Cell.* **34,** 213-226.
Debus, E., Weber, K., and Osborn, M. (1981). *Europ. J. Cell Biol.* **24,** 45-52.
De Mey, J., Joniau, M., De Brabander, M., Moens, W., and Gueuens, G. (1978). *Proc. Natl. Acad. Sci. U.S.A.* **75,** 1339-1343.
De Mey, J., Moeremans, M., Gueuens, G., Nuydens, R., and De Brabander, M. (1981). *Cell Biol. Int. Reports* **5,** 889-899.
Edelman, G. M., and Yahara, I. (1976). *Proc. Natl. Acad. Sci. U.S.A.* **73,** 2047-2051.
Franke, W. W., Seib, E., Osborn, M., Weber, K., Herth, W., and Falk, H. (1977). *Cytobiologie* **15,** 24-45.
Franke, W. W., Schmid, E., Osborn, M., and Weber, K. (1978). *Proc. Natl. Acad. Sci. U.S.A.* **75,** 5034-5038.
Franke, W. W., Applehans, B., Schmid, E., Freudenstein, M., Osborn, M., and Weber, K. (1979). *Differentiation* **15,** 7-25.
Frankel, F. R. (1976). *Proc. Natl. Acad. Sci. U.S.A.* **73,** 2798-2802.
Frankel, F. R., Tucker, R. W., Bruce, J., and Stenberg, R. (1978). *J. Cell Biol.* **79,** 401-408.
Fujiwara, K., and Pollard, T. D. (1978). *J. Cell Biol.* **77,** 182-195.
Fuller, G. M., Brinkley, B. R., and Boughter, M. J. (1975). *Science* **187,** 948-950.
Fulton, A., Wan, K. M., and Penman, S. (1980). *Cell* **20,** 849-858.
Fulton, C., and Simpson, P. (1979). *In* "Microtubules" (K. Roberts and J. S. Hyams, eds.), pp. 279-313. Academic Press, New York.
Gallagher, J. G. (1973). *In* "Tissue Culture: Methods and Applications" (P. F. Kruse and M. K. Patterson, eds.), pp. 102-106. Academic Press, New York.
Geiger, B. (1979). *Cell* **18,** 193-205.
Gordon, W. E., III, Bushnell, A., and Burridge, K. (1978). *Cell* **13,** 249-261.
Gottlieb, A. I., Heggeness, M. H., Ash, J. F., and Singer, S. J. (1979). *J. Cell. Physiol.* **100,** 563-578.
Harris, P., Osborn, M., and Weber, K. (1980a). *J. Cell Biol.* **84,** 668-679.
Harris, P., Osborn, M., and Weber, K. (1980b). *Exp. Cell Res.* **126,** 227-236.
Heggeness, M. H., and Ash, J. F. (1977). *J. Cell Biol.* **73,** 783-788.
Heggeness, M. H., Wang, K., and Singer, S. J. (1977). *Proc. Natl. Acad. Sci. U.S.A.* **74,** 3883-3887.
Heimer, G. V., and Taylor, C. E. D. (1974). *J. Clin. Pathol.* **27,** 254-256.
Henderson, D., and Weber, K. (1979). *Exp. Cell Res.* **124,** 301-316.
Heuser, J. E., and Kirschner, M. W. (1980). *J. Cell Biol.* **86,** 212-234.
Hiller, G., Weber, K., Schneider, L., Parajsz, C., and Jungwirth, C. (1979). *Virology* **98,** 142-153.
Hökfelt, T., Johannson, O., Ljungdahl, A., Lindberg, J. M., and Schultzberg, M. (1980). *Nature* **284,** 515-521.

Hynes, R. O., and Destree, A. I. (1978a). *Cell* **13,** 151-163.
Hynes, R. O., and Destree, A. I. (1978b). *Cell* **15,** 875-886.
Jongsma, A. P. M., Hijmans, W., and Ploem, J. S. (1971). *Histochemie* **25,** 329-343.
Karsenti, E., Guilbert, B., Bornens, M., and Avrameas, S. (1977). *Proc. Natl. Acad. Sci. U.S.A.* **74,** 3997-4001.
Karsenti, E., Guilbert, B., Bornens, M., Avrameas, S., Whalen, R., and Pantaloni, D. (1978). *J. Histochem. Cytochem.* **26,** 934-947.
Köhler, G., and Milstein, C. (1975). *Nature* (*London*) **256,** 495-497.
Lazarides, E. (1975). *J. Cell Biol.* **65,** 549-561.
Lazarides, E. (1980). *Nature* (*London*) **283,** 249-256.
Lazarides, E., and Balzer, D. R. (1978). *Cell* **14,** 429-438.
Lazarides, E., and Burridge, K. (1975). *Cell* **6,** 289-298.
Lazarides, E., and Weber, K. (1974). *Proc. Natl. Acad. Sci. U.S.A.* **71,** 2268-2272.
Lenk. R., Ramson, L., Kaufman, Y., and Penman, S. (1977). *Cell* **10,** 67-78.
Lloyd, C. W., Slobas, A. R., Powell, A. J., MacDonald, G., and Badley, R. A. (1979). *Nature* (*London*) **279,** 239-241.
Lockwood, A. H. (1978). *Cell* **13,** 613-627.
Luduena, R. F., and Woodward, D. O. (1973). *Proc. Natl. Acad. Sci. U.S.A.* **70,** 3594-3598.
Marchisio, P. C., Weber, K., and Osborn, M. (1979). *Eur. J. Cell Biol.* **20,** 45-50.
Matus, A., Bernhardt, R., and Hugh-Jones, T. (1981). *Proc. Natl. Acad. Sci. U.S.A.* **78,** 3010-3014.
Mazia, D., Schatten, G., and Sale, W. (1975). *J. Cell Biol.* **66,** 198-200.
Miller, C. L., Fuseler, J. W., and Brinkley, B. R. (1977). *Cell* **12,** 319-331.
Mullins, J. M., and Biesele, J. J. (1977). *J. Cell Biol.* **73,** 672-684.
Murphy, D. B., and Borisy, G. G. (1975). *Proc. Natl. Acad. Sci. U.S.A.* **72,** 2696-2700.
Nairn, R. C. (1976). "Fluorescent Protein Tracing." Livingstone, Edinburgh.
Oliver, J. M. (1976). *Am. J. Pathol.* **85,** 395-412.
Osborn, M. (1981). "Techniques in Cellular Physiology, Vol. 1." Elsevier/North-Holland Biomedical Press (in press).
Osborn, M., and Weber, K. (1976). *Proc. Natl. Acad. Sci. U.S.A.* **73,** 867-871.
Osborn, M., and Weber, K. (1977a). *Exp. Cell Res.* **106,** 339-350.
Osborn, M., and Weber, K. (1977b). *Cell* **12,** 561-571.
Osborn, M., and Weber, K. (1979). *Eur. J. Cell Biol.* **20,** 28-36.
Osborn, M., and Weber, K. (1980). *Exp. Cell Res.* **129,** 103-114.
Osborn, M., Franke, W. W., and Weber, K. (1977). *Proc. Natl. Acad. Sci. U.S.A.* **74,** 2490-2494.
Osborn, M., Webster, R. E., and Weber, K. (1978a). *J. Cell Biol.* **77,** R27-R34.
Osborn, M., Born, T., Koitzsch, H.-J., and Weber, K. (1978b). *Cell* **14,** 477-488.
Osborn, M., Franke, W. W., and Weber, K. (1980). *Exp. Cell Res.* **125,** 37-46.
Rodriguez, J., and Deinhardt, F. (1960). *Virology* **12,** 316-317.
Sato, H., Ohnuki, Y., and Fujiwara, K. (1976). *In* "Cell Motility" (R. D. Goldman, T. Pollard, and J. Rosenbaum, eds.), pp. 419-433. Cold Spring Harbor Lab., Cold Spring Harbor, New York.
Sharp, G., Osborn, M., and Weber, K. (1981). *J. Cell Sci.* **47,** 1-24.
Shay, J. W., and Fuseler, J. W. (1979). *Nature* (*London*) **278,** 178-180.
Shelanski, M. L., Gaskin, F., and Cantor, C. R. (1973). *Proc. Natl. Acad. Sci. U.S.A.* **70,** 765-768.
Sherline, P., and Schiavone, K. (1977). *Science* **198,** 1038-1040.
Sloboda, L. D., Dentler, W. L., and Rosenbaum, J. L. (1976). *Biochemistry* **15,** 4497-4505.
Small, J. V., and Celis, J. E. (1978). *J. Cell Sci.* **31,** 393-409.
Spiegelman, B. M., Lopata, M. A., and Kirschner, M. W. (1979). *Cell* **16,** 253-263.
Sternberger, L. A. (1979). "Immunocytochemistry." Wiley, New York.
Tucker, R. W., Sanford, K. K., and Frankel, F. R. (1978). *Cell* **13,** 629-642.
Vandekerckhove, J., and Weber, K. (1979). *Differentiation* **14,** 123-133.

Wang, K., Ash, J. F., and Singer, S. J. (1975). *Proc. Natl. Acad. Sci. U.S.A.* **72,** 4483–4486.
Weber, K., and Groeschel-Stewart, U. (1974). *Proc. Natl. Acad. Sci. U.S.A.* **71,** 4561–4564.
Weber, K., and Osborn, M. (1979). *In* "Microtubules" (K. Roberts and J. Hyams, eds.), pp. 279–313. Academic Press, New York.
Weber, K., and Osborn, M. (1981a). *In* "Cell Surface Reviews" (G. Poste and G. L. Nicolson, eds.). North-Holland Publ., Amsterdam. In press.
Weber, K., and Osborn, M. (1981b). *In* "Muscle and Non-Muscle Motility" (A. Stracher, ed.). Academic Press, New York. In press.
Weber, K., Pollack, R., and Bibring, T. (1975a). *Proc. Natl. Acad. Sci. U.S.A.* **72,** 459–463.
Weber, K., Bibring, T., and Osborn, M. (1975b). *Exp. Cell Res.* **95,** 111–120.
Weber, K., Wehland, J., and Herzog, W. (1976). *J. Mol. Biol.* **102,** 817–829.
Weber, K., Osborn, M., Franke, W. W., Seib, E., Scheer, U., and Herth, W. (1977). *Cytobiologie* **15,** 285–302.
Weber, K., Rathke, P. C., and Osborn, M. (1978). *Proc. Natl. Acad. Sci. U.S.A.* **75,** 1820–1824.
Webster, R. E., Henderson, D., Osborn, M., and Weber, K. (1978a). *Proc. Natl. Acad. Sci. U.S.A.* **75,** 5511–5515.
Webster, R. E., Osborn, M., and Weber, K. (1978b). *Exp. Cell Res.* **117,** 47–61.
Wehland, J., Osborn, M., and Weber, K. (1979). *J. Cell Sci.* **37,** 257–273.
Weingarten, M. D., Lockwood, A. H., Hwo, S.-Y., and Kirschner, M. W. (1975). *Proc. Natl. Acad. Sci. U.S.A.* **72,** 1858–1862.
Weisenberg, R. C. (1972). *Science* **177,** 1104–1105.
Wick, S. M., Seagull, R. W., Osborn, M., Weber, K., and Gunning, B. S. (1981). *J. Cell Biol.* **89,** 685–690.
Willingham, M. C., and Yamada, S. (1979). *J. Histochem. Cytochem.* **27,** 947–960.
Willingham, M. C., Yamada, S., and Pastan, I. (1980). *J. Histochem. Cytochem.* **5,** 453–461.
Yahara, I., and Kakimoto-Sameshima, F. (1978). *Cell* **15,** 251–260.

Chapter 8

An Automated Method for Defining Microtubule Length Distributions

DAVID KRISTOFFERSON, TIMOTHY L. KARR, THOMAS R. MALEFYT, AND DANIEL L. PURICH

Department of Chemistry
University of California, Santa Barbara
Santa Barbara, California

I. Introduction

The polymerization and depolymerization properties of various cytoskeletal components (e.g., actin, tubulin, and troponin) may be characterized in part by the polymer number concentration, the average polymer length, and the shape of the polymer length distribution. In terms of the condensation equilibrium model, Oosawa (1970) proposed that such self-assembly reactions proceed stepwise: first, nucleation; next, elongation; then, attainment of a monomer–polymer concentration ratio consistent with the critical concentration; and finally, adjustment to the equilibrium length distribution of polymer. Even in cases where significant length redistribution is an unlikely event, the shape of the polymer length distribution can provide considerable insight about nucleation and elongation.

ISBN 0-12-564124-9

Likewise, Kristofferson *et al.* (1980) recently offered a detailed kinetic model for endwise disassembly of linear protein polymers, and again the length distribution is an important factor. Finally, a measure of average polymer length is also of considerable value in defining the stoichiometry and mode of action of inhibitors of self-assembly reactions. For these reasons, there is much interest in methods to translate electron micrographic data into the previously mentioned numerical values. Generally, such methods are tedious and of limited accuracy, and it is desirable to consider reliable, fast alternatives that also minimize experimental error. In this chapter we present two examples of automated data acquisition and evaluation. We also examine various statistical problems, provide examples of actual experiment data, and present a detailed program listing for implementing this automated method.

II. Experimental Procedures

A. Sample Preparation and Electron Microscopy

Selection of the appropriate grid mesh is the first step in preparing to do accurate length distribution measurements. One of the main statistical biases in these measurements results from the tendency of longer polymers to run off the edges of electron micrographs. The measured average polymer length is then in error if these polymers are excluded from the measurement. The most straightforward solution to this problem is to take overlapping micrographs of a large area on a grid. If the grid mesh is too fine, the grid bars become obstructions to the overlapping of photographs. We have determined that 75-mesh copper grids are appropriate for our microtubule work, which is normally done at low magnification (1300×). Parllodian is used to coat the grids (Bils, 1974).

Unfortunately, a second problem results when the sample is applied to the coated grids. Our first attempts at microtubule length measurements were thwarted by the excessively high density of polymers on the grid. Because it is impossible to make accurate length measurements when the sample is extensively aggregated, it would be convenient to be able simply to dilute the sample first and then apply it to the grid. However, biological polymers unfortunately tend to depolymerize upon dilution. Thus any dilution of the sample must be done in a medium that maintains the integrity of the length distribution. We have found that a solution of 50% (w/v) sucrose in MEM (0.1 *M* 4-morpholineethanesulfonic acid, 1 m*M* $MgSO_4$, 1 m*M* EGTA, pH 6.8) satisfies this criterion as illustrated in Fig. 1. We normally dilute 10-μl aliquots of assembled microtubules in 90 μl of isothermal sucrose solution prior to applying the sample to the grid. Concentrated microtubule samples (>2 mg/ml) may require up to a 20 times dilution to obtain good results. The diluted sample is

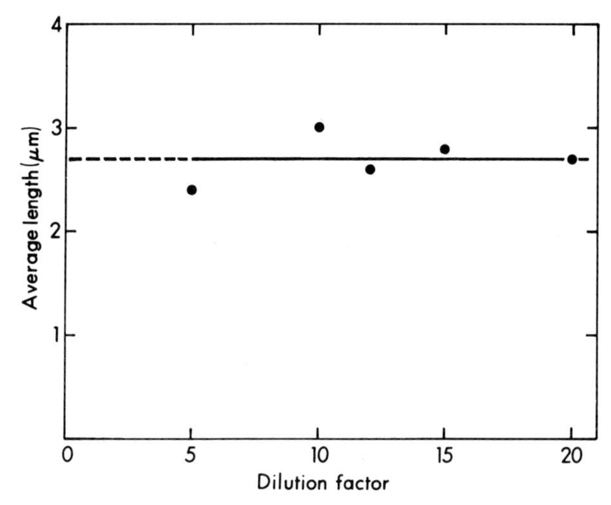

FIG. 1. Effect of dilution into a 50% sucrose in MEM solution on the average microtubule length. Identical aliquots of a 1.1-mg/ml microtubule sample (assembled to steady state and sheared by two passes through a 22 gauge needle to increase the sample size) were diluted 5, 10, 12, 15, and 20 times in 50% sucrose in MEM. Length determinations show little change in the average length with dilution. Sample size is 500.

normally applied to a grid as soon as possible after sucrose dilution. Undue haste is apparently unnecessary, however, as it has been reported (Margolis and Wilson, 1978) that sucrose-diluted samples maintain the same average length for up to 24 hr afterward.

Sucrose dilution leads to further complications during the staining of grids. We normally stain grids first with cytochrome c after drawing off the sample with a Kimwipe. The grid is then rinsed with a drop of water and uranyl acetate solution (1% in water) is applied for 20 sec. After the uranyl acetate, a final drop of water is used to wash the grid. We have found it necessary to draw off carefully as much of the sucrose solution as possible prior to staining. Failure to do so appears to result in occasionally heavy overstaining of the grid. The final drop of water after staining with uranyl acetate also appears to be necessary. Seventy five-mesh parllodian-coated grids tend to rip easily under the electron beam. Application of a single drop of water for as little as 10 sec following the uranyl acetate greatly enhances the stability of the parllodian. Of course, extensive rinsing poses the hazard of destaining the sample.

We usually examine our grids on a Siemens Elmiskop II at 1300× magnification. The microtubules are easily distinguishable at this low magnification and one benefits by obtaining a larger statistical sample in a single photograph. Final enlargement of the negatives gives a finished micrograph at 4000× magnification, allowing for easy length measurements. To minimize bias due to long polymers running off the pictures, we usually make a 2 × 3 mosaic by overlapping three pictures in the horizontal direction and two vertically ("vertically" referring to the longer side of our individual micrographs).

B. Microprocessor-Based Digitizer

We recently developed several computer methods for accumulating and processing microtubule length distribution data. The principal objective of these approaches was the rapid acquisition of length distribution data from electron micrographs. One program was written for use with a Hewlett-Packard microcomputer and the listing is provided in an appendix. The second approach used a Digital Equipment Corporation PDP-11/03 microprocessor interfaced with a Houston Instruments HiPad "bombsite" digitizer. All command statements and data output were made with the use of a Tektronix graphics terminal and permanent copy unit. The latter method has several advantages over the Hewlett-Packard microcomputer program, the chief advantage being the capability for the operator to interact with and modify the data acquisition and processing steps. Both methods rely on the use of a "bombsite" tracing unit with which the operator traces along each tubule in the electron micrographs. The start and finish of each tubule are controlled by depressing a record button on the "bombsite." The unit measures the length of each microtubule as the sum of line segments whose lengths are determined by the repeated application of the Pythagorean theorem. Each segment length is approximately 0.01 cm, and the data acquisition can proceed at a rate of 2.5 cm/sec. Any error in tracing a microtubule is easily recognized by the operator, and such entries may be erased by striking the appropriate one-letter command on the graphics terminal keyboard.

As the data are accumulated, the operator may at any time preview the corresponding histogram to determine the number of tubules measured, the length distribution, and the average length. The latter two are defined relative to a magnification factor that the operator enters at the beginning of the data accumulation phase. The program is designed to spread the distribution such that full scale on the histogram length axis includes the longest tubule traced by the operator. Nonetheless, one may define the length axis by a simple command, and the number of any overscale microtubules is also presented on the screen of the graphics terminal. Another valuable feature of the program is the provision for numerically listing the histogram data. This is especially useful if one wishes to employ the histogram data in a computer program for microtubule disassembly (Kristofferson *et al.*, 1980). With the listing the operator also gains the opportunity to collect part of the distribution data at one sitting and subsequently to input this data on the next occasion without the need to retrace the already processed tubules.

Details on implementing the microcomputer-based length distribution scheme are beyond the scope of this report, but the authors would be pleased to help other interested investigators with such technical aspects. Nonetheless, it is a relatively easy matter to employ the program provided in the appendix, and the number of minicomputers using HP-BASIC is considerable. This program may also be adapted to other computer languages, and again the authors would be pleased to consult on these matters.

C. Polymer Length Measurements

We employ the following two rules when measuring microtubule lengths: only measure distinct microtubules, and do not measure microtubules extending off the edges of the photographs. The first rule was adopted to minimize guesswork as to where a particular microtubule may begin or end, particularly if it is entangled with several others. Obviously this rule would tend to bias the results against longer polymers, which have a greater chance of entanglement. If one finds aggregates in one's micrographs, the remedy is to increase the sucrose dilution. The method of overlapping photographs is used to reduce the bias against long polymers inherent in the second rule. Some groups (e.g., Johnson and Borisy, 1977) have suggested that the bias could be overcome simply by doubling back over a polymer that goes off an edge. They argue that, on the average, polymers will tend to hit the edge at their midpoint. Unfortunately, such a technique would only be appropriate for average-length determinations but not for measurements that determine the distribution shape. For example, consider a sample containing polymers of equal length. If this doubling-back technique is used, the measured distribution will contain polymers that are very short, since only a small fraction of their length was on the picture, as well as polymers that are almost double their normal length. However, given a large enough sample to overcome statistical fluctuations, at least the average length would still be correct. In a recent paper (Kristofferson *et al.,* 1980) we showed that experimental microtubule depolymerization kinetics could be derived from the shape of the length distribution. Our calculations closely matched the experimental results as shown by Karr *et al.* (1980). Had we employed a "doubling-back" technique, the success of our experiments would have been jeopardized by inaccurate distribution shapes. Making a mosaic of photographs appears to be the most desirable technique.

In Fig. 2 we show the results obtained from a single set of photographs by using the three following techniques: (a) mosaics with exclusion of polymers exiting the outer boundary; (b) no mosaics but still excluding exiting polymers; (c) doubling back on microtubules that leave the pictures and not overlapping the micrographs. In this case it turned out that the area represented in a single photo was sufficiently wide so as not to bias the sample extensively. Accordingly, only a small increase in average length was obtained using mosaicked photos. The need for mosaics and the appropriate dimensions must be determined by each investigator for each polymer system.

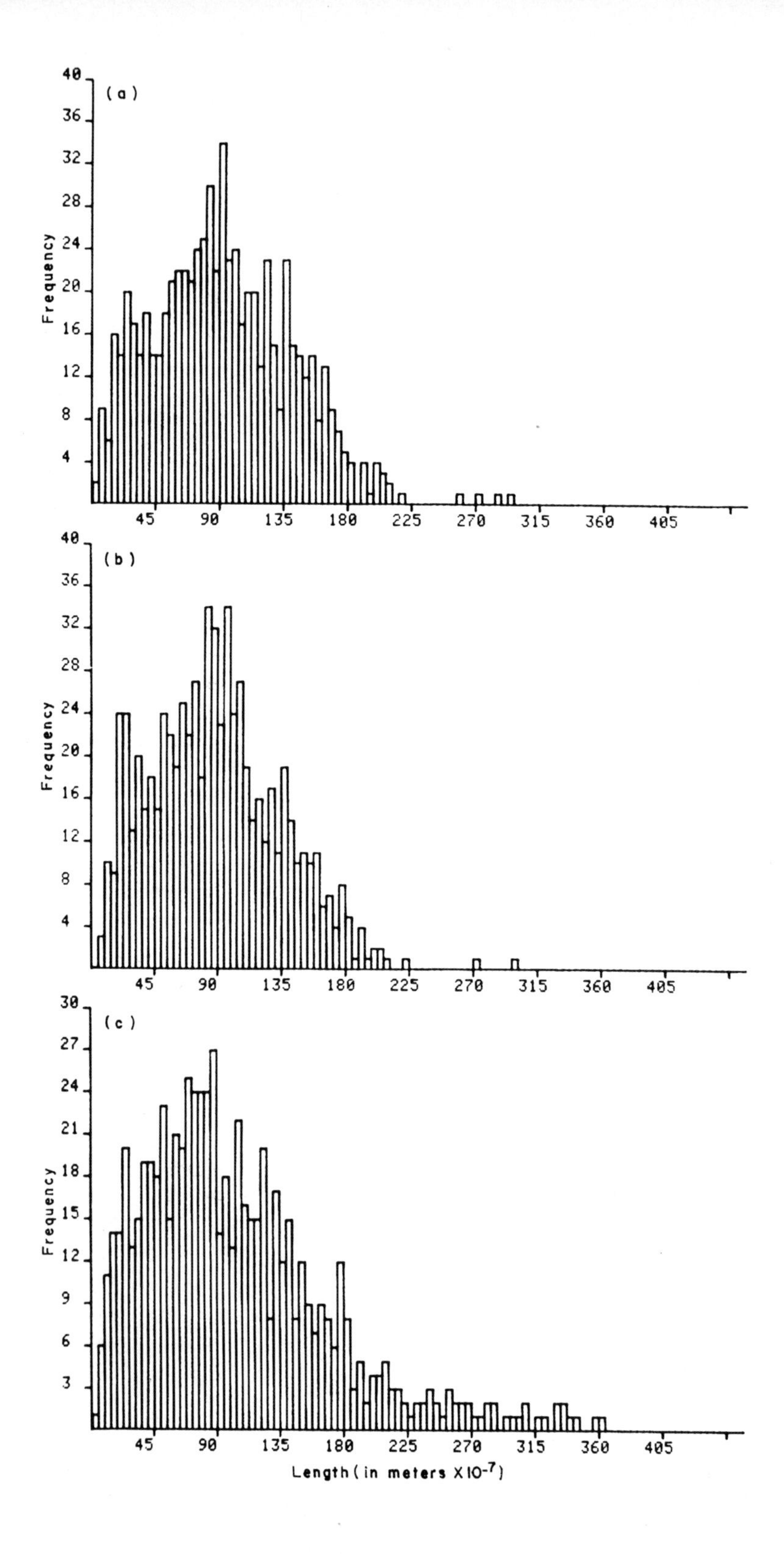
(a)
(b)
(c)
Frequency
45
90
135
180
225
270
315
360
405
Length (in meters X10⁻⁷)

TABLE I[a]

REPRODUCIBILITY OF THE AVERAGE LENGTH IN VARIOUS SAMPLES

Sample no.	Sample size	Average length (μm)
	250	4.1
1	250	4.3
	250	4.2
	250	7.4
2	250	6.9
	250	7.3
	125	8.7
3	125	7.2
	125	10.9

[a] Three length distributions of the indicated sample sizes were obtained from three sets of micrographs taken on the same EM grid for each of the three experiments. A single 1.5-mg/ml microtubule sample was used and the grids were taken at different times during assembly.

Differential expansion of the parllodian film by the electron beam causes overlapping micrographs to fit inexactly. However, no significant error in determining the lengths of single polymers occurs if the photos are frequently repositioned prior to measuring. Background debris on the film as well as adjacent polymers are an aid in this matter.

The final question to be considered in making length distribution measurements concerns the reproducibility and accuracy of the results. Table I shows experiments in which average lengths were determined for various sample sizes. The number of polymer measurements necessary to obtain an accurate average length is dependent on the breadth of the distribution. Since distribution shapes vary with experimental conditions, each investigator must determine the necessary sample size under the appropriate circumstances.

The standard statistical z test for the difference between two means may be usefully employed here. A measured difference greater than

$$1.96[(s_1^2/n_1) + (s_2^2/n_2)]^{1/2}$$

FIG. 2. Effect of measuring procedures on the microtubule length distribution. A 3 × 3 overlapping set of electron micrographs was used to make the following length distribution measurements. (a) Pictures were overlapped. Microtubules running off the outer boundary were not measured. Mosaics; average length = 9.1 μm. (b) The micrographs in (a) were used again. The pictures were not overlapped and any microtubules running off the photograph edges were excluded. No mosaics, no doubling; average length = 8.8 μm. (c) Five nonadjoining micrographs from the 3 × 3 set were measured. No mosaics, doubling-back; average length = 10.3 μm. Lengths of microtubules running off the micrographs were determined by "doubling back" along the tubule with the digitizer "bombsite." The sample size was kept at 690 in each case by doing the distribution in (b) first and using the number of microtubules in that sample as the limit.

between the means of two identical samples will occur by random error only 1 time in 20 determinations. Here s_1 and s_2 refer to the standard deviations of the two samples and n_1 and n_2 are the respective sample sizes. This test may be used for nonnormally distributed data (such as microtubule histograms) provided both sample sizes are greater than 30. A similar test can be employed to calculate a 95% confidence interval around a single measured mean. Thus, one will be correct 95% of the time in asserting that the true mean is within $\pm 1.96\ s/(n)^{1/2}$ of the measured mean. To use this test for nonnormal data, the sample size n must be greater than 30.

In Table I the average lengths in sample 1 are the same within experimental error as judged by the z test for the difference between two means. The same is true for sample 2. However, sample 3 shows significant deviations by the same test. This reflects the fact that small sample sizes such as 125 may be influenced by systematic errors such as nonuniformities in the distribution of microtubules on the surface of a grid. This is a particular problem if the distribution is broad and/or the sucrose dilution factor is small, resulting in areas of aggregation on the grid. These problems are readily apparent on micrographs and can be overcome by increasing the sucrose dilution and taking larger samples over a broader area on the grid. Our experience with fully polymerized bovine brain microtubules (Karr *et al.*, 1980) indicates that a sample size of 500 is usually adequate for average length and distribution shape determinations. Reproducibility of these results is dependent on the distribution width (this is related to s in the z tests), which is affected by protein concentration and other variables. Generally speaking, results obtained from distributions with under 500 microtubules may be suspect; even larger sample sizes may be necessary for more refined work. Use of the appropriate statistical tests can decide these issues.

III. Discussion

Length distribution measurements often provide the only available way to determine important parameters in the study of biological polymers. For example, since most polymerization mechanisms tend to be endwise, it is necessary to know the number of polymer ends in solution in order to study polymerization or depolymerization kinetics. This can be determined from a knowledge of the total protein in polymer form (usually determined by turbidity) and the average polymer length (Karr *et al.*, 1980). Using the automated techniques described herein, Karr *et al.* recently succeeded in correlating microtubule depolymerization rates with the number of microtubule ends in solution. The initial rates determined by turbidity agreed with the electron microscopic number concentration measurements to within a few percent. Accurate determinations of microtubule distribution shapes enabled us to predict theoretically the entire time course of the depolymerization reaction on the basis of these electron microscopic measurements. These results clearly demonstrate that electron microscopy can be used to determine accurately microtubule lengths in solution. In addition, one may utilize length distribution techniques to determine the mode of action of other microtubule assembly effectors. The ability to determine accurate average lengths and number concentrations as well as to monitor distribution shape changes as a function of time is invaluable in these efforts.

Finally, the use of a variety of methods for determining polymer length distributions causes a lack of uniformity and reproducibility in experimental results. Adoption of a standard automated method by researchers in the area of biological polymers would help to resolve this problem. Hopefully this paper will serve as a step toward that goal.

ACKNOWLEDGMENT

We would like to thank Mr. Michael K. McBeath from the Masters of Scientific Instrumentation program in the Department of Physics at UCSB for designing the microprocessor-based digitizer.

REFERENCES

Bils, R. F. (1974). "Electron Microscopy." Western Publ. Co., Los Angeles, California.
Johnson, K. A., and Borisy, G. G. (1977). *J. Mol. Biol.* **117,** 1-31.
Karr, T. L., Kristofferson, D., and Purich, D. L. (1980). *J. Biol. Chem.* **255,** 8560-8566.
Kristofferson, D., Karr, T. L., and Purich, D. L. (1980). *J. Biol. Chem.* **255,** 8567-8572.
Margolis, R. L., and Wilson, L. (1978). *Cell.* **13,** 1-8.
Oosawa, F. (1970). *J. Theor. Biol.* **27,** 69-86.

Appendix: Length Distribution Program for the HP 9825A

A. User Instructions and Program Listing

The following program is designed for use with an HP 9825A calculator equipped with the following features: an Advanced Programming ROM, a 9862A plotter, an I/O ROM for the plotter, and a 9864A digitizer. Instructions for using the program follow.

1. Enter the sample identification number. This may consist of letters, words, or numbers up to 32 characters.
2. Press CONTINUE.

3. Enter the scale factor, i.e., the magnification of the electron micrograph. This number converts the length directly measured on the micrograph to the true microtubule length.

4. Press CONTINUE. The digitizer is ready for use.

5. Set the origin at the lower left corner of the micrograph by positioning the crosshairs of the cursor at that point and pressing the red "O" button on the cursor. The red light on the cursor should be on at this stage.

6. To measure a microtubule, press and let up on cursor button "C" at the beginning of the tubule. The red light on the cursor will flash while the cursor is measuring lengths. Run the crosshairs along the microtubule and then press and let up on button "C" at the tubule's end. The light stops flashing, indicating that the measurement is completed.

MICROTUBULE LENGTH PROGRAM

```
Ø: dsp "microtubule length program"; dsp" "
1: dim A[2ØØØ],A$[32],L$[1Ø],K$[5],
   T$[15]
2: "start":
3: ent "sample identification number",A$
4: Ø→K
5: ent "scale factor",F;1/F→F
6: "2":
7: on err "error"
8: Ø→L
9: red 4,X,Y
1Ø: if X<Ø;beep;gto "15"
11: "5":
12: red 4,P,Q
13: if Q<Ø;beep;gto "1Ø"
14: ((X−P)↑2+(Y−Q)↑2)↑.5→A
15: 2.54AF+L→L
16: flt 3;str(L)→L$
17: fxd Ø;str(K+1)→K$
18: L$&K$→T$
19: dsp T$
2Ø: P→X;Q→Y
21: gto "5"
22: "1Ø":
23: if L>Ø;L→A[K+1→K]
24: gto "2"
25: "error":
26: dsp "last point erased . . . proceed"
27: K−1→K
28: gto "2"
29: "15":
3Ø: dsp "wait a minute, I'm
   thinking"
31: max(A[*])→M
47: if B>H;W/2→W;Ø→B;gto "2Ø"
48: next I
49: scl −.1M,M,−.1H,H
5Ø: axe Ø,Ø,.Ø5M,.1H
51: csiz 1,1.5,.7,Ø;fxd 1
52: for I=.9M to .1M by −.1M
53: plt I, −.Ø3H,−1;cplt −3,Ø
54: lbl DI/M
55: next I
56: plt .5M,−.Ø7H;lbl "length(cm)* 1Ø"
57: fxd Ø;plt .62M,−.Ø62H,−1;lbl −E
58: csiz 1,1,1,Ø; fxd Ø
59: for I=H−.1H to .1H by −.1H
6Ø: plt −.Ø7M,I,−1
61: lbl I
62: next I
63: csiz 1,1,1,9Ø;plt −.Ø6M,.35H,−1;lbl
   "frequency"
64: ent "list frequency vs. length ?",K$
65: if K$="yes";prt "length : freq.";spc 2;
   sfg 1
66: for I=WM to M by WM
67: Ø→B
68: for J=1 to K
69: if A[J]<I and A[J]>I−WM;B+1→B
7Ø: next J
71: plt I−WM,B;plt I,B;plt I,Ø
72: if not flg1 or B=Ø;gto "3Ø"
73: "from"→K$;flt 3;str(I−WM)→L$;K$&L$
   →T$;prt T$
74: "to"→K$;str(I)→L$;K$&L$→T$;prt T$
75: "freq"→K$;fxd Ø;str(B)→L$;K$&L$→T$;
   prt T$;spc 1
```

```
32: log(M)→N
33: int(N)→E
34: M*1Ø↑(−E)→C
35: if C<1Ø;1Ø→D
36: if C<5;5→D
37: if C<2;2→D
38: D*1Ø↑E→M
39: .Ø5→W;1Ø→H
4Ø: "2Ø":
41: if W<5e−3;H+1Ø→H;.Ø5→W;gto "2Ø"
42: for I=WM to M by WM
43: Ø→B
44: for J=1 to K
45: if A[J]<I and A[J]>I−WM;B+1→B
46: next J
```

```
76: "3Ø":
77: next I
78: pen
79: csiz .75,1,.75,Ø;fxd Ø
8Ø: plt .6M, .9H,−1;1b1 "sample size",K
81: plt .6M,.85H,−1;1b1 "I.D.",A$
82: Ø→G
83: for J=1 to K
84: A[J]+G→G
85: next J
86: G/K→O
87: flt 3;plt .6M,.8H,−1;1b1 "average
    length",O
88: cfg 1;ent "more data (1) or new
    data (2)",B
89: if B=1;gto "2"
9Ø: if B=2;gto "start"
91: end
```

7. Slide the cursor below the *positive* x-axis (defined in reference to the lower-left-hand corner origin) and press the "S" button on the cursor. A beep should be heard indicating that the measured length has been stored. The calculator display shows the length of the last microtubule in centimeters as well as the number of microtubules counted.

8. If a mistake is made while measuring a length, complete the process through step 7. After hearing the beep, press the STOP button on the calculator. Wait for the number "26:" to be displayed and then press CONTINUE. A message "last point erased . . . proceed" will be displayed. One may then continue as in steps 6 and 7.

9. Should the cursor be accidentally lifted off the digitizer pad, a beep will be heard and the red light on the cursor will go out. No data will be lost by doing this. Simply redefine the origin as in step 5 and then continue as in steps 6 and 7.

10. After the last microtubule has been measured, slide the cursor to the left of the origin (second quadrant, above the x axis and left of the y axis) and then press the "S" button. A beep will signal the end of the process and the display will indicate "Wait a minute, I'm thinking." Be sure the pen and paper are properly positioned on the plotter before pressing "S!" For large microtubule samples (>500) the wait between pressing the "S" button and the final plot may take several minutes. This time increases with the sample size. The program is currently set to handle a maximum of 2000 microtubules, but this may be easily changed by changing the number in the dimension statement for A in line 1 of the program.

11. Upon completion of the calculations, the plotter will draw the histogram axes and pause. The display asks if a numerical listing of lengths and frequencies by histogram interval is desired. Type in "yes" or "no" and press CONTINUE.

12. The list and/or the histogram will be printed.

13. Following the completion of step 12, the display will show "more data (1) or new data (2)." If one desired to add more microtubules to the same sample press "1" and then CONTINUE. Proceed as in steps 6 and 7. If one wishes to begin a new sample, press "2" and then CONTINUE. Proceed with step 1.

Chapter 9

Measurement of Steady-State Tubulin Flux

ROBERT L. MARGOLIS

The Fred Hutchinson Cancer Research Center
Seattle, Washington

I. Introduction

A. Microtubules as Dynamic Structures

Beginning almost 30 years ago, Inoué reported (for review, see Inoué and Sato, 1967) that the major birefringent elements of the eucaryotic spindle were rapidly and reversibly labile to pressure, cold temperature, and colcemid, a drug known for its ability to disrupt mitosis. Each of these treatments simultaneously reduced the presumptive fibrous network and brought the process of mitosis to a halt. Once it was known that microtubules were the major fibrous structural elements of the mitotic apparatus, and that the microtubule 6S subunit protein, tubulin, was a unique high-affinity receptor for colcemid (Shelanski and Taylor, 1967; Borisy and Taylor, 1967), Inoué and Sato (1967) proposed that microtubules in the mitotic spindle were in a state of dynamic equilibrium with their constituent subunits.

ISBN 0-12-564124-9

High pressure, cold temperature, and drugs that block mitosis (colchicine and its derivatives, the Vinca alkaloids, podophyllum derivatives, and other drugs) (for review, see Wilson and Bryan, 1974) all shift the microtubule equilibrium toward the subunits. Various glycols (Mazia *et al.*, 1961; Kane, 1965) that may act by changing solvent characteristics (Lee *et al.*, 1978; Lee and Timasheff, 1977) force the equilibrium toward microtubules and act to stabilize the polymeric structures. It is of interest to note that cells maintained in glycols will also remain blocked in mitosis indefinitely (Rebhun and Sawada, 1969), but with a fully formed mitotic spindle.

From these observations one may tentatively conclude that it is not merely the presence of microtubules, but rather their existence in a state of free equilibrium that allows for the completion of mitosis. If it is the equilibrium state itself that determines the behavior of microtubules, especially in mitosis, then it becomes of paramount importance to understand thoroughly what it is about the equilibrium state that establishes a functioning machine. The purpose of this chapter is to outline some of the methodology that we and others have used to elucidate the nature of the equilibrium state, and, in addition, to present some of the more interesting conclusions thus far drawn using the technology at hand.

Following the discovery by Weisenberg (1972) of the conditions for *in vitro* assembly of microtubules, it became clear that microtubules assembling outside the cell exhibited the same equilibrium behavior in response to solvent conditions, to drugs, or to temperature as did microtubules in the cell (Olmsted and Borisy, 1973; Olmsted *et al.*, 1974; Gaskin *et al.*, 1974). *In vitro* assembly thus could be used as a model system to describe at least some aspects of the microtubule equilibrium present in the living cell.

Shortly after Weisenberg's discovery, two laboratories were able to establish independently that microtubules contained an intrinsic polarity with respect to their assembly properties (Olmsted *et al.*, 1974; Dentler *et al.*, 1974). Each microtubule contains two assembly sites, one at either end. The important conclusion drawn was that, under assembling conditions (approach to equilibrium), one of these sites accumulated subunits much more rapidly than did the other site, thus establishing a biased polar assembly. We will call the more rapidly assembling site the A or assembly end.

The polarity of microtubule structure is expressed in the architecture of the microtubule itself. The α- and β-polypeptides of tubulin appear to align with the longitudinal axis of the microtubule (Amos and Klug, 1974), so that one microtubule end contains exposed α subunits and the other, β subunits. This polarity of structure can further be distinguished through the length of the microtubule (Mandelkow *et al.*, 1979). Where dynein arms protrude from the microtubule surface, they point with unique handedness with respect to the A end, since they bind to the microtubule in only one orientation at a unique binding site (Haimo, in this volume).

B. Opposite-End Assembly and Disassembly

These two aspects of the cytoplasmic microtubule, the intrinsic polarity and the continuous equilibrium state that is somehow linked to the microtubule mechanism of action, are joined by a third aspect of microtubule behavior that complements and is anticipated by the polarity and the constant equilibrium. There is also a constant opposite-end assembly/disassembly process going on (Margolis and Wilson, 1978). Microtubules, at their apparent equilibrium state, are adding subunits at the A end more rapidly than they are losing subunits at this site. Conversely, at the "D" or disassembly end, subunits are being lost more rapidly than they are gained. The two equilibria counterbalance each other so that net loss at the D site just matches the net gain of subunits at the A site when the microtubules are in overall equilibrium.

As a result of net subunit addition at the assembly end, microtubules at apparent equilibrium exhibit a constant flow of subunits from one end (the A end) to the other (the D end). A constant treadmilling machinery is thus established which can easily serve to transport materials that might attach on the microtubule surface from place to place in the cell. Similar observations with respect to F-actin behavior have also been made (Wegner, 1976; Brenner and Korn, 1979).

It is our belief that this mechanism may be of importance or may indeed be fundamental to the function of microtubules in the cell cytoplasm and mitotic apparatus. The methodologies that have been devised to study this phenomenon of subunit flow both in our lab and elsewhere follow.

II. Use of GTP as a Probe of the Equilibrium State

As stated earlier, GTP exchanges rapidly at the tubulin E site when the subunit is free in solution (Weisenberg *et al.*, 1968) but the guanine nucleotide becomes totally nonexchangeable following addition of the subunit to the microtubule with accompanying GTP hydrolysis (Weisenberg *et al.*, 1976; Arai and Kaziro, 1977; Margolis and Wilson, 1978). Only upon loss of the subunit from the polymer by disassembly is the guanine nucleotide again available for exchange.

The reality of this guanine nucleotide behavior in the microtubule is important to establish, for it is at the core of the rationale for using GTP as a probe of microtubule assembly reactions at steady state. A guanine nucleotide that enters the microtubule with a subunit, and remains in the microtubule as long as the subunit to which it is bound remains, is able, if it is radioactive, to act as a pulse-chase marker for the continued presence of that subunit in the microtubule. That the E-site guanine nucleotide behaves in this manner has been established.

The details of the methodology that we have used to establish the nonex-

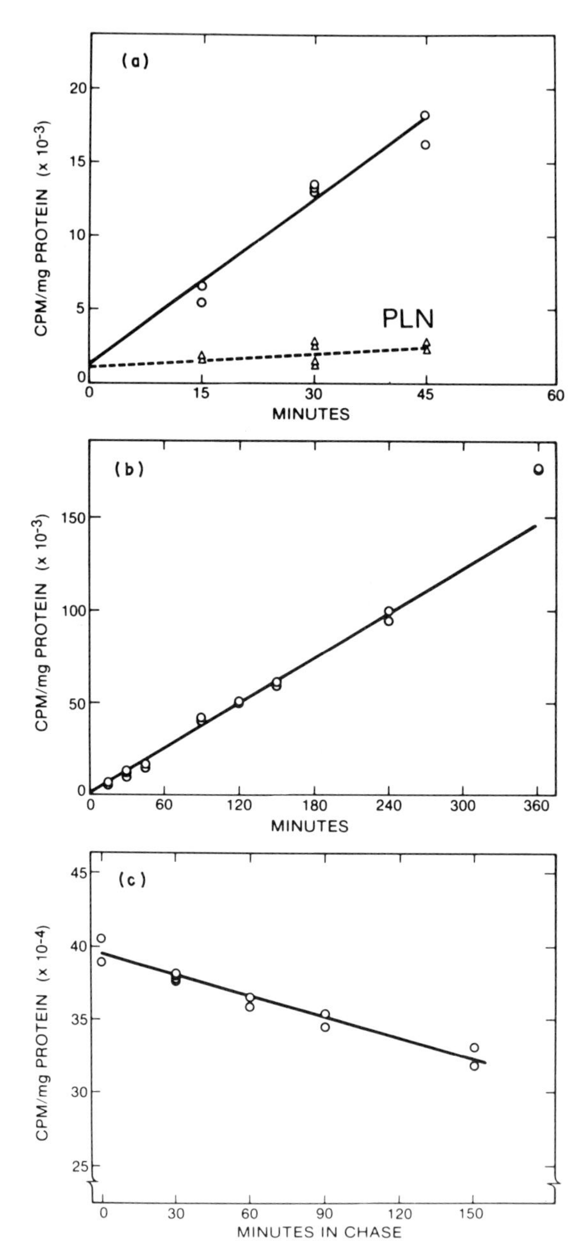

FIG. 1. Steady-state exchange of dimers with microtubules. (a) Microtubules were polymerized to steady state and then tested for their ability to incorporate [^{3}H]GTP in the presence or absence of podophyllotoxin. To one aliquot 50 μM podophyllotoxin was added, while the other remained drug free. The microtubules were then exposed to [^{3}H]GTP (10 μCi/ml) for the times indicated. (△---△),

changeability of E-site GDP in the microtubule constitute also the method we used to exploit this tool as a probe of microtubule steady-state behavior.

Briefly, we are able to show that microtubules, at an apparent equilibrium state, were able to incorporate linearly [^{3}H]GTP, eventually to a point of nearly molar saturation (one mole of [^{3}H]GTP per mole of tubulin dimer); but would incorporate no label if microtubule assembly were blocked with podophyllotoxin before introduction of the label (i.e., Fig. 1). GTP therefore adds to microtubules only during assembly, and does not freely exchange on subunits within microtubules.

A. Sucrose Cushion Method

1. Procedure Summary

For these experiments, we used beef brain tubulin prepared, as described previously (Margolis and Wilson, 1979), by three cycles of assembly and disassembly. Microtubule protein used in our experiments contains 75% tubulin and 25% microtubule-associated proteins (MAPs). Microtubule pellets that had been stored in liquid nitrogen were solubilized in a buffer containing 20 m*M* sodium phosphate, 100 m*M* sodium glutamate (pH 6.75) (also containing 0.02% NaN_3 to prevent bacterial contamination). The protein was Dounce-homogenized in this buffer (designated PG), and after 10 min on ice was centrifuged at 30,000 *g* for 10 min. The resulting supernatant was mixed with 0.1 vol of an assembly mix to yield final concentrations of 1–2 mg/ml total protein, 0.08 m*M* GTP, 1.0 m*M* $MgSO_4$, 1.0 m*M* EGTA, 20.0 m*M* acetyl phosphate (although this is sometimes added in smaller amounts in stages during the experiment, see below), and 0.05 IU/ml acetate kinase in either 3.3 or 6.6 ml total volume. The acetate kinase system (MacNeal *et al.*, 1977) served to replenish GTP constantly in the reaction, enabling long-term maintenance of steady state in the low GTP concentrations necessary both to keep [^{3}H]GTP specific activity high and to permit effective label chase conditions.

Microtubules were assembled by incubation of samples at 30°C. The time course for microtubules to reach steady state was determined by light-scattering measurement at 350 nm, using a recording spectrophotometer with a constant-

podophyllotoxin present; (○——○) podophyllotoxin absent. Podophyllotoxin prevents [^{3}H]GTP uptake. (b) [^{3}H]GTP uptake was measured over a prolonged time course at steady state in the absence of drug. Label was taken up linearly at 6.25% per hour. (c) Microtubules were uniformly prelabeled by polymerization in the presence of [^{3}H]GTP and were chased at steady state with a 30-fold excess of unlabeled GTP. Label was lost linearly at 6.7% per hour. The stoichiometry of GTP at zero time was 0.50 mole/mole tubulin. The ordinate is cpm/mg total protein incorporated into microtubules.

temperature cuvette chamber. Routinely, an apparent steady state was attained in 60 min. Experiments were begun after 100 min at 30°C had elapsed.

[^{3}H]GTP (ICN, 17.1 Ci/mmole), used in a final ratio of 10 μCi/ml, was dried down in test tubes and resuspended in the experimental solution at appropriate times for [^{3}H]GTP chase. Chase concentrations of GTP were 2.5 m*M*, or approximately 30 times the pulse concentrations.

At the end of incubation, 450 μl samples were layered on 5 ml of warm (30°C) PG buffer containing 50% sucrose and centrifuged at 25°C for 2 hr at 200,000 *g* to pellet microtubules. The supernatant was carefully aspirated and the pellet prepared as described in detail below. Finally, each pellet was resuspended in 1 ml of 0°C PG buffer and samples were taken for Lowry *et al.* (1951) protein assay and for scintillation counting. The end point of each timed experiment was designed to be the start of the sucrose centrifugation. Since all incubations terminated at a single time, time points were generated by initiating incubations at steady state in a reversed time sequence (longest time point first).

All experiments include electron microscopy of microtubules to determine lengths, and to determine physical constants for flow rates.

2. Procedure Detail

Various glycols stabilize microtubules in solution outside the cell, as first described by Kane (1965). Glycerol (Shelanski *et al.*, 1973), hexylene glycol (Kane, 1965) and polyethylene glycol (Lee and Lee, 1979) have all been used to prevent microtubule disassembly under conditions where the subunit population is so low in concentration that a standard equilibrium cannot be maintained. Sucrose shares this property.

We have shown that microtubules, maintained at low protein concentration in 50% sucrose, do not shorten at 30°C in a 24-hr period. Further, it can be shown that radioactive vinblastine, which ordinarily binds to either end of a microtubule and rapidly exchanges with free vinblastine at these binding sites, is retained quantitatively under chase conditions at these sites in 50% sucrose over a period of hours (Wilson *et al.*, submitted). Therefore, we conclude that, in 50% sucrose, subunit loss to disassembly is minimal.

Centrifugation of microtubules into 50% sucrose cushions is therefore an excellent means to stop equilibrium reactions simultaneously; to remove quantitatively microtubules from subunits and other materials in free solution; and to retain quantitatively the microtubules present at the beginning of the centrifugation. Experiments were therefore constructed to measure [^{3}H]GTP uptake into microtubules (one may, as explained earlier, use [^{3}H]GTP as a marker in order to make measurements of tubulin entry and exit from a microtubule at steady state) by allowing the label to incorporate for different periods of time and then terminating the period of incubation by beginning a centrifugation run into a sucrose cushion.

A rough calculation can show that particles the size of microtubules will quantitatively penetrate into a sucrose cushion in a Beckman SW50.1 at 40,000 rpm in approximately 0.8 min if they are applied to the cushion in a 0.5-ml sample. Since it takes approximately 2 min to accelerate to full speed, a very conservative estimate of the time from the beginning of the centrifugation run until microtubules are quantitatively removed from the incubation solution is 3 min.

Thus, for an experiment in which time points are taken over a period of hours, the data points can be expected to plot back to zero time, which is taken as the time at which the centrifuge run began. The data collected meets this expectation (Fig. 1b). For convenience, time points in sucrose cushion experiments are produced in reverse order, since all time points are defined by the time elapsed until the beginning of a centrifuge run in which all samples are simultaneously spun. Thus, if one were pulsing microtubules with [^{3}H]GTP, with time points every 15 min, up to 90 min, the 90-min time point would be exposed to label first, and the 15-min point last. In actual practice a 15-min time point would be pulsed just prior to loading the samples on sucrose cushions, since it takes approximately 15 min to layer samples on sucrose cushions, load the rotor, and pump down the vacuum on a preparative ultracentrifuge before a centrifuge run can begin.

Since experimental procedures described here involve [^{3}H]GTP pulses and since GTP is constantly hydrolyzed during microtubule assembly and therefore is constantly changing in concentration, a method must be used to maintain GTP at constant concentration. We have used acetyl phosphate and acetate kinase to regenerate GTP, a method of maintaining constant GTP introduced by MacNeal *et al.* (1977). It is not always a convenient system to use. At warm temperatures, acetyl phosphate is highly unstable. Typically, we introduce 10–20 m*M* acetyl phosphate to the reaction mixture just prior to assembly. An alternate reported method uses phosphoenolpyruvate and pyruvate kinase (Kobayashi and Simizu, 1976).

For long time course experiments, other strategies work better. If acetyl phosphate begins to deplete, the GTP concentrations will drop and pulses will have higher and higher specific activity and will be characterized by increasing levels of label incorporation per unit time. To keep GTP concentrations constant in long time course experiments, periodic boosters of acetyl phosphate may be given. Alternately, the intrinsic transphosphorylase activity of microtubules may be taken advantage of, and CTP or UTP may be used in relatively high concentrations to sustain constant GTP levels. Since we have shown ATP profoundly affects parameters of the microtubule steady state (Margolis and Wilson, 1979), this nucleotide cannot be used for the purpose of maintaining constant GTP level.

Another, more difficult, strategy has worked for us in long time course pulse-chase experiments (Fig. 2). Microtubules are assembled in 0.1 m*M* GTP with acetyl phosphate and acetate kinase, pulsed at steady state in one batch, and then

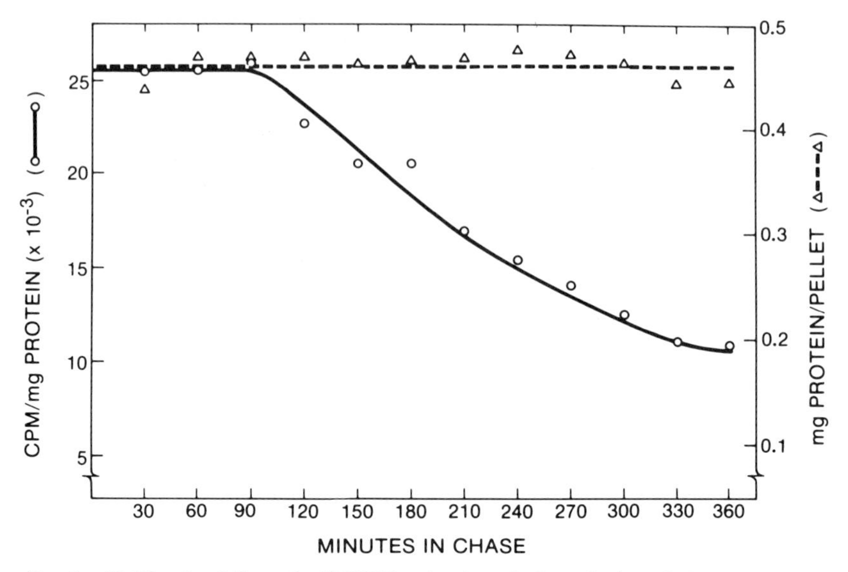

FIG. 2. Unidirectional flow of a [^{3}H]GTP pulse through sheared microtubules at steady state. Microtubules were polymerized to steady state and sheared by passage through a syringe to an average but disperse length of 3.5 μm, then exposed to [^{3}H]GTP for a 0.5-hr pulse. Pulse-labeled microtubules were split into 12 individual aliquots and pelleted through sucrose cushions. Upon resuspension at the appropriate time in PG buffer containing 2.5 m*M* GTP, and the time course of label loss from microtubules was determined after terminating the experiment with a second centrifugation through sucrose. Dashed line = mg protein/microtubule pellet; solid line = cpm/mg total microtubule pellet. Label incorporated at the A ends was completely retained for the first 90 min of chase. Label then was lost from the microtubules as the tubulin containing the guanine nucleotide pulse migrated to the D end. The total quantity of protein in microtubules did not change during the 6-hr experiment.

split into individual aliquots, which are then pelleted through sucrose cushions. These microtubule pellets are then gently resuspended at different times into warm chase buffer, and at the end of the experiment are pelleted through sucrose again and then analyzed. So long as the microtubule concentration remains high and microtubules need only minimally disassemble before the critical subunit concentration for assembly is reattained, the experiment will work quite well.

One reason for using sucrose cushions is to minimize contamination of microtubule pellets by unincorporated label. The amount of contamination can be determined by experiments in which microtubules are exposed to podophyllotoxin, which prevents further assembly, and then are exposed to [^{3}H]GTP (Fig. 1). The level of label contamination is, under the best circumstances, 10^{-4} of the label present during incubation. This low level of contamination is achieved only by attention to details such as choice of rotor type and cleaning procedure for the final pellet. We have used both swinging-bucket and fixed-angle rotors and find that swinging-bucket rotors consistently give one-tenth the level of pellet contamination that one obtains with fixed-angle rotors. The rotor we find most convenient to use is the Beckman SW50.1 since the small volume of its tubes is compatible with the 0.5-ml sample size we use.

Our protocol for cleaning of pellets may appear to be intricate, but the low levels of contamination achieved are worth the trouble.

After centrifugation, the supernatant is aspirated slowly downward until about one-third of the 50% sucrose layer remains. A buffered 70% sucrose solution (in the original microtubule assembly buffer) is then pipetted down the side of the tube to float the 50% layer away from the pellet. Again, the 50% layer is aspirated off, and then the 70% layer, leaving only the pellet. A second 70% sucrose rinse is performed. Next a disposable tissue (Kimwipe) is used to swab the inside of the tube down to the level of the pellet and, finally, another 70% sucrose rinse is done. All solutions used are at room temperature. The microtubule pellet is then resuspended in cold (4°C) buffer and analyzed for protein content and radioactivity.

B. Filter Method

Another approach that can yield successful results is the filtration method introduced by Maccioni and Seeds (1978). This method is much quicker and easier to set up than the sucrose cushion method.

The method essentially is to trap microtubules on glass fiber filters much as one would trap spaghetti in a colander. The microtubules can then be washed free of the solution, of subunits, and of unincorporated radioactivity. The entire process of running microtubules through glass fiber filters and rinsing the trapped material takes only seconds; therefore time points can be taken in rapid succession. This experimental procedure is dealt with in detail elsewhere in this volume (Wilson *et al.*).

III. A Simple Approach to Measurement of Treadmilling Rates

As was indicated earlier (Fig. 1), podophyllotoxin blocks microtubule assembly at steady state, so that no new subunits add in a net manner to preexisting microtubules. Subunits continue to be lost from podophyllotoxin-blocked microtubules. By measuring the rate of loss of [^{3}H]GTP and of protein from these drug-poisoned microtubules, we determined that the rate of loss of protein in the presence of the drug was nearly identical to the rate of loss of [^{3}H]GTP from the microtubule disassembly end at continuing steady state, in the absence of drug (Fig. 3). It can be concluded that podophyllotoxin not only blocks subunit addition of the A end, but also prevents subunit loss at the A end; assuming that there

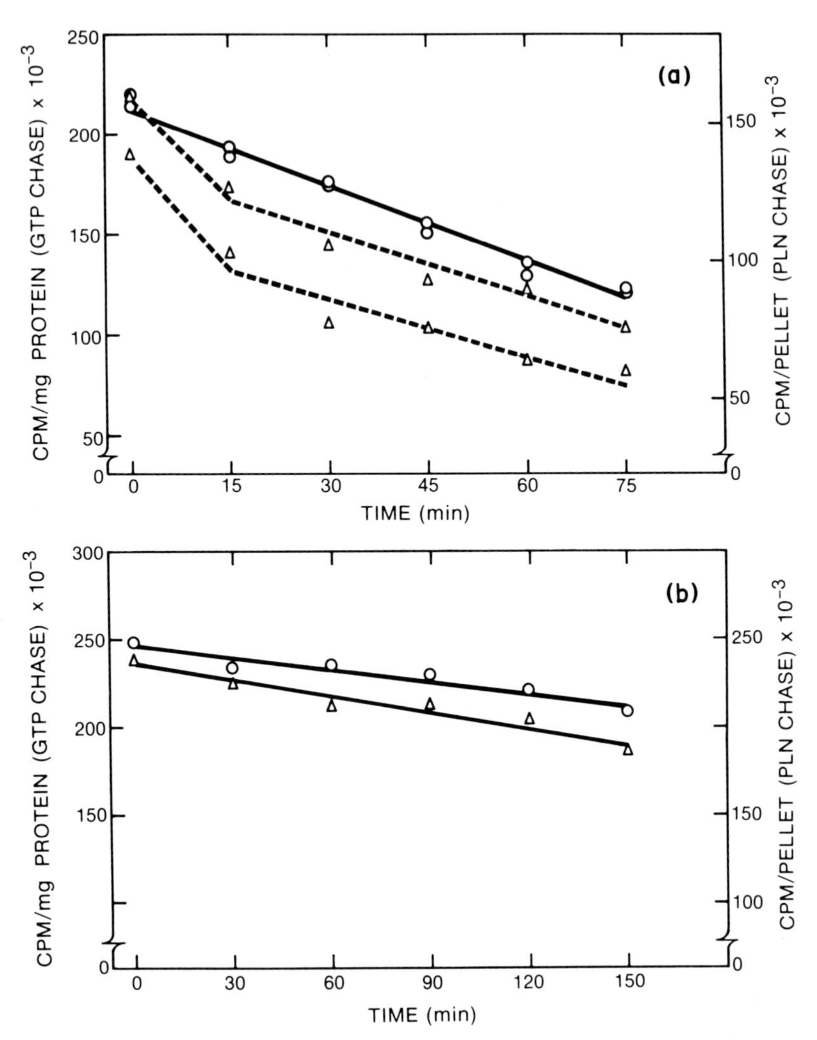

FIG. 3. Comparison of podophyllotoxin disassembly with the treadmilling rate. Microtubule protein (3.5 mg/ml) in two equal aliquots was assembled for 60 min (where ATP is present, this is a period of time sufficient to attain a stable steady state) in MEM buffer containing 0.1 m*M* GTP, 10 μCi/ml [^{3}H]GTP, 10 m*M* acetyl phosphate, and 0.1 IU/ml acetate kinase, either in the presence (a) or absence (b) of 1.0 m*M* ATP. At steady state, microtubules were either chased with 3.0 m*M* GTP or blocked in assembly with 50 μM podophyllotoxin. Label content and protein concentration were assayed for different time points in chase. (a) GTP chase, 35% per hour loss (○——○); podophyllotoxin chase, 35% per hour loss (upper curve), 38% per hour loss (lower curve) (△----△). (b) GTP chase, 5.6% per hour loss (○——○); podophyllotoxin chase, 7.8% per hour loss (△----△).

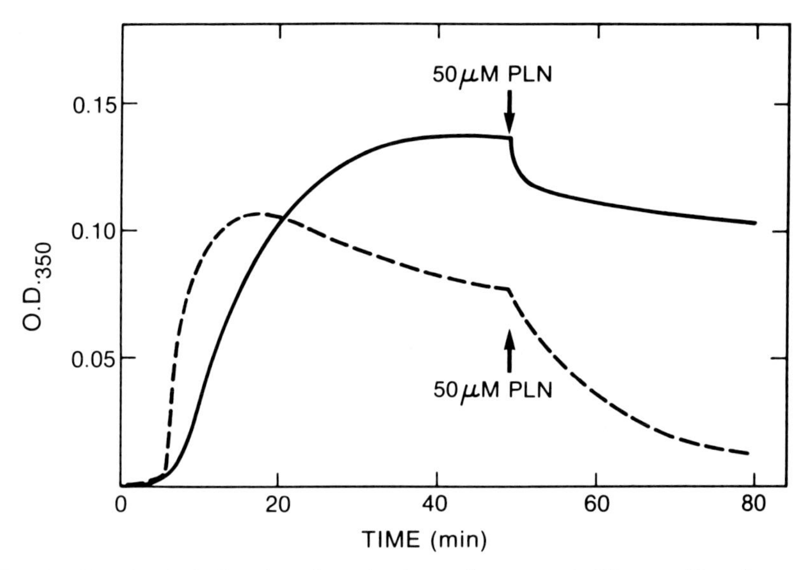

FIG. 4. Use of podophyllotoxin to determine the steady-state treadmilling rate. Two aliquots of microtubule protein (2 mg/ml) in MEM buffer were assembled at 30°C. One aliquot contained 1.25 m*M* ATP. Their assembly was monitored by turbidity increase. At 50 min, 50 μ*M* podophyllotoxin was added to each sample to poison further assembly and to make visible the intrinsic steady-state treadmilling rate. Both samples contain 0.1 m*M* GTP. (----) ATP present; (——) no ATP present.

is a natural loss to equilibrium at this site (Karr and Purich, 1979; Bergen and Borisy, 1980). Therefore the loss rate that one observes with podophyllotoxin is an accurate measure of the subunit treadmilling rate preexisting in the microtubule population.

The rate of loss of subunits from the microtubule need not be measured with [^{3}H]GTP. Since the extent of turbidity of microtubules directly reflects the mass concentration of the polymer, the rate of loss to podophyllotoxin, and therefore the treadmilling rate, can be measured quite simply by spectrophotometry.

To perform these measurements, we assemble microtubules to steady state in a recording spectrophotometer equipped with a constant temperature cuvette chamber, measuring the increase in turbidity at 350 nm that accompanies assembly. At steady state, podophyllotoxin is added at 50 μ*M* concentration, and after an initial rapid loss of polymer, a stable rate of decline is achieved (Fig. 4). We measure the apparent tangent to the second, more stable, rate of decline in order to determine the treadmilling rate. Experiments in which the rate of [^{3}H]GTP uptake into microtubules at steady state, and the rate of loss of polymer in podophyllotoxin by turbidity, were simultaneously measured show the values to be approximately equal under a variety of conditions (Margolis and Wilson, 1978, 1979; see also Fig. 3).

IV. Other Methods for Determination of Rates, Rate Constants, and Efficiency of Treadmilling

The preceding methods are limited. Although one can determine what amount of net subunit flow is generated at the microtubule's A end, and one can surmise that at steady state there is very little equilibrium tubulin addition at the D end compared to the loss rate (by showing that the rate of D end loss to PLN is the same as the rate of loss of [^{3}H]GTP to steady state at that end), one cannot estimate the relative rate constants at the A end. The problem of rate constants has been addressed by three independent methods that are in the literature. Actin polymer A-end rate constants have been determined by measuring the rate of actin uptake into the polymer at the apparent equilibrium and curve fitting this rate of uptake to mathematical predictions (Wegner, 1976).

Microtubule A-end rate constants have also been estimated based on two other methods. One involves kinetic analysis of the rate of loss of polymer to rapid dilution while monitoring microtubule concentration by turbidity (Karr and Purich, 1979). Another method involves independent measurement of the rates of assembly on the A and D ends of microtubules over a range of protein concentrations and curve fitting the results to computer-generated predictions based on different assembly end subunit addition efficiencies (the "efficiency" being roughly the net of the addition rate minus the loss rate at the A end) (Bergen and Borisy, 1980; see also Borisy and Bergen, in this volume). A and D ends were distinguished in this method by growing microtubules onto axonemal fragments, on which A and D ends can readily be distinguished. We have extended our methodology to permit the measurement of subunit addition efficiency at the microtubule A end.

We have previously shown that podophyllotoxin "caps" the A end so that gain and loss of subunits ceases at that site (Margolis and Wilson, 1978). Recently, we have been able to show that continued GTP hydrolysis maintains that cap (Margolis, 1981). If GTP is enzymatically hydrolyzed, the podophyllotoxin cap is lost and microtubules then disassemble at both their A and D ends, while podophyllotoxin prevents any further assembly addition.

Using drug and enzymatic manipulations (GTP is hydrolyzed quickly in this system by addition of phosphofructokinase and fructose-6-phosphate), one can quickly measure, under a variety of conditions, the efficiency of head-to-tail assembly of microtubules. The efficiency of treadmilling has been formalized in an expression derived by Wegner (1976):

$$S = (k_{11} \cdot c_c - k'_{21}) / (k'_{21} + k'_{22}) \tag{1}$$

where k_{11} is the rate constant for A-end subunit addition, c_c is the critical subunit concentration for assembly, k'_{21} is the rate of A end subunit loss, and k'_{22} is the rate of D-end subunit loss. S is the treadmilling efficiency, or the fraction of the population of subunits that binds at the A end and is not lost to equilibrium but proceeds unidirectionally through the microtubule.

The numerator in Eq. (1) expresses the treadmilling rate, which we have found equals the rate of podophyllotoxin induced disassembly (see Section III). The denominator expresses the rate of subunit loss from A and D ends in the absence of any assembly reaction. This rate is obtained by measuring the disassembly rate of microtubules in GDP and podophyllotoxin together. Details of this method may be found in Margolis, 1981.

Another finding, and potential methodology, worthy of note is that of Brenner and Korn (1979). They find that actin polymers are capped by cytochalasin D substoichiometrically in much the same way that microtubules are blocked by podophyllotoxin. The method involves application of a range of substoichiometric drug concentrations to actin filaments at steady state and measurement of polymer disassembly induced by these low drug levels. It is interesting to imagine that methodologies similar to those described here for microtubules (utilizing a capping drug and a nucleotide-depleting enzyme system) may now be applied to actin polymers as well. Conversely, we (Margolis and Rauch, in preparation) have recently applied the methodology of Brenner and Korn to microtubules and find it to be an alternative and valid method for determining whether and at what rate treadmilling occurs in a microtubule preparation.

REFERENCES

Amos, L. A., and Klug, A. (1974). *J. Cell Sci.* **14,** 523-537.

Arai, T., and Kaziro, Y. (1977). *J. Biochem. (Tokyo)* **82,** 1063-1071.

Bergen, L., and Borisy, G. G. (1980). *J. Cell Biol.* **84,** 141-150.

Borisy, G. G., and Taylor, E. W. (1967). *J. Cell Biol.* **34,** 525-533.

Brenner, S. L., and Korn, E. D. (1979). *J. Biol. Chem.* **254,** 9982-9985.

Dentler, W., Granett, S., Witman, G., and Rosenbaum, J. (1974). *Proc. Natl. Acad. Sci. U.S.A.* **71,** 1710-1714.

Gaskin, F., Cantor, C. R., and Shelanski, M. L. (1974). *J. Mol. Biol.* **89,** 737-758.

Inoué, S., and Sato, H. (1967). *J. Gen. Physiol.* **50,** 259-292.

Kane, R. E. (1965). *J. Cell Biol.* **25,** 137-144.

Karr, T. L., and Purich, D. L. (1979). *J. Biol. Chem.* **254,** 10885-10888.

Kobayashi, T., and Simizu, T. (1976). *J. Biochem. (Tokyo)* **79,** 1357-1364.

Lee, J. C., and Lee, L. L. Y. (1979). *Biochemistry* **18,** 5518-5526.

Lee, J. C., and Timasheff, S. N. (1977). *Biochemistry* **16,** 1754-1764.

Lee, J. C., Tweedy, N., and Timasheff, S. H. (1978). *Biochemistry* **17,** 2783-2790.

Lowry, O. H., Rosebrough, N. J., Farr, A. L., and Randall, R. J. (1951). *J. Biol. Chem.* **193,** 265-275.

Maccioni, R. B., and Seeds, N. W. (1978). *Arch. Biochem. Biophys.* **185,** 262-271.

MacNeal, R. K., Webb, B. C., and Purich, D. L. (1977). *Biochem. Biophys. Res. Commun.* **74,** 440-447.

Mandelkow, E. M., Mandelkow, E., and Schultheiss, R. (1979). *J. Mol. Biol.* **135,** 293-299.
Margolis, R. L. (1981). *Proc. Nat. Acad. Sci. U.S.A.* **78,** 1586-1590.
Margolis, R. L., and Rauch, C. T. (1981). In preparation.
Margolis, R. L., and Wilson, L. (1978). *Cell* **13,** 1-8.
Margolis, R. L., and Wilson, L. (1979). *Cell* **18,** 673-679.
Mazia, D., Mitchison, J. M., Medina, H., and Harris, P. (1961). *J. Biophys. Biochem. Cytol.* **10,** 467-474.
Olmsted, J. B., and Borisy, G. G. (1973). *Biochemistry* **12,** 4282-4289.
Olmsted, J. B., Marcum, J. M., Johnson, K. A., Allen, C., and Borisy, G. G. (1974). *J. Supramol. Struct.* **2,** 429-450.
Rebhun, L. I., and Sawada, N. (1969). *Protoplasma* **68,** 1.
Shelanski, M. L., and Taylor, E. W. (1967). *J. Cell Biol.* **34,** 549-554.
Shelanski, M. L., Gaskin, F., and Cantor, C. R. (1973). *Proc. Natl. Acad. Sci. U.S.A.* **70,** 765-768.
Wegner, A. (1976). *J. Mol. Biol.* **108,** 139-150.
Weisenberg, R. C. (1972). *Science* **177,** 1104-1105.
Weisenberg, R. C., Borisy, G. G., and Taylor, E. W. (1968). *Biochemistry* **7,** 4466-4479.
Weisenberg, R. C., Deery, W. J., and Dickinson, P. J. (1976). *Biochemistry* **15,** 4248-4254.
Wilson, L., and Bryan, J. (1974). *Adv. Cell Mol. Biol.* **3,** 21-72.
Wilson, L., Jordan, M. A., Morse, A., and Margolis, R. L., submitted.

Chapter 10

A Rapid Filtration Assay for Analysis of Microtubule Assembly, Disassembly, and Steady-State Tubulin Flux

LESLIE WILSON, K. BRADFORD SNYDER, AND
WILLIAM C. THOMPSON

Department of Biological Sciences
University of California, Santa Barbara
Santa Barbara, California

AND

ROBERT L. MARGOLIS

The Fred Hutchinson Cancer Research Center
Seattle, Washington

I. Introduction

When microtubules are assembled *in vitro* from tubulin or a mixture of tubulin and microtubule-associated proteins in the presence of GTP, the total quantity of polymer increases to a plateau, which is now recognized to reflect the summation of different reversible reactions that occur at each microtubule end (Margolis and

ISBN 0-12-564124-9

Wilson, 1978; Wilson and Margolis, 1978; Farrell *et al.*, 1979; Karr and Purich, 1979; Bergen and Borisy, 1980). Thus the overall reaction is not a simple equilibrium, but is a steady-state summation of the two distinct reactions at each microtubule end. Under steady-state conditions *in vitro,* net tubulin addition onto the microtubule occurs at one end of the microtubule (the *net assembly* or *primary assembly* end) and net tubulin loss occurs at the opposite end (the *net disassembly* or *primary disassembly* end[1]). Thus, a unidirectional flux of tubulin from one end of the microtubule to the other, or "treadmilling," occurs under steady-state conditions *in vitro*.

The opposite end assembly–disassembly behavior of microtubules, if it occurs within cells, could be fundamentally linked to the function of microtubules, as, for example, in the translocation of chromosomes during the process of mitosis (Margolis *et al.*, 1978).

The method currently employed for quantitating and characterizing steady-state tubulin flux makes use of the fact that GTP bound reversibly at the exchangeable site on a tubulin dimer when it is free in solution becomes bound irreversibly in the form of GDP upon or very shortly after addition of the tubulin onto the net assembly end of a steady-state microtubule (Weisenberg *et al.*, 1976; Margolis and Wilson, 1978). Thus, [^{3}H]GTP can be used as a marker for tubulin addition and loss at microtubule ends. The procedures employed until now have involved collection of the labeled microtubules by centrifugation through 50% sucrose cushions, which serve to free the microtubules of all unassembled tubulin and unincorporated tritium (for description of these methods, see Margolis, in this volume).

There are four limitations to the use of the centrifugation method for collection of labeled microtubules. First, each centrifuge rotor holds a maximum of 12 tubes, so the number of simultaneous data points that can be obtained in an experiment is limited by the number of available centrifuges and rotors. Second, sufficient quantities of microtubule protein must be applied to each sucrose cushion so that measurable yields of protein can be obtained in the final pellet of microtubules. Third, the assays are tediously slow. For example, an experiment with just 12 time points takes approximately 4–5 hr to complete. Fourth, and very importantly, it is not possible to monitor the early events of a pulse or chase experiment. It takes approximately 5 min to load the samples into the centrifuge rotor and to begin centrifugation under the most ideal conditions, so events that occur within seconds or even minutes of a pulse or a chase cannot be monitored using this procedure.

[1]By "net assembly," "primary assembly end," or "An end," we mean that end of the microtubule at which net tubulin addition occurs under steady-state conditions *in vitro*. Conversely, "net disassembly end," "primary disassembly end," or "D end" refers to that end of the microtubule at which net tubulin loss occurs under steady-state conditions *in vitro*. We are not using these terms to describe all the specific molecular tubulin association and disassociation events that reflect the reversible equilibrium reactions occurring at each end of the microtubule.

We have developed a rapid filtration procedure that we have modeled after the technique originally developed by Maccioni and Seeds (1978), in which aliquots of a steady-state suspension of labeled microtubules are diluted into a microtubule stabilizing solution, followed by collection of the microtubules on glass fiber filters by vacuum filtration. The method is known to be accurate with as little as 16 μg of total microtubule protein as microtubules in a single assay, and a time course of [^{3}H]GTP uptake or loss can be followed by taking time points at intervals as short as a few seconds, and beginning within a few seconds of a pulse or a chase. The procedure eliminates the experimental confinements of the centrifugation method for collection of labeled microtubules.

II. Rapid Filtration Assay: Details of the Procedure

A. Chemicals, Buffer Solutions, and Equipment

The basis of this assay is the rapid separation of radioactively labeled microtubules from labeled unincorporated nucleotides by vacuum filtration using glass fiber filters, under conditions in which the microtubules are maintained in a stabilized fashion by suspension in a stabilizing buffer. We usually use an assembly buffer consisting of 100 m*M* MES (2[*N*-morpholino]ethanesulfonic acid, Sigma Chemical Company); 1 m*M* EGTA (ethyleneglycolbis[β-aminoethyl ether]-*N,N*′-tetraacetic acid, Sigma Chemical Company); 1 m*M* $MgSO_4$, pH 6.75, containing 100 μM GTP (Sigma Chemical Company, Type I) and a GTP regenerating system consisting of 10 m*M* acetyl phosphate and 0.1 IU/ml acetate kinase (Sigma Chemical Company) (MacNeal *et al.*, 1977). When treadmilling experiments are carried out over long periods of time, we occasionally use 200 μM GTP, 20 m*M* acetyl phosphate, and 0.2 IU/ml acetate kinase, in order to maintain sufficient quantities of GTP. We routinely employ purified bovine brain microtubules in our treadmilling studies (Asnes and Wilson, 1979) and labeling of the microtubules is accomplished either by assembling to steady state in the presence of [^{3}H]GTP, or pulsing with a small volume of labeled GTP once steady state has been reached, as described by Margolis (in this volume). In a typical experiment, the ratio of unlabeled GTP to commercially purchased radioactive GTP is 100:1.

Quenching of the assembly or disassembly reaction is accomplished by pipetting the desired volume of reaction mixture into a microtubule stabilizing solution composed of 100 m*M* MES, 1 m*M* EGTA, 1 m*M* $MgSO_4$, 25% glycerol, 10% dimethylsulfoxide (DMSO) (see Himes *et al.*, 1976), and 5 m*M* ATP (Sigma Chemical Company, Grade II), pH 6.75. The purpose of the ATP is to reduce nonspecific association of GTP with the glass fiber filter during filtration of the microtubule suspension and subsequent washing of the trapped microtubules.

Filtration is accomplished using Whatman 2.4-cm diameter GF/F glass microfiber filters on a Millipore 25-mm glass microanalysis S.S. support, equipped with a stainless screen filter support (catalog number XX10 025 30), coupled to a model DD 20 vacuum pump (Precision Scientific Company). Filtration can also be carried out with a multiple filter manifold, such as the Hoeffer Scientific Instruments 10-place filtration manifold and collection box (catalog number 224V or 225V).

B. Procedure

Aliquots of a steady-state suspension of microtubules at 30°C (or any temperature desired) are diluted into 4.0 ml of prewarmed (30°C) stabilizing buffer contained in disposable glass tubes. The entire 4 ml of solution is pipetted into the filter holder at room temperature with the vacuum on. The filter is prewet by brief immersion in stabilizing buffer prior to assembly of the filter apparatus. The filter is washed three times after filtration of the original 4 ml of buffer containing the labeled microtubules with 5.0 ml vol of fresh stabilizing buffer. Filters are then removed and prepared for analysis of radioactivity, or radioactivity and protein content.

An important advantage of diluting the microtubule suspension into a microtubule stabilizing buffer prior to filtration is that it is unnecessary to filter the suspension immediately after dilution. There is little or no loss of radioactivity from labeled and stabilized microtubules within the first hour or so (see Section III,A). Thus it is possible to accumulate many samples taken at frequent intervals and hold them in a stabilized manner until it is feasible to filter them.

We have found an efficient method for analysis of the filters to be as follows. Filters are air-dried for approximately 15–30 min, then cut into quarters with a pair of scissors. If a protein determination is not required, the filter pieces can be placed directly in a liquid scintillation vial along with 1.5 ml of 0.1 *N* NaOH and incubated for 30 min to release the radioactivity from the filter prior to addition of scintillation fluid. If a protein determination is required, all protein and radioactivity may be released from the filter disks by immersion in 0.5 ml of 0.5 *M* NaOH and incubation at 37°C for 2 hr, essentially as described for protein analysis by Maccioni and Seeds (1978).

III. Characterization of the Assay System

A. Filters, Wash Procedure, Quantities of Protein Required

We have compared the relative filtration efficiencies of several types of glass filter fibers (Table I) and find the GF/F filters to be the most efficient for

TABLE I

EFFICIENCY OF MICROTUBULE COLLECTION BY VARIOUS TYPES OF GLASS FIBER FILTERS[a]

Filter type (Whatman)	Blank (no microtubules)	[^{3}H]Nucleotide in collected microtubules	Relative efficiency
	(CPM)	(CPM)	(% of GF/F)
GF/A	743	42,992	70.0
GF/B	2033	39,563	64.5
GF/C	1210	50,901	82.9
GF/F	1633	61,381	100.0

[a] Bovine brain microtubule protein (2 mg/ml) was polymerized as described in Section II,B for 1 hr at 37°C in the presence of 0.25 m*M* [^{3}H]GTP (specific activity 197 Ci/mole), but without a GTP-regenerating system. The reaction mixture was diluted into stabilizing buffer and then aliquots of the solution of stabilized microtubules were filtered through the various types of glass fiber filters. Each value represents the mean of three individual filters. To obtain blank values in the absence of microtubules, aliquots of a similar reaction mixture containing 50 μM podophyllotoxin were diluted with stabilizing buffer and filtered as above. The blank values vary with the thickness and pore size of the glass fiber filters.

collection of microtubules. Both the GF/C and GF/F filters are more efficient than the GF/A filters originally utilized by Maccioni and Seeds (1978). Although the relative efficiencies of the various filter types probably depends upon the distribution of microtubule lengths, the GF/F filters are quite efficient at collecting even the extremely short microtubules formed in the presence of the drug taxol (Schiff *et al.*, 1979). No microtubules have been detected in the filtrates of GF/F filters when examined by electron microscopy. The use of two or more GF/F filters in a stack does not increase the collection of microtubules, and only serves to slow the rate of filtration and increase the background adherence of unincorporated labeled nucleotide. Some free tubulin is bound by the GF/F filters, but this poses no problem when using [^{3}H]GTP as the marker, because such dimers do not bind [^{3}H]GTP tightly. However, in a similar assay for the covalent incorporation of [^{14}C]tyrosine into free tubulin and microtubules (Thompson, in this volume) GF/C filters were utilized because of their low nonspecific binding of free tubulin dimers.

The volume of stabilizing buffer used for the initial dilution of the microtubule suspension appears to be relatively uncritical. Similarly, the volume of each wash, and the number of washes used, is not critical, except that a minimum of three washes is necessary with 5-ml wash volumes. In one experiment, we washed with six 5-ml volumes of stabilizing buffer. A plateau was reached after the third wash, and the plateau was maintained for an additional three washes.

One important advantage of the rapid filtration assay procedure is that the microtubules with the incorporated label can be "frozen" at any instant by dilution of the microtubules into an appropriate stabilizing buffer. Thus, if the rate of sample accumulation in stabilizing buffer exceeds the rate of filtration capability, the stabilized samples can be retained for a while until filtration becomes possible. The results shown in Fig. 1 were obtained by assembling bovine brain microtubule protein to steady state in the presence of labeled GTP as described in Section II,A. Labeled microtubules were diluted into stabilizing wash buffer, and aliquots were filtered after further incubation in stabilizing buffer as indicated in the figure. Essentially no label loss occurred during the first few minutes, and the minimal loss of label that occurred as incubation continued was quantifiable.

It is important to emphasize that the ability to employ a stabilizing buffer in this way requires either that the microtubules be fully stable in the buffer, or that, if they are unstable, the rate of polymer loss be quantifiable. We have found that one class of microtubules, those reassembled from sea urchin outer doublet tubulin, is very unstable in the buffer employed for stabilization of purified bovine brain microtubules. Thus, for these microtubules, and possibly micro-

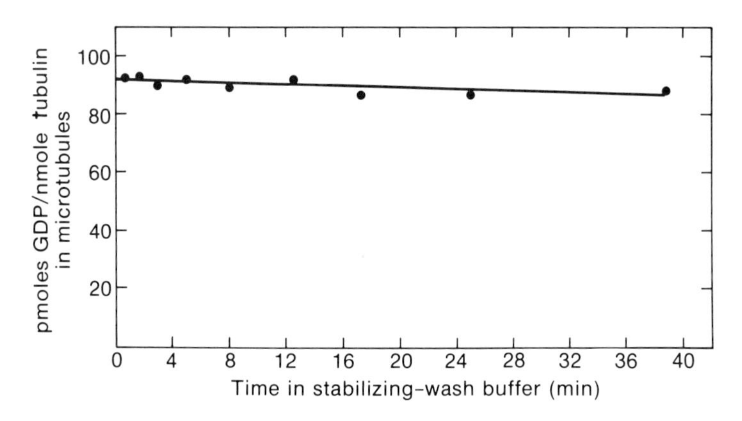

FIG. 1. Stability of labeled microtubules in stabilizing buffer. A solution of three-times-cycled bovine brain microtubule protein (Asnes and Wilson, 1979) in reassembly buffer (0.5 ml, 2.12 mg total microtubule protein/ml, in 100 mM MES, 1mM EGTA, 1 mM $MgSO_4$, 100 μm GTP, 10 mM acetyl phosphate, and 0.1 IU/ml acetate kinase, pH 6.75) was assembled to steady state by incubation for 45 min at 30°C. [^{3}H]GTP (20 μCi in 20 μl water, 13.5 Ci/mmole, ICN) was added and incubation was continued for an additional 50 min to label the primary assembly ends of the microtubules. At this time (zero time) the entire microtubule suspension was diluted into 40 ml of stabilizing buffer at 30°C, and at the times indicated in the figure, 4.0 ml aliquots, each containing 106 μg of total microtubule protein and 64 μg of tubulin as microtubules, were filtered as described in Section II,B. The first sample taken, at 0.6 min, contained 15,700 cpm on the filter. Background was determined by filtering a microtubule suspension that had been incubated with a 100-fold excess of unlabeled GTP prior to addition of the labeled GTP (609 cpm/filter). Data are expressed in terms of the number of picomoles of nucleotide, considered to be GDP, incorporated into the microtubules per nanomole of tubulin in the microtubules.

tubules from other sources and compositions as well, specific stabilizing buffers will have to be designed.

Another significant benefit of the filtration assay procedure is that accurate quantitation of label incorporation can be obtained with very small amounts of protein. The results shown in Fig. 2 were obtained by assembling bovine brain microtubule protein as described in the experiment of Fig. 1 in the presence of labeled GTP to label the microtubules uniformly. Then different volumes of the labeled steady-state microtubules were diluted into stabilizing wash buffer and filtered. The quantity of label retained on the filter was proportional to the quantity of total protein applied, between protein values of 16 μg per filter and 240 μg per filter. Although we have not carried the experiment out with lower or higher quantities of microtubule protein, the assay would likely be accurate with

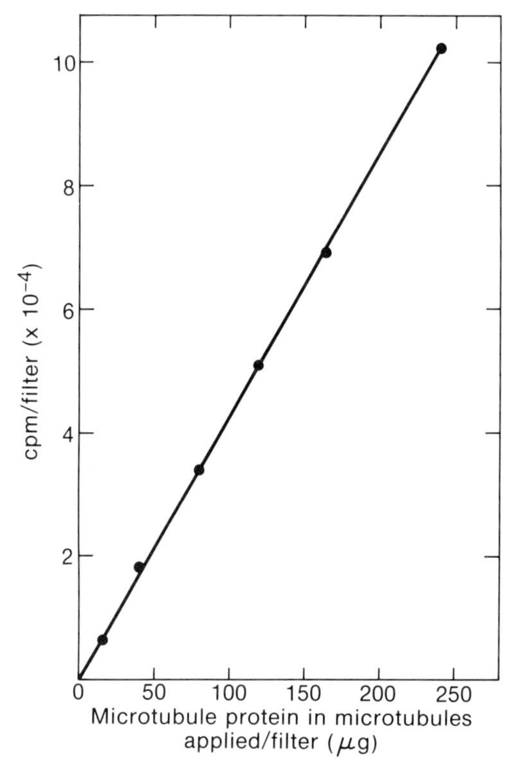

FIG. 2. Linearity of the filter assay. A solution of bovine brain microtubule protein in reassembly buffer (0.5 ml, 2.33 mg/ml total protein, see legend to Fig. 1) was assembled to steady state at 30°C for 1.5 hr in the presence of [^{3}H]GTP (8 μCi in 8 μl water, 13.5 Ci/mmole) to label the microtubules uniformly. Aliquots of the labeled microtubule suspension were filtered as described in Section II,B, except that four 2.5-ml washes were employed instead of three 5-ml washes. Background radioactivity was determined as described previously (see legend to Fig. 1).

quantities of microtubule protein both below and above those used in the experiment; and the ultimate limitation with small quantities of protein would probably be the ability to detect radioactivity.

Evidence that the rapid filtration assay procedure using [^{3}H]GTP as a marker for tubulin addition to microtubules provides a quantitative estimate of polymer formation is shown in Fig. 3. Bovine brain microtubule protein was assembled to steady state in the presence of [^{3}H]GTP, and the assembly reaction was monitored independently by light scattering and by the filter assay. Polymerization as determined by [^{3}H]nucleotide incorporation and trapping on glass fiber filters very closely mirrored polymerization as determined by light-scattering measurements at 350 nm. Addition of 50 μM podophyllotoxin after steady state had been reached resulted in an initial rapid loss of polymer, followed by a slow rate of loss; and the only discrepancy between the two methods was that the loss of polymer as judged by light scattering was slightly slower than the loss as judged by the filter assay. It is conceivable that the filter assay more accurately reflects

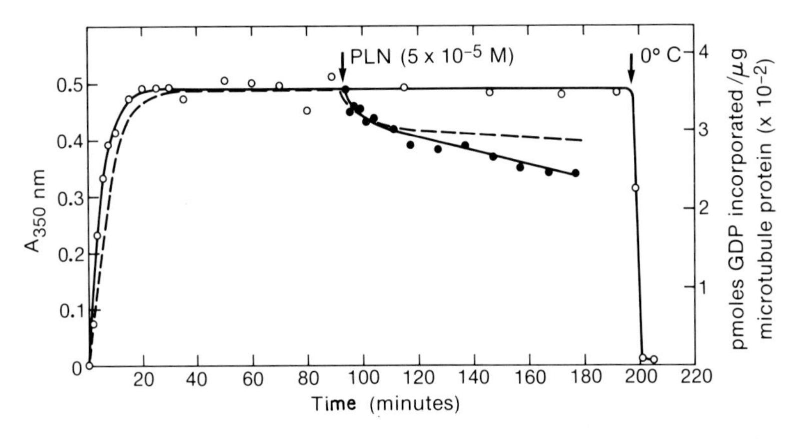

FIG. 3. Comparison of the filter assay procedure witha light-scattering assay. A solution of three-times-cycled bovine brain microtubule protein in reassembly buffer (2.8 ml, 2.62 mg/ml total protein) was assembled to steady state at 30°C in the presence of [^{3}H]GTP (50 μCi in 50 μl of water, 13.5 Ci/ mmole). At the times indicated, 50-μl aliquots were assayed by the rapid filter procedure as described in Section II,B (open circles, solid line). Podophyllotoxin (PLN, final concentration, 5×10^{-5} M) was added to a 900-μl aliquot of the microtubule suspension after 101 min of assembly (arrow) and 50-μl volumes were assayed by the filter procedure at the indicated times (closed circles, solid line). The remaining untreated suspension was cooled to 0°C at 191 min (arrow) by immersion in an ice-water slurry, and 50-μl aliquots of the cooled (depolymerized) microtubule protein solution were assayed at the indicated times (open circles, solid line). Background radioactivity was determined as described in the legend to Fig. 1. Data are expressed as picomoles of nucleotide, considered to be GDP, incorporated per microgram total microtubule protein applied per filter. A 1-ml aliquot of the original microtubule protein suspensions was assembled at 30°C in the absence of labeled GTP and assayed by light scattering as described previously (Asnes and Wilson, 1979) (dashed line, no symbols). Podophyllotoxin (5×10^{-5} M) was added at 101 min, and the assay was continued (dashed line, no symbols).

polymer quantity because the assay involves the direct biochemical measurement of polymer; however, this remains to be determined experimentally. Cooling to 0°C results in a rapid loss of polymer, as expected.

One important difficulty with the filter assay concerns nonspecific adherence of labeled nucleotide on the filters. The inclusion of 5 m*M* ATP in the stablizing wash buffer reduces background counts very significantly. GTP would be a better nucleotide to use, but its use would be prohibitively expensive.

B. Applications of the Procedure

The rapid filter assay procedure may be used to quantitate polymerization in a manner similar to that depicted in Figure 3. It is important to emphasize that in this experiment, and the one that follows, much more protein and radioactivity were used than the minimum required for the assay. The procedure would be most useful when insufficient protein is available for measurement of assembly

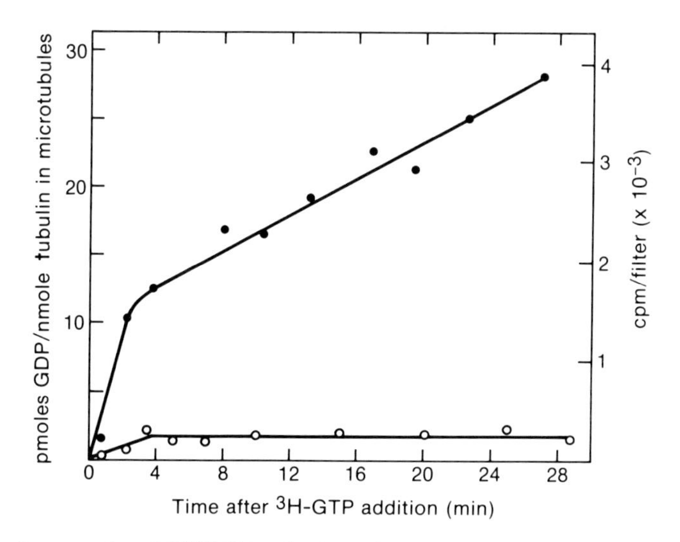

FIG. 4. Incorporation of [^{3}H]GTP into bovine brain microtubules at steady state. A solution of three-times-cycled bovine brain microtubule protein (1.5 ml, 2.12 mg/ml of total microtubule protein) was assembled to steady state by incubation for 2.5 hr at 30°C as described in the legend to Fig. 1. The suspension of steady-state microtubules was divided into two 600-μl aliquots, and one aliquot was incubated for an additional 10 min with 1×10^{-4} *M* colchicine, to block further polymerization. [^{3}H]GTP (20 μCi, in 20 μl of water, 13.5 Ci/mmole) was added to each aliquot at zero time, and at the indicated times, 50-μl samples, each containing 64 μg of tubulin as microtubules, were removed and assayed for incorporation by the rapid filtration technique as described in Section II,B. The data are presented as described in the legend to Fig. 1. An initial burst of labeled nucleotide incorporation occurs during the first several minutes, which is followed by a slower incorporation rate (solid circles). Addition of 10^{-4} *M* colchicine, which blocks all further assembly, prevented all incorporation of nucleotide into the microtubules.

by light scattering, viscometry, or sedimentation. This method is uniquely useful for monitoring tubulin polymerization in the presence of drugs like taxol, where the extremely short microtubules cannot be adequately monitored by light scattering or even viscometry.

The most important application of the rapid filtration assay procedure is for quantitation and characterization of steady-state tubulin flux, and characterization and quantitation of the specific association and disassociation events that occur at the microtubule ends. An interesting example of the usefulness of the procedure is shown in Fig. 4. Bovine brain microtubule protein was assembled to steady state in the absence of labeled GTP and divided into two aliquots, and one aliquot was incubated briefly with colchicine. [^{3}H]GTP was then added (zero time) and the incorporation of label into the steady-state microtubules was monitored during the following 28–29 min period. It is evident that there is an initial burst of label incorporation, which is prevented by the drug, and which precedes the normal slow incorporation rate characteristic of this microtubule system (Margolis and Wilson, 1978). The burst was not detected using the centrifugation method (Margolis and Wilson, 1978). This phenomenon may be related to the utilization of GTP at the primary assembly end of the microtubule, and is currently being investigated in our laboratory.

IV. Concluding Remarks

In summary, we have adapted the rapid filter assay, developed initially by Maccioni and Seeds (1978), for the analysis of steady-state tubulin flux, using [^{3}H]GTP incorporation into steady-state microtubules. The filtration procedure for collection of labeled microtubules offers several significant advantages over the centrifugation method which has been employed until now. Most notably, early events during assembly or disassembly, or after perturbation of microtubules, as, for example, after drug treatment, may be investigated. In addition, a large number of data points may be obtained within a short period of time, with small quantities of microtubule protein. Thus, it is now possible to investigate events occurring at the microtubule ends that could not be studied using previously available techniques, and the procedure enables study of microtubule polymerization with proteins available in very limited supply.

Acknowledgments

We wish to thank Mr. Thanh Diec and Mr. Scott Komoto for technical assistance. Support was given by the American Cancer Society (grant CD-3F), and by the United States Public Health

Service (grant NS13560). W.C.T. was supported by a Muscular Dystrophy Association grant to L.W. and Dr. Daniel Purich.

References

Asnes, C. F., and Wilson, L. (1979). *Anal. Biochem.* **98,** 64–73.
Bergen, L. G., and Borisy, G. G. (1980). *J. Cell Biol.* **84,** 141–150.
Farrell, K. W., Kassis, J. A., and Wilson, L. (1979). *Biochemistry* **18,** 2642–2647.
Himes, R. H., Burton, P. R., Kersey, R. N., and Pierson, G. B. (1976). *Proc. Natl. Acad. Sci. U.S.A.* **73,** 4397–4399.
Karr, T. L., and Purich, D. L. (1979). *J. Biol. Chem.* **254,** 10885–10888.
Maccioni, R. B., and Seeds, N. W. (1978). *Arch. Biochem. Biophys.* **185,** 262–271.
MacNeal, R. K., Webb, B. C., and Purich, D. L. (1977). *Biochem. Biophys. Res. Commun.* **74,** 440–447.
Margolis, R. L., and Wilson, L. (1978). *Cell* **13,** 1–8.
Margolis, R. L., Wilson, L., and Kiefer, B. (1978). *Nature (London)* **272,** 450–452.
Schiff, P. B., Fant, J., and Horowitz, S. B. (1979). *Nature (London)* **277,** 665–667.
Weisenberg, R. C., Deery, W. J., and Dickinson, P. J. (1976). *Biochemistry* **15,** 4248–4254.
Wilson, L., and Margolis, R. L. (1978). *ICN-UCLA Symp. Mol. Cell. Biol.* **12,** 241–258.

Chapter 11

A Direct Method for Analyzing the Polymerization Kinetics at the Two Ends of a Microtubule

GARY G. BORISY AND LAWRENCE G. BERGEN

Laboratory of Molecular Biology
University of Wisconsin
Madison, Wisconsin

I. Introduction

Microtubules have an intrinsic structural polarity (Amos and Klug, 1974; Crepeau *et al.*, 1977; Amos, 1979), which means that the two ends of a microtubule are physically and chemically distinct. This structural polarity is reflected in the polarity of microtubule growth (Olmsted *et al.*, 1974; Binder *et*

ISBN 0-12-564124-9

al., 1975; Dentler *et al.*, 1974; Allen and Borisy, 1974). Another manifestation of growth polarity is that in the steady state, net assembly occurring at one end of a microtubule is balanced by net disassembly at the other end such that a flux of subunits traverses the tubule. This phenomenon has been referred to as opposite end assembly–disassembly or head-to-tail polymerization (Wegner, 1976; Margolis and Wilson, 1978; Bergen and Borisy, 1980a; Cote *et al.*, 1980).

The solution biochemical techniques used to investigate microtubule polymerization cannot directly differentiate the reactions at the two ends, but only permit inferences as to the relative behavior of these ends. To analyze directly the growth and depolymerization of microtubule ends, we introduced a quantitative morphological method based on the structure and template-nucleating capacity of flagellar axonemes.

The flagellar axoneme has a number of properties that make it a useful tool for the analysis of microtubule polarity. The seeds have a defined polarity, all of the doublets being oriented in the same direction (Amos and Klug, 1974). The two ends of the seed can be distinguished morphologically, which in turn defines the absolute structural polarity of the seeded extensions (Allen and Borisy, 1974). The seed can be distinguished from the added cytoplasmic microtubules, thereby making clear in static electron micrographs what lengths correspond to the seed and to the added polymer. Under our standard *in vitro* conditions the flagellar axoneme is quite stable; the length of the seed does not change upon either polymerization or depolymerization of the cytoplasmic tubules. Thus, the flagellar axoneme serves to separate the two ends of the labile cytoplasmic microtubules and renders them accessible for independent analysis.

In this report we will outline the axoneme-subunit system in detail and describe three applications of the system that illustrate how it can be used to study microtubule polarity at the molecular level.

II. Materials and Methods

A. Microtubule Protein and Subunit Preparation

Microtubule protein was prepared from porcine brain tissue by two cycles of polymerization and depolymerization with differential centrifugation according to the procedure of Borisy *et al.* (1975). The buffer generally used was PEMG [0.1 *M* piperazine-*N,N'*-bis(2-ethanesulfonic acid) (PIPES), 1 m*M* ethyleneglycolbis(aminoethyl ether)-*N,N'*-tetraacetic acid (EGTA), 0.1 m*M* $MgCl_2$, 1 m*M* GTP, pH 6.94]. For storage, protein was then frozen in liquid nitrogen and kept at −80°C. Protein determinations were done by the method of Lowry *et al.* (1951), using bovine serum albumin as a standard.

Solutions of nucleation-supressed subunits were prepared by a high speed centrifugation of fresh cycled protein (185,000 *g*, 90 min, 0–4°C). For a typical preparation, 2 ml of 4–5 mg/ml microtubule protein was centrifuged in a Beckman type 65 rotor (2-ml delrin adapter tubes were used). After centrifugation, the top 1.0–1.4 ml was carefully pipetted off, and the remainder was discarded. This procedure strongly attenuated the ability of the protein to self-assemble, but did not alter the subunits' competence to elongate microtubules (Olmsted *et al.*, 1974; Allen and Borisy, 1974). These subunit preparations were either used immediately after the high-speed centrifugation or frozen in liquid nitrogen and stored at −80°C without detectable loss of activity (Bergen and Borisy, 1980a). Frozen samples were thawed immediately before use.

B. *Chlamydomonas* Culture Conditions

Chlamydomonas reinhardtii strain 21 GR (wild type) was grown in the defined algal medium developed by Sager and Granick (1953). The cultures were aerated continuously and maintained at 25 ± 1°C. Cell synchrony was induced by growing the cultures with a 14-hr light, 10-hr dark day–night cycle (Allen and Borisy, 1974; Bernstein, 1960). For a typical preparation of flagellar axonemes, one flask of 300 ml medium was grown to an approximate density of 5×10^6 cell/ml.

C. Preparation of Axonemes

Detergent treatment of *Chlamydomonas* culture induced both flagellar amputation and demembranation of the flagella. Axonemes were then separated from the cell bodies by differential centrifugation. The following is our routine protocol, which is a modification of the Allen and Borisy (1974) procedure.

Cells were harvested between the third and fifth hour of the light cycle. This gave the best yield of full-length axonemes. Harvesting was done by low-speed centrifugation (3000 rpm, Sorval SS-34 rotor; 1100 *g*, 1 min). The cells were washed twice by resuspension in distilled water and subsequent centrifugation. The second pellet was washed once with the flagellar isolation buffer (PEMG diluted 1:10 with 1 m*M* dithiothreitol in distilled water) and again pelleted. The cells were then mixed with 25 ml of isolation buffer containing 0.04% (w/v) Non-Idet, NP40, which is a nonionic detergent. The detergent solution induced amputation of virtually 100% of the flagella. Once the detergent was applied to the cells, they were subsequently handled only with plastic pipettes and tubes, since the demembranated flagella adhered well to glass. The cell bodies were separated from the axonemes by the same low-speed centrifugation used in the washes. The supernatant was then carefully removed, and the centrifugation repeated. the axonemes were then concentrated by a higher speed centrifugation

(16,000 rpm, Sorvall SS-34 rotor; 31,000 *g*, 30 min). This supernatant was discarded and the pellet was resuspended in 0.2–0.5 ml PEMG. Residual cell bodies were removed by centrifugation in a clinical centrifuge adapted to hold 1.5 ml polypropylene conical tubes (Eppendorf micro-test-tubes). The axoneme solution was then divided into aliquots of 50 μl, frozen in liquid nitrogen, and stored at −80°C. Freezing did not alter the ability of these seeds to nucleate growth. Frozen axonemes were stable for several months, but were routinely used within one month of preparation.

D. Microtubule Assembly onto Flagellar Axonemes

Solutions of microtubule subunits were thawed immediately before an assembly experiment. Preparations of flagellar seeds were also thawed just before an experiment and diluted with PEMG to an appropriate concentration as determined by electron microscopic examination. Microtubule elongation was initiated by mixing these subunit and axoneme solutions at polymerizing temperatures, usually 30°C. To ensure that the temperature of a sample did not fluctuate, experiments were performed in a constant-temperature room, and the solutions were in tubes that were held by an aluminum block. The temperature of the experiment was recorded as the temperature of the block. The centrosome experiments were performed in a table-top gravity convection incubator that contained the aluminum block. Polymerization reactions were quenched and the structures fixed for further analysis by removing a 20-μl sample and placing it in 10–50 μl of warm (30°C) 2% glutaraldehyde (in polymerization buffer without GTP).

E. Tannic Acid Staining Procedure for Visualization of Microtubule Protofilaments

Self-nucleated microtubules or axoneme-seeded microtubules were fixed for 5 min (at room temperature) with glutaraldehyde as performed for all other experiments. After fixation, 3 vol of tannic acid solution (8% in 0.05 *M* sodium phosphate, pH 6.8) were added to the fixed solution. Tannic acid staining required at least 5 hr, and the sample was often left overnight at this stage. Stained material was collected by centrifugation (100,000 *g*, 30 min) and the pellet was postfixed with 1% osmium tetroxide. The hardened and somewhat brittle pellet was then scraped from the centrifuge tubes and dehydrated with a series of acetone solutions (from 30% acetone in water to anhydrous acetone), then infiltrated with Araldite 502 resin (Ladd). The resin was then polymerized in a 60°C oven (48 hr). This material was trimmed with an LKB Pyramitome and then sectioned with a Reichert OMU-3 ultramicrotome. The thin sections (500–700 Å) were counterstained with Reynold's (1963) lead citrate and 2% uranyl acetate.

F. Mammalian Cell Culture

Chinese hamster ovary (CHO) cells were grown as monolayers in Ham's F-10 medium, supplemented with 10% fetal bovine serum, antibiotics and 15 mM HEPES (*N*-2-hydroxyethyl piperazine-*N*′-2 ethanesulfonic acid), pH 7.2, and were maintained at 37°C in a humid atmosphere with 5% CO_2.

G. Preparation of Centrosomes from Mitotic Cells

Centrosomes were prepared from mitotic cells by the method of Gould and Borisy (1977), with some modification. Mitotic cells were obtained by a two-step procedure. First, the cells (estimated to be at a density of approximately two divisions prior to confluence) were synchronized at S phase by treatment with 2 m*M* thymidine (for 12–16 hr). The cells were incubated with fresh medium for 3–5 hr, and then collected at mitosis by further incubation with medium containing 0.1 μg/ml Colcemid (4–5 hr). The mitotic cells were collected by centrifugation in a clinical centrifuge (1,100 *g*, 4 min). The medium was carefully aspirated and the pellet resuspended in 0.1 ml distilled water. After 1 or 2 min the cells were lysed by addition of an equal volume of 2× lysis buffer (10 mM PIPES, 1 mM EGTA, 0.1 mM $MgCl_2$, 0.25% Triton X-100, pH 6.7). Lysis and release of centrosomes were checked by phase contrast microscopy. These lysates were then used to study centrosome-nucleated assembly.

H. Microtubule Assembly onto Centrosomes

Solutions of microtubule subunits were thawed immediately before an assembly experiment and ranged from 0.7 to 2.0 mg/ml in protein concentration. Many centrosome experiments contained flagellar axonemes as internal nucleation standards; These axonemes were also thawed just before an experiment and diluted with PEMG to an appropriate concentration as determined by electron microscopic examination. The proportions of subunits, flagellar seeds, and mitotic lysate were varied, but typically 500 μl of subunits and 20 μl of flagellar seeds were added to 200 μl of mitotic lysate in a 1.5-ml polypropylene tube and incubated at 30°C. Aliquots were removed at intervals thereafter. Polymerization reactions were quenched by the addition of glutaraldehyde (in polymerization buffer, final concentration 1%).

I. Purification of Tubulin and Tubulin–Colchicine Complex (TC)

Microtubule protein was separated into tubulin and microtubule-associated proteins by cation exchange chromatography (Murphy and Borisy, 1975). The peak fractions of tubulin were pooled and dialysed against 100 vol of PEMG for

1 hr (0–4°C). This purified tubulin (10–30 μM) was incubated with colchicine (100–200 μM) at 37°C for 30–45 min. The resultant tubulin–colchicine complexes (TC) were then separated from the excess colchicine by gel filtration on a G-25 coarse Sephadex column. The bed volume equaled six times the applied volume (which was usually 1 ml). The column flow was controlled by a Gilson Minipuls-2 peristaltic pump at 0.2 ml/min, and 0.5 ml fractions were collected with a Gilson Microfractionator. The column was monitored at 280 nm with an Altex analytical optical cell coupled with an Altex model 153 monitor. To show that all the unbound colchicine was removed, the peak fraction of TC was then run on a second G-25 column, and the fractions were analyzed at OD_{350} with a Gilford spectrophotometer as described by Sternlicht and Ringel (1979). By comparing the OD_{350} and OD_{280} readings of the peak TC fractions, we found that virtually all the dimer was complexed. The TC solutions were used immediately after preparation.

J. Electron Microscope and Computer Analyses of Microtubule Length

All samples were fixed with glutaraldehyde (final concentration 0.5–2%). When the samples contained a high enough concentration of axonemes, an aliquot (10 μl) was placed directly onto an ionized, carbon- and Formvar-coated, electron microscope grid. These grids were negatively stained with either 1% uranyl acetate of 1% phosphotungstic acid. However, for the centrosome experiments, the concentrations of flagella and centrosomes were too low for this procedure. Therefore, we sedimented these organizing centers onto grids by the method of Gould and Borisy (1977; Gould, 1975). For this procedure, we used lucite disks (25 mm diameter, 7 mm thick) with a hole (4 mm diameter) bored through the center as adapters to sediment the material. For a typical sedimentation, 100 μl solution was layered on top of the electron microscope grid and then centrifuged (3200 *g*, 5 min).

Data were recorded on 70-mm roll film (Kodak) using a Philips 300 electron microscope. The fixed microtubules were frequently curved, rendering it impractical to determine their length with a ruler. For this reason, and because of the large number of determinations required, we used a computer-assisted length determination system. The system contained a customized projector that cast the image of a 70-mm negative upon a ground glass screen. On top of this screen rested the arm of a Numonics digitizer (model 264s, Numonics Corp., Lansdale, PA) which was interfaced to a Hewlett-Packard 9825A computer. A program, developed by R. K. Littlewood, allowed the length of a line traced by the digitizer arm to be calculated and stored for subsequent statistical analyses. This apparatus was originally designed to determine the length of DNA molecules and has been used extensively for denaturation and heteroduplex mapping studies.

This length-histogram program has proved to be equally useful in analyzing the length distributions of microtubules. After magnification factors were calculated and entered, the data were recorded as lengths in micrometers. Different classes of microtubules, such as those grown onto the plus and minus ends of axonemes, were stored separately. For each data file, data were stored in one of ten categories (called plot codes). The data were then routinely expressed as the mean and standard deviation of the determinations in one plot code. These means were often plotted against time and/or concentration of protein. Best-fit lines were calculated by a least-mean-squares equation, which was also computer-assisted.

III. Results and Discussion

A typical flagellar axoneme is shown in Fig. 1a. Under our *in vitro* conditions, axonemes tend to fray at one end while remaining intact at the other. The frayed end was shown by Allen and Borisy (1974) to be the one originally distal to the cell body. Following our previous convention (Borisy, 1978), the frayed (distal) end is designated plus (+), while the intact (proximal) end is referred to as minus (−). When axonemes such as this one were incubated with greater than 0.4 mg/ml microtubule subunits under polymerizing conditions, net growth occurred at both ends of the axonemes (Fig. 1b). The flagellar microtubules apparently served as seeds upon which porcine brain subunits could polymerize, thereby forming hybrid structures. Extensions of the axonemal microtubules by the brain subunits were readily distinguishable from the axonemal microtubules themselves, and thus permitted us to analyze separately the polymerization properties of the two microtubule ends.

Several authors have called attention to the heterologous nature of the flagellar seed/brain subunit system as if the heterologous nature *per se* indicated a questionable degree of significance of this *in vitro* polymerization system. However, we have evidence that this heterologous system may, in fact, mimic the *in vivo* condition more closely than any other available *in vitro* experimental system. Our evidence comes from determinations of the number of protofilaments of self-nucleated and axoneme-nucleated microtubules (Scheele *et al.,* 1980).

The microtubules in mammalian brain have been reported to consist of precisely 13 protofilaments (Burton *et al.,* 1975; Pierson *et al.,* 1978; Scheele *et al.,* 1980). However, when microtubule protein is isolated from brain, and self-assembled *in vitro,* the resulting microtubules most often have 14 or 15 protofilaments, though the number may range from 11 to 17 (Pierson *et al.,* 1978). Figure 2a shows a gallery of electron micrographs of self-assembled brain microtubules stained with tannic acid and cross-sectioned to reveal protofilament

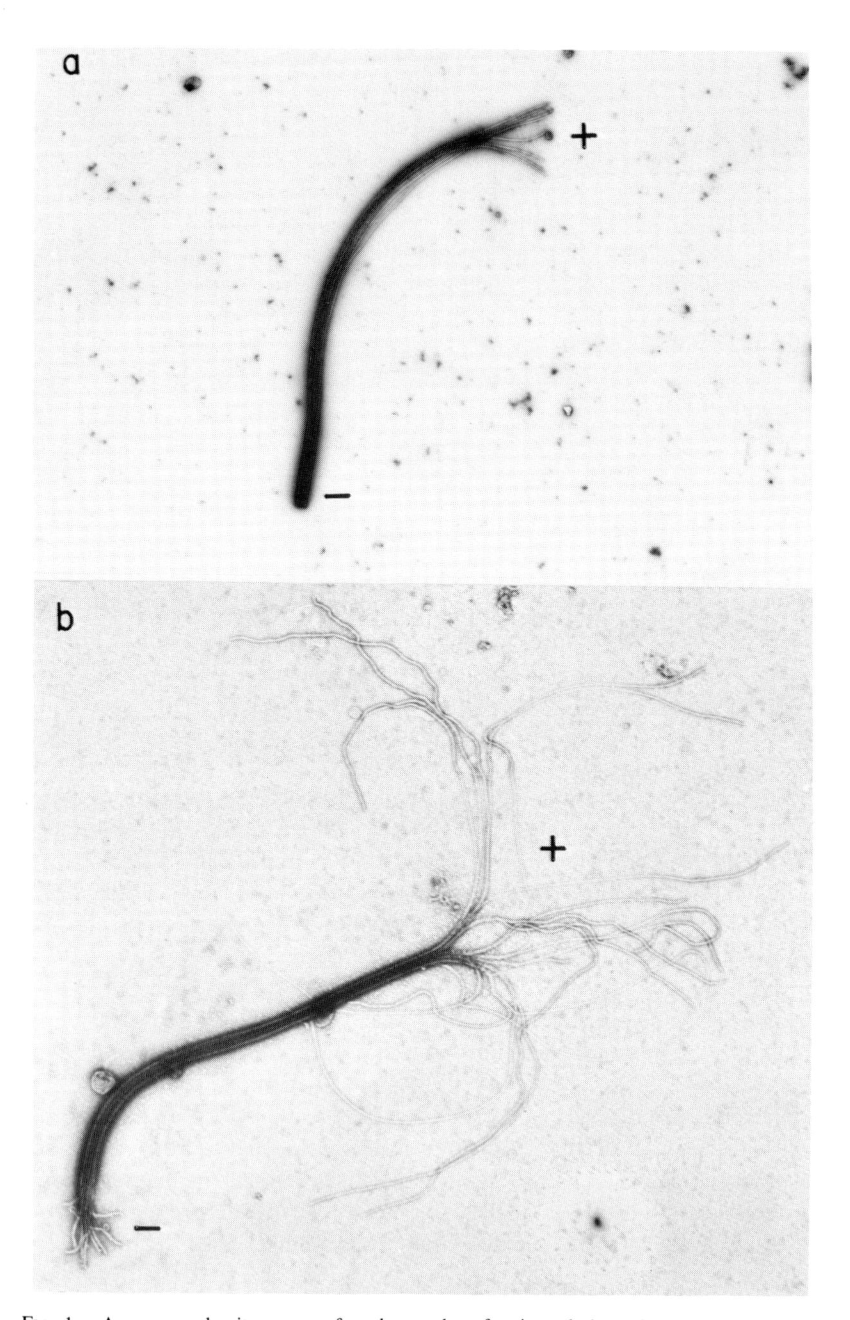

FIG. 1. Axoneme-subunit system for the study of microtubule polymerization. Flagellar axonemes were isolated from the green alga *Chlamydomonas reinhardtii*. Microtubule subunits were prepared from porcine brain tissue. (a) Without incubation with microtubule protein; (b) after incubation at 30°C with 0.60 mg/ml microtubule protein for 6 min. The microtubules grown at the plus end are longer than those at the minus end. 7300×.

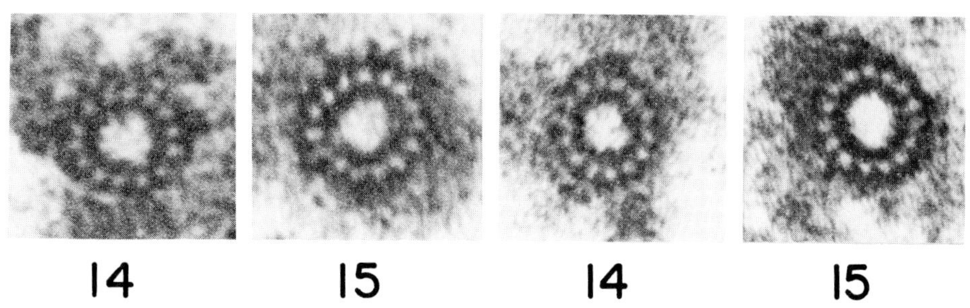

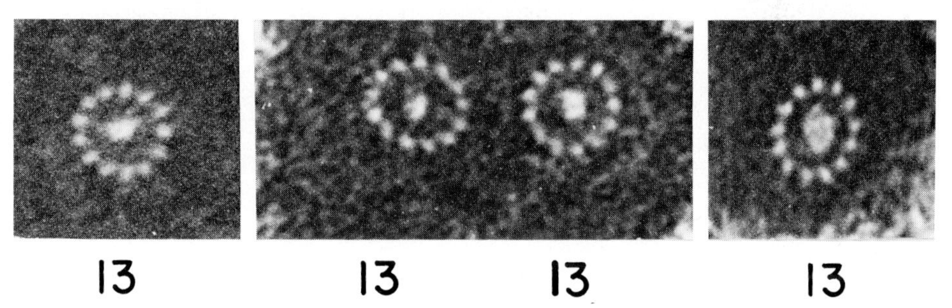

FIG. 2. Electron micrographs of microtubule cross sections. Porcine brain microtubule protein was purified by four cycles of polymerization and then either (a) self-assembled at 30°C, or (b) subjected to high-speed centrifugation and then polymerized onto the 13-protofilament axoneme templates. Both samples were fixed with glutaraldehyde, tannic acid stained, embedded in Araldite 502, and sectioned. The cross sections are representative of the microtubules examined. Self-nucleated microtubules usually contain 14 or 15 protofilaments, while the A subfiber nucleates microtubules that have 13 protofilaments. 486,000×.

composition. Out of 59 microtubules, we found 57 to consist of either 14 or 15 protofilaments, and only once did we encounter a 13-protofilament microtubule. Apparently, some component ensuring fidelity of microtubule structure is missing from the *in vitro* self-assembly system. As a consequence, most biochemical analyses that have been performed on *in vitro* polymerized microtubules have been studies on structures that differ detectably from the ones found *in vivo*.

On the other hand, we have determined the structure of axoneme-nucleated microtubules and found that the A subfiber and central pair microtubules, which consist of 13 protofilaments, nucleate the formation of microtubules that also have 13 protofilaments (Scheele *et al.*, 1980). Figure 2b is a gallery of cross sections of brain protein microtubules that have been nucleated by these 13-protofilament templates. Quantitative analysis showed that greater than 90% of these microtubules were composed of 13 protofilaments, and hence were indistinguishable from native microtubules.

Thus, the axoneme–subunit system, although heterologous in nature, possesses two important advantages: It lets one study the polymerization properties of 13-protofilament microtubules *in vitro,* and it allows separate analyses of growth and depolymerization at the two ends. We usually study growth and depolymerization as the time-dependent changes in the lengths of the plus and minus extensions. The length analysis of a typical growth experiment is shown in Fig. 3. The mean lengths and standard deviations of the microtubules nucleated by either end are plotted versus time. As is clear from the figure, the time points generally fall on straight lines that intersect at the origin. In this experiment the growth at the plus end occurred approximately three times as fast as at the minus end. However, this ratio of rates is not a constant, but rather a complex function of several variables, most notably the concentration of monomer. The concentration of monomer at which a microtubule end is in equilibrium is called the critical monomer concentration for that end. As shown in Fig. 4, which is the plot of rate of growth versus the concentration of monomer, the critical monomer concentrations for the plus and minus ends are not the same (under our buffer

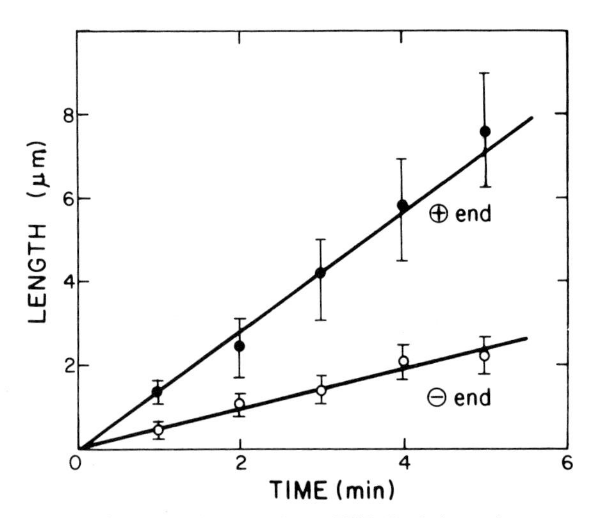

FIG. 3. Time course of microtubule elongation at 30°C. Each time point represents the mean and standard deviation of greater than 20 length determinations. Protein concentration was 0.66 mg/ml. Rates of elongation were 1.33 ± 0.17 μm/min at the plus end (solid circles) and 0.45 ± 0.07 μm/min at the minus end (open circles).

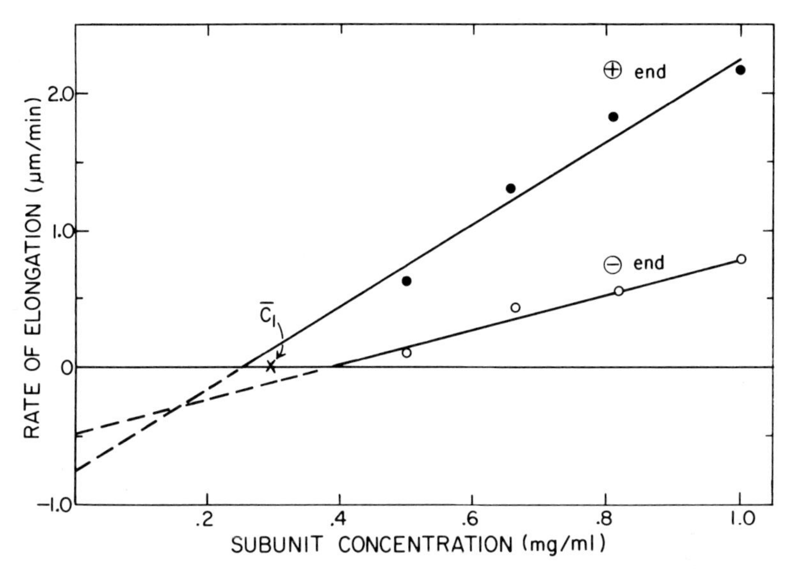

FIG. 4. Dependence of rate of elongation on subunit concentration. Each point represents the slope of a time course such as is shown in Fig. 3. The data were fit by a least squares linear regression analysis (solid lines) and the intercepts on the ordinate were obtained by extrapolation (dotted lines). The slope for growth at the plus end (solid circles) was 3.0 μm/min $(mg/ml)^{-1}$; the intercept was −0.77 μm/min. The slope for the minus end (open circles) was 1.3 μm/min $(mg/ml)^{-1}$; the intercept was −0.48 μm/min. The steady-state monomer concentration ($\bar{c}_1$ = 0.3 mg/ml) is denoted by *X*. The correlation coefficients for the plus and minus ends are 0.97 and 0.98, respectively.

conditions, at 30°C). Therefore, the ratio of plus to minus end lengths will vary with monomer concentration.

By plotting the rates of growth versus monomer concentration (see Fig. 4), we are also able to calculate the association and dissociation rate constants for addition and loss of subunits at the two ends of a microtubule. This information is not available from conventional solution biochemical studies that average the properties of the two ends. As shown previously by Johnson and Borisy (1977), the slope of this kind of plot provides the association rate constant, while the *Y*-intercept provides the dissociation rate constant. In the studies of Johnson and Borisy (1977) the reported rate constants were necessarily the sum of the values for the two ends. However, with the axoneme-subunit system, we were able to apply this rate analysis to each end separately (Bergen and Borisy, 1980a). Typical values of the rate constants obtained under our standard conditions at 30°C are given in Table I.

Several implications follow from these values. First, they are a chemical manifestation of the structural polarity intrinsic to the microtubule. The two ends may be distinguished operationally by their kinetic properties. Second, since the ratio of the on and off rate constants is different at each end, the critical monomer concentrations also differ. It is evident that the plus end, with the lower critical concentration, can elongate at a monomer concentration insufficient to support growth at the minus end. As a consequence, the steady-state monomer concentration is intermediate between the critical concentrations for the plus and minus ends. Third, this difference in critical concentrations allows us to infer that in the steady state a flux of subunits traverses the microtubules from their plus to minus ends. Therefore, in the steady state there is net growth at the plus ends and equal net loss at the minus ends. However, there are both association and dissociation reactions occurring at both ends, thereby reducing the efficiency of the flux. From our rate constants we have calculated that at 30°C there is a flux of 2 subunits per second, and the efficiency of this flux is only 0.07. This value of 0.07 signifies that only one of every 14 association and dissociation events actually contributes to the flux.

The polarity of depolymerization can also be studied with the axoneme-subunit system. This is done by incubating axonemes with subunits as before, and then diluting the solution isothermally with buffer to a concentration of subunits

TABLE I

HEAD-TO-TAIL POLYMERIZATION CONSTANTS[a]

k_2^+	$7.2 \pm 0.9 \times 10^6\ M^{-1}\ \text{sec}^{-1}$
k_2^-	$2.2 \pm 0.15 \times 10^6\ M^{-1}\ \text{sec}^{-1}$
k_{-1}^+	$16.0 \pm 5.0\ \text{sec}^{-1}$
k_{-1}^-	$7.0 \pm 0.9\ \text{sec}^{-1}$
c_1^+	$2.2 \pm 0.6\ \mu M$
c_1^-	$3.2 \pm 0.6\ \mu M$
$\bar{c}_1$	$2.5 \pm 0.6\ \mu M$
s	0.07 ± 0.04
ϕ	1.6 ± 0.8 subunits sec^{-1}

[a] The conditions for determination of the head-to-tail constants are given in Section II. The association constants for the (+) and (−) ends are denoted k_2^+, k_2^-; the dissociation constants are k_{-1}^+, k_{-1}^-. The critical monomer concentrations for each end, c_1^+, c_1^-, are given by the ratio of the rate constants for that end. The steady-state monomer concentration $\bar{c}_1$ is given by

$$\bar{c}_1 = (k_{-1}^+ + k_{-1}^-) / (k_2^+ + k_2^-)$$

The efficiency of head-to-tail polymerization s is calculated from

$$s = \left(\frac{k_2^+}{(k_2^+ + k_2^-)}\right) - \left(\frac{k_{-1}^+}{(k_{-1}^+ - k_{-1}^-)}\right)$$

The flux ϕ of subunits through the tubule is calculated from

$$\phi = s(\bar{c}_1)(k_2^+ + k_2^-) = s(k_{-1}^+ + k_{-1}^-)$$

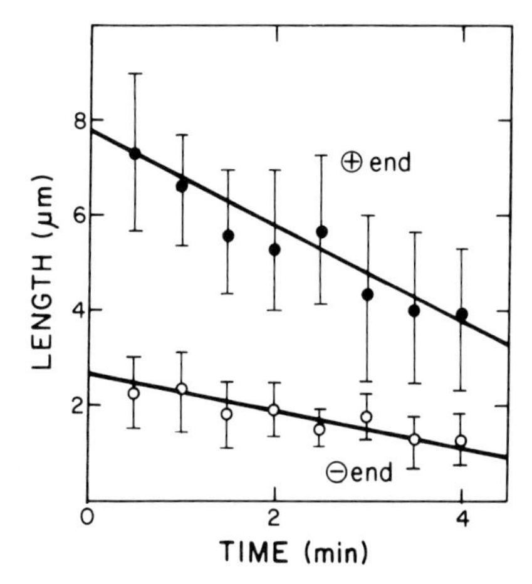

FIG. 5. Time course of dilution-induced depolymerization. Microtubules were polymerized onto axonemes as described previously. After growth had proceeded to an extent of 6–9 μm on the plus end and 2–3 μm on the minus end, an aliquot was removed and diluted 11-fold with buffer at 30°C, reducing the protein concentration to 0.06 mg/ml. This is noted as time zero. Then, at intervals of 30 sec, aliquots were removed and fixed to monitor depolymerization. Rates of depolymerization were 0.98 ± 0.10 μm/min for the plus end (solid circles) and 0.30 ± 0.06 μm/min for the minus end (open circles). The correlation coefficients for the plus and minus end lines are −0.96 and −0.88, respectively.

below the critical monomer concentration of either end. The microtubules of both ends then shorten at constant rates (Fig. 5), and as predicted by the straight-line extrapolation of the plot of rates versus concentrations (Fig. 4), the plus ends depolymerize faster than the minus ends.

After establishing the parameters of growth for the two microtubule ends, we were able to investigate the growth polarity of microtubules nucleated by other microtubule organizing centers. This was done by using flagellar axonemes as internal standards for the comparison with the growth occurring on the organizing centers.

To illustrate this type of analysis we selected the centrosomes of mitotic mammalian cells. Centrosomes isolated from Chinese hamster ovary cells have been shown to nucleate microtubules *in vitro* (Gould and Borisy, 1977). We incubated axonemes and these centrosomes together with subunits and the resulting polymerization was monitored by whole-mount electron microscopy of aliquots fixed at specified time intervals. At zero time there were no detectable microtubules on the isolated centrosomes, and therefore we attributed all growth to the addition of brain subunits.

Figure 6 shows the growth that occurred on an axoneme and centrosome in one of these experiments. As is evident by inspection, the microtubules nucleated by the centrosome are approximately the same length as those grown onto the plus

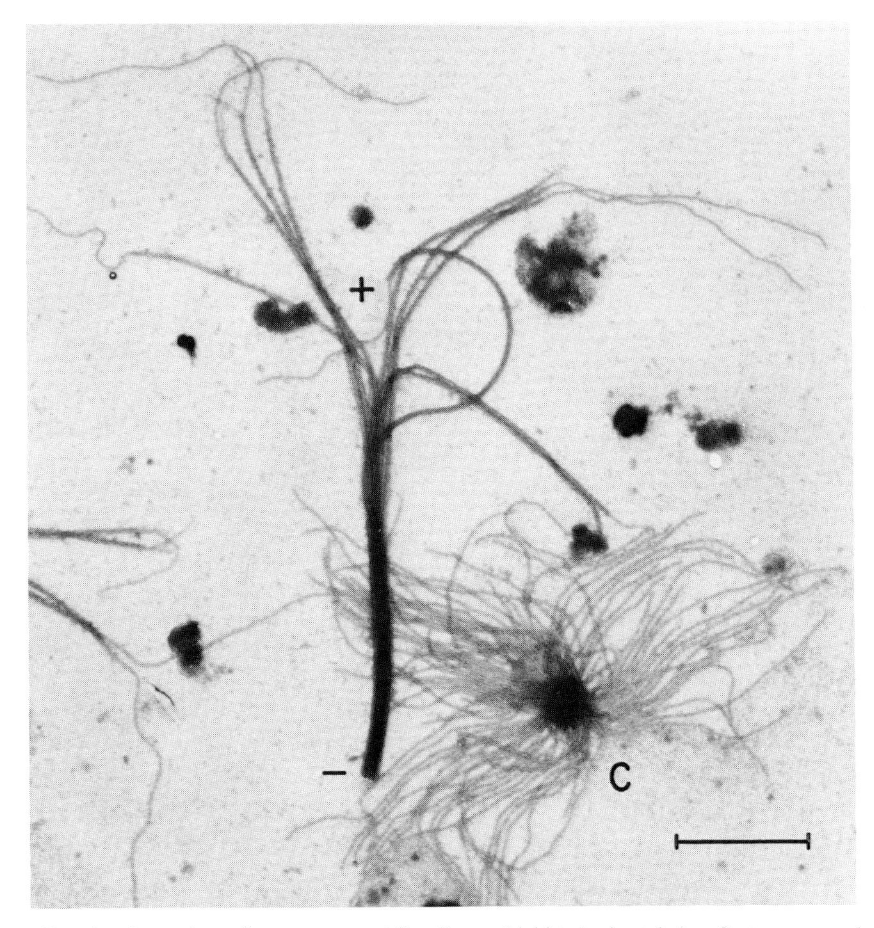

FIG. 6. Comparison of centrosome and flagellar seed-initiated microtubules. Centrosomes and axonemes were mixed together and incubated with 0.35 mg/ml microtubule subunits at 30°C for 8 min. Note that the added tubules at the plus ends of the flagellar seeds are approximately the same length as the centrosomal microtubules, and that under these conditions there was no minus end growth. C, centrosome. Bar = 2 μm.

end of the axoneme. Quantitative data showed that growth of the centrosomal microtubules was linear with time and occured at the rate characteristic of the plus end (Bergen *et al.*, 1980). From these and other experiments at higher concentrations, we were able to conclude that only one end of a centrosomal microtubule was able to add subunits and that end was the plus end. Isothermal dilution experiments resulting in depolymerization of centrosomal microtubules showed that loss of subunits also occurred only from the plus end. Thus by using axonemes as internal polarity standards, we were able to draw conclusions not attainable by other analytical methods.

Another application of the axoneme-subunit system is its ability to test the polarity of drug–tubulin interaction. To illustrate this application, we tested the effects of colchicine on growth and depolymerization of axoneme-nucleated microtubules. Colchicine has been shown to inhibit microtubule polymerization substoichiometrically by first forming a tubulin dimer–colchicine complex (TC), which is the active inhibitory species (Margolis and Wilson, 1978; Sternlicht and Ringel, 1979). By using the axoneme-subunit system we have demonstrated that at high concentrations of TC all four polymerization reactions were inhibited (Bergen and Borisy, 1980b). However, at lower concentrations, TC preferentially inhibited the depolymerization of the minus end. This result is shown in Fig. 7. In this experiment, microtubules were polymerized onto axonemes as before, and depolymerization was then induced by an isothermal dilution into buffer with or without added TC. The dashed lines show the characteristic time course of microtubule depolymerization as previously shown in Fig. 5. The data points and solid lines represent the changes of microtubule length in the presence of 1.2 μM TC: depolymerization of the plus end was slightly attenuated; however, depolymerization of the minus end was totally inhibited. Therefore, under appropriate conditions the inhibition of microtubule depolymerization by colchicine is seen to be a polar phenomenon. The power of the axoneme-subunit system for this type of analysis is that it allows direct identification of the end affected by the reagent. In a similar fashion this technique can be applied to analyze the effects of a wide range of solution conditions in terms of alterations of the four reactions that define the polymerization system.

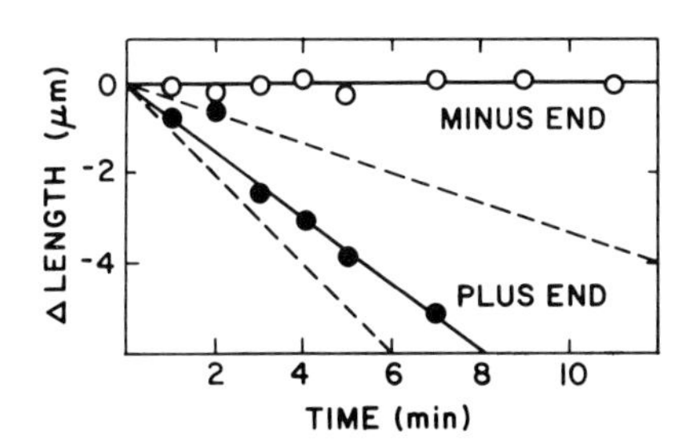

FIG. 7. Effect of tubulin–colchicine (TC) complex on microtubule depolymerization. The changes of length at the plus (closed circles) and minus (open circles) ends are plotted versus time after an 11-fold dilution into PEMG containing 1.2 μM tubulin–colchicine complex (solid lines) at 30°C. The dashed lines represent the corresponding length changes in the absence of TC. At $t = 10$ min there were very few plus end microtubules, while the minus end microtubules were still at the t_0 length. Each point corresponds to the mean length for that time.

IV. Perspective

The axoneme-subunit system is a relatively simple technique that is well suited to studying several aspects of microtubule polarity *in vitro*. The main feature of this system is the direct visualization of changes in microtubule length that can be attributed to the sum of addition and loss of subunits at a defined end. There is an absolute identification of the two ends. Another feature of this system is the small volume required for the studies, an experiment can be performed with 10–20 μl of sample. Also, one can study changes in microtubule length that occur relatively quickly, since the time points can be taken every 20 sec. Furthermore, the axoneme-subunit system is the only *in vitro* polymerization system in which the microtubules have the *in vivo* protofilament structure.

The axoneme-subunit system has so far been applied to three different problems:

1. Measurement of the association and dissociation rate constants for the two ends under defined solution conditions.
2. Comparison of the growth of microtubules nucleated by organizing centers with the plus and minus end standards.
3. Evaluation of the effects of microtubule drugs on the polymerization or depolymerization at the two ends.

Acknowledgments

This study was supported by National Institutes of Health (NIH) grant GM25062 to G. G. Borisy. L. G. Bergen was supported by an NIH predoctoral traineeship.

References

Allen, C., and Borisy, G. (1974). *J. Mol. Biol.* **90,** 381–402.
Amos, L. (1979). *In* "Microtubules" (K. Roberts and J. Hyams, eds.), pp. 1–64. Academic Press, New York.
Amos, L., and Klug, A. (1974). *J. Cell Sci.* **14,** 523–549.
Bergen, L., and Borisy, G. (1980a). *J. Cell Biol.* **84,** 141–150.
Bergen, L., and Borisy, G. (1980b). *J. Cell Biol.* **87,** 249a.
Bergen, L., Kuriyama, R., and Borisy, G. (1980). *J. Cell Biol.* **84,** 151–159.
Bernstein, E. (1960). *Science* **313,** 1528–1529.
Binder, L., Dentler, W., and Rosenbaum, J. (1975). *Proc. Natl. Acad. Sci. U.S.A.* **72,** 1122–1126.
Borisy, G. (1978). *J. Mol. Biol.* **124,** 565–570.
Borisy, G., Marcum, J., Olmsted, J., Murphy, D., and Johnson, K. (1975). *Ann. N.Y. Acad. Sci.* **253,** 107–132.
Burton, P., Hinkley, R., and Pierson, G. (1975). *J. Cell Biol.* **65,** 227–233.

Cote, R., Bergen, L., and Borisy, G. (1980). *In* "Microtubules and Microtubule Inhibitors" (M. De Brabander and J. De Mey, eds.), pp. 325-338. Elsevier/North-Holland, Amsterdam.
Crepeau, R., McEwen, B., Dykes, G., and Edelstein, S. (1977). *J. Mol. Biol.* **116,** 301-315.
Dentler, W. L., Granett, S., Witman, G. B., and Rosenbaum, J. L. (1974). *Proc. Natl. Acad. Sci. U.S.A.* **71,** 1710-1714.
Gould, R. (1975). *J. Cell Biol.* **65,** 65-74.
Gould, R., and Borisy, G. (1977). *J. Cell Biol.* **73,** 601-615.
Johnson, K., and Borisy, G. (1977). *J. Mol. Biol.* **117,** 1-31.
Lowry, O., Rosebrough, N., Farr, A., and Randall, R. (1951). *J. Biol. Chem.* **193,** 265-275.
Margolis, R., and Wilson, L. (1978). *Cell* **13,** 1-8.
Murphy, D., and Borisy, G. (1975). *Proc. Natl. Acad. Sci. U.S.A.* **72,** 2696-2700.
Olmsted, J., Marcum, J., Johnson, K., Allen, C., and Borisy, G. (1974). *J. Supramol. Struct.* **2,** 429-450.
Pierson, G., Burton, P., and Himes, R. (1978). *J. Cell Biol.* **78,** 223-228.
Reynolds, E. (1963). *J. Cell Biol.* **17,** 208-212.
Sager, R., and Granick, S. (1953). *Ann. N.Y. Acad. Sci.* **56,** 831-838.
Scheele, R., Bergen, L., and Borisy, G. (1980). *J. Cell Biol.* **87,** 249a.
Sternlicht, H., and Ringel, I. (1979). *J. Biol. Chem.* **254,** 10540-10550.
Wegner, A. (1976). *J. Mol. Biol.* **108,** 139-150.

Chapter 12

Dynein Decoration of Microtubules—Determination of Polarity

LEAH T. HAIMO

Division of Biology
California Institute of Technology
Pasadena, California

I. Introduction

The directional assembly of microtubules is a manifestation of their intrinsic structural polarity. For example, the axoneme assembles primarily at its distal tip during flagellar regeneration (Rosenbaum and Child, 1967; Witman, 1975). Moreover, assembly of tubulin *in vitro* occurs predominantly onto one end of microtubule pieces (Dentler *et al.*, 1974; Olmsted *et al.*, 1974) and the distal end of axonemes (Allen and Borisy, 1974; Binder *et al.*, 1975), basal bodies (Snell *et al.*, 1974), and centrioles (McGill and Brinkley, 1975; Gould and Borisy,

ISBN 0-12-564124-9

1977; Telzer and Rosenbaum, 1979). The preferred and nonpreferred ends for assembly of a microtubule have been termed "plus" and "minus," respectively (Borisy, 1978). While microtubules free in solution exhibit net polymerization and depolymerization at opposite ends (Margolis and Wilson, 1978), kinetic anaylsis indicates that those bound to chromosomes (Summers and Kirschner, 1979; Bergen *et al.*, 1980) or centrosomes (Bergen *et al.*, 1980) assemble and disassemble *in vitro* only at their plus end. Direct observations suggest that the site of this assembly is the free distal end of these microtubules (Heidemann *et al.*, 1980; Summers and Kirschner, 1979). Accordingly, it has been proposed that microtubules emanating from the mitotic poles *in vivo* are antiparallel to those projecting from chromosomes within the same half spindle (Bergen *et al.*, 1980). It is necessary to demonstrate, however, that the rate and direction of microtubule assembly *in vitro,* in fact, reflect the structural polarity of microtubules present within the cell.

We have recently developed techniques to determine directly the polarity of a microtubule independent of its assembly (Haimo *et al.*, 1979). By decorating *in vitro* assembled microtubules with flagellar dynein, structural polarity is revealed in negatively stained preparations and in both longitudinal and transverse thin sections of microtubules. Because dynein can bind to preexisting microtubules, an examination of the polarity of native microtubules can be conducted. Application of this technique *in situ* will demonstrate if adjacent microtubules possess the same or opposite polarity.

This chapter describes the methods used and results obtained when microtubules polymerized *in vitro* from calf brain tubulin are incubated with dynein from flagella of *Chlamydomonas*. Recent studies have demonstrated that dynein from *Tetrahymena* cilia will bind to the microtubules present in isolated spindles of *Spisula solidissima* revealing the half spindle to be composed of microtubules possessing a single polarity (Telzer and Haimo, 1980, 1981). Accordingly, the interaction of dynein with cytoplasmic microtubules can now be investigated and the polarity of the microtubules determined.

II. Methods

A. Dynein Preparation

Chlamydomonas reinhardii, strain 21 gr, is grown with continuous aeration on an alternating 14-hr light/10-hr dark cycle in medium I (Sager and Granick, 1953) supplemented with 22 m*M* sodium acetate and five times the normal phosphate concentration. After reaching a density of 3–5 $\times$ 10^6 cells/ml, 32 liters of cells are harvested in a DeLaval cream separator operated at half maximum speed. This procedure results in the deflagellation of the majority of cells, which are then resuspended in approximately 2 liters of medium I and aerated under lights for 1.5 hr to permit flagellar regeneration.

Cells are collected by centrifugation at 600 *g* for 5 min and washed by resuspension twice in 10 m*M* HEPES, pH 7.4, and centrifuged at 1550 *g* for 7 min. The packed cell volume is approximately 150 ml. Flagella are then isolated at 4°C by minor modifications of the method of Witman *et al.* (1978). The pellet of cells is resuspended in 10 m*M* HEPES, 5 m*M* $MgSO_4$, 1 m*M* dithiothreitol (DTT), and 4% sucrose at pH 7.4 and mixed on a magnetic stirrer while approximately 50 ml (or ⅓–½ the packed cell volume) of 50 m*M* dibucaine (Nupercaine HCl, Ciba-Geigy Co.) is added. Flagella detach in 1–2 min, and EGTA (ethyleneglycolbis(β-aminoethyl ether)-*N, n'*-tetraacetic acid) is then added to a final concentration of 0.5 m*M*. The suspension is centrifuged at 1300 *g* for 4 min in order to sediment cell bodies. The supernatant containing the flagella is centrifuged as before, and the resulting supernatant is then underlayed with 25% sucrose in 10 m*M* HEPES, 5 m*M* $MgSO_4$, pH 7.4, and centrifuged at 1550 *g* for 10 min to remove residual cell bodies. This step is then repeated if the flagella preparation is not sufficiently pure as indicated by light microscopy. Flagella are subsequently sedimented by centrifugation of the supernatant at 27,000 *g* for 10 min, demembranated by resuspension in HMDEK (10 m*M* HEPES, 5 m*M* $MgSO_4$, 1 m*M* DTT, 0.5 m*M* EDTA [ethylenediaminetetraacetic acid], 25 m*M* KCl, pH 7.4) containing 0.04% Nonidet P-40, and centrifuged again at 12,000 *g* for 10 min. The final pellet containing the axonemes is resuspended and washed twice with HMDEK.

The dynein preparation is obtained by extracting axonemes in high salt by a modification of the method of Gibbons and Gibbons (1973). Approximately 20 mg of axonemes are resuspended in 2–3 ml of 0.6 *M* NaCl, 10 m*M* HEPES, 4 m*M* $MgSO_4$, 0.2 m*M* EDTA, and 1 m*M* DTT at pH 7.4. After 15 min at 4°C the axonemes are centrifuged at 27,000 *g* for 15 min and the supernatant containing dynein is retained. The pellet is resuspended, reextracted, and centrifuged as above. The two supernatants containing the dynein are combined and centrifuged at 27,000 *g* for 15 min. The resulting supernatant is then desalted by centrifugation through Sephadex G-25 (Neal and Florini, 1973) into 50 m*M* PIPES (piperazine-*N,N'*-bis[2-ethanesulfonic acid]), 1 m*M* EGTA, 0.5 m*M* $MgSO_4$, pH 6.9, and concentrated to 3–6 mg/ml by placing Aquacide outside a dialysis membrane containing the dynein solution. The dynein preparation is then clarified by centrifugation at 100,000 *g* for 1 hr and can be stored up to one week at 4°C. This procedure yields approximately 5 mg of a dynein preparation.

B. Microtubule Preparation

Tubulin is obtained from calf brain by cycles of assembly/disassembly according to modification (Sloboda *et al.*, 1976) of the method of Shelanski *et al.*

(1973). Brains are suspended in 0.1 *M* PIPES, 2 m*M* EGTA, 1 mM $MgSO_4$, 4 *M* glycerol, pH 6.9, and homogenized at 4°C in a Sorval Omnimixer for 50 sec at setting #3 and for 10 sec at setting #10. The homogenate is centrifuged at 10,000 *g* for 15 min at 4°C, and the supernatant is centrifuged at 130,000 *g* for 75 min at 4°C. The resulting supernatant is then adjusted to 1 m*M* GTP and incubated at 37°C for 45 min to promote microtubule assembly. Microtubules are sedimented at 130,000 *g* for 60 min at 30°C, the pellet resuspended in 0.1 *M* PIPES, 2 m*M* EGTA, 1 m*M* $MgSO_4$, 1 m*M* GTP, pH 6.9, at 4°C with a Dounce homogenizer and incubated at 0°C for 30 min to depolymerize microtubules. Residual aggregates are then removed by centrifugation of the solution at 130,000 *g* for 45 min at 4°C. The supernatant containing 1×-cycled tubulin is then adjusted to 4 *M* glycerol and stored frozen at −80°C.

Aliquots of 1×-cycled tubulin are thawed and then warmed to 37°C for 30 min. Polymerized microtubules are collected by centrifugation at 100,000 *g* for 1 hr and resuspended in a Dounce homogenizer with cold column buffer (CB) containing 50 m*M* PIPES, 0.5 m*M* $MgSO_4$, 1 m*M* EGTA, and 0.1 m*M* GTP at pH 6.9. After incubation at 0°C for 30 min the twice cycled (2×) tubulin is clarified by centrifugation at 100,000 *g* for 45 min.

By phosphocellulose chromatography of the 2× tubulin 6 S tubulin dimers are obtained (Weingarten *et al.*, 1975; Sloboda *et al.*, 1976). Approximately 300–400 mg protein are loaded on a 22 × 3-cm phosphocellulose column equilibrated in CB. Purified 6 S tubulin, collected in the void volume, is adjusted to 1 m*M* $MgSO_4$ (Williams and Detrich, 1979) and concentrated to 5–7 ml (20–25 mg/ml) by ultrafiltration using an Amicon PM-30 membrane. The 6 S tubulin is subsequently desalted into CB, clarified by centrifugation at 100,000 *g* for 1 hr, and stored as frozen droplets in liquid N_2.

C. Decoration of Microtubules with Dynein

Decoration of microtubules with dynein is accomplished by mixing the dynein and tubulin preparations at 0°C and then raising the temperature. In order to avoid possible activation of dynein's latent ATPase activity at elevated temperature (Gibbons and Fronk, 1979) incubations are performed at 30°C rather than 37°C. Alternatively, the tubulin can first be assembled and subsequently incubated with dynein. In studies where it is necessary to ensure that dynein binds only to preformed microtubules, microtubules are assembled, and then 10^{-5} *M* colchicine is added to prevent additional microtubule assembly that would occur upon dynein addition. Glycerol can also be included in the tubulin preparation to stabilize microtubules. The tubulin and dynein preparations are incubated at 30°C for approximately 45 min in PME (50 m*M* PIPES, 2.5 m*M* $MgSO_4$, 1 m*M* EGTA, pH 6.9) containing 1 m*M* GTP. Both microtubule assembly and dynein binding to the microtubules can be readily observed by dark-field microscopy and by monitoring absorbance changes at 350 nm. A protein ratio of the dynein–tubulin preparations of approximately 1:3 results in extensive dynein arm decoration of the microtubules.

To obtain purified preparations of microtubules suitable for electron microscopy, gel electrophoresis, or ATPase analysis it is necessary to separate the microtubules from unbound dynein and unpolymerized tubulin by sedimentation into a sucrose gradient. A microtubule preparation is layered on a discontinuous gradient consisting of 0.5 ml each of 40–50–60–70% and 2 ml of 80% sucrose in PME. The gradients are centrifuged in an SW 50.1 rotor (Beckman) at 35,000 *g* for 30 min at 30°C (Telzer and Rosenbaum, 1979). Microtubules containing bound dynein band as a tight mass at the 60–70% sucrose interface. On the other hand, microtubules assembled from purified 6 S tubulin alone form a loose band at the 70–80% sucrose interface. Microtubules removed from the sucrose gradients are negatively stained for electron microscopy or are diluted 1:1 with PME at 30°C and centrifuged at 27,000 *g* for 20 min at 30°C. The resulting pellets are then fixed and embedded for ultrastructural analysis or are resuspended for ATPase assays or gel electrophoresis.

D. Microscopy

1. Dark-field Microscopy

Dark-field microscopy of *in vitro* assembled microtubules is performed as described by MacNab (1976). The light source consists of a 500-W xenon arc (Oriel Corp.) fitted with a UV, yellow, and heat filter. Observations are made with a Zeiss microscope equipped with a universal mirror, an oil immersion dark-field Ultracondenser (NA 1.2/1.4), and a Neofluar 40×/0.75 objective. For photography, microtubules can be suspended in 20% sucrose in PME to decrease Brownian motion. Photographs are taken using Tri-X film exposed for 1/15–1 sec and developed with Acufine Acu-I developer at ASA 1200.

2. Electron Microscopy

a. Negatively Stained Preparations. Microtubules are fixed with 1% glutaraldehyde in PME prior to negatively staining with phosphotungstic acid. Microtubules obtained from sucrose gradients are diluted with PME subsequent to fixation because high sucrose concentrations interfere with optimal staining. A drop of material is incubated on a grid for about 1 min and then drawn off with filter paper. The grid is then stained by inverting it for several seconds each on a drop of 0.02% cytochrome c in 1% *n*-amyl alcohol and on a drop of neutralized 4% phosphotungstic acid in 0.4% sucrose (Rosenbaum *et al.*, 1975).

b. Thin Sectioned Preparations. Material is prepared for ultrastructural analysis according to a modification (Kim *et al.*, 1979) of the method of Begg *et al.* (1978). The use of tannic acid greatly enhances the appearance of dynein arms. Microtubule pellets are fixed for at least 2 hr at 20°C with 1% glutaraldehyde containing 1% tannic acid (Sigma) buffered at pH 7.0 with 10 m*M* sodium phosphate. The pellets are rinsed with several changes of the same buffer and then postfixed at 4°C for ½ hr with 1% OsO_4 buffered at pH 7.0 with 10 m*M* sodium phosphate. The pellets are rinsed with H_2O and stained en bloc with 1% aqueous uranyl acetate for ½ hr. The material is subsequently dehydrated through a graded ethanol series and embedded in Epon 812. Sections are stained for 30–60 min with 1% uranyl acetate in 12.5% methanol, 35% ethanol, then for 30 min in 0.4% lead citrate in 0.1 *M* NaOH. Microscopy is performed using a Philips 201 electron microscope operated at 60 or 80 kV.

III. Analysis of Dynein Binding to Microtubules

The simplest and most rapid method for determining that dynein has bound to microtubules is observation of the preparations by dark-field microscopy. Using a 500-Watt xenon arc as a light source, the light scattered by microtubules can be readily discerned. Microtubules assembled from purified 6 S tubulin alone (Fig. 1a) are quite distinct from those assembled in the presence of dynein (Fig. 1b). The latter are extremely bright and appear clustered into large, beaded cables. These aggregates are, in fact, groups of microtubules cross-bridged together by dynein arms that have bound to them (Haimo *et al.*, 1979). Less than an hour incubation with dynein is sufficient time for most microtubules in a preparation to become cross-bridged together into large bundles.

The dynein cross-bridges can be relaxed by ATP addition. However, ATP causes microtubules containing dynein to depolymerize rapidly. To overcome this depolymerization so that the effect of ATP on cross-bridging alone can be analyzed, it is necessary to add an ATP generating system to the microtubules. 16 m*M* creatine phosphate and 0.1 mg/ml creatine phosphokinase will delay microtubule depolymerization caused by ATP addition for approximately 20 min (Haimo *et al.*, 1979). ADP also causes depolymerization of microtubules containing dynein, but neither ATP nor ADP causes depolymerization of microtubules assembled from purified 6 S tubulin alone. It is not clear why microtubules containing dynein depolymerize as a result of ATP addition nor what implication this might have for microtubules interacting with dynein within the cell.

Addition of 1 m*M* ATP and the ATP generating system to cross-bridged microtubules results in the dispersal of the microtubule aggregates (Fig. 1c). The

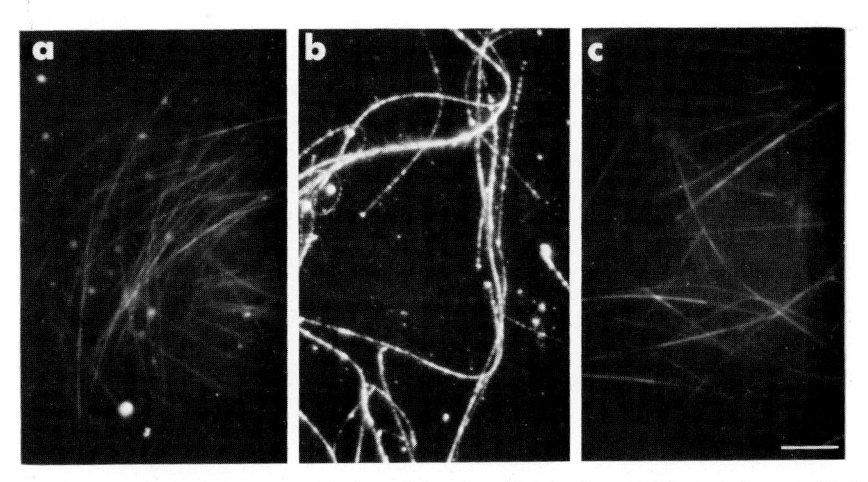

FIG. 1. Dark-field micrographs of microtubules assembled *in vitro*. (a) Microtubules assembled from 6 S tubulin alone. (b) Microtubules assembled from 6 S tubulin and *Chlamydomonas* flagellar dynein. (c) Microtubules assembled as in (b) and then incubated with 1 m*M* ATP and an ATP generating system. Bar = 5 μm.

light intensity varies along the lengths of some of these dispersed microtubules suggesting that they may still be partially associated. The observation that flagellar dynein is capable of both cross-bridging *in vitro* assembled microtubules and releasing them upon ATP addition indicates that the B-subfiber binding end of the dynein arm does not require the specific B-subfiber surface lattice, which is distinct from that of assembled microtubules (Amos *et al.*, 1976), in order to interact with a microtubule. It will be important to determine if these assembled microtubules are actually sliding along each other or, rather, are simply separating from each other. ATP-induced sliding of these microtubules would be analogous to ATP-induced outer doublet sliding of trypsin-treated axonemes (Summers and Gibbons, 1971) and might have substantial implications for understanding the mechanism by which movements occur *in vivo* in association with cytoplasmic microtubules.

The fact that dynein causes microtubules to aggregate permits an immediate determination by dark-field microscopy that dynein has bound to them. Indeed, one of the more remarkable properties of the interaction of flagellar dynein with cytoplasmic microtubules is that cross-bridging occurs spontaneously between assembled microtubules containing dynein.

Analysis of these preparations by electron microscopy provides additional information about the extent of microtubule cross-bridging. When microtubules are present at relatively high concentration, most microtubules that possess dynein arms become cross-bridged. Two groups of microtubules cross-bridged by dynein are illustrated in Fig. 2. Present in the same micrograph are several

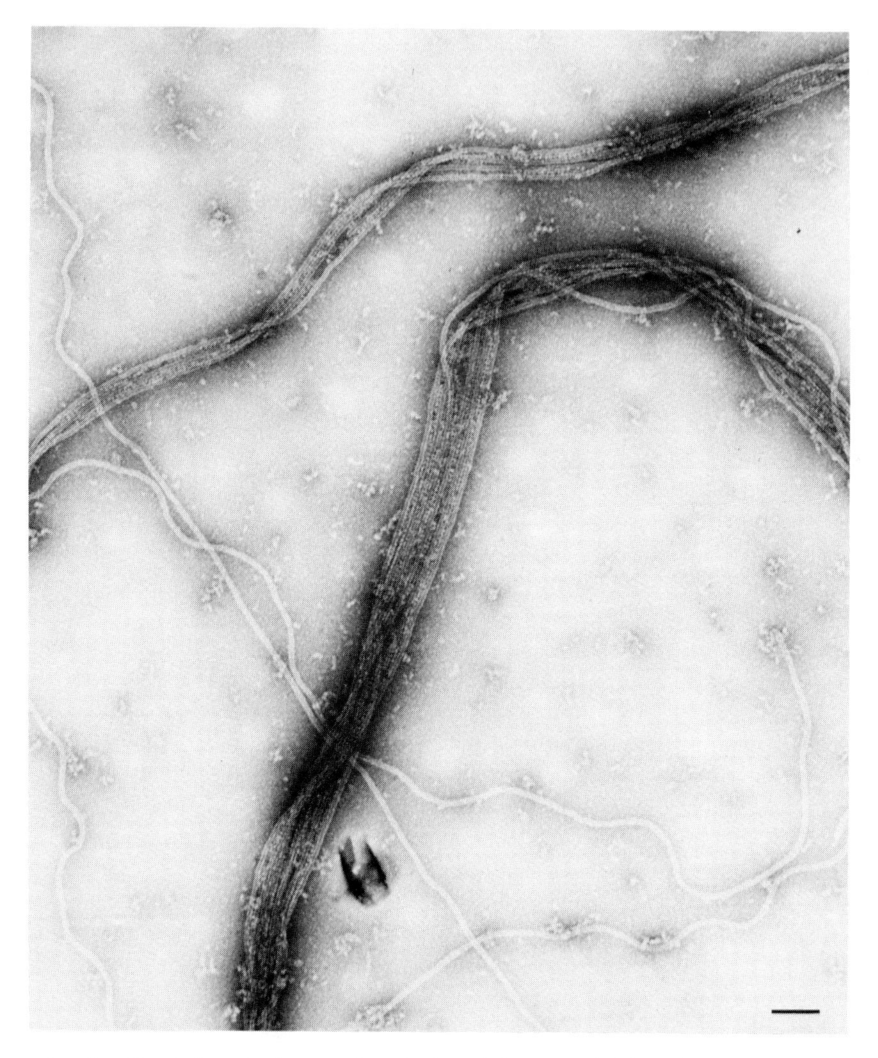

FIG. 2. Negatively stained preparation of microtubules assembled from 6 S brain tubulin and flagellar dynein. Microtubules are cross-bridged into groups of four and six. Microtubules that are not cross-bridged possess few bound dynein arms. Bar = 0.2 μm.

microtubules that are not cross-bridged and have very few bound dynein arms. Thin sections of these preparations, viewed at low magnification, demonstrate that the majority of microtubules are cross-bridged (Fig. 3). The presence of GTP in the polymerization solution does not interfere with the formation or maintenance of these cross-bridges. Moreover, microtubules are often cross-bridged along their entire lengths (Fig. 4).

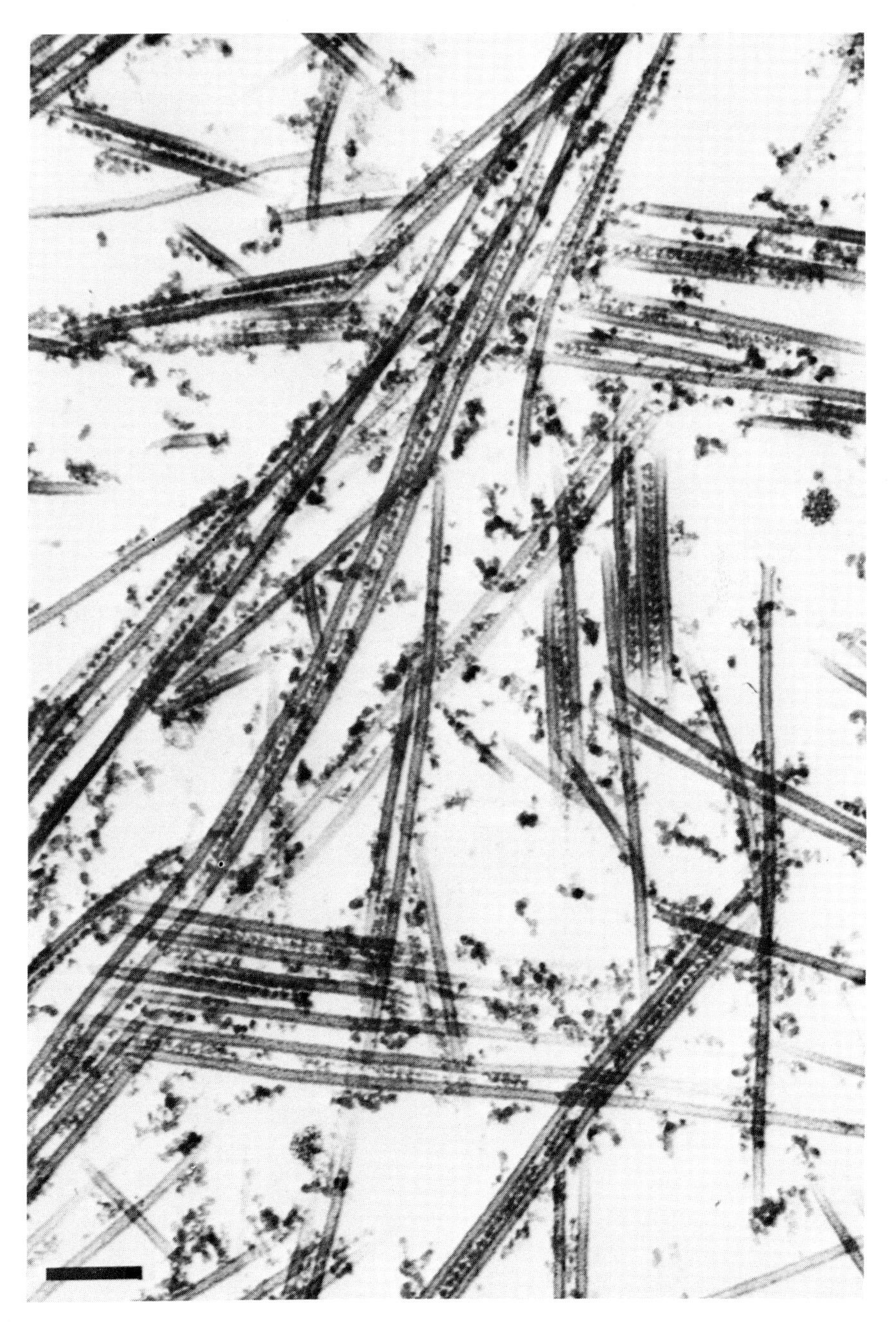

FIG. 3. Electron micrograph of a section through a pellet of microtubules assembled from 6 S tubulin and dynein. The majority of microtubules are cross-bridged. Bar = 0.2 μm.

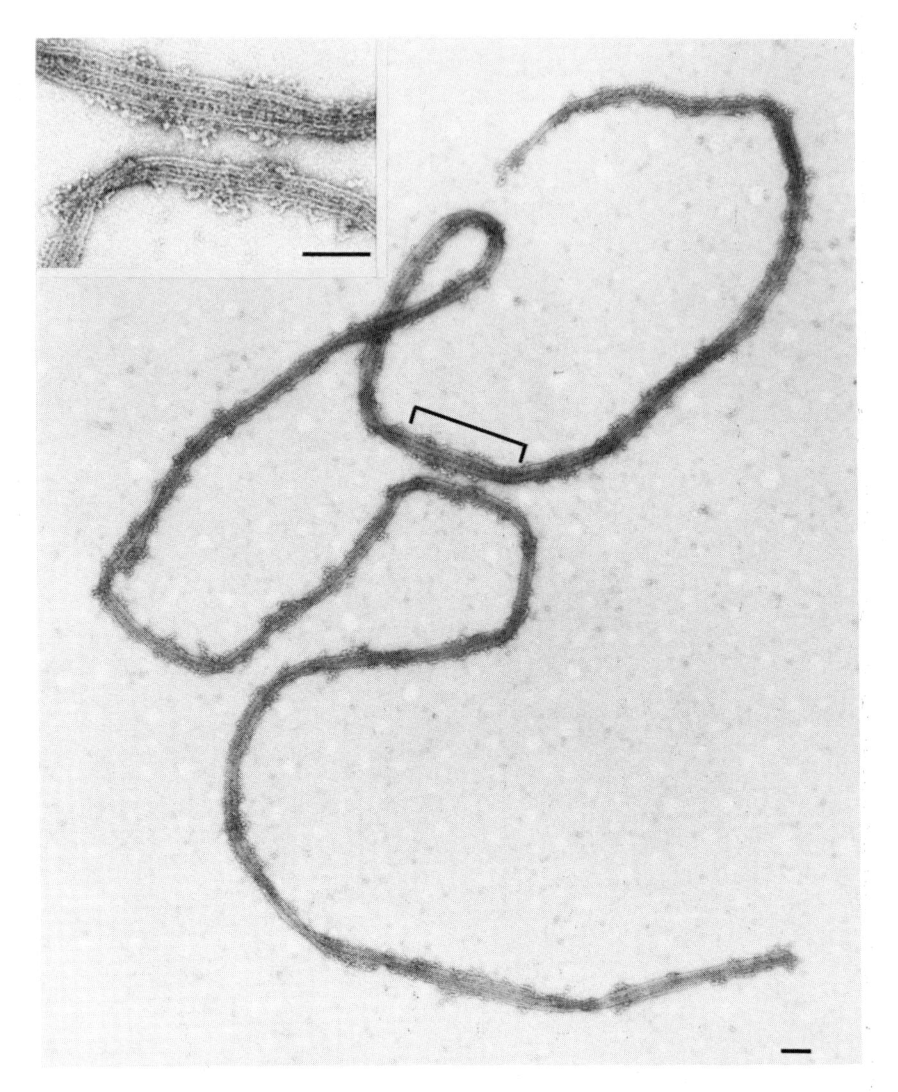

FIG. 4. Negatively stained preparation showing microtubules cross-bridged along their entire lengths by dynein. Bracket indicates region shown enlarged in inset. Bar = 0.2 μm.

These observations, together with studies showing that reactivating axonemes enter rigor when the ATP concentration of the reactivation solution is rapidly lowered (Gibbons and Gibbons, 1974; Gibbons, 1975) and that individual outer doublets containing dynein aggregate into large groups upon ATP depletion (Takahashi and Tonomura, 1978), suggest that the interaction of dynein with brain microtubules and axonemal outer doublets is similar to that of myosin with actin in that rigor complexes form spontaneously in the absence of ATP.

Previous studies based on electron microscopic observations of axonemes viewed in transverse sections suggest that the ability of dynein to cross-bridge adjacent outer doublet microtubules within the axoneme is dependent on particu-

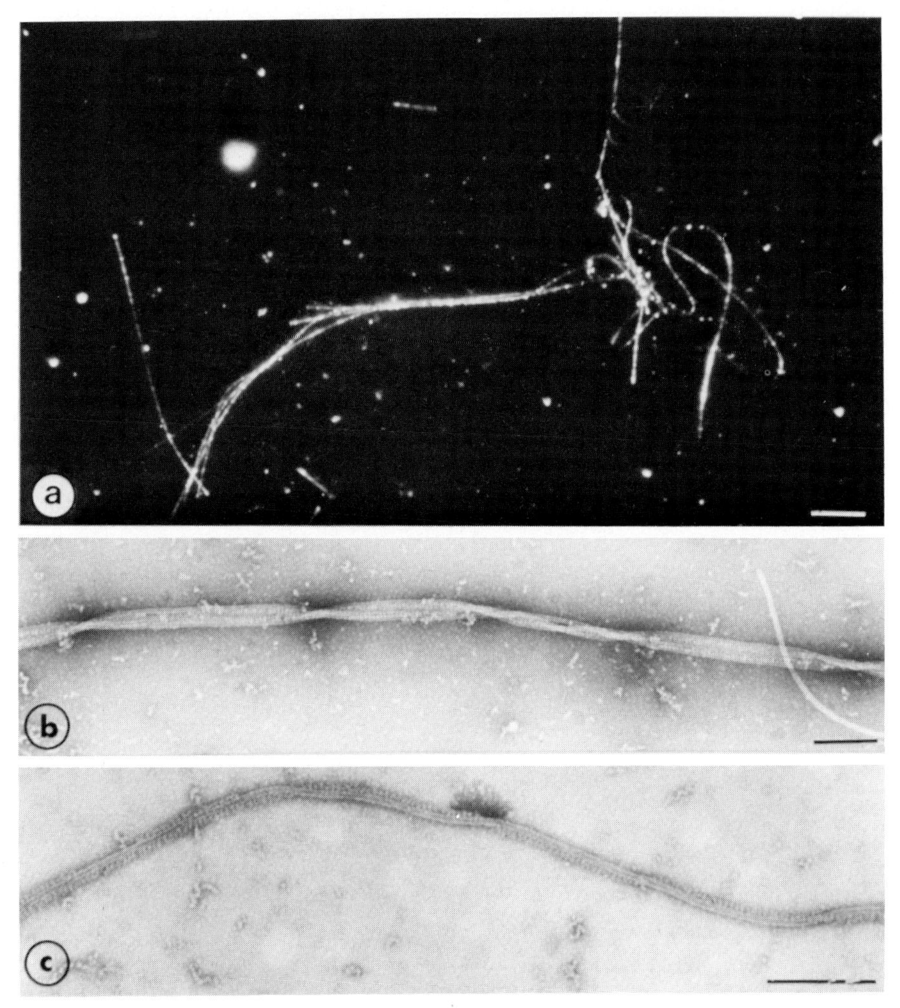

FIG. 5. (a) Dark-field and (b and c) electron micrographs of microtubules polymerized from 6 S tubulin and dynein. (a) Microtubules appear beaded by dark-field microscopy. (b) Cross-bridged microtubules twist over each other with a period of approximately 1 μm. (c) After ATP addition to cross-bridged microtubules, dynein arms can be visualized twisting around the individual microtubules. The twist has a period of 0.6–1 μm. Bar = 5 μm (a), 0.5 μm (b and c).

lar buffers and cations (Zanetti *et al.,* 1979). The fact that the *in vitro* assembled brain microtubules in the present study are cross-bridged despite the observation that the dynein arms do not span the entire distance between any two bridged microtubules (see Fig. 7) suggests that fixation artifacts may be responsible for the B-subfiber binding end of the dynein arm not directly attaching to the adjacent microtubule. Thus, caution should be used in interpreting micrographs of axonemes in which the attached or relaxed state of the dynein arm is being studied.

Microtubules cross-bridged by flagellar dynein have a strong tendency to twist over each other. These twists often have a regular period of approximately 1 μm (Fig. 5b). The beaded appearance of microtubules containing dynein when viewed by dark-field microscopy (Fig. 1b, 5a) is probably a reflection of twisting, cross-bridged microtubules. When the bridged microtubules are separated from each other by addition of ATP, it can be observed that the dynein arms, themselves, twist around the microtubules and do so with a helical pitch similar to the twisting of cross-bridged microtubules (Fig. 5c). In contrast, dynein arms on the A subfiber of axonemes are present in a straight, linear array. *In vitro* assembled microtubules often have more than 13 protofilaments (Binder and Rosenbaum, 1978; Pierson *et al.,* 1978; Kim *et al.,* 1979), a property that would affect both the structure of the microtubule surface and the lattice of the bound dynein arms. It is possible that the helical twist of dynein arms on assembled microtubules will be useful in providing information about the surface lattice of a microtubule.

IV. Analysis of Microtubule Polarity Using Bound Flagellar Dynein

Dynein arms present on assembled microtubules provide a cytological tool by which microtubule polarity can be directly resolved. In negatively stained preparations or longitudinal thin sections, microtubule polarity is defined by the direction of the tilt of the dynein arms, and in transverse sections the asymmetric cross-sectional orientation of the dynein arm serves as a marker for defining microtubule polarity (Haimo *et al.,* 1979).

A. Polarity of Microtubules Containing Dynein Viewed in Longitudinal Sections

In longitudinal thin sections dynein arms appear to be composed of a large, rhombous-shaped head attached to the microtubule by a thin stalk (Fig. 6; see

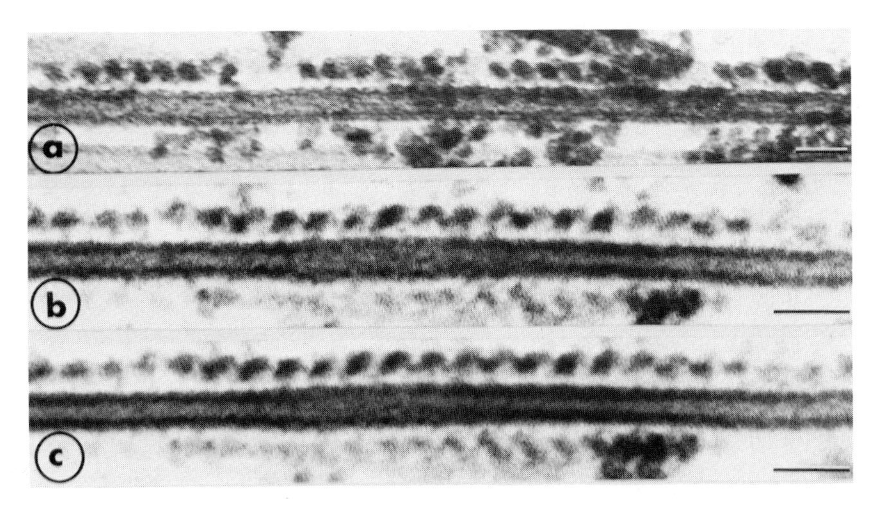

FIG. 6. Longitudinal sections through microtubules polymerized from 6 S tubulin and dynein showing that all dynein arms on a given microtubule tilt in the same direction. (a) Microtubule contains several clusters of dynein arms, all of which tilt toward the left. (b) Microtubule contains two rows of dynein arms. Arms in each row tilt toward the right. (c) One-step linear translation of micrograph shown in (b). The picture was photographed twice and translated a distance equivalent to 24 nm between exposures. The dynein arms have been reinforced, and the tilt of those in the bottom as well as the top row can be more clearly discerned. Bar = 50 nm.

also Fig. 3 in Haimo *et al.*, 1979). The head tilts with respect to the wall of the microtubule, and the angle subtended by lines drawn bisecting the head and along the length of the microtubule is approximately 55° (Haimo *et al.*, 1979).

Examination of microtubules assembled *in vitro* from 6 S brain tubulin and *Chlamydomonas* dynein demonstrate that all dynein arms on a given microtubule tilt in the same direction (Fig. 6). Figure 6a is a longitudinal section through a microtubule containing several clusters of dynein arms, all of which uniformly tilt toward the left. Moreover, when a microtubule contains more than one row of dynein arms as is illustrated in Fig. 6b (also see Fig. 5c), the dynein arms in both rows tilt in the same direction. The bottom row of arms is not in the plane of section but is reinforced when the micrograph is linearly translated (Fig. 6c).

In *Chlamydomonas,* the dynein arms present on intact axonemes tilt toward the distal tip of the flagellum in the absence of ATP (Allen and Borisy, 1974; Witman and Minervini, 1979). Because flagellar microtubules preferentially assemble at their distal tips (Rosenbaum and Child, 1967; Witman, 1975), it is assumed that dynein arms on assembled microtubules tilt in the direction of preferred or "plus" assembly. Thus, microtubule assembly would preferentially occur onto the left end of the microtubule in Fig. 6a and onto the right end of the microtubule in Fig. 6b.

B. Polarity of Microtubules Containing Dynein Viewed in Transverse Sections

The dynein arms that bind to microtubules assembled *in vitro* appear identical to the outer row of dynein arms on *Chlamydomonas* axonemes (Haimo *et al.*, 1979). In transverse section, the arm has a very distinctive, asymmetric ultrastructure; a hook or spur projects from one side of the arm (see Fig. 7). When a given cross section of a microtubule contains more than one dynein arm, the arms are oriented so that these hooks all point either clockwise or counterclockwise around the microtubule. Thus, microtubule polarity can be defined.

Figure 7 illustrates transverse sections of microtubules that contain two or more bound dynein arms. In Fig. 7a, one microtubule has three dynein arms, all oriented counterclockwise around the microtubule. Another microtubule in the same micrograph contains two dynein arms, both pointing clockwise around that microtubule. Figures 7b and c each contain a microtubule possessing two dynein arms that point counterclockwise, while a microtubule possessing two clockwise oriented dynein arms is illustrated in Fig. 7d.

Gibbons and Grimstone (1960) have demonstrated that when an axoneme is

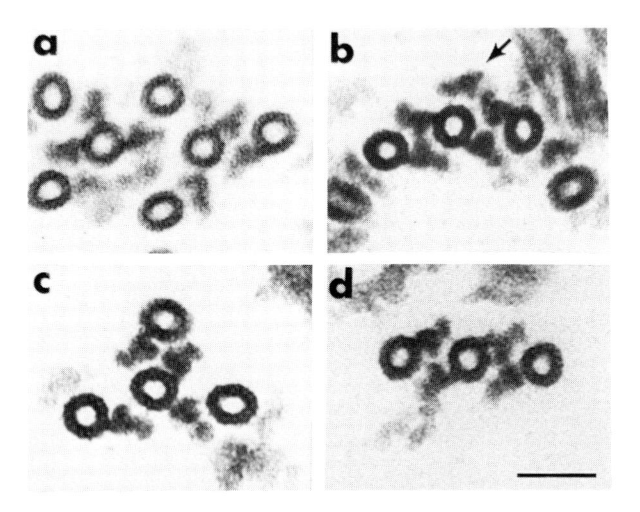

FIG. 7. Transverse sections of microtubules containing bound dynein showing that arms on a given microtubule are oriented in the same direction and that cross-bridged microtubules possess the same polarity. (a) One microtubule possesses three dynein arms, all oriented counterclockwise. A second microtubule possesses two dynein arms, both oriented clockwise. (b) The center microtubule possesses two dynein arms both oriented counterclockwise as are the other dynein arms within the cross-bridged group. A dynein arm, bridged to the center microtubule but containing no microtubule at what would normally comprise its A-subfiber binding site within the axoneme (see arrow), also exhibits a counterclockwise orientation. (c and d) One microtubule in each micrograph possesses two dynein arms, both pointed in the same direction. All dynein arms within the cross-bridged group of microtubules are oriented counterclockwise (c) or clockwise (d). Bar = 50 nm.

viewed in an orientation so that the dynein arms point clockwise around the axoneme, the observer is looking toward its distal tip. Extending this observation, the hook on the outer dynein arm of *Chlamydomonas* points clockwise around each outer doublet microtubule when the axoneme is viewed in this clockwise orientation (see Fig. 4 in Haimo *et al.*, 1979). Accordingly, it is proposed that when dynein arms point clockwise around microtubules assembled *in vitro* the preferred end of microtubule assembly is away from the observer. Conversely, microtubules preferentially assembling toward the observer would be decorated with dynein arms pointing counterclockwise around the microtubule. The orientation of dynein arms can be used to determine the polarity of those microtubules illustrated in Fig. 7.

C. Polarity of Cross-Bridged Microtubules

An analysis of microtubule polarity reveals that only microtubules possessing the same polarity are cross-bridged by flagellar dynein. Dynein arms within a given group of cross-bridged microtubules are all oriented in the same direction (Fig. 7). Moreover, in Fig. 7b a dynein arm appears to be cross-bridged to the center microtubule but this arm contains no microtubule at what would normally comprise its A-subfiber binding site on the axoneme (see arrow). Yet, if an imaginary microtubule were drawn in at this site, the dynein arm would be pointed counterclockwise around that microtubule, thus exhibiting the same orientation as the other dynein arms within this group of cross-bridged microtubules. Accordingly, the cross-bridging end of the dynein arm is, in itself, capable of recognizing microtubule polarity.

Both the A and B subfibers within the axoneme should possess the same polarity as it is probable that they both assemble in the same direction. The fact that flagellar dynein binds to and cross-bridges cytoplasmic microtubules possessing the same polarity is evidence that the A and B subfibers do, indeed, possess the same polarity.

Recently, Dr. B. R. Telzer and I have studied the interaction of *Tetrahymena* dynein with microtubules in isolated meiotic spindles of *Spisula solidissima*. In transverse sections, the spindle microtubules are surrounded by four to seven dynein arms, all of which bind to the microtubules by the end of the arm that would normally cross-bridge the B subfiber within the axoneme (Fig. 8, inset). Moreover, the ATPase activity of dynein is stimulated by the presence of the spindles, and dynein either contributes to or stabilizes their birefringence (Telzer and Haimo, 1980, 1981). An analysis of microtubule polarity in any given region of the spindle has revealed that greater than 90% of the microtubules possess a uniform polarity (Fig. 8). These observations are in agreement with a study of microtubule polarity in the mitotic apparatus of PtK_1 cells using microtubule hooks (Euteneuer and McIntosh, 1981) and demonstrate that chromosomes cannot move to the poles as a result of sliding between microtubules possessing the

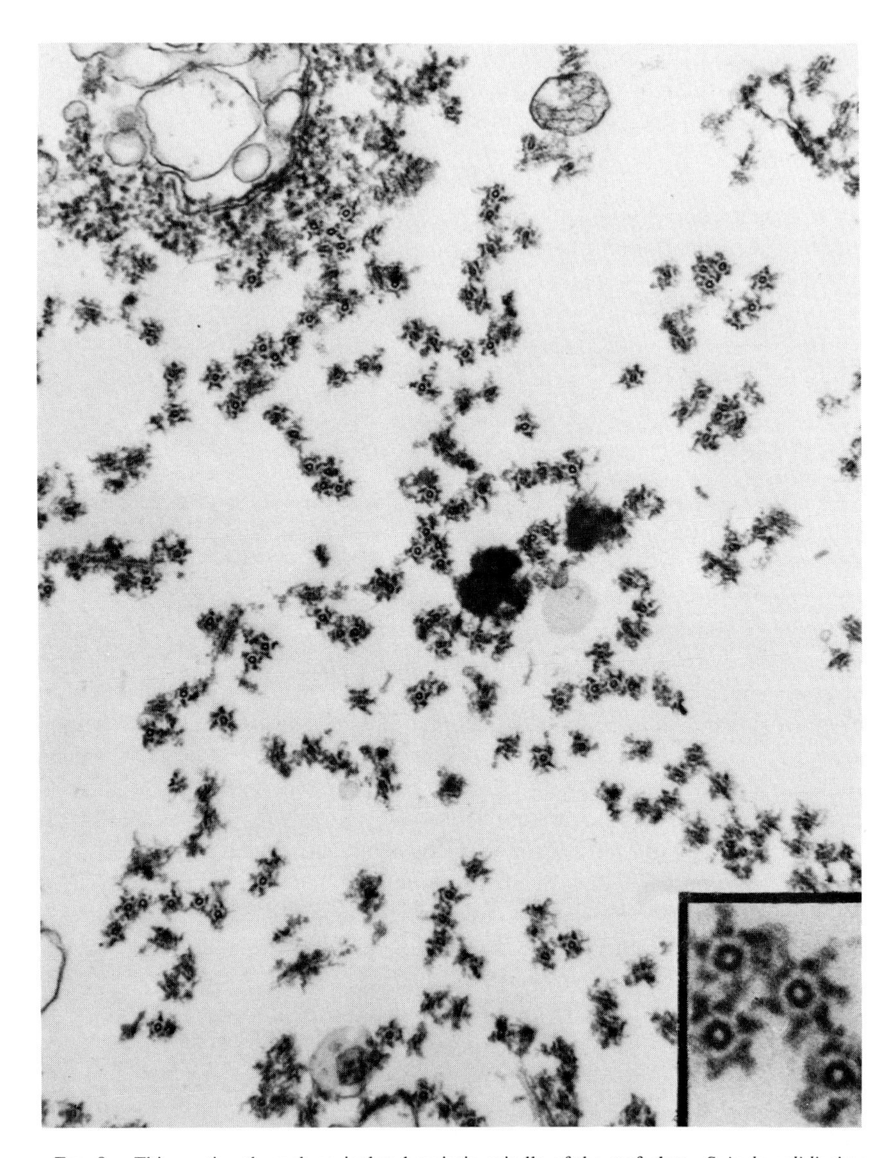

FIG. 8. Thin section through an isolated meiotic spindle of the surf clam, *Spisula solidissima,* incubated with *Tetrahymena thermophila* ciliary dynein. Each microtubule in the spindle is surrounded by 4 to 7 dynein arms attached to the microtubules by their end which would normally cross-bridge the B subfiber of the adjacent outer doublet microtubule within the axoneme (see inset). The dynein arms are pitched, and the decorated microtubules look like a pinwheel or rotary blade. The arms are oriented counterclockwise around the microtubules in the main figure and clockwise around the microtubules in the inset. In any given section through the spindle, 95% of the microtubules possess dynein arms with the same clockwise or counterclockwise pitch. Accordingly, the half spindle may be composed of microtubules possessing a uniform polarity.

opposite polarity. Proteins with properties similar to but distinct from ciliary dynein have been identified in the cytoplasm of unfertilized sea urchin eggs (Pratt, 1980) and associated with isolated mitotic apparatuses (Pratt *et al.*, 1980). Further experimentation should reveal if these proteins are, in fact, cytoplasmic dyneins, and, if so, if they produce sliding between cytoplasmic microtubules possessing the same or opposite polarity.

V. Perspectives

It is significant that dynein can be used to determine the polarity of a microtubule. This technique is analogous to that of using heavy meromyosin to determine the polarity of actin filaments (Huxley, 1963; Ishikawa *et al.*, 1969) and has the advantage that polarity can be resolved in transverse as well as in longitudinal sections of microtubules. Furthermore, dynein arms can be visualized on negatively stained preparations of microtubules, thus precluding the necessity to embed and section material that is amenable to whole mount analysis.

In addition to a determination of microtubule polarity, the observation that cytoplasmic microtubules can be cross-bridged by flagellar dynein and that the cross-bridges can be relaxed by ATP addition raises the possibility that movements associated with cytoplasmic microtubules within the cell may be produced by a dyneinlike protein interacting with the microtubules. Thus, the specialized movements exhibited by ciliary and flagellar beating may serve as a model for understanding other motile systems based on microtubules.

Acknowledgments

I am most grateful to Dr. Joel L. Rosenbaum and Dr. Bruce R. Telzer for their help and collaboration on this project. This research was supported by U.S. Public Health Service Grant GM 14642-12 to J.L.R., by a Steps Towards Independence Fellowship to B.R.T., and by an American Cancer Society Postdoctoral Fellowship to L.T.H.

References

Allen, C., and Borisy, G. G. (1974). *J. Mol. Biol.* **90,** 381–402.
Amos, L. A., Linck, R. W., and Klug, A. (1976). *In* "Cell Motility" (R. Goldman, T. Pollard, and J. L. Rosenbaum, eds.), pp. 847–867. Cold Spring Harbor Lab., Cold Spring Harbor, New York.
Begg, D. A., Rodewald, R., and Rebhun, L. I. (1978). *J. Cell Biol.* **79,** 846–852.
Bergen, L. G., Kuriyama, R., and Borisy, G. G. (1980). *J. Cell Biol.* **84,** 151–159.
Binder, L. I., and Rosenbaum, J. L. (1978). *J. Cell Biol.* **79,** 500–515.

Binder, L. I., Dentler, W. L., and Rosenbaum, J. L. (1975). *Proc. Natl. Acad. Sci. U.S.A.* **72,** 1122-1126.
Borisy, G. G. (1978). *J. Mol. Biol.* **124,** 565-570.
Dentler, W. L., Granett, S., Witman, G. B., and Rosenbaum, J. L. (1974). *Proc. Natl. Acad. Sci. U.S.A.* **71,** 1710-1714.
Euteneuer, U., and McIntosh, J. R. (1981). *J. Cell Biol.* **89,** 338-345.
Gibbons, B. H., and Gibbons, I. R. (1973). *J. Cell Sci.* **13,** 337-357.
Gibbons, B. H., and Gibbons, I. R. (1974). *J. Cell Biol.* **63,** 970-985.
Gibbons, I. R. (1975). *In* "Molecules and Cell Movement" (S. Inoué and R. E. Stephens, eds.), pp. 207-232. Raven, New York.
Gibbons, I. R., and Fronk, E. (1979). *J. Biol. Chem.* **254,** 187-196.
Gibbons, I. R., and Grimstone, A. V. (1960). *J. Biophys. Biochem. Cytol.* **7,** 697-716.
Gould, R. R., and Borisy, G. G. (1977). *J. Cell Biol.* **73,** 601-615.
Haimo, L. T., Telzer, B. R., and Rosenbaum, J. L. (1979). *Proc. Natl. Acad. Sci. U.S.A.* **76,** 5759-5763.
Heidemann, S. R., Zieve, G. G., and McIntosh, J. R. (1980). *J. Cell Biol.* **87,** 152-159.
Huxley, H. E. (1963). *J. Mol. Biol.* **7,** 281-308.
Ishikawa, H., Bishoff, R., and Holtzer, H. (1969). *J. Cell Biol.* **43,** 312-328.
Kim, H., Binder, L. I., and Rosenbaum, J. L. (1979). *J. Cell Biol.* **80,** 266-276.
McGill, M., and Brinkley, B. R. (1975). *J. Cell Biol.* **67,** 189-199.
MacNab, R. M. (1976). *J. Clin. Microbiol.* **4,** 258-265.
Margolis, R., and Wilson, L. (1978). *Cell* **13,** 1-8.
Neal, M. W., and Florini, J. R. (1973). *Anal. Biochem.* **55,** 328-330.
Olmsted, J. B., Marcum, J. M., Johnson, K. A., Allen, C., and Borisy, G. G. (1974). *J. Supramol. Struct.* **2,** 429-450.
Pierson, G. B., Burton, P. R., and Himes, R. H. (1978). *J. Cell Biol.* **76,** 223-228.
Pratt, M. M. (1980). *Dev. Biol.* **74,** 364-378.
Pratt, M. M., Otter, T., and Salmon, E. D. (1980). *J. Cell Biol.* **86,** 738-745.
Rosenbaum, J. L., and Child, F. M. (1967). *J. Cell Biol.* **34,** 345-364.
Rosenbaum, J. L., Binder, L. I., Granett, S., Dentler, W. L., Snell, W. J., Sloboda, R. D., and Haimo, L. (1975). *Ann. N.Y. Acad. Sci.* **253,** 147-177.
Sager, R., and Granick, S. (1953). *Ann. N.Y. Acad. Sci.* **56,** 831-838.
Shelanski, M. L., Gaskin, F., and Cantor, C. R. (1973). *Proc. Natl. Acad. Sci. U.S.A.* **70,** 765-768.
Sloboda, R. D., Dentler, W. L., and Rosenbaum, J. L. (1976). *Biochemistry* **15,** 4498-4505.
Snell, W. J., Dentler, W. L., Haimo, L. T., Binder, L. I., and Rosenbaum, J. L. (1974). *Science* **185,** 357-360.
Summers, K. E., and Gibbons, I. R. (1971). *Proc. Natl. Acad. Sci. U.S.A.* **68,** 3092-3096.
Summers, K. E., and Kirschner, M. W. (1979). *J. Cell Biol.* **83,** 205-217.
Takahashi, M., and Tonomura, Y. (1978). *J. Biochem. (Tokyo)* **84,** 1339-1355.
Telzer, B. R., and Haimo, L. T. (1980). *Biol. Bull. (Woods Hole, Mass.)* **159,** 446-447.
Telzer, B. R., and Haimo, L. T. (1981). *J. Cell Biol.* **89,** 373-377.
Telzer, B. R., and Rosenbaum, J. L. (1979). *J. Cell Biol.* **81,** 484-497.
Weingarten, M. D., Lockwood, A. H., Hwo, S.-Y., and Kirschner, M. W. (1975). *Proc. Natl. Acad. Sci. U.S.A.* **72,** 1858-1862.
Williams, R. C., Jr., and Detrich, H. W. (1979). *Biochemistry* **18,** 2499-2503.
Witman, G. B. (1975). *Ann. N.Y. Acad. Sci.* **253,** 178-191.
Witman, G. B., and Minervini, N. (1979). *J. Cell Biol.* **83,** 181a. (Abstr.)
Witman, G. B., Plummer, J., and Sander, G. (1978). *J. Cell Biol.* **76,** 729-747.
Zanetti, N. C., Mitchell, D. R., and Warner, F. D. (1979). *J. Cell Biol.* **80,** 573-588.

Chapter 13

Microtubule Polarity Determination Based on Conditions for Tubulin Assembly In Vitro

STEVEN R. HEIDEMANN

Department of Physiology
Michigan State University
East Lansing, Michigan

AND

URSULA EUTENEUER

Department of Molecular, Cellular, and Developmental Biology
University of Colorado
Boulder, Colorado

I. Introduction

Microtubules are intrinsically polar fibers. This has been observed directly in structural studies of sufficient resolution (Amos and Klug, 1974; Crepeau *et al.*, 1977; Baker and Amos, 1978). This polarity is manifested in the different assembly rates at the two ends of microtubules *in vitro* (Allen and Borisy, 1974; Rosenbaum *et al.*, 1975; Margolis and Wilson, 1978; Summers and Kirschner,

ISBN 0-12-564124-9

1979; Bergen and Borisy, 1980; Bergen *et al.*, 1980). Given both a structural and growth asymmetry, it is not surprising that microtubule polarity plays an important role in models for the mechanism of microtubule-mediated transport processes and the control of microtubule assembly *in vivo* (Subirana, 1968; McIntosh *et al.*, 1969; Nicklas, 1971; Margolis *et al.*, 1978; Schulz and Jarosch, 1980; Filner and Yadav, 1979; Kirschner, 1980). All of these proposals make strong predictions for the polarity arrangements of different subpopulations of cytoplasmic and spindle microtubules. Visualization of microtubule polarity orientation in various cell structures should contribute to our understanding of microtubule activity and provide an experimental test for many ideas of microtubule function.

We describe here a convenient method for the visualization of the polarity of cellular microtubules. An alternative method is described in this volume and elsewhere by Haimo (Haimo *et al.*, 1979; Haimo, 1981). Our method uses conditions for the assembly of brain microtubule protein *in vitro* by which the walls of the cytoplasmic microtubules become decorated with curved, protofilament appendages we call hooks. Two groups have pointed out that such junctions between microtubule walls would contain polarity information if only one handedness of hook were formed (Mandelkow and Mandelkow, 1979; Burton and Himes, 1978). We have shown that the curvature of hooks formed with our method is a dependable display of the intrinsic polarity of microtubules (Heidemann and McIntosh, 1980). Clockwise-curving hooks are observed in cross section if one is looking from the plus toward the minus end of the microtubule (Borisy, 1978), counterclockwise hooks if one is looking in the other

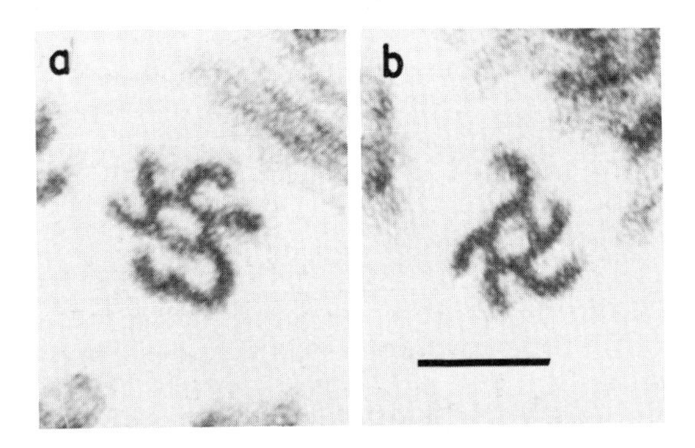

FIG. 1. Microtubule polarity is revealed by the curvature of the protofilament hooks formed on the walls of cellular microtubules by incubation in the microtubule assembly mixture described in the text. (a) Clockwise-curving hooks are observed in cross section on microtubules viewed from their plus end toward their minus end. (b) Counterclockwise hooks are observed on microtubules viewed from their minus end toward their plus end. Bar = 50 nm.

direction (Fig. 1) (Heidemann and McIntosh, 1980; Euteneuer and McIntosh, 1980).

II. Overview of the Method

The cell or tissue whose microtubules are to be assayed for polarity orientation is lysed and incubated in an *in vitro* microtubule assembly mixture. During the incubation, assymetric microtubule protofilament sheets (hooks) assemble onto the walls of the cellular microtubules. The material is then fixed, embedded, and thin sectioned for electron microscopy by standard methods. However, special precautions with respect to the orientation of the material must be taken during sample processing in order to interpret the handedness of hooks. The polarity of cellular microtubules is determined from the handedness of hooks as seen in cross sections of microtubules in electron micrographs.

III. Conditions for Microtubule Assembly *in Vitro* That Promote Hook Formation

We purify microtubule protein from bovine or porcine brain in the presence of glycerol by the method of Shelanski *et al.* (1973). After 2 or 3 cycles of polymerization the protein is resuspended in 0.5 piperazine *N,N'*-bis(2-ethanesulfonate) pH 6.9 (PIPES), 1 m*M* $MgCl_2$, 1 m*M* ethylenediaminetetraacetic acid (EDTA), and 1 m*M* GTP. We call this buffer "PB." The resuspended protein is spun at 250,000 *g* for 3 hr. The supernatant, called tubulin, is stored frozen in small aliquots in liquid nitrogen.

This PB is an *in vitro* microtubule assembly buffer similar to that reported by Himes *et al.* (1979). We have confirmed and extended these workers studies (Heidemann, 1980). We found that brain microtubule protein will assemble to the same mass of polymer in this high concentration of PIPES as in 0.1 *M* PIPES buffer. However, the polymers initiated *de novo* in PB are primarily ribbons of laterally associated protofilaments, not true microtubules. Addition of brain tubulin onto microtubule seeds produces microtubules with an evident lumen and ribbons of protofilaments on the microtubule wall that appear as hooks in cross section. It is this latter kind of assembly that is exploited in our polarity assay. Assembly of microtubule protein in PB requires GTP, is poisoned by substoichiometric concentrations of colchicine, and is rapidly reversed by cold and millimolar concentrations of calcium ions. Pure 6 S tubulin purified by phosphocellulose chromatography (Weingarten *et al.*, 1975) polymerizes efficiently in PB. Another sulfonic acid buffer, MES (Good *et al.*, 1966), routinely used for microtubule assembly at 0.1 *M* does not support microtubule assembly at 0.5 *M* without the addition of 10% dimethylsulfoxide (DMSO).

IV. Formation of Hooks on Cytoplasmic Microtubules

Although cellular microtubules can be elongated efficiently in PB without added DMSO, we found that 2.5% DMSO added to PB (PBD) greatly increases the frequency of hook formation on cellular microtubules. The PIPES concentration, pH, and DMSO concentration in PBD is likely to be close to the optimum conditions for the procedure. Hook frequency diminished at PIPES concentrations less than 0.4 M or at DMSO concentrations less than 1% or greater than 2.5% (McIntosh *et al.*, 1980).

Cell lysis is required during hook formation to allow access of the *in vitro* tubulin assembly mixture to the cellular microtubules. A number of detergents have been used for this purpose: a strong detergent mixture containing 1% Triton X-165, 0.5% deoxycholate, and 0.2% (or 0.02%) sodium dodecyl sulfate (SDS) has proved unusually useful. This mixture causes complete cell lysis while still maintaining cytoplasmic and spindle microtubules (Heidemann *et al.*, 1980). Moreover, brain tubulin polymerizes to the same extent in the mixture and actually shows a shortened lag time relative to PBD without detergents.

Three different reactions take place under the conditions described: elongation of cellular microtubules, decoration with hooks of the preexisting cellular microtubules and their elongated parts, and spontaneous initiation of protofilament polymers from the brain tubulin. By varying tubulin concentration, incubation time, and incubation temperature one can optimize hook decoration of microtubules and minimize the other reactions. With the particular material under investigation these factors and the detergents must be varied by trial and error to obtain a high proportion of hooks on cellular microtubules. For example, the polarity of microtubules in HeLa cells has been determined by lysing these cells with PBD containing a detergent mixture of 1% Triton X-165, 0.5% deoxycholate, and 0.2% sodium dodecyl sulfate. These lysed cells were incubated for 3 min with 2 mg/ml tubulin for determining the polarity of aster microtubules (Heidemann and McIntosh, 1980); midbody microtubule polarity (Fig. 2) was determined through a 30-min incubation with 0.5 mg/ml tubulin in the same mixture (Euteneuer and McIntosh, 1980). *Haemanthus* endosperm cells were incubated in PBD with 0.5 mg/ml tubulin and a mild detergent, either 0.04% saponin or 0.5% Brij 58, for 30 min at 37°C to determine the polarity of phragmoplast microtubules (Euteneuer and McIntosh, 1980). For axopodia of the heliozoan *Actinosphaerium nu-*

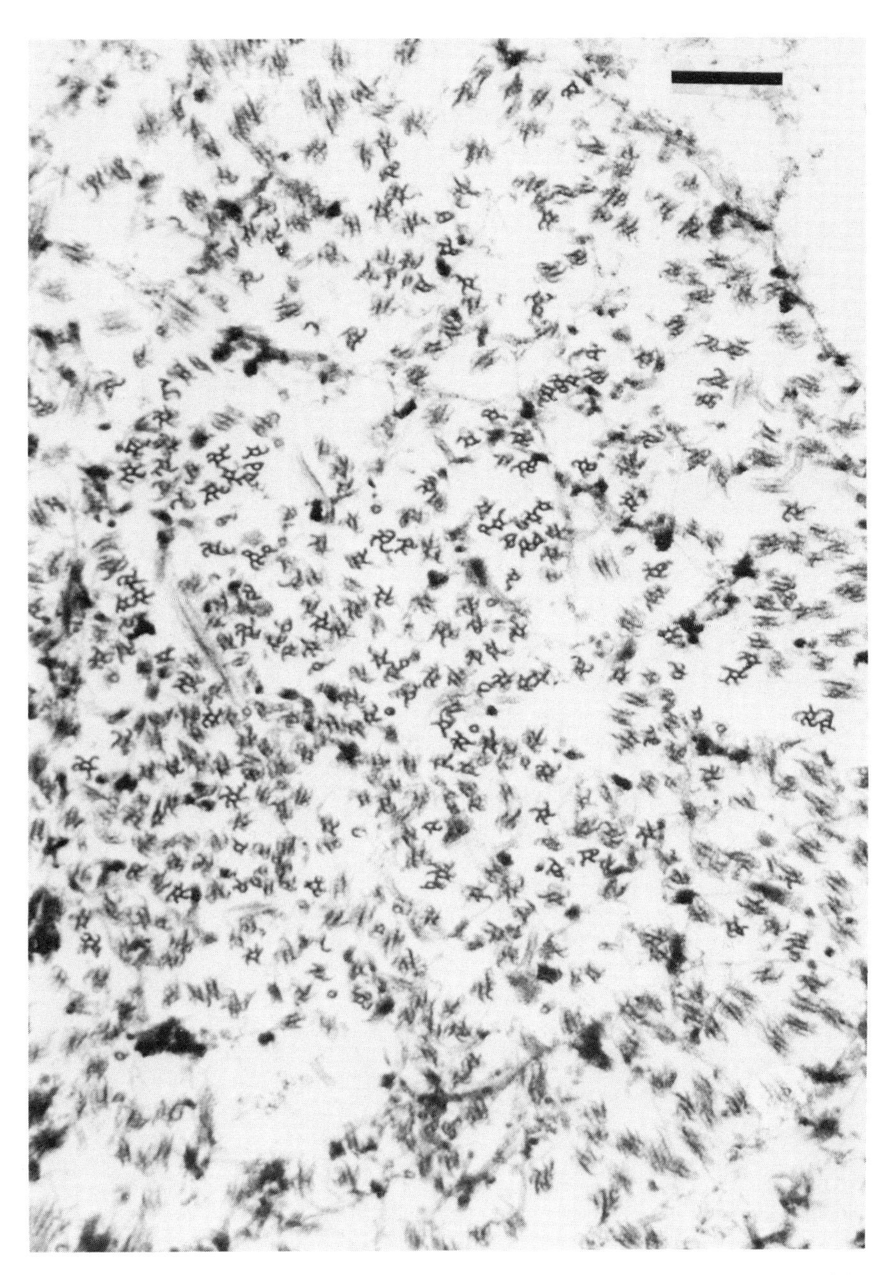

FIG. 2. Cross section through a midbody of a PtK_1 cell incubated for 30 min in PBD containing 0.5 mg/ml tubulin and the strong detergent mixture described in the text. This section was taken before the zone of overlap was reached during serial sectioning. Given this apparent vantage point, the predominantly counterclockwise hooks observed here indicate that it is the minus ends of the microtubules that are proximal to the cell nucleus. Bar = 0.25 μm.

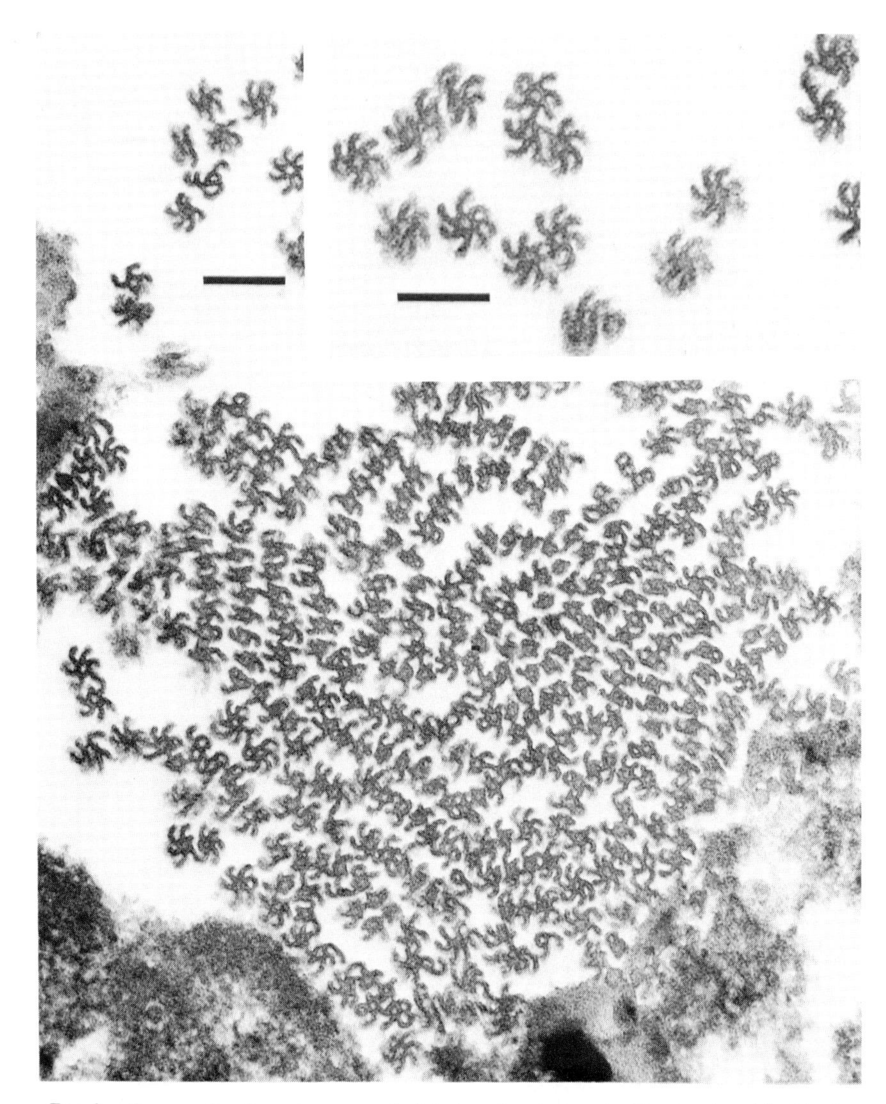

FIG. 3. Cross section through an axopod of *Actinosphaerium nucleofilum* incubated for 20 min at room temperature in PBD containing 1.5 mg/ml tubulin and 0.5% Brij 58. This section was taken from a series sectioned toward the cell center. The clockwise hooks indicate that the microtubules in the axopod are oriented with their plus ends distal to the cell center. Bar = 0.125 μm. (Inset) Bar = 0.1 μm.

cleofilum (Fig. 3) a combination of high tubulin concentration (1.5 mg/ml) and long incubation time in mild detergent at room temperature was found to be optimal (Euteneuer and McIntosh, 1981). Suitable conditions for lysis can often be determined with trials assayed by polarization or Nomarski microscopy. Regrettably, we have found no assay for hook formation other than time consuming thin sectioning.

V. Electron Microscopy

Fixation, dehydration, embedment, and so on, of specimens for electron microscopy can generally be carried out by standard procedures. However, in the case of metaphase-arrested HeLa cells lysed in buffer containing 1% TX-165, 0.5% deoxycholate, 0.2% SDS we found it necessary to dehydrate very slowly with methoxyethanol in order to avoid wavy microtubules (Heidemann *et al.*, 1980). Application of our method does require unusual awareness of the orientation of the specimens, the thin sections, and the photographic negatives since the handedness of hooks decorating microtubules depends on the direction from which they are viewed. Moreover, the polarity of microtubules must be determined relative to some cellular referencc point such as the nucleus or mitotic center. Such complete structure determination often requires serial thin sectioning and/or an awareness of the orientation of the material relative to its original orientation *in situ* from specimen preparation prior to incubation through the printing of electron micrographs. Some examples of precautions we have taken and of common procedures that might reverse polarity are presented here. The neuron has an intrinsic morphological polarity; the cell body lies at one end of the axon. In order to determine microtubule polarity relative to the polarity of the cell, the cell polarity was permanently marked by tying a ligature with suture at the end of the nerve segment nearest the ganglion. All sectioning proceeded toward the ligature, that is, from the axon toward the cell body. Thin sections placed onto grids from above will show the opposite handedness from those in sections picked up from below, other things being equal. The grids used should allow an unequivocal identification of the side on which the sections are placed. Another possible site of inversion is the electron microscope itself. The Phillips 300 microscope that we use inverts the specimen during its movement into the beam path. Also during printing, care must be taken to avoid producing mirror images of the original hook handedness by inverting the negative. In general, care must be taken while working to avoid mirror-image reversal of hook handedness through unnoted inversions.

VI. Interpretation of Hook Handedness

The handedness of hooks is scored on prints of electron micrographs with a final magnification of about 50,000×. Figure 4 shows some of the images we

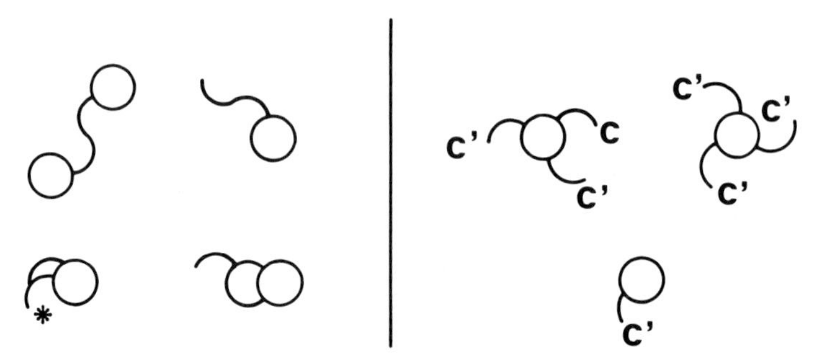

FIG. 4. Diagram illustrating some of the observed hook images. All images on the right side were interpretable as either clockwise (c) or counterclockwise (c′) hooks. Images on the left side, with the exception of the image marked with an asterisk, were regarded as uninterpretable.

have observed and our classification of them. If the original handedness of hooks is preserved during electron microscopy, then predominantly clockwise hooks are seen on microtubules that are oriented with their fast-growing or "+" end (Borisy, 1978) toward the observer. Naturally, counterclockwise hooks are seen on microtubules with their slow-growing or "−" end oriented toward the observer. It is clear, however, that hooks of the wrong sense can form. Several images we have obtained show hooks of both curvatures on one microtubule. Our evidence also indicates that as many as 10% of the microtubules with a given polarity may show hooks of the wrong curvature. For example, in experiments on the polarity of microtubules grown from basal bodies (Heidemann and McIntosh, 1980), only 90% of the microtubules with hooks were consistent with the orientation determined by growth rates (Allen and Borisy, 1974; Rosenbaum *et al.*, 1975). On the other hand, hooks attached to preexisting, cellular microtubules produced more consistent data. The fraction of hooks with one curvature on microtubules in axopods of heliozoa was 98–99%. *All* hooks on ciliary microtubules indicated the same polarity for these microtubules (Euteneuer and McIntosh, 1981a). A result of 98% hooks of one curvature is convincing evidence for a single polarity of microtubules. Given the data from the microtubules grown from basal bodies, however, 10% hooks of opposite polarity need not mean that some of the microtubules are antiparallel. It may be that certain conditions influence the fidelity of the relationship between hook curvature and microtubule polarity. Alternatively, we may be observing a case of biological variability or a case in which microtubules of opposite polarity not grown from a basal body were included in the fields under study.

Thus far useful results have been obtained with this method in several microtubule systems: asters of mammalian spindles (Heidemann and McIntosh, 1980), cilia, axopodia of heliozoa and melanophores (Euteneuer and McIntosh, 1981a), midbodies of mammalian cells and phragmoplasts of plants (Euteneuer and McIntosh, 1980), kinetochore fibers in PtK_1 spindles (Euteneuer and McIntosh, 1981b), and axons of the cat renal nerve (Heidemann, 1980).

In some cases the method has confirmed results obtained with other techniques, lending confidence to observations (Heidemann and McIntosh, 1980). In others, there is a direct disagreement between results from hook curvature and the polarity established by other methods. Although it is unclear which method is giving the right answer, the consistency of hook formation and the clarity of the result makes this method appear dependable as well as convenient.

REFERENCES

Allen, C., and Borisy, G. G. (1974). *J. Mol. Biol.* **90,** 381-402.

Amos, L., and Klug, A. (1974). *J. Cell Sci.* **14,** 523-549.

Baker, T. S., and Amos, L. A. (1978). *J. Mol. Biol.* **123,** 89-106.

Bergen, L. A., and Borisy, G. G. (1980). *J. Cell Biol.* **84,** 141-150.

Bergen, L. A., Kuriyama, R., and Borisy, G. G. (1980). *J. Cell Biol.* **84,** 151-159.

Borisy, G. G. (1978). *J. Mol. Biol.* **124,** 565-570.

Burton, P. R., and Himes, R. H. (1978). *J. Cell Biol.* **77,** 120-133.

Crepeau, R., McEwen, B., Dykes, G., and Edelstein, S. J. (1977). *J. Mol. Biol.* **116,** 301-315.

Euteneuer, U., and McIntosh, J. R. (1980). *J. Cell Biol.* **87,** 509-515.

Euteneuer, U., and McIntosh, J. R. (1981a). *Proc. Natl. Acad. Sci. U.S.A.* **78,** 372-376.

Euteneuer, U., and McIntosh, J. R. (1981b). *J. Cell Biol.* **89,** 338-345.

Filner, P., and Yadav, N. S. (1979). *Encycl. Plant Physiol., New Ser.* **7,** 95-113.

Good, N. E., Winget, G. D., Winter, W., Connolly, T. N., Izawa, S., and Singh, R. M. W. (1966). *Biochemistry* **5,** 467-477.

Haimo, L. T. (1981). *Methods Cell Biol.* **24a** (in press).

Haimo, L. T., Telzer, B. R., and Rosenbaum, J. R. (1979). *Proc. Natl. Acad. Sci. U.S.A.* **76,** 5759-5763.

Heidemann, S. R. (1980). *In* "Microtubules and Microtubule Inhibitors" (M. De Brabander and J. De Mey, eds.), pp. 341-355. Elsevier/North-Holland, Amsterdam.

Heidemann, S. R., and McIntosh, J. R. (1980). *Nature (London)* **286,** 517-519.

Heidemann, S. R., Zieve, G. W., and McIntosh, J. R. (1980). *J. Cell Biol.* **87,** 152-159.

Himes, R. H., Newhouse, C. S., Haskins, K. M., and Burton, P. R. (1979). *Biochem. Biophys. Res. Commun.* **87,** 1031-1038.

Kirschner, M. W. (1980). *J. Cell Biol.* **86,** 330-334.

McIntosh, J. R., Hepler, P. K., and Van Wie, D. G. (1969). *Nature (London)* **224,** 659-663.

McIntosh, J. R., Euteneuer, U., and Neighbors, B. N. (1980). *In* "Microtubules and Microtubule Inhibitors" (M. De Brabander and J. De Mey, eds.), pp. 357-371. Elsevier/North-Holland, Amsterdam.

Mandelkow, E. M., and Mandelkow, E. (1979). *J. Mol. Biol.* **129,** 135-148.

Margolis, R., and Wilson, L. (1978). *Cell* **13,** 1-10.

Margolis, R. L., Wilson, L., and Kiefer, B. J. (1978). *Nature (London)* **272,** 450-452.

Nicklas, R. B. (1971). *Adv. Cell Biol.* **2,** 225-297.

Rosenbaum, J. L., Binder, L. I., Granett, S., Dentler, W. L., Snell, W., Sloboda, R., and Haimo, L. (1975). *Ann. N.Y. Acad. Sci.* **253,** 147-177.

Schulz, D., and Jarosch, R. (1980). *Eur. J. Cell Biol.* **20,** 249-253.

Shelanski, M. L., Gaskin, I., and Cantor, C. R. (1973). *Proc. Natl. Acad. Sci. U.S.A.* **70,** 765–768.
Subirana, J. A. (1968). *J. Theor. Biol.* **20,** 117–123.
Summers, K., and Kirschner, M. W. (1979). *J. Cell Biol.* **83,** 205–217.
Weingarten, M. D., Lockwood, A. H., Hwo, S.-Y., and Kirschner, M. W. (1975). *Proc. Natl. Acad. Sci. U.S.A.* **72,** 1858–1862.

Chapter 14

The Use of Tannic Acid in Microtubule Research

KEIGI FUJIWARA AND RICHARD W. LINCK

Department of Anatomy
Harvard Medical School
Boston, Massachusetts

I. Introduction

Our present knowledge concerning the morphology and the distribution of microtubules comes from the application of conventional electron microscopy, optical diffraction of electron micrographs of negatively stained microtubules (Erickson, 1974; Amos and Klug, 1974; Linck and Amos, 1974; Amos, 1979; McEwen and Edelstein, 1980), and immunohistochemistry (Goldman *et al.*, 1976; De Brabander *et al.*, 1977). These morphological approaches have established the universal occurrence and structural features of microtubules in eukaryotic cells, although some exceptions to this general statement have been reported, as we will point out later. Because the aim of this chapter is not to discuss the detailed structure and function of microtubules, readers interested in such information should refer to more comprehensive recent review articles (Stephens and Edds, 1976; Kirschner, 1978; Roberts and Hyams, 1979; Linck, 1982).

METHODS IN CELL BIOLOGY, VOLUME 24

ISBN 0-12-564124-9

In this chapter we will describe a fixation method using tannic acid with glutaraldehyde for the study of microtubule structure. In addition to describing the methodology, we will consider the possible mechanism of the effect tannic acid provides. Various applications of the use of tannic acid to other ultrastructural studies will also be discussed.

II. Microtubule Substructure—Brief Review

The success of electron microscopy depends largely on the fixation method used, and in the study of microtubules there have been two major breakthroughs. One was the introduction of glutaraldehyde as an ultrastructural fixative (Sabatini *et al.*, 1963). The use of this fixative quickly established the presence of microtubules in all eukaryotic cells, not to mention superb preservation of the general cell morphology.

The second milestone was set by Mizuhira and Futaesaku (1971, 1972), who added tannic acid to the glutaraldehyde fixative. In cross sections of rat sperm tails fixed with glutaraldehyde and tannic acid, these investigators observed a clear substructure within the walls of the axonemal microtubules. These subunits are protofilaments of the microtubules and appear as unstained electron translucent areas surrounded by an electron dense precipitate. This appearance of the microtubule wall is commonly referred to as "negative staining," although the term *negative staining* more properly refers to the heavy metal contrasting procedures for electron microscopy (Huxley, 1963).

The new fixative provided the means to settle the controversy concerning the exact number of protofilaments that make up the wall of a microtubule. The protofilaments were originally observed by André and Thiéry (1963) and by Pease (1963) in the negatively stained images of microtubules of the flagellar axoneme, and since then similar subunit structure has been recognized by many other investigators (Grimstone and Klug, 1966; Gall, 1966; Behnke and Zelander, 1966; Kiefer *et al.*, 1966; Warner and Satir, 1973). By negative staining procedures these authors were unable to determine the exact number of these subunits per microtubule because the dried protofilaments overlapped on a grid. Neither image reinforcement techniques (Markham *et al.*, 1963; Agrawal *et al.*, 1965; Friedman, 1970) nor X-ray diffraction (Cohen *et al.*, 1971) yielded unequivocal counts of protofilaments. It seemed that the only way to determine the number of protofilaments within the wall of a microtubule was to count directly from a cross section of a microtubule properly stained so that such a direct counting was possible. Prior to the discovery of the tannic acid technique of Mizuhira and Futaesaku (1971), there were two examples that clearly demonstrated 13 protofilaments in the wall of a microtubule: the "negatively stained" plant microtubules (Ledbetter and Porter, 1964), and central pair microtubules of the sperm flagellum of the black scavenger fly (Phillips, 1966).

Tilney and co-workers (1973) used this new fixative to investigate the protofilament arrangement in a variety of microtubules including the axoneme of *Echinosphaerium,* sea urchin mitotic spindles, the axostyles of certain protozoa, the axoneme of sea urchin sperm tails, *in vitro* polymerized brain microtubules, and the sea urchin sperm basal body. They found that the wall of a complete microtubule consisted of 13 protofilaments. There were less than 13 protofilaments in incomplete "microtubules," such as the B subfibers in the axoneme (10–11 subunits[1]) and the B and C subfibers in the basal body (10 subunits). Later studies also found 13 protofilaments in the spindle microtubules of cranefly spermatocytes (LaFountain, 1975; LaFountain and Thomas, 1975).

Several investigators have found, however, that there are exceptions to this "universal" rule of the composition of 13 protofilaments within the wall of a microtubule. Burton and Hinkley (1974) reported that the axonal microtubules of the crayfish ventral nerve cord consisted of 12 protofilaments, although those of the supporting cells had 13. Burton *et al.* (1975) investigated further the number of protofilaments in microtubules from other neuronal and nonneuronal sources. They counted 13 protofilaments without exception in microtubules of the frog lung fluke sperm cortex, the small axons of the olfactory nerve of the frog, the axons of goldfish brain, and the pleurovisceral connective nerves of *Aplysia californica*. However, most microtubules of the crayfish sperm consisted of 15 protofilaments while those of the nerve cords of the lobster had 12, as was the case in the crayfish nerve cord. Fifteen protofilaments were also found in the microtubules in the epidermal cells of the cockroach (Nagano and Suzuki, 1975).

The number of protofilaments in microtubules assembled *in vitro* vary, depending on the purity of the polymerizing system and the conditions of polymerization. Initial investigations using high-speed tissue extracts of chick brain (Tilney *et al.*, 1973), bovine brain (Burton *et al.*, 1975), and fruit-fly embryo and egg (Green *et al.*, 1975) all yielded 13 protofilaments. However, using bovine brain tubulin, Pierson *et al.* (1978) found that microtubules with 14 and more protofilaments became more abundant as the number of polymerization–depolymerization cycles increased. This observation implies that highly purified tubulin assembles into microtubules with 14 protofilaments. Indeed, purified tubulin from bovine brain (Burton and Himes, 1978), calf brain (Kim *et al.*, 1979), and sea urchin sperm tail flagella (Binder and Rosenbaum, 1978; Linck and Langevin, 1981) formed predominantly microtubules with 14 protofilaments. The

[1]The B subfiber of doublet microtubules appears to be composed of 10 protofilaments; the eleventh subunit often appears smaller or altogether absent and may correspond to a unique protein structure responsible for the association of the A and B subfibers (Linck, 1976).

presence of microtubule-associated proteins (MAPs) apparently causes no effect on these results (Kim *et al.*, 1979).

Pierson *et al.* (1978, 1979) studied systematically the effects of pH, different buffers and buffer concentrations on the number of protofilaments in the *in vitro* assembled microtubules and found that the protofilament number is influenced by all these factors. They found microtubules with 9 to 17 protofilaments, although the majority of microtubules had 13 or 14 protofilaments. Microtubules assembled in 20 m*M* MES (pH 6.5), 70 m*M* NaCl, 0.5 m*M* $MgCl_2$, 1 m*M* EGTA, 1 m*M* GTP, and 4 *M* glycerol most consistently showed 13 protofilaments.

A correlation between the diameter of microtubules and the number of protofilaments has been noted by many investigators (Nagano and Suzuki, 1975; Burton *et al.*, 1975; Kim *et al.*, 1979): the more protofilaments, the greater the microtubule diameter. Tyson and Bulger (1973) summarized various conditions that induced microtubules of larger diameter (macrotubules) in many cell types. Attempts have been made to investigate the arrangement of protofilaments in some of these macrotubules using the tannic acid technique. Fujiwara and Tilney (1975) failed to observe clear subunit arrangement in the walls of two types of such macrotubules: the 34-nm-diameter tubules induced by low temperature in *Echinosphaerium* (Tilney and Porter, 1967) and the "tubules" of tubulin paracrystals induced by vinblastine in sea urchin eggs. Burton *et al.* (1975) also detected no clear substructure in the wall of vinblastine-induced tubulin paracrystals. These observations indicate that some macrotubular walls do not contain protofilaments arranged parallel to the axis of the tubule. Both Tilney and Porter (1967) and Fujiwara and Tilney (1975) suggested a helical arrangement of protofilaments in these macrotubules.

The tannic acid method has provided, in addition to the exact number of protofilaments in various microtubules, more precise morphology of the axoneme and other morphological features of microtubules. Tilney *et al.* (1973) showed that the B subfiber of the sea urchin sperm tail outer doublet was a partial tubule consisting of 10–11 protofilaments and that the eleventh subunit appeared both smaller and less obvious than the other protofilaments (see footnote 1). They also found that the connection of protofilaments at the junction of A and B subfibers was from subunit to subunit, not from subunit to groove, and that there are three protofilaments of the subfiber A between the two protofilaments to which the subfiber B attaches. A similar subunit arrangement is observed in the triplets of the basal body except that the eleventh subunit is missing from the B and C fragments. Linck (1976) revealed the existence of yet another subunit structure applied to the inside wall of the A tubule. Furthermore, he suggested a systematic nomenclature to identify the individual protofilaments of the doublet microtubule.

Fujiwara and Tilney (1975) demonstrated a remarkably precise arrangement of protofilaments in the flagellar axoneme using the Markham rotation technique (Markham *et al.*, 1963) on demembranated sea urchin sperm tail axonemes fixed in the presence of tannic acid. The position of each protofilament in the outer doublet microtubule is so precisely determined that all the protofilaments in each of the nine doublets superimpose when an axoneme is rotated by multiples of 40°. Because the exact position of protofilaments in the axoneme can be observed and because arguments exist to favor precise attachment sites of the arms and bridges, one might expect to identify precisely the protofilaments to which these structures attach in the electron micrographs of cross-sectioned tannic-acid-treated axonemes. However, no serious attempt has been made to resolve this question.

When microtubules of both axonemal and nonaxonemal origins are observed longitudinally, periodic projections can be detected. In the case of axonemal microtubules the presence and arrangement of these accesory structures has been studied by whole mount negative stain electron microscopy (for review, see Linck, 1979). However, tannic acid methods can also be applied to demonstrate the radial spoke components (Huang *et al.*, 1981; Piperno *et al.*, 1981). Also using tannic acid, Kim *et al.* (1979) showed a periodic arrangement of the HMW-MAP_2 (high-molecular-weight–microtubule-associated protein) component along microtubules assembled *in vitro* from calf brain tubulin. Under saturating conditions, the MAP_2 decorate the circumference of the singlet microtubules and the MAP_2 projections are reported to be arranged along the microtubule with an axial spacing of 32 nm (Murphy and Borisy, 1975; Kim *et al.*, 1979). However, a controversy exists as to the three-dimensional arrangement of the MAP_2 components (Amos, 1977).

Although it is tempting to believe that the observed periodicity reflects the position of MAP_2 binding sites of *in vivo* microtubules, it should be kept in mind that 95% of the microtubules used in these studies consisted of 14 protofilaments rather than 13 found in most *in vivo* microtubules. It is possible that the arrangement of MAP_2 binding sites on native microtubules composed of 13 protofilaments may be different from those on microtubules synthesized *in vitro*.

III. Use of Tannic Acid in Microtubule Research

A. Tannic Acid

Tannic acid can be purchased from many companies. Because it is a natural product and is a mixture of many types of polyphenolic compounds, the exact chemical nature is difficult, if not impossible, to obtain. In addition, it should be realized that one product may be different from others, largely because of the source from which the product is made. Both Chinese and Turkish nutgalls are the common sources of commercial tannic acids. According to Simionescu and Simionescu (1976a) the tannic acid of the Chinese nutgall (Chinese gallotannin) contains about 80% hepta- and decagalloylglucoses and 9% penta- and hexagalloyl esters, whereas the tannic acid from Turkish nutgalls (Turkish gallotannin) contains roughly 51% penta- and hexagalloylglucoses and 30% hepta- and octagalloylglucoses. The molecular weights assigned to these tannic acids are approximately 1400 for Chinese gallotannin and 1000 for Turkish gallotannin.

Most chemical companies (e.g., Fisher Scientific Co., Eastern Organic Chemicals, Matheson Coleman and Bell Manufacturing Chemists, Hach Chemical Co., and Galland-Schlesinger Chemical) assign tannic acid an empirical formula $C_{76}H_{52}O_{46}$ and the formula weight 1701.18, and some others (for example, Merck Chemical Co., Sigma, and Baker Chemicals) provide no information. Although tannic acid sold by Polysciences carries the same empirical formula as that of the other companies, its product is described as "galloylglucose-low M.W." Mallinckrodt is the only manufacturer that provides tannic acid characterized as chiefly $C_{14}H_{10}O_9$, or digallic acid.

Simionescu and Simionescu (1976a,b) empirically came to the conclusion that Aleppo tannin consisting mostly of penta- and hexagalloylglucoses gave the most satisfactory results on the ultrastructure of animal tissues. Among the commercial products they tested, tannic acid distributed by Mallinckrodt produced results similar to those obtained with Aleppo tannin.

Although Mallinckrodt tannic acid can be used to demonstrate protofilaments of microtubules, many investigators insisted on using tannic acid sold by Merck Chemical Co. because some workers failed to obtain satisfactory results with tannic acids from other sources. However, we as well as other workers used tannic acids from companies other than Mallinckrodt or Merck and encountered no serious problem. It is not clear what causes these inconsistencies. Nevertheless, it seems safe to conclude that the products from Mallinckrodt and Merck Chemical give the most consistent results.

Precipitation may result when tannic acid and certain compounds are present in the same solution (Table I). First, we tested the compatibility of various buffer solutions with tannic acid. A 2% solution of tannic acid (Mallinckrodt, Type AR 1764) was made using buffers listed in the table at the concentration of 100 m*M* at a given pH. No precipation was detected with any of the buffers. We then studied the effect of common salts, detergents, and other compounds that are used frequently in the microtubule research. Typically, an appropriate amount of concentrated stock solution of a test compound was added to 5 ml of 2% tannic acid solution made with 100 m*M* Na-phosphate buffer and the pH was adjusted to the desired value. The mixture was immediately vortexed. Where it was possible, we determined the highest concentration of reagents that gave no

TABLE I

COMPATIBILITY OF REAGENTS WITH 2% TANNIC ACID SOLUTION AT ROOM TEMPERATURE

Reagents[a]	pH	Concentration	Precipitate[b]
Na-phosphate	6.0–7.0	100 mM	–
Na-cacodylate	6.0–7.0	100 mM	–
Tris-HC1	6.0–7.0	100 mM	–
PIPES	7.0	100 mM	–
MES	7.0	100 mM	–
NaC1	6.0–7.0	200 mM	–
KC1	6.1	75 mM	±
KCl	7.0	70 mM	±
CaC_2[c]	6.1	6 mM	±
$CaC1_2$	7.0	2 mM	±
$MgC1_2$	6.1	50 mM	–
$MgC1_2$	7.0	25 mM	±
Triton X-100	7.0	0.01%	+
Brij	6.5	0.05%	+
Tween	6.5	0.05%	+
Digitonin	6.5	1%	–
Glycerol	6.0–7.0	50%	–
EDTA	6.5	10 mM	–
EGTA	6.5	10 mM	–

[a] All reagents except buffers are tested using 100 mM Na-phosphate buffer solution.

[b] Presence or absence of precipitate is determined after the mixture has been vortexed (see text for details). –: No precipitate. +: Precipitate forms. ±: Precipitate begins to form at the given concentration of salts.

[c] $CaCl_2$ may form a precipitate with phosphate ion; much higher $CaCl_2$ (15 mM, at pH 6.1) concentration without precipitation can be obtained with Na-cacodylate buffer.

precipitate. Mildly acidic pH allows slightly higher concentrations of salts to be present without forming precipitate. All the detergents we tested cause extensive formation of a white precipitate even at very low concentrations. However, digitonin, which is practically insoluble in water, becomes soluble in the presence of tannic acid. Glycerol (50%), EDTA (10 mM), and EGTA (10 mM) cause no precipitation.

B. Mechanism of Tannic Acid Effect

Since the introduction by Mizuhira and Futaesaku (1971) of tannic acid as an additive to conventional glutaraldehyde fixatives, several mechanisms have been proposed to explain various effects observed in electron micrographs. The Japanese workers have suggested that tannic acid acts as a fixative because it causes polypeptides with more than nine amino acids to precipitate, provided that the pH is optimally controled (Mizuhira and Futaesaku, 1972; Futaesaku, 1979). Most proteins precipitate at a slightly acidic pH, whereas phosphoproteins and basic proteins precipitate under alkaline pH. It is interesting to note that these polypeptides that precipitate with tannic acid are probably not covalently cross-linked, because Futaesaku (1979) has reported that the precipitates can be resolubilized by altering pH or by removing tannic acid by dialysis.

Although covalent cross-linking may not occur in the biological specimen treated with tannic acid, it has been noted that tannic acid can stabilize certain biological structures, such as many types of membranes (Simionescu and Simionescu, 1976a,b; Kalina and Pease, 1977a,b), interstitial structures (Rodewald and Karnovsky, 1974; Van Deurs, 1975), and isolated tubulin paracrystals induced by vinblastine (Fujiwara and Tilney, 1975). The most dramatic example is the stabilization of isolated tubulin paracrystals by tannic acid. When conventional aldehyde fixatives are used to fix isolated paracrystals, they quickly dissolve. However, if tannic acid is added to the fixative, the dissolution of the paracrystal is prevented, although certain morphological changes do take place. Such alteration can be easily observed in a polarization microscope. Isolated paracrystals are highly birefringent, but their birefringence weakens in the glutaraldehyde fixative containing tannic acid. On the other hand, when the isolated paracrystals are placed in buffered tannic acid solution, the level of birefringence does not change, suggesting that the structural integrity of the isolated paracrystal is maintained. In fact, they can be further processed for electron microscopy without aldehyde fixation (Fujiwara and Tilney, 1975).

Whether or not tannic acid should be regarded as a fixative may be a question of semantics. If one defines a fixative as an agent that covalently cross-links macromolecules within a biological structure, tannic acid is not a fixative; however, most workers define fixation in less chemical terms. For example, fixation is to kill living cells quickly without damage and to *stabilize* cellular organization so that its ultrastructural relations are preserved during dehydration, embedding, and microscopy (Hayat, 1970). According to Bloom and Fawcett (1975, p. 10), it is "to render the structural components of cells *insoluble,* and this is accomplished by the use of various chemicals that *precipitate* the proteins and certain other classes of compounds." As we have described earlier, tannic acid can stabilize certain biological structures. Thus it is not unreasonable to consider tannic acid as a fixing agent.

The idea that tannic acid is a mordant was suggested. Simionescu and Simionescu (1976a,b) studied systematically various effects of tannic acid on glutaraldehyde-fixed animal tissues. They concluded that tannic acid should not be thought of as a fixative, but that it should be regarded as a mordanting agent. They proposed that the primary role of tannic acid was to act as a mordant between osmicated structure and lead. The mordanting effect does not depend on the residual aldehyde groups in the fixed tissue but *does* depend on OsO_4.[2] The functional groups required for the mordanting effect are found to be a carboxyl group and at least one hydroxyl group on the tannic acid (Simionescu and Simionescu, 1976b).

There is no doubt that tannic acid acts as a mordanting agent that enhances dramatically the contrast of electron micrographs of various biological structures (see Section IV) and also helps to reveal protofilaments of microtubules. However, there exist equally strong arguments and evidence to asign a fixative role to tannic acid. It is, therefore, safe to conclude that tannic acid has both mordanting and fixing effects (Begg *et al.*, 1978).

Many investigators have used the term *tannic acid staining* to describe the effect of tannic acid (Tilney *et al.*, 1973; Burton *et al.*, 1975; Green *et al.*, 1975; LaFountain and Thomas, 1975; Burton and Himes, 1978; Pierson *et al.*, 1979). This terminology may draw some support because of the fact that electron micrographs of tannic-acid-treated specimens have better contrast. However, based on what is known regarding its mechanism of action, there appears to be no justification for considering tannic acid to be a "stain." For this reason the use of "tannic acid staining" should be avoided.

C. Methods

1. *In Vitro* MICROTUBULES

The category of *in vitro* microtubules includes both isolated cell organelles that contain microtubules (such as spindles and axonemes of various types) and *in vitro* assembled microtubules. Because in these cases there is no plasma membrane to interfere with tannic acid penetration, it is not necessary to use tannic acid of smaller molecular weight. Rather, it may be of advantage to use larger species of tannic acid because they may have a better mordanting effect. To handle these specimens it is most convenient to form a small pellet by centrifugation. If a sample is easily sedimentable (e.g., isolated spindles), the pellet may be resuspended in the fixative and the washing solution. However, purified microtubules are best dealt with as pellets throughout the entire procedure. A typical step-by-step procedure is outlined here. However, the exact condition may vary depending on the specimen, the nature of tannic acid, and other parameters.

1. Obtain specimen in the form of a pellet in a centrifuge tube.
2. Prepare fixative containing tannic acid immediately before use as follows:

[2]Our observation that the isolated vinblastine-induced tubulin paracrystal can be processed for electron microscopy by treating paracrystals with tannic acid and OsO_4 without glutaraldehyde fixation (Fujiwara and Tilney, 1975) supports this conclusion.

Dissolve tannic acid (1-8%) in Na-phosphate buffer (100 m*M*, pH 7.0) or other appropriate buffer solution determined empirically. Stir and/or heat to dissolve and clear by centrifugation or filtration on cooling. Add glutaraldehyde to the final concentration of 1-3% and readjust pH. The freshly made fixative should be used within 6 hr.

3. Fix the specimen by either resuspending the pellet in the fixative or overlaying the pellet with the fixative and incubate for 1-18 hr at 0-24°C. The choice of temperature and the duration of fixation must be determined empirically. The higher concentrations of tannic acid become insoluble at 0-4°C.
4. Wash the specimen with the buffer to remove unbound tannic acid.
5. Postfix with OsO_4 in a manner similar to a conventional procedure for electron microscopy.
6. Dehydrate, embed, and cut thin sections.

Although a Na-phosphate buffer is most commonly used, other buffers may also be used (Tables I and II). Potassium phosphate should not be used, since K^+ causes a white precipitate when mixed with tannic acid. Depending on one's goals, the investigator is wise to establish empirically the optimal conditions for fixation. Tannic acid concentration may vary from 1% to 8% for use in the determination of protofilament counts in microtubules, and higher concentrations produce more electron-dense precipitates around microtubules. However, as we will discuss in the following section, lower concentrations (0.2-1%) may be optimal for some ultrastructural analyses. The pH of the fixative should be adjusted so that the specimen becomes most stable. For isolated microtubule-containing organelles, the pH should be that of isolation media, and for *in vitro* assembled microtubules it is the pH of the assembly buffer. Table II summarizes conditions for various microtubule systems.

It is not necessary to expose microtubules to glutaraldehyde and tannic acid simultaneously in order to observe protofilaments. Microtubules may be either fixed in glutaraldehyde and then treated with tannic acid (Tilney *et al.*, 1973) or treated first with tannic acid and then fixed with glutaraldehyde (Tilney *et al.*, 1973; Green *et al.*, 1975). In an unusual case, specimens can be treated with tannic acid without glutaraldehyde, followed by osmication, in order to preserve the structure for electron microscopy (Fujiwara and Tilney, 1975).

2. *In Situ* MICROTUBULES

Ultrastructure of microtubules in cells and tissues may also be studied by the tannic acid method. The general procedure is similar to that for *in vitro* microtubules; again, however, the precise conditions must be determined empirically for each cell and tissue type. Table III summarizes published procedures. The greatest difficulty involved in this approach is the penetration of tannic acid through the plasma membrane into the cytoplasm. This problem can be

TABLE II

SUMMARY OF CONDITIONS USED TO FIX ISOLATED MICROTUBULE ORGANELLES AND *in Vitro* REASSEMBLED MICROTUBULES WITH GLUTARALDEHYDE AND TANNIC ACID

Specimen	Tannic acid concentration (%)	Glutaraldehyde concentration (%)	Buffer and pH	Fixation condition	Reference
Mitotic apparatus	8	2	Na-PO_4 (100 mM) 6.3	1 hr 0°C	Tilney *et al.* (1973)
Axostyle	4	2	Na-PO_4 (100 mM) 6.8	1 hr 0°C	Tilney *et al.* (1973)
Axostyle	1	1	Na-PO_4 (10 mM) 7.0	24 hr room temperature	Woodrum and Linck (1980)
Axoneme	2, 4, and 8	2	Na-PO_4 (100 mM) 7.0–7.5	1 hr 0°C	Tilney *et al.* (1973)
Centriole	8	2	Na-PO_4 (100 mM) 7.0	1 hr 0°C	Tilney *et al.* (1973)
Vinblastine-induced tubulin paracrystal	4	2	Na-PO_4 (100 mM) 7.0	1 hr 0°C	Fujiwara and Tilney (1975)
Vinblastine-induced tubulin paracrystal	4	–	Na-PO_4 (100 mM) 7.0	1 hr 0°C	Fujiwara and Tilney (1975)
Microtubule	8	2	Na-PO_4 (100 mM) 6.5	1 hr 0°C	Tilney *et al.* (1973)
Microtubule	8	3	Na-PO_4 (200–500 mM) 6.2 or 6.8	1 hr room temperature	Burton *et al.* (1975)
Microtubule	8	2.5	Na-cacodylate (100 mM) 7.2	—	Green *et al.* (1975)
Microtubule	8	3	Na-PO_4 (50 mM) 6.8	1 hr room temperature	Pierson *et al.* (1978); Burton and Himes (1978)
Microtubule	6	1	Na-PO_4 (150 mM) 6.7	Overnight	Binder and Rosenbaum (1978)
Microtubule	1	1	Na-PO_4 (10 mM) 7.0	Overnight	Kim *et al.* (1979)

TABLE III

SUMMARY OF CONDITIONS USED TO FIX *In Situ* MICROTUBULES WITH GLUTARALDEHYDE AND TANNIC ACID

Specimen	Tannic acid concentration (%)	Glutaraldehyde concentration (%)	Buffer and pH	Fixation condition	Agents added	Reference
Rat sperm tail	2	2.5	Na-PO_4(50 m*M*) 6.8	1.5–2.0 hr 0°C	Digitonin (saturated)	Mizuhira and Futaesaku (1971)
Rat sperm tail	2	2.5	Veronal-acetate (70 m*M*) 6.8	1.5 hr 4°C	Na_2SO_4(1.7%)	Mizuhira and Futaesaku (1972)
Echinosphaerium	8	2	Na-PO_4 (50 m*M*) 6.8	1 hr 0°C	$CaCl_2$ (15 m*M*) digitonin (1%)	Tilney *et al.* (1973)
Crayfish nerve cord	8	3	Na-PO_4 (50 m*M*) 6.5–6.8	1 hr room temperature	$CaCl_2$ (15 m*M*)	Burton and Hinkley (1974)
Cockroach epidermal cells	2–4	2.5	Na-PO_4 (100 m*M*) 7.2	Overnight	—	Nagano and Sazuki (1975)
Crane-fly spermatocyte	8	4	Cacodylate (100 m*M*)	30 min	—	LaFountain and Thomas (1975)
Frog lung fluke sperm Crayfish sperm Frog olfactory axon Goldfish brain	2–8	3	Na-PO_4 (200–500 m*M*) 6.2–7.0	1 hr room temperature	$CaCl_2$ (15 m*M*) digitonin (0.5–1%)	Burton *et al.* (1975)

minimized (1) by using tannic acid that contains higher amounts of low-molecular-weight species and (2) by using detergents to permeabilize the plasma membrane. D. A. Begg (personal communication) has found that tannic acid penetrates poorly into cells in the presence of glutaraldehyde and thus recommends that tannic acid be used after glutaraldehyde fixation. High concentrations of tannic acid tend to obscure the ultrastructure of the cytoplasm, and for this reason lower concentrations should be tried first.

As we recall from Table I, most detergents form precipitates when added to solutions of tannic acid. However, of possible importance is the fact that a mixture of digitonin–tannic acid remains soluble. Digitonin has been used in the past to permeabilize or solubilize cell membranes (Gibbons, 1965; Mizuhira and Futaesaku, 1971; Tilney *et al.*, 1973; Burton *et al.*, 1975). Thus, a combined digitonin–tannic acid solution may provide a means to apply the tannic acid procedure to the study of *in situ* cytoplasmic microtubules and other cytoskeletal elements. Table I also shows that glycerol is compatible with tannic acid.

IV. Use of Tannic Acid in Other Ultrastructural Studies

In this section we review briefly studies unrelated to microtubules but in which tannic acid has been used to demonstrate various ultrastructural features of normal and experimentally treated biological specimens. Special effects yielded by tannic acid include improved contrast, better preservation of the extracellular matrix and the membrane, and improved demonstration of microfilaments and their polarity after decoration with myosin fragments.

Simionescu and Simionescu (1976a,b) obtained electron micrographs of mammalian tissues with much better contrast when they treated fixed tissues with tannic acid. They also reported ultrastructurally improved preservation of the general cell morphology, the extracellular matrix, and the membrane. It is important to note that the tannic acid treatment comes after fixing the specimen with glutaraldehyde and OsO_4, yet better morphological preservation results. This is because tannic-acid-treated tissues appear to be more resistant to extraction during dehydration and embedding (Simionescu and Simionescu, 1976a). For the purpose of general ultrastructural studies, tannic acid must be of the low-molecular-weight type, which minimizes the problem of penetration and the formation of large precipitates that interfere with high-resolution microscopy.

Tannic acid has proved to be very useful in demonstrating many morphological features that are unclear or impossible to see in conventionally fixed specimens. Rodewald and Karnovsky (1974) showed detailed extracellular structures of the mammalian renal glomerulus present in the space between the podocyte processes and the endothelial cells of the glomerular capillary. Tannic acid provided increased contrast of the plasma membrane, the slit diaphragm, the basement membrane, and cell surface coats, and allowed them to visualize fine structure of the slit diaphragm, which conventionally fixed materials did not show. The ability of tannic acid to give a highly contrasted image of the plasma membrane yielded a clear demonstration of gap and tight junctions of the mouse liver tissue (Van Deurs, 1975), the asymmetry of the membrane of the sarcoplasmic reticulum (Saito *et al.*, 1978), and a discovery of a new type of membrane specialization in the triad of the skeletal muscle (Somlyo, 1979).

Another example, which is highly relevant to the subject of this volume, is the usefulness of tannic acid in visualizing microfilaments. Leblond and co-workers noted at the light microscope level that many cells processed by tannic acid–phosphomolybdic acid–amido black (TPA) technique revealed fibrillar structures in their cytoplasm (Puchtler and Leblond, 1958; Leblond *et al.*, 1960, 1966). Unfortunately, the chemical characterization of these fibrillar structures was unknown. However, we now know ultrastructurally that the particular areas of those cells and tissues (e.g., the terminal web region of the intestinal epithelium) or the particular cell type (e.g., smooth muscle, skin cells, and pillar cells of the organ of Corti) contain an abundant amount of microfilaments and/or intermediate filaments.

To identify a cytoplasmic filament as a microfilament at the ultrastructural level, it is most common to employ HMM or S_1 decoration techniques (Ishikawa *et al.*, 1969). Perhaps more important, this method also provides the polarity of each microfilament because of the fact that the specific binding of HMM or S_1 forms an arrowhead configuration along an actin filament (Huxley, 1963). It is, however, not easy to demonstrate unambiguously the direction of arrowheads in thin-sectioned material. Recently, Begg *et al.* (1978) overcame this technical difficulty by adding 0.2% tannic acid to the glutaraldehyde fixative, which both improved the preservation of the S_1-microfilament complex and enhanced the contrast, and obtained better visualization of the arrowhead. These investigators also noted that the addition of tannic acid improved dramatically the preservation of all types of cytoplasmic filaments. Several investigators capitalized on this latter advantage of tannic acid to study various aspects of microfilament organization in the cytoplasm of various cells. Goldman *et al.* (1979) discovered alternating light and dark electron densities along bundles of actin filaments in BHK-21 cells and suggested that these banding patterns might reflect nonuniform localization of various contractile proteins within stress fibers. Using plant material, Seagull and Heath (1979) showed improved preservation of cytoplasmic filaments and attempted to investigate morphologically possible sites of microtubule–microfilament interaction.

As has been mentioned, in order to obtain satisfactory tannic acid effects on the ultrastructure of intact cells, it is extremely important to use low-molecular-weight tannic acid at low concentrations, usually less than 1%. Concentrations higher than this level cause extensive precipitation inside and outside of cells, and this in turn obscures the fine structural detail.

V. Conclusion

It was 10 years ago that Mizuhira and Futaesaku (1971) published magnificent electron micrographs of tannic acid treated axonemes of the rat sperm tail, and its usefulness was immediately recognized by many investigators studying the morphology of microtubules. Tannic acid can most dramatically demonstrate the number of protofilaments in the wall of a microtubule. However, it is important to note that it can also accentuate existing structures associated laterally along microtubules of various origin. Perhaps it is this latter property that will allow tannic acid to become a pivotal factor in understanding various functions of microtubules. For example, one of the important morphological questions that needs to be answered is whether or not there is a structural basis to support various interactions among several types of cytoplasmic filamentous elements—including microtubules, microfilaments, and intermediate filaments—and among these filaments and cell organelles. Given a relatively short history of the use of tannic acid for ultrastructural studies and given the fact that is has yielded many new pieces of morphological information, it is justified to expect that tannic acid will provide means to visualize ultrastructural details that presently are not known.

Acknowledgments

We thank Dr. D. A. Begg for his helpful discussions and suggestions. We are also happy to acknowledge technical and other assistance provided by Dottie Rowe, Joan Nason, Mike Smith, Elena McBeath, Randy Byers, and Glen White. The original work mentioned in the text was supported by NIH grant GM 25637 to K. F.

References

Agrawal, H. O., Kent, J. W., and Mackay, D. M. (1965). *Science* **148,** 638-640.

Amos, L. A. (1977). *J. Cell Biol.* **72,** 642-654.

Amos, L. A. (1979). *In* "Microtubules" (K. Roberts and J. S. Hyams, eds.), pp. 1-64. Academic Press, New York.

Amos, L. A., and Klug, A. (1974). *J. Cell Sci.* **14,** 523-549.

André, J., and Thiéry, J. P. (1963). *J. Microsc. (Paris)* **2,** 71-80.

Begg, D. A., Rodewald, R., and Rebhun, L. I. (1978). *J. Cell Biol.* **79,** 846-852.

Behnke, O., and Zelander, T. (1966). *Exp. Cell Res.* **43,** 236-239.

Binder, L. I., and Rosenbaum, J. L. (1978). *J. Cell Biol.* **79,** 500–515.
Bloom, W., and Fawcett, D. W. (1975). "A Textbook of Histology," 10th ed. Saunders, Philadelphia, Pennsylvania.
Burton, P. R., and Himes, R. H. (1978). *J. Cell Biol.* **77,** 120–133.
Burton, P. R., and Hinkley, R. E. (1974). *J. Submicrosc. Cytol.* **6,** 311–326.
Burton, P. R., Hinkley, R. E., and Pierson, G. B. (1975). *J. Cell Biol.* **65,** 227–233.
Cohen, C., Harrison, S. C., and Stephens, R. E. (1971). *J. Mol. Biol.* **59,** 375–380.
De Brabander, M., De Mey, J., Joniau, M., and Gueuens, G. (1977). *Cell Biol. Int. Rep.* **1,** 177–183.
Erickson, H. P. (1974). *J. Cell Biol.* **60,** 153–167.
Friedman, M. H. (1970). *J. Ultrastruct. Res.* **32,** 226–236.
Fujiwara, K., and Tilney, L. G. (1975). *Ann. N.Y. Acad. Sci.* **253,** 27–50.
Futaesaku, Y. (1979). *Seitai no Kagaku* **30,** 470–482.
Gall, J. C. (1966). *J. Cell Biol.* **31,** 639–643.
Gibbons, I. R. (1965). *Arch. Biol.* **76,** 317–352.
Goldman, R., Pollard, T., and Rosenbaum, J. (1976). "Cell Motility." Cold Spring Harbor Lab., Cold Spring Harbor, New York.
Goldman, R. D., Chojnacki, B., and Yerna, M. (1979). *J. Cell Biol.* **80,** 759–766.
Green, L. H., Brandis, J. W., Turner, F. R., and Raff, R. A. (1975). *Biochemistry* **14,** 4487–4491.
Grimstone, A. V., and Klug, L. R. (1966). *J. Cell Sci.* **1,** 351–362.
Hayat, M. A. (1970). "Principles and Techniques of Electron Microscopy," Vol. 1. Van Nostrand-Reinhold, New York.
Huang, B., Piperno, G., Ramanis, Z., and Luck, D. J. L. (1981). *J. Cell Biol.* **88,** 80–88.
Huxley, H. E. (1963). *J. Mol. Biol.* **7,** 281–308.
Ishikawa, H., Bischoff, R., and Holtzer, H. (1969). *J. Cell Biol.* **43,** 312–328.
Kalina, M., and Pease, D. C. (1977a). *J. Cell Biol.* **74,** 726–741.
Kalina, M., and Pease, D. C. (1977b). *J. Cell Biol.* **74,** 742–746.
Kiefer, B. I., Sakai, H., Solaris, A. J., and Mazia, D. (1966). *J. Mol. Biol.* **20,** 75–79.
Kim, H., Binder, L. I., and Rosenbaum, J. L. (1979). *J. Cell Biol.* **80,** 266–276.
Kirschner, M. W. (1978). *Int. Rev. Cytol.* **54,** 1–71.
LaFountain, J. R., Jr. (1975). *BioSystems* **7,** 363–369.
LaFountain, J. R., Jr., and Thomas, H. R. (1975). *J. Ultrastruct. Res.* **51,** 340–347.
Leblond, C. P., Puchtler, H., and Clermont, Y. (1960). *Nature (London)* **186,** 784–788.
Leblond, C. P., Sarkar, K., Kallenbach, E., and Clermont, Y. (1966). *Cancer Res.* **26,** 2259–2266.
Ledbetter, M. C., and Porter, K. R. (1964). *Science* **144,** 872–874.
Linck, R. W. (1976). *J. Cell Sci.* **20,** 405–439.
Linck, R. W. (1979). *In* "The Spermatozoon" (D. W. Fawcett and J. M. Bedford, eds.), pp. 99–115. Urban & Schwarzenberg, Munich.
Linck, R. W. (1982). *Ann. N.Y. Acad. Sci.* **383,** in press.
Linck, R. W., and Amos, L. A. (1974). *J. Cell Sci.* **14,** 551–559.
Linck, R. W., and Langevin, G. L. (1981). *J. Cell Biol.* **89,** 323–337.
McEwen, B., and Edelstein, S. J. (1980). *J. Mol. Biol.* **139,** 123–145.
Markham, R., Frey, S., and Hills, G. J. (1963). *Virology* **20,** 88–125.
Mizuhira, V., and Futaesaku, Y. (1971). *Proc.—Annu. Meet., Electron Microsc. Soc. Am.* **29,** 494–495.
Mizuhira, V., and Futaesaku, Y. (1972). *Acta Histochem. Cytochem.* **5,** 233–236.
Murphy, D. B., and Borisy, G. G. (1975). *Proc. Natl. Acad. Sci. U.S.A.* **72,** 2696–2700.
Nagano, T., and Suzuki, F. (1975). *J. Cell Biol.* **64,** 242–245.
Pease, D. C. (1963). *J. Cell Biol.* **18,** 313–326.
Phillips, D. M. (1966). *J. Cell Biol.* **31,** 635–638.
Pierson, G. B., Burton, P. R., and Himes, R. H. (1978). *J. Cell Biol.* **76,** 223–228.

Pierson, G. B., Burton, P. R., and Himes, R. H. (1979). *J. Cell Sci.* **39,** 89-99.
Piperno, G., Huang, B., Ramanis, Z., and Luck, D. J. L. (1981). *J. Cell Biol.* **88,** 73-79.
Puchtler, H., and Leblond, C. P. (1958). *Am. J. Anat.* **102,** 1-31.
Roberts, K., and Hyams, J. S. (1979). "Microtubules." Academic Press, New York.
Rodewald, R., and Karnovsky, M. J. (1974). *J. Cell Biol.* **60,** 423-433.
Sabatini, D. D., Bensch, K., and Barrnett, R. J. (1963). *J. Cell Biol.* **17,** 19-58.
Saito, A., Wang, C. T., and Fleischer, S. (1978). *J. Cell Biol.* **79,** 601-616.
Seagull, R. W., and Heath, I. B. (1979). *Eur. J. Cell Biol.* **20,** 184-188.
Simionescu, N., and Simionescu, M. (1976a). *J. Cell Biol.* **70,** 608-621.
Simionescu, N., and Simionescu, M. (1976b). *J. Cell Biol.* **70,** 622-633.
Somlyo, A. V. (1979). *J. Cell Biol.* **80,** 743-750.
Stephens, R. E., and Edds, K. T. (1976). *Physiol. Rev.* **56,** 709-777.
Tilney, L. G., and Porter, K. R. (1967). *J. Cell Biol.* **34,** 327-343.
Tilney, L. G., Bryan, J., Bush, D. J., Fujiwara, K., Mooseker, M. S. Murphy, D. B., and Snyder, D. H. (1973). *J. Cell Biol.* **50,** 267-275.
Tyson, G. E., and Bulger, R. E. (1973). *Z. Zellforsch. Mikrosk. Anat.* **141,** 443-458.
Van Deurs, B. (1975). *J. Ultrastruct. Res.* **50,** 185-192.
Warner, F. D., and Satir, P. (1973). *J. Cell Sci.* **12,** 313-326.
Woodrum, D. T., and Linck, R. W. (1980). *J. Cell Biol.* **87,** 404-414.

Chapter 15

The Cyclic Tyrosination/Detyrosination of Alpha Tubulin

WILLIAM C. THOMPSON

Departments of Biological Sciences and Chemistry
University of California, Santa Barbara
Santa Barbara, California

I. Introduction

Of primary interest in the study of the cytoskeleton are the mechanisms whereby the assembly, disassembly, interactions, and functions of the various cytoskeletal elements are controlled in the living cell. In view of the rapid assembly and disassembly of microtubules during the cell cycle, as well as the

ISBN 0-12-564124-9

speed of intracellular translocations dependent on microtubules, the control elements should be capable of rapid changes. The recent demonstration (Thompson *et al.*, 1979) that the terminal tyrosine residue of alpha tubulin exchanges rapidly with free tyrosine in living muscle cells makes this posttranslational modification an interesting candidate for such a control element. The detyrosination phase of the rapid turnover of terminal tyrosine appears rather tightly linked *in vivo* to the presence of intact microtubules. *In vitro* experiments (Arce *et al.*, 1978; Thompson *et al.*, 1980) indicate that the tyrosination phase of the turnover of terminal tyrosine occurs only on free tubulin dimers, and not on microtubules. Therefore, the rapid cyclic tyrosination/detyrosination observed in living cells seems to be linked to the dynamic exchange of tubulin dimers between the free dimer pool and the microtubules.

Since the literature concerning this posttranslational modification of tubulin occurs in diverse journals and abstracts and has not been reviewed, the first section of this chapter will be a brief review of the literature through mid-1980. In the following methods section several methods deemed useful to the study of the cellular aspects of this molecular modification will be given. Basic biochemical methods, such as the elegant purification of tubulin:tyrosine ligase (TTLase) by Murofushi (1980), will not be presented here. Possible implications of the linkage of the rapid tyrosine turnover with the dynamic exchange of free tubulin dimers and microtubules will be discussed at the end of this chapter.

II. Review

A. Characterization of Posttranslational Tyrosine Incorporation

Posttranslational incorporation of tyrosine into protein by a mechanism not requiring RNA was described by Barra *et al.* (1973a) utilizing a rat brain supernatant system. In addition to the brain supernatant proteins and tyrosine, the enzymatic reaction required ATP, Mg^{2+}, and K^{+}. This enzymatic system was also shown to catalyze the incorporation of phenylalanine and L-3,4-dihydroxyphenylalanine into the same acceptor protein, although tyrosine is by far the best amino acid substrate (Barra *et al.*, 1973b; Rodriguez *et al.*, 1975; Deanin and Gordon, 1976). Barra and co-workers (1974) soon noted similarities between the endogenous tyrosine acceptor protein of rat brain supernatants and tubulin, such as reversible polymer formation, vinblastine precipitation and co-chromatography with the colchicine-binding protein. The penultimate residue of the tyrosinated protein was identified as either glutamine or glutamic acid (Arce *et al.*, 1975a).

B. Characterization of Tubulin: Tyrosine Ligase Substrates

A significant advance in the study of this enzymatic reaction was made possible by the use of purified tubulin as a substrate instead of crude supernatant systems (Arce *et al.*, 1975b; Raybin and Flavin, 1975). The endogenous tyrosine acceptor activity of the supernatant protein decays on storage; therefore the cell-free tyrosine incorporation can be made entirely dependent on the addition of purified tubulin. Raybin and Flavin (1975) reported that the presence or absence of C-terminal tyrosine apparently did not affect the ability of cold-depolymerized tubulin to assemble into microtubules when warmed. Since subclasses of tubulin exist *in vivo,* the exact nature of the TTLase substrate has been of continuing interest. The tyrosine incorporation was shown to occur exclusively into the alpha tubulin subunit (Raybin and Flavin, 1975). Free tubulin dimers definitely are a substrate for TTLase, as has been demonstrated by the lack of any inhibition of tyrosine incorporation in the presence of colchicine (Arce *et al.*, 1975b; Raybin and Flavin, 1975), in the presence of podophyllotoxin (Thompson *et al.*, 1979), or in the presence of vinblastine (Nath and Flavin, 1978). A clear-cut answer as to whether microtubules could be a substrate for TTLase has not been so simple to obtain. Arce *et al.* (1978) presented an equilibrium mixture of tubulin dimers and microtubules to the TTLase system and showed that the free dimer pool became much more heavily labeled with radioactive tyrosine than did the microtubule pool. Although the results were ambiguous because of a simultaneous detyrosination of microtubular tubulin in their system, these authors suggested that tyrosine incorporation could have occurred exclusively into free tubulin dimers, some of which were subsequently taken into the microtubules because of the dynamic equilibrium between the two tubulin pools. Based on our increased understanding of the nature of that dynamic equilibrium in cell-free systems (Margolis and Wilson, 1978) and the methods given in Section III,B of this chapter, it has been possible to demonstrate that this scenario is correct (Thompson *et al.*, 1980). Nath and Flavin (1979a) briefly note that T. Kobayashi has found that intact doublet microtubules from sea urchin sperm are unable to accept tyrosine, but the tubulin dimers released from them are capable of being tyrosinated by the TTLase system.

There have been some unsuccessful attempts to separate tyrosinated tubulin physically from nontyrosinated or nonsubstrate alpha tubulin. Lu and Elzinga (1977) separated alpha tubulin into two peaks by hydroxylapatite chromatography in the presence of sodium dodecyl sulfate. Raybin and Flavin (1977b) showed that both alpha tubulins separated by this method were tyrosine acceptors in the TTLase system. Gozes and Littauer (1978) labeled chick brain tubulin with radioactive tyrosine using the TTLase system and then separated the alpha tubulin into five bands by isoelectric focusing. Each of the five species of alpha tubulin contained labeled tyrosine.

C. Studies of Tubulin:Tyrosine Ligase Activity Levels in Tissues

The use of purified tubulin as an exogenous tyrosine acceptor in cell-free systems made possible the accurate determination of tubulin:tyrosine ligase activity in extracts of tissues with low endogenous tubulin concentrations or very high enzyme activity levels. Utilizing these methods tubulin:tyrosine ligase activity was shown not to be confined to brain, but was present in all tissues of the rat or chick examined (Deanin and Gordon, 1976; Raybin and Flavin, 1977b). In adult animals, however, brain tissue has the highest levels of TTLase activity, and that activity is apparently not confined to one cell type. Preparations of synaptosomes have been shown to incorporate labeled tyrosine into endogenous synaptosomal alpha tubulin *in vitro* (Deanin and Gordon, 1976).

Developmental patterns of changes in TTLase activity are tissue specific. Barra *et al.* (1973a) showed that TTLase activity of rat brain decreases during the first 30 days of neonatal development and then remains fairly constant. Chick brain TTLase activity levels are substantially higher in the 12 to 18-day-old embryo than in week-old hatched chicks (Deanin *et al.*, 1977; Rodriguez and Borisy, 1978). TTLase activity of chick skeletal muscle reaches a peak of activity in 12–14-day embryos that is much higher than that of even embryonic brain, but the muscle enzyme activity decreases rapidly with further development to a level only a fraction of that of brain in week-old chicks (Deanin *et al.*, 1977). The highest TTLase activity reported for any tissue source is that of embryonic chick sympathetic ganglia (0.12 nmoles tyrosine fixed/min mg supernatant protein) (Pierce *et al.*, 1978).

Attempting to clarify the role of changes in TTLase activity in processes of differentiation, several systems of *in vitro* differentiation of nerve and muscle have been studied. Levi *et al.* (1978) used nerve growth factor to induce pheochromocytoma cells to differentiate along a neuronal pathway and noted that the morphological change was accompanied by a two- to threefold increase in TTLase specific activity. Nath and Flavin (1979a), on the other hand, report very high TTLase specific activity levels (0.40 nmoles of tyrosine fixed/min mg supernatant protein) in rat sympathetic ganglia after 2 days of culture *in vitro* either in the presence or absence of nerve growth factor. These authors also report an increase of only 20% in TTLase specific activity when neuroblastoma/glioma cells are induced to differentiate in culture by cAMP. Muscle differentiation from myoblasts to spontaneously contracting myotubes is characterized *in vitro* by a sustained high specific activity of TTLase, unlike embryonic muscle differentiation (W. C. Thompson, unpublished observations).

Tubulin:tyrosine ligase from many species and tissue sources can utilize exogenous brain tubulin as the tyrosine acceptor. Preston *et al.* (1979), in a study of the phylogenetic distribution of TTLase, were able to demonstrate the incorporation of tyrosine into beef brain tubulin catalyzed by supernatant proteins from all vertebrate species tested. These workers were unable to demonstrate TTLase activity in the supernatant proteins of invertebrates or protochordates. Kobayashi and Flavin (1977) were able to demonstrate low levels of TTLase activity in at least one invertebrate species, the sea urchin, but only after several steps of blind purification of the enzyme, since no TTLase activity was detected in crude supernatant fractions (Raybin and Flavin, 1977b).

Forrest and Klevecz (1978) reported changes in TTLase activity synchronized with the cell cycle in CHO cells and transformed CHO cells growing in culture. Unfortunately, this study was seriously flawed by the use of undialyzed cell extracts and the addition of only 22.5 n*M* tyrosine to the assay systems. Since the added tyrosine concentration was only one-thousandth that required for half-maximal tyrosine incorporation in the TTLase system, authentic changes in TTLase activity cannot be distinguished from apparent changes in TTLase activity due to variations in tyrosine concentration and specific activity in the assay mixtures. Therefore, a definitive correlation of TTLase activity with the cell cycle awaits further work.

D. Purification and Characterization of Tubulin:Tyrosine Ligase

Rabin and Flavin (1977a) reported a partial purification of TTLase from beef brain by a method involving ammonium sulfate fractionation and DEAE cellulose chromatography. This method had the disadvantage that much of the enzymatic activity was lost during ammonium sulfate treatment and during dialysis. However, these authors were able to demonstrate that the apparent molecular weight of the DEAE-purified enzyme was about 35,000 daltons, whereas in the presence of the tubulin substrate the enzymatic activity was found in a 150,000-dalton complex, presumably containing a molecule of TTLase and a tubulin dimer. Based on these findings, an extensive purification (8100-fold) of TTLase from porcine brain has recently been reported by Murofushi (1980). Murofushi avoided any ammonium sulfate fractionation by subjecting a postmicrotubule supernatant directly to DEAE chromatography to free the ligase from tubulin. The DEAE fraction containing free TTLase was then subjected to two successive affinity chromatography steps, the first utilizing an ATP-Sepharose 4B matrix and the second utilizing a tubulin-Sepharose 4B matrix. The resulting protein is monodisperse on SDS gel electrophoresis migrating with an apparent molecular weight of 46,000. On gel filtration the apparent molecular weight of the TTLase activity was 37,000. The difference in apparent molecular weight might be due to some interaction of the enzyme with the gel filtration medium. The purified enzyme has a K_m for ATP of 8.5 μM and a K_m for tyrosine of 30 μM. The optimum pH for TTLase activity is around 8, with a second peak of activity at pH 6.5.

E. Enzymatic Removal of Carboxyl-Terminal Tyrosine from Tubulin

The tyrosine incorporated at the carboxyl terminus of alpha tubulin can be removed by several enzymes. Rodriguez *et al.* (1973) reported that some of the terminal tyrosine was released in the presence of ADP and inorganic phosphate. Raybin and Flavin (1976) demonstrated that arsenate could substitute for inorganic phosphate in this reaction. Raybin and Flavin (1977a) also reported that the ability to release terminal tyrosine in the presence of ADP and inorganic phosphate copurified with TTLase, indicating that a partial reversal of the TTLase reaction coupled with hydrolysis of an activated intermediate is probably responsible for that tyrosine release.

Removal of the terminal tyrosine residue *in vitro* by the use of pancreatic carboxypeptidase A has been utilized by workers in several laboratories to determine the extent to which alpha tubulin has been tyrosinated *in vivo* (see Section II,G). Argaraña *et al.* (1978) reported an enzyme activity present in brain extracts that is capable of catalyzing the removal of the terminal tyrosine by a carboxypeptidase-like activity, which was physically separable from TTLase. Even heat-denatured tubulin served as a substrate for tyrosine release by the carboxypeptidase-like activity, unlike the TTLase-catalyzed tyrosine release. This carboxypeptidase-like brain enzyme has been partially purified from chicken brain and shown to differ from the pancreatic enzyme in substrate specificity and susceptibility to inhibitors (Argaraña *et al.*, 1980). Martensen and Flavin (1979, p. 612) partially purified a similar enzyme from sheep brain. In addition to complete inhibition of the activity by 0.1 *M* KCl or NaCl, the enzymatic activity was inhibited by colchicine and by a nucleoside triphosphate regenerating system. Thompson (1977b) reported that a regeneration of ATP in a crude chick muscle supernatant system inhibited the turnover of the terminal tyrosine residue of the endogenous muscle tubulin. High salt concentrations and/or nucleoside triphosphate regeneration have been utilized to inhibit the detyrosinating activity in cell-free systems while allowing tyrosine incorporation.

Perhaps another type of tyrosine release, and one that seems to be important *in vivo* (Thompson *et al.*, 1979), utilizes intact microtubules as the substrate for detyrosination. When tubulin that had incorporated labeled tyrosine at the C-terminal position *in vivo* was allowed to polyermize *in vitro*, much of the terminal tyrosine was lost during the warm incubation (Thompson, 1977a). The specific activity of the resultant microtubular tubulin was found to be approximately one half that of the free tubulin dimers. Hallak *et al.* (1977) demonstrated that those conditions that allowed for microtubule polymerization were also best for tyrosine release. Arce *et al.* (1978) separated tyrosinated microtubules and dimers and demonstrated that the removal of labeled tyrosine was much greater from the microtubule pool than from the dimer pool during subsequent warm incubation. Rodriguez and Borisy (1979a) demonstrated that in neonatal rat brain the free dimer tubulin pool was tyrosinated to about twice the extent of the microtubules *in vivo*. This difference could result from a microtubule-dependent detyrosination reaction.

F. Studies of the Cyclic Tyrosination/Detyrosination *in Vivo*

Posttranslational incorporation of labeled tyrosine at the C-terminus of alpha tubulin *in vivo* has been demonstrated in cells and tissues treated with inhibitors of protein synthesis (Thompson, 1977a; Argaraña *et al.*, 1977; Raybin and Flavin, 1977b; Nath and Flavin, 1979a). Tyrosine incorporation under these conditions indicates that tubulin substrate is being made available at a rate much in excess of the residual tubulin synthetic rate, either because the drug treatment makes available previously untyrosinated tubulin as a suitable substrate or by detyrosination of previously tyrosinated tubulin. Thompson *et al.* (1979) made a study of the rate of turnover of the terminal tyrosine residue of alpha tubulin in living muscle cells in culture. Rapid turnover of the terminal tyrosine residue was demonstrated in living muscle cells by two types of pulse/chase experiment, in which the radioactivity of the terminal tyrosine residue of alpha tubulin could be determined. In the first method the cells were preincubated with hypertonic culture medium to inhibit the initiation of protein synthesis (Wengler and Wengler, 1972), then pulse-labeled under these conditions with tritiated tyrosine, and subsequently chased with fresh culture medium. Cultures were analyzed by dissolving the cells *in situ* in a buffer containing sodium dodecyl sulfate (SDS) and proteolytic inhibitors. On separation of the cellular proteins by polyacrylamide gel electrophoresis in the presence of SDS, a protein comigrating with alpha tubulin was the major labeled protein. The radioactivity of the terminal tyrosine residue decayed during the chase incubation with a half-life of 37 min in the interval of 30–90 min after the pulse. For the second method (see Section III,C) the muscle cultures were not treated to inhibit protein synthesis, but were given a pulse of tritiated tyrosine directly into the culture medium at least 18 hr after they had last been fed. Under these conditions the posttranslational tyrosine incorporation accounts for only about 5% of the total tyrosine incorporated. However, when tubulin was isolated from such pulse-labeled cultures by vinblastine sulfate treatment in the cold, the bulk of the radioactivity of the alpha tubulin was due to the terminal tyrosine residue (Fig. 1A). This could be demonstrated by susceptibility of the C-terminal tyrosine radioactivity to release by carboxypeptidase A (Fig. 1B). During a chase incubation in fresh medium containing unlabeled tyrosine the radioactivity of the terminal tyrosine residue decreased with an apparent first-order decay and a half-life of 37 min (Fig. 2). Preincubation of the muscle cells with antimicrotubular drugs inhibited the rapid turnover of the terminal tyrosine as evidenced by a decrease of tyrosine incorporation during the pulse and by an increase in the half-life of the terminal tyrosine radioactivity during the chase.

In a study of the effect of podophyllotoxin on the kinetics of labeled tyrosine incorporation into tubulin in cultured muscle cells, Thompson (1979) reported that "podophyllotoxin-capped" microtubules continue to provide detyrosinated tubulin substrate at a normal rate for 10–15 min. However, preincubation of cultures for that length of time brought about a severe decrease in tyrosine incorporation during a subsequent labeled tyrosine pulse incubation. These results were considered consistent with a rapid opposite-end net assembly and disassembly of microtubules occuring *in vivo,* with 10–15 min approximating the time required for a newly incorporated tubulin subunit to "transit" an average-length microtubule and be released from the opposite end. These studies also indicated the presence on the microtubules of a pool of detyrosinated subunits, which became available for tyrosination when muscle cultures were chilled.

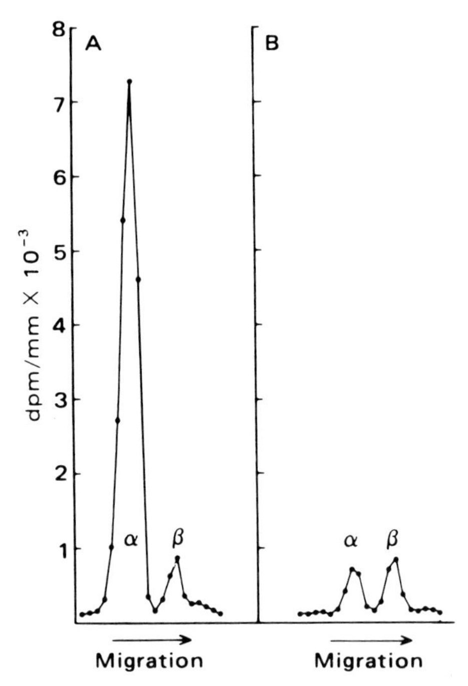

FIG. 1. Radioactivity of alpha- and beta-tubulin subunits isolated from muscle cultures that had been pulse-labeled for 15 min with [^{3}H]tyrosine. Tubulin was isolated from a muscle cell homogenate by vinblastine precipitation in the presence of carrier brain tubulin. Alpha and beta tubulins were resolved by SDS electrophoresis. (A) Radioactivity of the two tubulin subunits is shown. (B) A second aliquot of the same labeled tubulin was digested with carboxypeptidase A before electrophoresis. Radioactivity of the beta subunit approximates the translational contribution to the radioactivity of the alpha subunit. Reprinted from Thompson *et al.*, 1979.

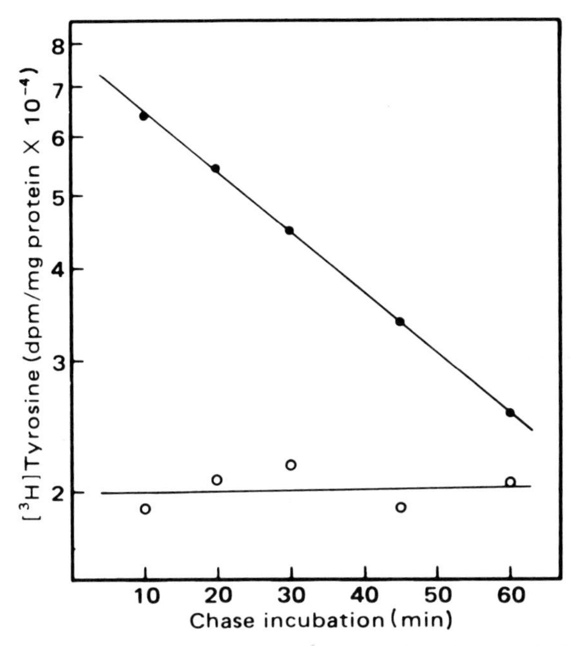

FIG. 2. Loss of labeled tyrosine from the carboxyl terminus of alpha tubulin in unperturbed muscle cultures. Cultures pulse-labeled as in Fig. 1 were incubated for the indicated times in chase medium containing excess unlabeled tyrosine. Tubulin was then isolated and the radioactivity of the alpha- and beta-tubulin subunits was determined as in Fig. 1A. The radioactivity of the terminal tyrosine residue (●) was calculated as the difference between the radioactivities of the two tubulin subunits. The radioactivity of beta tubulin is also shown (○). The data are normalized to protein in the supernatant before vinblastine precipitation. Reprinted from Thompson *et al.*, 1979.

G. Studies of the Extent to Which Alpha Tubulin Is Tyrosinated *in Vivo*

Several laboratories have attempted to determine the extent to which alpha tubulin is tyrosinated *in vivo*. Since the terminal tyrosine content can change during the warm incubations necessary for tubulin polymerization (Thompson, 1977a; Hallak *et al.*, 1977), it is necessary for these determinations that tubulin be purified by some method that can be carried out in the cold, such as chromatography or vinblastine precipitation. Lu and Elzinga (1978) isolated the C-terminal peptide of beef brain tubulin after cyanogen bromide cleavage and consistently found an intermediate value of between one and two tyrosine residues in the peptide. Treatment of the peptide with carboxypeptidase A released 0.3 mole of tyrosine for each mole of peptide. Ponstingl *et al.* (1979) reported the carboxy-terminal amino acid sequence of alpha tubulin from pig brain and find tyrosine as the ultimate residue in 15% of the polypeptides. Rodriguez and Borisy (1978) demonstrated that the extent of tyrosination of tubulin of chick brain varies as a function of development and is found to be greatest in embryonic stages characterized by highest TTLase activity levels. These same authors also separated free dimer tubulin from microtubular tubulin in rat brain and determined the extent of tyrosination in each type (Rodriguez and Borisy, 1979a). In newborn and week-old rat brain tyrosine is the C-terminal residue of 45–50% of the free dimer alpha tubulin, but only 20–25% of the microtubular alpha tubulin. In young adult rat brain the extent of tyrosination of free tubulin dimers is lower (about 22%) and approximates that of the microtubular alpha tubulin. Although TTLase can catalyze the incorporation of phenylalanine into alpha tubulin (Barr *et al.*, 1973b), in normal brain tissue tubulin containing C-terminal phenylalanine represents only a very small fraction of the total tubulin (Rodriguez and Borisy, 1978). Rodriguez and Borisy (1979b) have shown that the induction of experimental phenylketonuria in rats changes the percentage of the total aromatic C-terminal residues of tubulin that are phenylalanine from 4% to as high as 31%. The possible consequences of such an alteration of the C-terminal residue as a result of disease remain unknown.

Raybin and Flavin (1977b) developed another method for estimating the extent of prior tyrosination of tubulin *in vivo,* which consists of adding tyrosine maximally to tubulin before and after a treatment with carboxypeptidase A. The increase in tyrosine acceptor activity of the tubulin after CPA digestion is considered a measure of the tyrosine removed from the tubulin. Utilizing this methodology to analyze brain tubulin, Raybin and Flavin (1977b) were able to demonstrate an increase of tubulin capacity from about 0.25 mole of labeled tyrosine per mole alpha tubulin to about 0.36 mole of labeled tyrosine per mole alpha tubulin. When Nath and Flavin (1978) analyzed a fraction of ''membrane-bound'' tubulin from brain by this method, they found no increase in tyrosine acceptor capacity after digestion with carboxypeptidase A, indicating that this fraction contained no terminal tyrosine *in vivo*. In a later abstract (Nath and Flavin, 1979b) these authors report that incorporation of labeled tyrosine by living cells and tissues indicates, on the contrary, that ''membrane-bound'' tubulin does contain some C-terminal tyrosine. Therefore the method of determining C-terminal tyrosine by difference in tyrosine acceptance before and after carboxypeptidase A digestion may have some deficiencies for some tubulin types. Nevertheless, the method has proved useful to demonstrate changes in terminal tyrosine content during differentiation in a neuroblastoma/glioma cell line (Nath and Flavin, 1979a). Accompanying the cyclic AMP-induced differentiation was an increase in the extent of tyrosination from 30–40% in the undifferentiated cells to nearly 85% saturation of the carboxyl-terminus. This increase was at the expense of the pool of tubulin previously considered ''nonsubstrate tubulin.'' Bulinski *et al.* (1980) purified tubulin from interphase or mitotic populations of

HeLa cells. They reported that no differences were apparent in tyrosine acceptance capacity between these two tubulin preparations, either before or after digestion with carboxypeptidase A.

H. Possible Roles of Tyrosinated Tubulin *in Vivo*

High levels of TTLase activity have been found in embryonic tissues whose elongating cellular morphologies are based on a scaffolding of microtubules (Deanin *et al.*, 1977). This might indicate that tyrosination of tubulin or the rapid turnover of terminal tyrosine is important for the development or maintenance of these morphologies.

Two recent reports (Matsumoto and Sakai, 1979; Matsumoto *et al.*, 1979) indicate that tyrosination of tubulin might have an important role in maintaining membrane excitability in the giant axon of the squid. Excitability of the squid membrane is lost after perfusion with a buffer containing Ca^{2+}, which causes the loss of microtubules. Addition of purified tubulin, TTLase and 300k axon protein, together with tyrosine, ATP, KCL, Mg^{2+} was found to restore the membrane excitability. In the presence of all of these components except TTLase, the membrane excitability was not restored. Therefore, the TTLase activity seems to be important for interaction of tubulin with the membrane or with integral proteins at the membrane surface.

III. Methods

Since the optimal pH for TTLase activity *in vitro* is not optimal for stabilizing the acceptor activity of tubulin, the systems to be described here are based on the fact that a second peak of TTLase activity is found at a lower pH, where the tubulin substrate should be much more stable. The first system is useful for assaying extracts for TTLase activity levels, and the second supports microtubule polymerization and function as well as tyrosination of tubulin.

A. Assay of Tubulin:Tyrosine Ligase Activity

When assaying TTLase activity in cell-free extracts of various tissues, it is desirable to remove any endogenous tyrosine acceptor activity in the extracts. The tyrosine acceptor activity of alpha tubulin decays during storage or dialysis at pH 7.4 or greater (Raybin and Flavin, 1975). Therefore tissues are homogenized in buffer A: 0.05 *M* Tris (adjusted to pH 7.4 with HCl), 0.1 *M* KCl, 5 m*M* Mg-acetate, 1 m*M* dithiothreitol, and 10% glycerol. The homogenate is centrifuged for 1 hr at 200,000 *g* at 2°C, and the resulting supernatant fraction is dialyzed overnight against buffer A. Assay of TTLase activity can be made under essentially these buffer conditions (Deanin *et al.*, 1977); however, the purified tubulin substrate will decay somewhat during the assay incubation. This is a particular problem when maximal tyrosination of a small amount of tubulin is desired. Therefore the assay conditions presented here are based on the buffer conditions of Murofushi (1980). Following the dialysis against buffer A to inactivate endogenous tubulin, the cell extracts are dialyzed against the TTLase buffer B containing: 0.15 *M* KCl, 25 m*M* MES (adjusted to pH 6.8 with KOH at room temperature), 12.5 m*M* Mg-acetate, 1 m*M* dithiothreitol, and 10% glycerol. The standard assay mixtures contain in addition to the supernatant enzyme, 2.5 m*M* ATP, 1.0 mg/ml microtubule protein, 0.1 m*M* labeled tyrosine and 0.1 m*M* colchicine, all in buffer B. Tritiated tyrosine is diluted with unlabeled tyrosine to a convenient specific activity of about 1 Ci/mmole before use. Labeled tyrosine is dried under a stream of nitrogen and taken up in buffer B just before use. Microtubule protein substrate can be prepared by three cycles of polymerization according to the method of Shelanski *et al.* (1973), the method of Karr *et al.* (1979), or that of Asnes and Wilson (1979). Colchicine is added (Raybin and Flavin, 1977a) because of the nonsubstrate nature of assembled microtubules. The enzyme assay mixtures are prepared at 0°C and incubated at 37°C for a convenient time (usually 10 min) before an aliquot is removed for the determination of labeled tyrosine incorporation into protein by the method outlined below. By staggering the introduction of individual assay samples to the 37°C water bath at 30-sec intervals, up to 20 samples can be assayed at a time by a single worker. Even more samples can be assayed in a group if a second person is available to continue adding samples to the water bath while aliquots of the first samples are being taken. To insure that no reagents become limiting during the 10-min assay incubation, the time course of tyrosine incorporation of the sample containing the highest enzyme activity should be determined. If the incorporation of tyrosine is not linear for at least 10 min, the more active enzyme samples should be diluted with buffer B and reassayed until the rate of tyrosine incorporation is in the linear range. Appropriate controls should include unincubated assay mixtures to detect any nonspecific absorption of labeled tyrosine and reaction mixtures lacking the enzyme fraction to detect any residual TTLase activity in the microtubule protein preparation. One unit of TTLase activity catalyzes the synthesis of 1 nmole of tyrosinated tubulin per minute in the standard assay conditions.

The method of determination of tyrosine incorporation into protein is based on that of Mans and Novelli (1961). Paper disks (Whatman 3 MM, 2.1-cm diameter) are saturated with 10% trichloroacetic acid (TCA) and dried in advance of the experiment. Before the experiment the paper disks are labeled with identifying marks with a #2 pencil and then individually pinned to a styrofoam block using brass straight pins. This author finds it convenient to prepare 30-μl assay mixtures and to transfer 20 μl at the appropriate time to the TCA-saturated paper disks using a Pipetteman micropipette. After the samples have been dried on the filters at room temperature under a stream of air, the filter disks together with the pins are dropped into a beaker of 10% TCA containing unlabeled tyrosine. After soaking for 15 min the solution is decanted off and replaced by 5% TCA. The beaker is covered with foil and heated to boiling for 15 min in a hot water bath. After three more rinses in 5% TCA and two washes with absolute ethanol, the filters are dried in a vacuum oven. Alternatively, the filters can be rinsed through 50% diethyl ether : 50% ethanol, then pure diethyl ether and finally air dried (Mans and Novelli, 1961). The dry filters are then counted in a toluene-based scintillator.

B. Cell-Free Tyrosination of Mixtures of Tubulin Dimers and Microtubules: Use of a Rapid Filtration Assay for Determining the Radioactivity of the Microtubule Pool

The second method was designed to allow for tyrosine incorporation under conditions that support tubulin polymerization and microtubule treadmilling (Margolis and Wilson, 1978, 1979). The buffer for this system, buffer C, contains 0.05 *M* MES (adjusted to pH 6.8 with KOH at room temperature), 0.05 *M* KCl, 2.5 m*M* $MgSO_4$, 1 m*M* EGTA, 1 m*M* dithiothreitol, and 10% glycerol. A pellet of microtubule protein prepared according to the method of Asnes and Wilson (1979) is taken up directly in buffer C and allowed to stand at 0°C for 10 min; then GTP is added to 0.5 m*M* and the microtubule protein stock is centrifuged at 2°C for 20 min at 30,000 *g* before use. Each milliliter of reaction mixture for polymerization contains in addition to 2 mg of tubulin, TTLase (purified by DEAE chromatography), 0.1 mg ribonuclease A, 0.07 units of acetate kinase (E.C. 2.7.2.1), 1 m*M* ATP, 0.5 m*M* GTP, and 10 m*M* acetyl phosphate, all in buffer C. The system is allowed to polymerize to steady state, and then the reaction mixture is transferred to a warm tube in which sufficient [^{14}C]tyrosine has been dried to bring the final tyrosine concentration to 0.04–0.1 m*M*. The incubation at 37°C is continued and from time to time aliquots are removed and assayed for total incorporation of labeled tyrosine into protein by the method described earlier except that the filters are counted in Aquasol (New England Nuclear). In order to assay rapidly the contribution of that incorporation due to microtubules, microtubules are collected by a variation of the method of Maccioni and Seeds (1978). Aliquots of the reaction mixture (usually 20 μl) are taken and immediately diluted with 4 ml of a warm microtubule stabilizing buffer containing 5 *M* glycerol, 50 m*M* PIPES (adjusted to pH 6.8 with KOH at room temperature), 1 m*M* $MgSO_4$, 1 m*M* EGTA, and 1 m*M* unlabeled tyrosine. This solution is poured through a glass fiber filter (Whatman GF/C, previously wetted with the stabilizing buffer). The filter is rinsed first with 15 ml of stabilizing buffer at room temperature and then with 25 ml of 5% TCA. The filter is sucked dry and transferred to 10 ml of Aquasol and then the radioactivity is determined in a liquid scintillation spectrometer. Incorporation of labeled tyrosine into the tubulin dimer fraction is estimated as the difference between the total labeled protein and that binding to the glass fiber filters. Podophyllotoxin-containing reaction mixtures, in which no microtubules were ever allowed to form, are used to determine the extent of any nonspecific binding of tubulin dimers to the glass fiber filters. Utilizing these methods, it has been possible to show that no radioactive tyrosine enters microtubules previously blocked with colchicine or podophyllotoxin, even though the pool of tubulin dimers becomes rapidly labeled (Thompson *et al.*, 1980). In the absence of drugs some newly tyrosinated tubulin dimers enter the microtubules by the treadmilling process. This methodology should prove quite useful in correlating the dynamics of turnover of the microtubule and dimer pools in cell-free systems with that occurring *in vivo*.

C. Determination of the Rate of Turnover of the C-Terminal Tyrosine Residue of Alpha Tubulin in Living Cells

This methodology was developed for use with cultured muscle cells (Thompson *et al.*, 1979) but should be applicable to other cell types that have an appreciable rate of posttranslational tyrosine incorporation compared to that due to translation.

1. Pulse Labeling of Cultured Cells

Primary cultures of chick muscle cells were grown on 60- or 100-mm-diameter gelatin-coated plastic tissue culture dishes. Cultures were pulse labeled at least 18 hr after the last feeding with F-10 medium supplemented with 10% horse serum and 3% embryo extract. It should be noted that F-10 medium contains 10 μM tyrosine, much lower than many other popular tissue culture media, and therefore causes less dilution of tyrosine specific activity. Tritiated tyrosine (50 μCi) is added directly to the culture medium on each plate. After 15 min of continued incubation at 37°C the radioactive medium is removed by aspiration, the plate is quickly washed with fresh medium, flooded with fresh medium containing unlabeled tyrosine, and returned to the incubator for the chase incubation.

2. Determination of the Radioactivity of the Terminal Tyrosine Residue of Tubulin

At each time point during the pulse/chase incubations when a determination of the radioactivity of the terminal tyrosine is to be made, the medium is removed from one of the cultures and rapidly replaced with ice-cold buffered saline (Dulbecco's PBS without calcium) to stop enzymatic activity. The culture plate is rinsed several times with ice-cold PBS and then twice with buffer D containing 0.1 *M* MES (pH adjusted to 6.8 at room temperature with KOH) 1 m*M* EGTA, 0.5 m*M* MgCl, 2 m*M* mercaptoethanol, 0.5 m*M* GTP, and 10% glycerol. Each plate is drained and the cells are scraped off with a rubber policeman into buffer D and then disrupted at 0°C using a sonicator equipped with a microprobe. The homogenate is centrifuged for 60 min at 100,000 *g* and the supernatant is saved. An aliquot of the supernatant is taken for protein determination and a second aliquot containing the bulk of the protein is used for tubulin purification. Purified beef brain tubulin (0.065 mg) is added to the muscle supernatant fraction, $MgCl_2$ is added to 10 m*M* and then vinblastine sulfate (1 m*M* final concentration) is added and the samples are incubated in the dark for 1 hr at 0°C. The samples are then centrifuged for 30 min to recover the vinblastine-precipitated tubulin. The supernatant fraction is discarded and the pellet is taken up directly into 0.2 ml of SDS electrophoresis sample buffer (Laemmli, 1970) and heated for 2 min in a boiling water bath. One drop of additional mercaptoethanol is added to each sample immediately before electrophoresis on 7.5% acrylamide gels (10 × 0.8-cm diameter) according to the method of Laemmli (1970). Other methods of gel electrophoresis that separate the alpha and beta tubulin bands well could be substituted. The gels are fixed and simultaneously stained at 50°C with 0.24% Coomassie Brilliant Blue R-250 in a solution of 50% isopropanol, 9% acetic acid. Destaining at 50°C is carried out overnight in a diffusion destainer containing 5% isopropanol, 7.5% acetic acid. The alpha and beta bands of the carrier tubulin are predominant and well separated on the destained gels. Each gel is sectioned into 1-mm-thick disks by a regular array of blades. The radioactivity of each disk is determined by first drying it in a glass vial, dissolving it in 0.05 ml of hydrogen peroxide, and then counting it in 4 ml of Aquasol. Figure 1A shows a typical electrophoretic pattern of radioactive vinblastine-precipitated protein from a pulse-labeled muscle culture. Only the central portion of the gel containing the stained alpha and beta tubulin bands is shown in this figure. Note that the incorporation of tyrosine into the alpha tubulin is greatly in excess of that incorporated into beta tubulin by protein synthesis alone, even though the two subunits contain an equivalent number of internal tyrosine residues (Bryan and Wilson, 1971). In order to correct the incorporation into alpha tubulin for radioactivity contributed by protein synthesis, radioactivity equal to that of the beta tubulin region is subtracted from the alpha tubulin radioactivity peak. Figure 1B shows that the assumed equivalence of the translational contribution to the radioactivity of alpha and beta tubulin is valid. For comparison of such data from different cultures the specific activity of the alpha tubulin terminus per milligram of protein in the high-speed supernatant is used. This corrects for any differences in the culture density, cell recovery, or efficiency of cell disruption from one culture to another. Comparison of the uncorrected values from one culture to another indicates that such corrections are usually not great. By determining the radioactivity of the terminal residue at several times during the chase incubation, the rate of turnover of the terminal tyrosine residue can be estimated from the steady-state loss of radioactive tyrosine (Fig. 2). It should be noted that the half-life of tubulin molecules in these cells is several days, as opposed to 37 min for the terminal tyrosine residue of those molecules.

IV. Perspectives

The largely nonmitotic muscle tissue of the 13 to 14 day-old chick embryo is characterized by high levels of TTLase activity (Deanin *et al.*, 1977). The terminal tyrosine residue of alpha tubulin turns over rapidly in myotubes developed from embryonic myoblasts in culture (Thompson *et al.*, 1979). Such rapid turnover of terminal tyrosine requires the presence of microtubules, apparently because microtubules are the substrate for detyrosination *in vivo*. Free tubulin dimers are apparently the only substrate for tyrosination, at least in cell-free systems (Thompson *et al.*, 1980). Based on these and other findings, a model of tubulin metabolism in embryonic muscle tissue has been developed and is summarized in Fig. 3. Briefly, muscle tubulin exists as a pool of free tubulin dimers and a smaller pool of microtubule tubulin. These two tubulin pools are considered to be in an equilibrium (Inouyé and Sato, 1967) maintained by the opposite-end assembly and disassembly mechanism demonstrated in cell-free systems by Margolis and Wilson (1978). Alpha tubulin is considered to be present in three forms: tyrosinated, substrate for immediate tyrosination, and a pool of detyrosinated subunits that are temporarily sequestered from the action of TTLase. A possible fourth type of alpha tubulin, which can never by tyrosinated, is ignored in this analysis, though it may exist *in vivo*. Because of the high levels of TTLase, the ATP cleavage during tyrosination, and tyrosine concentrations well in excess of the K_m value, most of the available alpha tubulin in the free dimer pool is considered to be tyrosinated. These tyrosinated tubulin dimer units are then randomly incorporated into microtubules at the ends undergoing net assembly. As the dimers "transit" the microtubule they tend to become detyrosinated, and are not subject to retyrosination until they are released from the opposite end of the microtubule as free tubulin dimers. At that time they would be quickly tyrosinated by TTLase at the expense of one molecule of ATP per cycle of detyrosination/tyrosination. Some experimental evidence for this proposed metabolism will be considered in more detail below.

The tendency for microtubules (MTs) to undergo opposite-end assembly and disassembly at steady-state, "treadmilling," was demonstrated under cell-free condi-

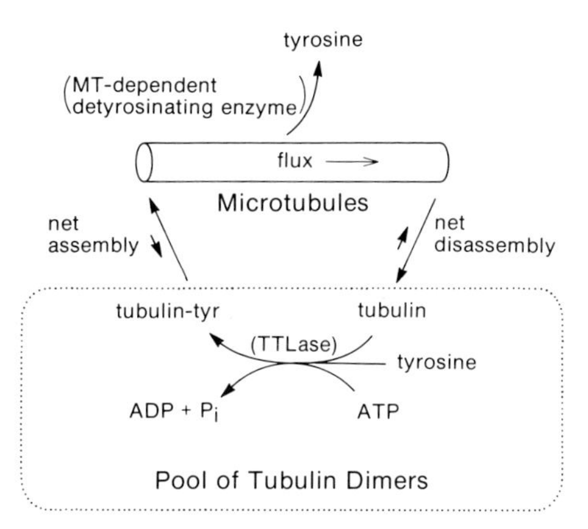

FIG. 3. Model for the rapid cyclic tyrosination/detyrosination of alpha tubulin in embryonic muscle.

tions that did not involve tyrosination (Margolis and Wilson, 1978). Although there is no solid demonstration that such a MT metabolism occurs *in vivo,* many observations are compatible with the hypothesis. In order for tubulin to undergo a complete cycle of tyrosination/detyrosination, it must evidently pass from the free tubulin dimer pool into the MTs and then back into the free dimer pool (Thompson *et al.,* 1979, 1980). The half-life of labeled tyrosine at the C-terminus of alpha tubulin in cultured muscle cells is about 37 min (Thompson *et al.,* 1979). Accordingly, half of the tubulin dimers must be incorporated in MTs at least once every 37 min. The phrase "at least once" indicates the possibility, according to the proposed model, that effective detyrosination might require more than one pass through average-length MTs. If the dimer pool is large with respect to the pool of MT subunits, then the time required for individual tubulin units to transit an average length MT would be considerably less than 37 min. When muscle cultures are treated with podophyllotoxin, the production of tyrosinable alpha tubulin subunits continues for 10–15 min at the control rate and is then drastically curtailed (Thompson, 1979). One interpretation of this observation is that 10–15 min represents the time required for "capped" MTs to depolymerize *in vivo*. According to the proposed model, that period should also approximate the average transit time *in vivo*.

The exact nature of the site of the major detyrosination event involving MTs *in vivo* is not yet clear. If detyrosination occurs at sites along the length of the MTs, then detyrosinated subunits would become available for retyrosination only when they are released from the end of the MT undergoing net depolymerization. Arce *et al.* (1978) demonstrated detyrosination of glycerol-stabilized MTs in a cell-free system, which might be similar to the *in vivo* detyrosination reaction. Such a rather nonspecific detyrosination, in conjunction with treadmilling, would tend to produce a MT that is heavily tyrosinated at its assembly end and detyrosinated at its disassembly end to an extent dependent on the rate of detyrosination, the rate of flux of tubulin units along the MTs, and the length of the individual MT. This additional asymmetry aspect of the MT might have important consequences for the partial reactions occurring at the two ends of the MT. The ends of the MTs are considered to be in rapid equilibrium with free tubulin dimers (Bergen and Borisy, 1979; Karr and Purich, 1979; see also Wegner, 1976). As an alternative to the proposed model, detyrosination could occur mainly at the MT ends, resulting in the immediate release of some tubulin dimers for retyrosination, and thus providing for a rapid tyrosine turnover that is not linked to the rate of flux of tubulin units along the MTs. If the major detyrosination event does occur at MT ends, it is not efficient, since Rodriguez and Borisy (1979a) have shown that about 20–25% of MT subunits contain a terminal tyrosine *in vivo*. However, the free tubulin dimer pool in the same neonatal rat brain tissue contained about twice as much tyrosine, confirming perhaps that some of the MT subunits had been detyrosinated but were not available for retyrosination *in vivo*. A pool of detyrosinated tubulin subunits, possibly derived from MTs, has been demonstrated in extracts of chilled muscle tissue (Thompson, 1977a) or chilled brain tissue (e.g., Rodriguez and Borisy, 1978). The only model that is probably not consistent with these findings is one in which detyrosination only occurs at the net disassembly end of the MT.

The rapid turnover of C-terminal tyrosine observed in embryonic muscle tissue might not be characteristic of other cell types. It would appear from the tyrosine uptake experiments of Raybin and Flavin (1977b) and Nath and Flavin (1979a) that a fairly rapid, but as yet undetermined, turnover rate is also characteristic of some cell lines derived from the nervous system. In both nerve and muscle tissue "treadmilling" microtubules could of course support intracellular transport of material along the major cellular axis by a simple attachment of the transported molecule to an integral microtubule subunit. The observed rapid tyrosine turnover, and the implied rapid "treadmilling" of the muscle microtubules, might be related also to the special role of microtubules in establishing and maintaining the extraordinarily elongated morphology of these cells, particularly in culture. The necessary role of microtubules in normal muscle morphogenesis has been demonstrated by the disruptive effects of colchicine on muscle development *in vivo* (Warren, 1968) and in tissue culture (Godman and Murray, 1953; Godman, 1955; Bischoff and Holtzer, 1968). Utilizing electron microscopy of serial sections, Warren (1974) demonstrated the pattern of the network of cytoplasmic microtubules in myoblasts and myotubes. The microtubules are formed into a parallel array aligned with the major axis of the cell. Although it appears that individual microtubules are much shorter than the length of the cell, close paraxial associations between individual microtubules were seen, which might provide for effective continuity of tension and or transport function along the entire length of the cell.

Since cytoplasmic microtubules are needed for normal muscle development, microtubule dysfunction might play a part in muscle diseases. The muscular dystrophy syndrome includes a defect in microtubule metabolism that is expressed in several cell types. Shay and Fuseler (1979) have shown by fluorescent staining with antitubulin antibodies that fibroblastic cells derived from explants of dystrophic chick muscle tissue lack the normal complex array of cytoplasmic microtubules during interphase, although mitosis is apparently normal. These original findings were challenged when Connolly and co-workers (1979) were unable to demonstrate any abnormalities in the microtubule complex of cells derived from another strain of dystrophic chicks. However, the microtubule defect has since been confirmed even in this second strain of chickens (J. W. Shay, personal communication), but the age of onset of the defect is more delayed. Shay *et al.* (1981) have recently reported observations of decreased complexes of microtubules in mononuclear cells collected from Duchenne muscular dystrophy patients and carriers. Although no direct observation of this microtubule defect in muscle cells has been yet reported, any such inability of muscle microtubules to form or function normally could seriously affect myogenesis, and thus contribute to the aborted muscle regeneration characteristic of muscular dystrophy. Examination of the metabolism of tubulin and microtubules in dystrophic tissues and extracts of those tissues is now underway in this laboratory.

Acknowledgments

The author wishes to thank Drs. L. Wilson and D. L. Purich for helpful discussions, for collaboration in the present aspects of this work, and for generously sharing their laboratory facilities. The author also thanks the Musuclar Dystrophy Association for its support.

References

Arce, C. A., Barra, H. S., Rodriguez, J. A., and Caputto, R. (1975a). *FEBS Lett.* **50,** 5–7.

Arce, C. A., Rodriguez, J. A., Barra, H. S., and Caputto, R. (1975b). *Eur. J. Biochem.* **59,** 145–149.

Arce, C. A., Hallak, M. E., Rodriguez, J. A., Barra, H. S., and Caputto, R. (1978). *J. Neurochem.* **31,** 205–210.

Argaraña, C. E., Arce, C. A., Barra, H. S., and Caputto, R. (1977). *Arch. Biochem. Biophys.* **180,** 264–268.

Argaraña, C. E., Barra, H. S., and Caputto, R. (1978). *Mol. Cell. Biochem.* **19,** 17–21.

Argaraña, C. E., Barra, H. S., and Caputto, R. (1980). *J. Neurochem.* **34,** 114–118.

Asnes, C. F., and Wilson, L. (1979). *Anal. Biochem.* **98,** 64–73.

Barra, H. S., Rodriguez, J. A., Arce, C. A., and Caputto, R. (1973a). *J. Neurochem.* **20,** 97–108.
Barra, H. S., Arce, C. A., Rodriguez, J. A., and Caputto, R. (1973b). *J. Neurochem.* **21,** 1241–1251.
Barra, H. S., Arce, C. A., Rodriguez, J. A., and Caputto, R. (1974). *Biochem. Biophys. Res. Commun.* **60,** 1384–1390.
Bergen, L. G., and Borisy, G. G. (1979). *J. Cell Biol.* **84,** 141–150.
Bischoff, R., and Holtzer, H. (1968). *J. Cell Biol.* **36,** 111–127.
Bryan, J., and Wilson, L. (1971). *Proc. Natl. Acad. Sci. U.S.A.* **68,** 1762–1766.
Bulinski, J. C., Rodriguez, J. A., and Borisy, G. G. (1980). *J. Biol. Chem.* **255,** 1684–1688.
Connolly, J. W., Kalnins, V. I., and Barber, B. H. (1979). *Nature (London)* **282,** 511–513.
Deanin, G. G., and Gordon, M. W. (1976). *Biochem. Biophys. Res. Commun.* **71,** 676–683.
Deanin, G. G., Thompson, W. C., and Gordon, M. W. (1977). *Dev. Biol.* **57,** 230–233.
Forrest, G. L., and Klevecz, R. R. (1978). *J. Cell Biol.* **78,** 441–450.
Godman, G. C. (1955). *Exp. Cell Res.* **8,** 488–499.
Godman, G. C., and Murray, M. R. (1953). *Proc. Soc. Exp. Biol. Med.* **84,** 668–672.
Gozes, I., and Littauer, U. Z. (1978). *Nature (London)* **276,** 411–413.
Hallak, M. E., Rodriguez, J. A., Barra, H. S., and Caputto, R. (1977). *FEBS Lett.* **73,** 147–150.
Inouyé, S., and Sato, H. (1967). *J. Gen. Physiol.* **50,** Suppl., 259–292.
Karr, T. L., and Purich, D. L. (1979). *J. Biol. Chem.* **254,** 10885–10888.
Karr, T. L., White, H. D., and Purich, D. L. (1979). *J. Biol. Chem.* **254,** 6107–6111.
Kobayashi, T., and Flavin, M. (1977). *J. Cell Biol.* **75,** 277a.
Laemmli, U. K. (1970). *Nature (London)* **227,** 680–685.
Levi, A., Castellani, L., Calissano, P., Deanin, G. G., and Gordon, M. W. (1978). *Bull. Mol. Biol. Med.* **3,** 42s–50s.
Lu, R. C., and Elzinga, M. (1977). *Anal. Biochem.* **77,** 243–250.
Lu, R. C., and Elzinga, M. (1978). *Biochim. Biophys. Acta* **537,** 320–328.
Maccioni, R. B., and Seeds, N. W. (1978). *Arch. Biochem. Biophys.* **185,** 262–271.
Mans, R. J., and Novelli, G. D. (1961). *Arch. Biochem. Biophys.* **94,** 48–53.
Margolis, R. L., and Wilson, L. (1978). *Cell* **13,** 1–8.
Margolis, R. L., and Wilson, L. (1979). *Cell* **18,** 673–679.
Martensen, T. M., and Flavin, M. (1979). *In* "From Gene to Protein: Information Transfer in Normal and Abnormal Cells" (T. R. Russell, K. Brew, H. Faber, and J. Schultz, eds.), p. 612. Academic Press, New York.
Matsumoto, G., and Sakai, H. (1979). *J. Membr. Biol.* **50,** 15–22.
Matsumoto, G., Kobayashi, T., and Sakai, H. (1979). *J. Biochem. (Tokyo)* **86,** 1155–1158.
Murofushi, H. (1980). *J. Biochem. (Tokyo)* **87,** 979–984.
Nath, J., and Flavin, M. (1978). *FEBS Lett.* **95,** 335–338.
Nath, J., and Flavin, M. (1979a). *J. Biol. Chem.* **254,** 11505–11510.
Nath, J., and Flavin, M. (1979b). *J. Cell Biol.* **83,** 335a.
Pierce, T., Hanson, R. K., Deanin, G. G., Gordon, M. W., and Levi, A. (1978). *In* "Maturation of Neurotransmission" (A. Vernadakis, E. Giacobini, and G. Filogamo, eds.), pp. 142–151. Karger, Basel.
Ponstingl, H., Little, M., Krauhs, E., and Kempf, T. (1979). *Nature (London)* **282,** 423–424.
Preston, S. F., Deanin, G. G., Hanson, R. K., and Gordon, M. W. (1979). *J. Mol. Evol.* **13,** 233–244.
Raybin, D., and Flavin, M. (1975). *Biochem. Biophys. Res. Commun.* **65,** 1088–1095.
Raybin, D., and Flavin, M. (1976). *In* "Cell Motility" (R. Goldman, T. Pollard, and J. Rosenbaum, eds.), pp. 1133–1138. Cold Spring Harbor Lab., Cold Spring Harbor, New York.
Raybin, D., and Flavin, M. (1977a). *Biochemistry* **16,** 2189–2194.
Raybin, D., and Flavin, M. (1977b). *J. Cell Biol.* **73,** 492–504.

Rodriguez, J. A., and Borisy, G. G. (1978). *Science* **206,** 463–465.
Rodriguez, J. A., and Borisy, G. G. (1979a). *Biochem. Biophys. Res. Commun.* **89,** 893–899.
Rodriguez, J. A., and Borisy, G. G. (1979b). *Science* **206,** 463–465.
Rodriguez, J. A., Arce, C. A., Barra, H. S., and Caputto, R. (1973). *Biochem. Biophys. Res. Co-mun.* **54,** 335–340.
Rodriguez, J. A., Barra, H. S., Arce, C. A., Hallak, M. E., and Caputto, R. (1975). *Biochem. J.* **149,** 115–121.
Shay, J. W., and Fuseler, J. W. (1979). *Nature* (*London*) **278,** 178–180.
Shay, J. W., Thomas, L. E., and Fuseler, J. W. (1982). *In* "Diseases of the Motor Unit" (D. L. Schotland, ed.), John Wiley and Sons, New York. In press.
Shelanski, M. L., Gaskin, F., and Cantor, C. R. (1973). *Proc. Natl. Acad. Sci. U.S.A.* **70,** 765–768.
Thompson, W. C. (1977a). *FEBS Lett.* **80,** 9–13.
Thompson, W. C. (1977b). *J. Cell Biol.* **75,** 324a.
Thompson, W. C. (1979). *J. Cell Biol.* **83,** 335a.
Thompson, W. C., Deanin, G. G., and Gordon, M. W. (1979). *Proc. Natl. Acad. Sci. U.S.A.* **76,** 1318–1322.
Thompson, W. C., Purich, D. L., and Wilson, L. (1980). *J. Cell Biol* **87,** 255a.
Warren, R. H. (1968). *J. Cell Biol.* **32,** 544–555.
Warren, R. H. (1974). *J. Cell Biol.* **63,** 550–566.
Wegner, A. (1976). *J. Mol. Biol.* **108,** 139–150.
Wengler, G., and Wengler, G. (1972). *Eur. J. Biochem.* **27,** 162–173.

Chapter 16

Tubulin–Tyrosine Ligase from Brain

MARTIN FLAVIN, TAKAAKI KOBAYASHI,
AND TODD M. MARTENSEN

Laboratory of Cell Biology
National Heart, Lung, and Blood Institute
National Institutes of Health
Bethesda, Maryland

I. Introduction

This procedure yields an enzyme that, although only partially purified, is sufficiently concentrated and stable to be useful for determining the tyrosinolation state of tubulin and for preparing maximally tyrosinolated tubulin. The procedure is a slight modification of that developed in this laboratory in 1976–1977, and briefly reported at that time (Kobayashi and Flavin, 1977). Several papers have appeared since then in which use has been made of the partially purified ligase (Nath and Flavin, 1978, 1979, 1980).

ISBN 0-12-564124-9

II. Assay Method

The standard reaction mixtures contain, in 60 μl final volume: 25 m*M* Tris-HCl, pH 7.2, 100 m*M* KCl, 12.5 m*M* $MgCl_2$, 2.5 m*M* ATP, 10 m*M* dithiothreitol, 0.1 m*M* [U-^{14}C]-L-tyrosine (50 mCi/mmole), 3 mg/ml of microtubule protein (purified from brain by three assembly cycles), and 0.01 to 0.1 unit/ml of ligase. After incubating for 20 min at 37°C, 50 μl is transferred to a Whatman 3MM paper disk, which is dropped into a beaker of cold 10% trichloroacetic acid and washed to remove free tyrosine (Mans and Novelli, 1961). Dried disks are counted in 12 ml of 0.5% (w/v) 2,5-diphenyloxazole in toluene in a scintillation counter. A unit of ligase is the amount fixing 1 nmole of tyrosine into protein in 1 min.

III. Purification Procedure

So that tubulin could be prepared from the same brain extract, we used as starting material the first warm supernatant obtained by centrifuging down microtubules assembled from a glycerol-supplemented extract (Shelanski *et al.*, 1973). This starting material could be stored for months at −70°C without loss of activity.

A. Step 1: DEAE-Cellulose Chromatography

Whatman DE-52, equilibrated in 25 m*M* K^+MES, pH 6.7, was sucked free of standing liquid on a Buchner funnel, and 140 g of moist cake was added to 545 ml of first warm supernatant (obtained from sheep or bovine brain), and stirred gently for 20 min. This and all subsequent operations were done at 0–4°C unless otherwise specified. The suspension was centrifuged for 10 min at 10,000 rpm. The pelleted DEAE-cellulose was suspended in 140 ml of 25 m*M* K^+MES containing 1 m*M* each of EDTA and dithiothreitol (ligase buffer), and poured into a 4-cm-diameter column. As the cellulose settled, the supernatant was overlaid with 200 ml of ligase buffer containing 50 m*M* KCl, maintaining a flow of 3 ml per minute and packing the bed to 15 cm. The flow-through + 50 m*M* salt fractions contained one-tenth of the protein and no ligase. The column was then eluted at 1.5 ml/min with 350 ml of 200 m*M* KCl in ligase buffer. Fractions of high specific activity were pooled, and then concentrated, from 240 to 40 ml, through an Amicon PM-30 membrane. Fractions were always centrifuged (40 min at 100,000 *g*) immediately prior to filtration. The concentrated fraction could also be stored at least several days at −70°C.

B. Step 2: Tyrosine Affinity Chromatography

Tyrosine was coupled through its carboxyl group to Affi-Gel 101,[1] a Bio-Gel A-5m matrix with *n*-propylamine side chains (Bio-Rad Laboratories), by the conventional procedure using the water-soluble carbodiimide, 1-ethyl-3(3-dimethylaminopropyl)carbodiimide hydrochloride. Forty-eight milliliters of the suspension of Affi-Gel 101 was placed on a 6-cm diameter coarse sintered glass filter and washed repeatedly at 25°C with deionized water, by gentle suction that was released as soon as the free-standing water had passed through. For convenient gentle agitation, the 25 g of moist cake was divided between two 15 × 2.5-cm plastic tubes. To each was added 35 ml of water, 0.5 ml of 2 *M* pyridine-HCl, pH 4.8, and 30 mg of solid L-tyrosine (total 340 μmoles). After warming the solution to 37°C for 5 min to help dissolve the tyrosine, 780 mg of the carbodiimide was added to each tube (total 7300 μmoles), and the stoppered tubes were agitated for 6 hr at 25°C by gentle reciprocal shaking. After 20 min the pH was adjusted with a few drops of *N* HCl from 5.1 to 4.8, where it remained. The tyrosinolated matrix was then washed repeatedly on the same filter with 25 m*M* K^+MES, pH 6.7, degassed *in vacuo* in a stoppered Buchner flask at 0°C, and packed with the same cold buffer to a 1.8 × 10.5-cm column bed (28-ml bed volume). If it was not to be used for some time, the column was equilibrated with the same buffer containing *M* KCl and stored at +4°C.

The step 1 fraction was diluted to 200 ml, to reduce the KCl below 50 m*M*, with ligase buffer containing also enough $MgCl_2$, GTP, and tubulin (purified by three assembly cycles) to give final concentrations of 1.0 m*M*, 0.1 m*M*, and 0.3 mg/ml, respectively. This solution was pumped through the bed of tyrosinolated Affi-Gel 101 at 3 ml/min. The column was next eluted at 1.2 ml/min with 100 ml of the same buffer containing 50 m*M* KCl, and with tubulin reduced to 0.2 mg/ml, and finally, while collecting 12-ml fractions, with the latter buffer but containing 100 m*M* KCl. Ligase activity was usually found in fractions 3 to 5. Almost all the protein in this fraction was now tuublin. If tubulin is not added to both the buffer in which ligase is applied and that with which it is eluted, no activity can be recovered. The pooled active fractions could also be stored at −70°C overnight at this stage.

C. Step 3: Phosphocellulose Chromatography

Whatman P-11, pretreated as by Weingarten *et al.* (1975), was suspended in 2 vol of 100 m*M* K^+MES, pH 6.7. The pH of the slurry was adjusted to 6.7 with

[1]Bio-Rad Laboratories' 1981 catalogue no longer lists Affi-Gel 101. It has been replaced by Affi-Gel 102, which has a different spacer chain. Ligase was not retained by the latter when it was tyrosinolated. Chris Siebert at the Richmond, California, office has agreed to prepare the Affi-Gel 101 on request.

TABLE I

PURIFICATION OF TUBULIN–TYROSINE LIGASE FROM BOVINE BRAIN[a]

Step	Volume (ml)	Total protein (mg)	Specific activity (nmole/min mg)	Total units	Yield (%)	Purification fold
First warm supernatant[b]	545	4470	0.045	202	(100)	—
1. DEAE-cellulose	240	430	0.350	150	74	8
2. Tyrosylated Affi-Gel 101	45	156	0.71	110	54	16
3. Phosphocellulose	10	8	11.0	88	44	240

[a] The starting material for enzyme purification was the supernatant obtained after centrifuging down assembled microtubules at 30°C. Ligase could also be purified directly from the extract.

[b] Brain extract was prepared, and tubulin assembled in the presence of glycerol by the Shelanski procedure (Shelanski *et al.*, 1973).

KOH, and it was degassed *in vacuo* at 0°C. It was packed to a bed of 1.9 × 13 cm by washing with 25 m*M* K^+MES, pH 6.7, and shortly before use was washed with the same buffer containing 1 m*M* each EDTA and dithiothreitol. The step 2 fraction was centrifuged for 15 min at 20,000 rpm, and the supernatant was pumped onto the column at maximum flow rate, 0.5 ml/min. All of the ligase was retained, but 80% of the protein (including the tubulin) was not. The column was next eluted with 35 ml of the preceding buffer containing 200 m*M* KCl at maximum rate (2.5 hr), and then, while collecting 4 ml fractions, with 35 ml of buffer with 400 m*M* KCl. The active fractions (2 to 6) were stable for at least a year at −70°C stored in the 400 m*M* KCl with protein about 1 mg/ml.

The results of a typical purification are shown in Table I.

IV. Properties

The partially purified ligase was completely stable for at least a year at a protein concentration of about 1 mg/ml. It was inactivated (90% in 20 min at 25°C) if further diluted 10:1 in its own or several other buffers with or without high salt, thiol, or ovalbumin. It could be stabilized to dilution only by addition of tubulin (0.1 mg/ml). It follows that whenever assaying or using the enzyme it should be added only after the tubulin substrate.

Murofushi (1980) has recently reported an 8000-fold purification of brain ligase, to homogeneity, in three steps: the first is similar to ours, the second and

TABLE II

RELATIVE CONCENTRATIONS OF PURIFIED LIGASE FRACTIONS

Ligase fraction	Protein (mg/ml)	Specific activity (units/mg)	Ligase concentration (units/ml)
Brain extract	12	0.056	0.67
Ligase purified by:			
This procedure	0.8	11.0	8.8
Murofushi procedure	0.005	240.	1.2

third are based on affinity for ATP and tubulin, respectively. We have been able to repeat his procedure, but have found that the final fraction loses all activity within weeks, and is therefore less suitable for the applications described below. The instability may be related to the extremely dilute protein in his final preparation, as illustrated in Table II, which compares the two final fractions as prepared in this laboratory. A variety of methods tried for further concentration of either final fraction have so far produced unacceptable activity losses.

A. Utilization of Ligase to Determine the State of Tyrosinolation of Tubulin

The success of this procedure rests on the availability of ligase that is active enough to tyrosinolate the tubulin sample maximally in a reasonable time (30 min at 37°C) before any significant denaturation has occurred. The details of the procedure have been described elsewhere (Nath and Flavin, 1980). In outline, the tubulin is divided into two portions, one of which is briefly incubated with sufficient carboxypeptidase A to remove all preexisting tyrosine. The penultimate glutamate of the α chain is not attacked. After adding β-phenylpropionate to inhibit the carboxypeptidase, each sample is incubated with radiolabeled tyrosine, the other reaction components (Nath and Flavin, 1980), and ligase. To show that tyrosinolation has been maximal, ligase is added in different amounts (usually 0.5 and 1.0 unit/ml) to aliquots of each sample. Typical results are shown in Table VII of ref. 4. The results indicate what porportion of α chains in a tubulin sample have C-terminal tyrosine, and what additional proportion can accept tyrosine; a portion of tubulin from every source so far examined can not accept tyrosine even after carboxypeptidase treatment. The structural alteration responsible for this "nonsubstrate" species has not been determined.

We will describe here two kinds of control experiment, done in this laboratory several years ago (Kobayashi and Flavin, 1977), to show that this analysis of tubulin tyrosinolation state is valid. Since the ligase reaction is reversible (Nath

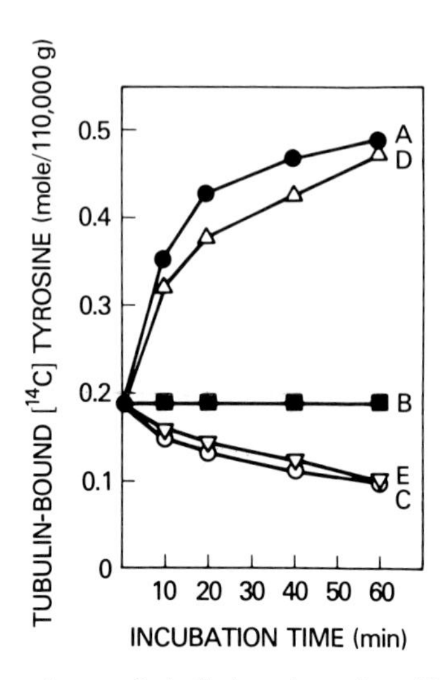

FIG. 1. Requirements for exchange of tubulin-bound tyrosine with free tyrosine. The incubation mixtures contained: 0.60 mg/ml of ^{14}C-tyrosinolated tubulin (0.19 mole tyrosine/110,000 g tubulin, 50 mCi/mmole tyrosine); 0.2 unit/ml ligase; 0.1 *M* K^+MES, pH 6.8; 5 m*M* $MgCl_2$; 10 μ*M* GTP; 1 m*M* DTT; and other components as follows: (A) 1 m*M* ATP and 0.1 m*M* [^{14}C]tyrosine (50 mCi/mmole); (B) 1 m*M* ATP and 0.1 m*M* [^{12}C]tyrosine; (C) 1 m*M* ADP and 10 m*M* phosphate; (D) 1 m*M* ATP, 1 m*M* ADP, 10 m*M* phosphate and 0.1 m*M* [^{12}C]tyrosine; (E) 1 m*M* ATP, 1 m*M* ADP, 10 m*M* phosphate and 0.1 m*M* [^{12}C]tyrosine. The mixtures were incubated at 37°C, and at intervals 50μl aliquots were taken to measure protein-bound radioactivity.

and Flavin, 1980), i.e., tyrosine is fixed in the presence of ATP and released in the presence of ADP + P_i, it was important to show that there was no exchange between free and bound tyrosine during the preceding tyrosinolations. If there were, the assay would give erroneously low values for preexisting tyrosine. Figure 1, curve B, shows that no exchange occurs during a 60-min incubation starting with ^{14}C-tyrosinolated tubulin, [^{12}C]tyrosine, and ATP. The other curves in Fig. 1 confirm that, however, if one starts with ATP + ADP + P_i, the reaction proceeds in both directions, with maximal tyrosinolation and exchange occuring simultaneously.

The following results indicate that the "nonsubstrate" tubulin is not an artifact of the analytical procedure. The β-phenylpropionate does not inhibit ligase and does inhibit carboxypeptidase completely, as shown by detecting the same proportions of nonsubstrate when carboxypeptidase was irreversibly inhibited with dithiothreitol (10 m*M* for 10 min at 37°C) or separated on DEAE-cellulose. Several observations show that carboxypeptidase does not create nonsubstrate by further digesting the tubulin; i.e., amino acid analyses do not show any glutamate (which would be the first amino acid released from either α or β chain), and carboxypeptidase concentration can be varied over orders of magnitude without changing the proportions of nonsubstrate. Finally, nonsubstrate tubulin is not simply the detyrosinolated moiety present at an equilibrium point of the reversible reaction. When the small molecule fraction is removed by gel filtration after maximally tyrosinolating tubulin, and the protein fraction supplemented with fresh ATP and [^{14}C]tyrosine, no more of the latter is fixed.

B. Utilization of Ligase to Prepare Maximally Tyrosinolated Tubulin

Tubulin is tyrosinolated by incubation with tyrosine, ATP, and large amounts of ligase for 60 min at 37°C. A typical incubation mixture contains 100 mg of microtubule protein (purified by three assembly cycles) and 40 units of ligase fraction in 20 ml of: 100 m*M* K^+MES, pH 6.7, 4 m*M* $MgCl_2$, 1 m*M* GTP, 3 m*M* ATP, and 1 m*M* *L*-tyrosine. After incubation the protein is equilibrated into reassembly buffer by Sephadex G-25 filtration and purified by a fourth assembly cycle. Pure tyrosinolated tubulin (free of associated proteins) may then be prepared by phosphocellulose chromatography, concentrated by adding ammonium sulfate to 50% of saturation, and again passed through a Sephadex column equilibrated with reassembly buffer.

References

Kobayashi, T., and Flavin, M. (1977). *J. Cell Biol.* **75,** 277a.

Mans, R. J., and Novelli, G. D. (1961). *Arch. Biochem. Biophys.* **94,** 48.

Murofushi, H. (1980). *J. Biochem. (Tokyo)* **87,** 979.

Nath, J., and Flavin, M. (1978). *FEBS Lett.* **95,** 335.

Nath, J., and Flavin, M. (1979). *J. Biol. Chem.* **254,** 11505.

Nath, J., and Flavin, M. (1980). *J. Neurochem.* **35,** 693.

Shelanski, M. L., Gaskin, F., and Cantor, C. R. (1973). *Proc. Natl. Acad. Sci. U.S.A.* **70,** 765.

Weingarten, M. D., Lockwood, A. H., Hwo, S.-Y., and Kirschner, M. W. (1975). *Proc. Natl. Acad. Sci. U.S.A.* **72,** 1858.

Chapter 17

Preparation of Brain Tyrosinotubulin Carboxypeptidase

TODD M. MARTENSEN

Laboratory of Cell Biology
National Heart, Lung, and Blood Institute
National Institutes of Health
Bethesda, Maryland

I. Introduction

Brain tubulin can be interconverted between tyrosinolated and detyrosinolated species by the action of tyrosine tubulin ligase and tyrosinotubulin carboxypeptidase, respectively. Tubulin detyrosinolation activity was found in brain extracts by Hallak *et al.* (1977), and subsequently (Argaraña *et al.*, 1978) this activity was shown to be distinct from the detyrosinolation activity of the ligase which required Mg^{2+}, ADP, and P_i (Raybin and Flavin, 1977; Rodriguez *et al.*, 1973).

II. Assay

The activity of the carboxypeptidase is measured by the release of radioactive tyrosine from ^{14}C- or ^{3}H-tyrosinolated tubulin. The procedure requires the prepa-

ISBN 0-12-564124-9

ration of radiolabeled tyrosinotubulin with the ligase. It is desirable to remove preexisting tyrosinotubulin species by carboxypeptidase A digestion prior to preparation of the radiolabeled substrate to achieve a substrate of defined specific radioactivity. Activity in fresh brain extract is not proportional to protein concentration above 2 mg/ml, which is in part due to the presence of endogenous tyrosinotubulin in the extract.

A. Preparation of Detyrosinolated Tubulin

Tubulin stored at $-80°$ in RA buffer [0.1 *M* Mes (4-morpholinoethanesulfonic acid), 1 m*M* EGTA, 0.5 m*M* $MgCl_2$, pH 6.8] containing 1 m*M* GTP is diluted to 10 mg/ml with RA buffer and stored on ice. Carboxypeptidase A obtained as a crystalline suspension (10–40 mg/ml) treated with diisopropylfluorophosphate is dissolved by diluting 0.2 mg into 0.15 ml of 1% $NaHCO_3$ on ice and then adding 0.15 ml of 0.1 *M* NaOH with mixing (Ambler, 1967). After solubilization RA buffer is added to 1.0 ml. The solution is used within an hour after preparation. Prior to tubulin digestion the solution is diluted 10-fold to 20 μg/ml, and an aliquot of this solution is diluted into the tubulin solution to give a final concentration of 0.2–0.5 μg/ml. The digestion mixture is incubated at 37°C for 15 min, then dithiothreitol (1 *M*) is added to 15 m*M*, and incubation is continued for 10–15 min. The thiol irreversibly inactivates carboxypeptidase A. [If radiolabeled tyrosinolated tubulin is available, the extent of the digestion can be quantified by the addition of tubulin (specific activity ~2000 cpm/μg) to 25 μg/ml in the digestion mixture. Aliquots (50 μl) are removed at the beginning and the end of the digestion to a 6 × 50-mm test tube containing 0.1 ml of serum albumin (5 mg/ml) followed by the addition of 0.5 ml of 15% trichloroacetic acid. After 10 min on ice the samples are centrifuged in an adapter at 2500 *g* and an aliquot (500 μl) removed for scintillation counting.] The digestion mixture is put on ice for 30 min to depolymerize tubulin, and centrifuged for 15 min in a Sorvall SS-34 rotor at 15,000 rpm to remove any aggregated protein. The supernatant is passed through a Sephadex G-25 fine column equilibrated with RA buffer and 0.1 m*M* GTP at a flow rate of 1–1.5 ml/min. The column volume is 10 times the sample volume with a height to diameter ratio of 3–5. Fractions containing protein are pooled to yield a final protein concentration of ~5 mg/ml and GTP is added to 1 m*M*. The detyrosinolated tubulin is polymerized by incubation at 37°C for 30 min and the microtubules are isolated by centrifugation at 10^5 *g* for 30 min at 26–28°C. After removal of the supernatant, RA buffer containing 1 m*M* GTP is added (1 ml/25 mg microtubule protein) and the sample is left on ice for 30 min. The tubulin sample is homogenized by repeated transfers into a Pasteur pipette, or with a loosely fitted Dounce homogenizer, and stored at $-80°$C.

Digestion of tubulin by carboxypeptidase A under the conditions specified appears to remove only tyrosine from the α subunit. Removal of glutamate which is pentultimate to tyrosine as measured by the loss of acceptor capacity of digested tubulin only occur when the carboxypeptidase A level is $>10\ \mu g/ml$. Normally, carboxypeptidase A digestions are carried out at alkaline pH (7.5–8.5). However, the rate of tubulin detyrosinolation is inversely related to pH from 6.8–8.8; addition of NaCl or KCl to 0.1 *M* is moderately inhibitory. Thiol inhibition of carboxypeptidase A activity appears irreversible, perhaps because of the reduction of a single disulfide bond and/or the chelation of Zn^{2+} by EGTA in the RA buffer.

B. Preparation of Radiolabeled Tyrosinolated Tubulin

Detyrosinolated tubulin is tyrosinolated by incubation at 10 mg/ml in RA buffer, pH 6.8, containing 0.20 *M* KCl, 10 m*M* dithiothreitol, 5 m*M* $MgCl_2$, 5 m*M* ATP, 0.5 m*M* GTP, and 0.15 m*M* [^{14}C]tyr of a specific activity >450 mCi/mmole. Ligase is added to this solution to ~1 unit/ml final concentration and the reaction is incubated at 37°C for 30–60 min. The progress of the reaction is followed by the filter disk assay. Approximately 1.8×10^3 cpm [^{14}C]tyr should be incorporated/μg tubulin. The procedure is regularly carried out within a vial containing 250 μCi of [^{14}C]tyr (~500 mCi/mmole). The [^{14}C]tyr solution is taken to dryness in a hood by carefully positioning a Pasteur pipette (connected to N_2 tank with the gas flow on [gently]) above the solution. To the dry vial is added 1.5 ml of reaction mix containing twice the concentrations listed above followed by 1.5 ml of tubulin (20 mg/ml), and lastly ligase (0.3 ml, 10 unit/ml). After incubation the reaction is put on ice for 30 min, then desalted, polymerized, centrifuged, and solubilized as described for the carboxypeptidase A digestion. An aliquot of the homogenized tubulin is removed for the measurement of protein as well as total and trichloroacetic acid soluble radioactivity before storage at −80°C. Normally, 1 μg of microtubule protein containing 80% tubulin is radiolabeled with ~2000 cpm [^{14}C]tyr (counting efficiency 85%) whose specific activity is 500 mCi/mmole. This quantity of tubulin is 7.2 pmole; the quantity of incorporated tyrosine is 2.12 pmole, and the extent of labeling is 30%. (Approximately 50% of polymeric brain tubulin is unable to be tyrosinolated irrespective of treatment with the brain or pancreatic carboxypeptidase A.) Release of counts from this substrate in the detyrosinolation assay (see below) is used to express activity in pmole/min.

C. Detyrosinolation Measurement

The standard assay is carried out in 6×50-mm test tubes containing 15 μl : 5 μl ^{14}C-tyrosinolated microtubule protein (3 mg/ml) in RA buffer containing 1 m*M* GTP and 10 μl tyrosinotubulin carboxypeptidase (2–30 u /ml) in 50 m*M*

Mes, pH 6.6. The reaction is initiated by the addition of the enzyme to the tubulin solution while on ice followed by incubation for 15–20 min at 37°C. The reaction is terminated by the addition of 0.1 ml of 0.1 *N* HCl. Protein is precipitated after the addition of 0.1 ml of albumin (5 mg/ml) with 0.4 ml of 15% trichloroacetic acid. The tubes are left on ice for 15 min prior to centrifugation for 10 min at 2500 *g*. [^{14}C]tyr released is measured by counting an aliquot of the supernatant (500 μl) by scintillation counting. One unit of activity releases 1 pmole of [^{14}C]tyr/min.

III. Purification of Brain Tyrosinotubulin Carboxypeptidase

Sheep brains from animals sacrificed within the same day are used for purification. After removal of meninges, chilled brains are homogenized in a blender for 60 sec in 1.2 times their weight of 50 m*M* Mes, pH 6.6. All operations are carried out at 4°C. The homogenate is centrifuged for 30 min at 10,000 rpm in a Sorvall GS-A rotor. The supernatant (specific activity = 1.0) is decanted and brought to 42% saturation with solid $(NH_4)_2SO_4$ (243 g/liter) added slowly with rapid mixing, and centrifuged as above. The supernatant is adjusted to 68% $(NH_4)_2SO_4$ saturation (168 g/L) and centrifuged in the same manner. The pellet is dissolved into a minimum volume of 50 m*M* Mes, pH 6.6, (~5% of the homogenate supernatant) and dialyzed against the same buffer for 8–10 hr with two changes of buffer. The dialyzed sample (specific activity = 3.7) is passed over a DEAE cellulose column equilibrated with the same buffer. The volume of DEAE cellulose is equivalent to the volume of the dialyzed sample. The unadsorbed protein fractions whose concentrations exceeded 0.5 mg/ml are pooled (specific activity = 6.8) and applied (1 ml/min) to a CM cellulose column equilibrated with the same buffer containing one-fourth the volume of the pooled DEAE eluate. After loading, the column is washed with 2 vol of buffer, followed by 1 vol of buffer containing 50 m*M* NaCl. Next, a linear gradient is applied of 50 m*M* → 250 m*M* NaCl; each chamber contains five times the volume of the CM cellulose column. Enzymatic activity reaches a maximum when the salt concentration of the eluate is 100 m*M*. Conductivity measurements are made of samples to be assayed. Fractions containing a salt concentration greater than 50 m*M* NaCl are diluted to yield ~50 m*M* [NaCl] prior to assay. The fractions that display a specific activity ⩾400 pmole/min mg are pooled and dialyzed versus 70% $(NH_4)_2SO_4$ in 0.1 *M* Mes buffer, pH 6.6. After ~6 hr the sample is removed and the precipitate collected by centrifugation, diluted in a minimal volume, and dialyzed versus 100 times its volume of buffer for 8–10 hr with two buffer changes. The contents of the dialysis bag are centrifuged to remove denatured protein and the clear supernatant is carefully removed with a Pasteur pipette and stored at −80°C.

IV. Remarks

This purification has been carried out with 0.2–1.5 kg of brain with 50% yield of enzyme with a specific activity >400 pmoles/min mg (Martensen and Flavin, 1978). The preparation is stable at −80°C (<2% loss of activity/month) Dilutions of the frozen stock solution to 50–100 μg/ml in the Mes buffer are stable on ice for a day. The stock solution is partitioned into several aliquots to minimize the number of freeze-thaw cycles. No observable endoprotease is seen during a detyrosinolation time course followed by sodium dodecylsulfate PAGE. The enzyme is distinct from pancreatic carboxypeptidase A by several criteria (Martensen and Flavin, 1978; Argaraña *et al.*, 1980). Both native and urea denatured iodoacetate alkylated tubulin are substrates for the enzyme (Martensen and Flavin, 1978). Using polymeric tubulin as substrate, the detyrosinolation activity is strongly inhibited by 0.1 *M* salt, $[Mg^{2+}]$ > 10 m*M*, and colchicine (Hallak *et al.*, 1977; Martensen and Flavin, 1978).

References

Ambler, R. P. (1967). *In* "Enzyme Structure" (C. H. W. Hirs, ed.), Methods in Enzymology, Vol. 11, pp. 155–166. Academic Press, New York.

Argaraña, C. E., Barra, H. S., and Caputto, R. (1978). *Mol. Cell. Biochem.* **19,** 17–21.

Argaraña, C. E., Barra, H. S., and Caputto, R. (1980). *J. Neurochem.* **34,** 114–118.

Hallak, M. E., Rodriguez, J. A., Barra, H. S., and Caputto, R. (1977). *FEBS Lett.* **73,** 147–150.

Martensen, T. M., and Flavin, M. (1978). *Miami Winter Symp.* **16,** 612.

Raybin, D., and Flavin, M. (1977). *Biochemistry* **16,** 2189–2194.

Rodriguez, J. A., Arce, C. A., Barra, H. S., and Caputo, R. (1973). *Biochem. Biophys. Res. Commun.* **54,** 335–340.

Chapter 18

Purification of Muscle Actin

JOEL D. PARDEE AND JAMES A. SPUDICH

Department of Structural Biology
Stanford University School of Medicine
Stanford, California

I. Introduction

The opportunity to study the molecular events responsible for muscle contraction and cell motility was made possible over 40 years ago by Banga and Szent-Györgyi (1941) and by Straub (1942), who discovered myosin and actin in the extracts of rabbit skeletal muscle. Actin was first isolated when Straub (1942)

ISBN 0-12-564124-9

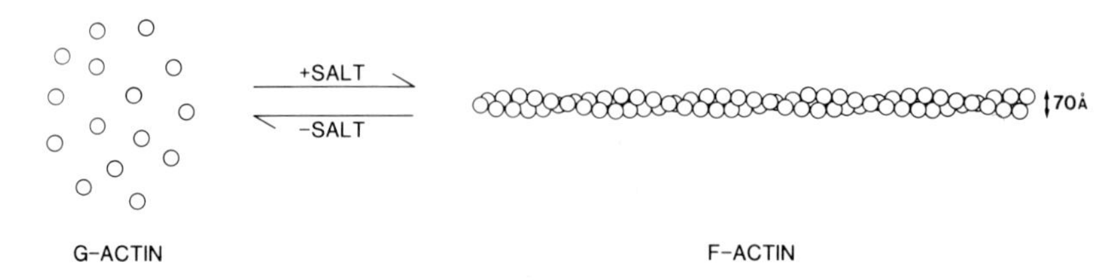

FIG. 1. Assembly of muscle actin. G-actin containing 1 ATP and 1 tightly bound divalent cation per monomer assembles in the presence of salt into 70-Å-diameter filaments of F-actin.

separated the viscous protein from an actomyosin preparation. Subsequent work (Straub, 1943) revealed that actin could be obtained in a nonviscous state (G-actin) by extracting muscle with a low ionic strength buffer, and that addition of salt induced conversion to a viscous form called F-actin (Fig. 1). An improved procedure by Straub and colleagues then incorporated a step that denatured muscle proteins not stabilized in an actomyosin complex; minced muscle was dehydrated with acetone before the actin extraction step (Straub, 1943; Feuer *et al.*, 1948). After the discovery that bound ATP was important for maintaining the functional integrity of actin (Laki *et al.*, 1950; Mommaerts, 1951; Straub and Feuer, 1950), the stability of isolated actin was enhanced by inclusion of ATP in the extraction buffers, but the purification protocol of Feuer *et al.* (1948) has remained essentially intact. With the advent of polyacrylamide gel electrophoresis as a highly resolving analytical tool for ascertaining protein purity, it became evident that muscle actin isolated by this classical procedure contained significant amounts of actomyosin-associated muscle proteins such as tropomyosin (Laki *et al.*, 1962) and α-actinin (Ebashi and Ebashi, 1965). These contaminants promote the gelation of F-actin and greatly affect the physical properties of filaments in solution (Maruyama *et al.*, 1974). Drabikowski and Gergely (1962) showed that an enhanced purification could be achieved by extraction of the actin from muscle acetone powder at 0°C, and Spudich and Watt (1971) devised a modification of this method designed to eliminate tropomyosin from muscle actin preparations. Their purification resulted in a single band on SDS–polyacrylamide gels and has met with widespread use as a general method for obtaining muscle actin.

A subtle problem in establishing methods for actin purification resides in the level of purity acceptable for the investigations at hand. Emerging experimentation in cell biology and, specifically, cytoskeletal biochemistry requires probing sensitive properties of actin itself and actin associations with other cell proteins. It is therefore the goal of this review to explore some of the pitfalls associated with actin purification and to clarify in some detail the correct usage and expected result from each step of the widely used muscle actin purification proce-

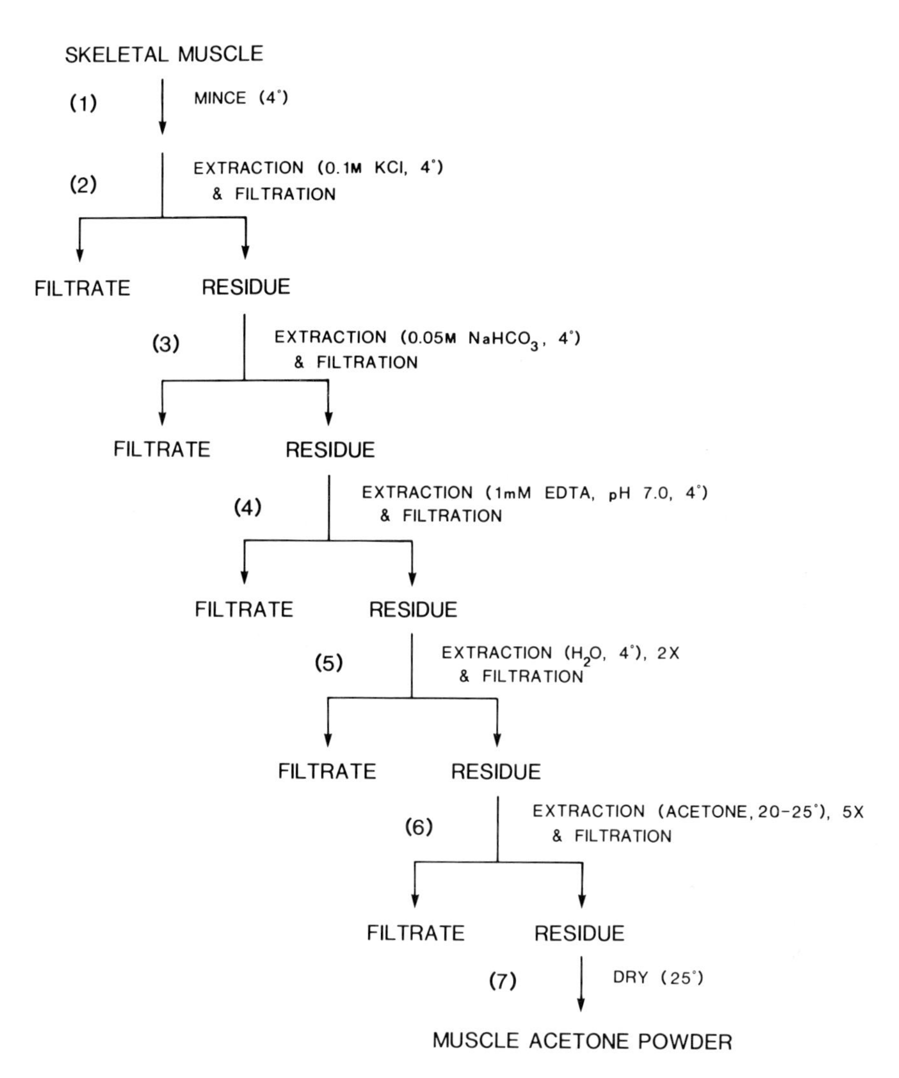

FIG. 2. Flow diagram for muscle acetone powder preparation. For complete explanation of the protocol, refer to steps in Section II of the text.

dure of Spudich and Watt (1971). Furthermore, additional steps to eliminate trace contaminants are described.

Flow diagrams for isolating rabbit skeletal muscle actin are given in Figs. 2 and 3. The detailed procedure for the acetone powder preparation of Feuer *et al.* (1948) with minor modifications is presented (Fig. 2) because of its importance in eliminating myosin and proteases from the final product and because the original reference may not be readily available. Adherence to requirements of temperature, buffer conditions, and incubation times are of prime importance for

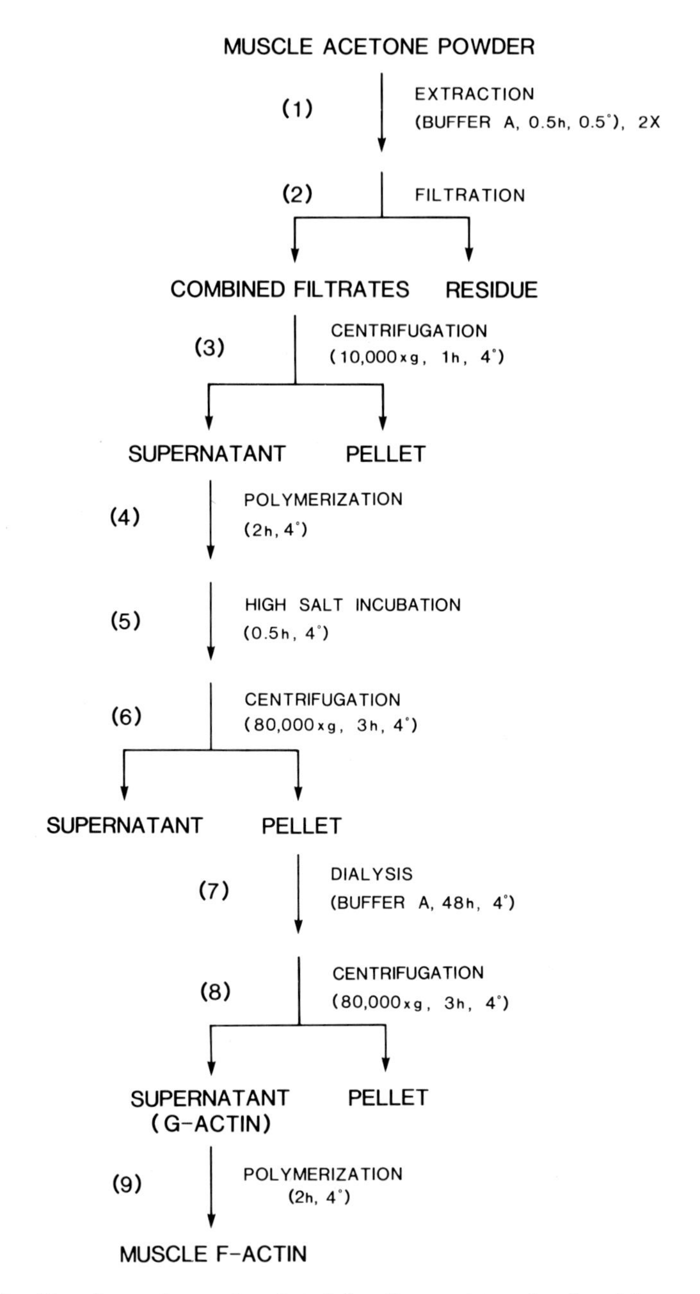

FIG. 3. Flow diagram for muscle actin isolation. For complete explanation of the protocol, refer to steps in Section III of the text.

obtaining high-purity actin. Steps at 4°C are preferably carried out in a cold room; cold extraction buffers and solvents are prechilled to 4°C before use, and buffer pH is determined at 25°C before chilling.

II. Acetone Powder Preparation (Fig. 2)

a. Preparation of Muscle Mince. The preferred way to kill the rabbit is to let it hang head down by grasping the hind legs with one hand while delivering a sharp blow to the back of the neck, followed by bleeding the animal completely. Immediately after sacrifice, the dorsal lateral skeletal muscles and the hind leg muscles are excised (about 350 g), chilled on ice, washed clear of blood with distilled H_2O, and minced at 4°C in a prechilled meat grinder.

b. Extraction with KCl. The mince is quickly extracted with stirring for 10 min in 1 liter of ice-cold 0.1 *M* KCl, 0.15 *M* potassium phosphate, pH 6.5. All extracts are filtered by squeezing through four layers of cheesecloth which have been previously boiled for approximately 20 min in distilled H_2O, drained, and brought to 4°C.

c. Extraction with $NaHCO_3$. The filtered muscle mince is extracted with stirring for 10 min at 4°C in 2 liters of prechilled 0.05 *M* $NaHCO_3$ and filtered. Longer extraction times at this stage cause appreciable extraction of actin and are to be avoided (J. R. Bamburg, personal communication).

d. Extraction with EDTA. The filtered residue is extracted with 1 liter of 1 m*M* EDTA, pH 7.0, by stirring for 10 min at 4°C.

e. Extraction with H_2O. The next two extractions are with 2 liters of 4°C distilled H_2O for 5 min with stirring.

f. Extraction with Acetone. The final five extractions are with 1 liter of acetone for 10 min each. All acetone extractions are performed at 20–25°C. Acetone should be cooled to below 20°C or the acetone–mince mixture becomes too warm. Clumps of residue are broken up by stirring during each extraction.

g. Drying. The filtered residue is placed in large glass evaporating dishes and air-dried overnight in a hood to obtain dried "acetone powder." The resulting acetone powder is stable for months if stored at −20°C.

III. Actin Isolation (Fig. 3)

Typical preparations use about 10 g of acetone powder. The minimal yield is approximately 10 mg actin per gram acetone powder, but can be as high as 30 mg actin per gram acetone powder.

A. Extraction

The acetone powder is extracted at 0–0.5°C for 30 min by stirring with 20 ml buffer A per gram acetone powder. The temperature must be kept low during the extraction (Drabikowski and Gergely, 1962). Buffer A consists of 2 m*M* Tris-Cl, 0.2 m*M* Na_2ATP, 0.5 m*M* 2-mercaptoethanol, 0.2 m*M* $CaCl_2$, 0.005% azide, final pH = 8.0 at 25°C.

Because actin is extracted under depolymerizing conditions, it is desirable to keep the concentrations of Mg^{2+}, K^+, and Na^+ in the buffer as low as possible. Measurable actin assembly occurs in 2 m*M* K^+ or Na^+, and in 0.2 m*M* Mg^{2+} (Pardee and Spudich, 1982). Therefore, it is useful to employ reagent powders of Tris-base, K_2ATP or Na_2ATP, and $CaCl_2$. To prepare the buffer, the reagents are dissolved in double distilled H_2O and titrated to pH 8.0 at 25°C with HCl. Add 2-mercaptoethanol after the pH determination, since 2-mercaptoethanol (and dithiothreitol) interferes with accurate pH determination by impairing the sensitivity of the pH electrode. The result is drifting pH readings.

An important caution is that actin is susceptible to proteolysis resulting from even minor bacterial contamination. Therefore, if buffers are prepared from stock solutions rather than from reagent powders, buffer stocks should be stored at 4°C with 0.1% sodium azide present to prevent bacterial or mold growth.

B. Filtration

The extract is separated from the hydrated acetone powder by squeezing through several layers of sterile cheesecloth; latex gloves are used to avoid contamination. Filtration through a coarse sintered glass filter under vacuum can also be used, but if sufficient care is not taken, foaming of the protein filtrate will occur, resulting in actin denaturation and reduced yields. If necessary, low-speed centrifugation at 5000–10,000 *g* for 10–20 min readily removes the bulk of the solids. Reextract the residue by stirring 10 min in 20 ml buffer A per gram acetone powder. Filter and combine extracts.

C. Centrifugation

The extract is centrifuged at 20,000 *g* for 1 hr at 4°C. Decant the supernatant by hand pipetting, leaving the turbid lower layer in the centrifuge tube.

D. Polymerization

The KCl concentration of the supernatant is brought to 50 m*M*, Mg^{2+} to 2 m*M*, and ATP to 1 m*M*. Inclusion of 1 m*M* ATP at this step ensures full polymerization. Allow to assemble for 2 hr at 4°C. At this stage of isolation a visible increase in the solution viscosity should be observed.

E. High Salt Wash: Tropomyosin Removal

Solid KCl is slowly added with stirring to a final concentration of 0.6 *M*, and the solution is stirred gently for 0.5 hr. Some investigators (MacLean-Fletcher and Pollard, 1980) have successfully employed 0.8 *M* KCl at 4°C in the wash step, which is useful in the event that the 0.6 *M* KCl treatment does not eliminate tropomyosin from the actin preparation. However, the actin monomer concentration increases with increasing salt concentration above 0.15 *M* (Kasai, 1969); thus lower yields of actin may result from washes with higher concentrations of salt.

F. Sedimentation of Filamentous Actin

The polymerized actin is centrifuged in 30-ml tubes at 80,000 *g* (ave) for 3 hr at 4°C. To obtain optimal purity it is advisable to remove contaminants trapped in the liquid phase of the F-actin pellet. This can be achieved by homogenizing the total pelleted F-actin into 150 ml of fresh wash buffer (buffer A + 0.6 *M* KCl, 2 m*M* $MgCl_2$, 1 m*M* ATP) and resedimenting the F-actin at 80,000 *g* (ave) for 3 hr at 4°C.[1] After discarding the supernatant, the intact F-actin pellet is rinsed thoroughly with buffer A.

G. Depolymerization

The pellets of F-actin are resuspended by gentle homogenization in 3 ml of cold buffer A per gram acetone powder originally extracted. Large actin losses can occur because of incomplete transfer to the homogenizer. A good technique to maximize recovery of F-actin from the centrifuge tube is to allow each pellet to stand on ice in 1 ml of buffer A for 1 hr before transferring to the homogenizer. The softened pellets can then be partially homogenized with a Teflon-coated rod and transferred with a plastic disposable pipette to the homogenizer without significant losses. Dialysis at 4°C against 1 liter of prechilled buffer A with one or two changes over a 3-day period gives complete depolymerization of actin, although dialysis times can be shortened considerably if vigorous stirring and large surface area dialysis bags are employed. One technique is to divide the homogenate into equal 6-ml aliquots (for 10 g acetone powder extracted) and place them into dialysis bags of ¼-in. diameter × 12 in. long. The bags are then either mounted on a rapid dialyser or tied to a magnetic stir bar in a 1-liter graduated cylinder. Rapid rotation of the dialysis bag permits nearly complete exchange of solutes in approximately 6 hr. Three buffer changes at 12-hr intervals results in >90% depolymerization of the F-actin (Fig. 4).

[1]This additional wash was not included as a step in the original Spudich–Watt report (Spudich and Watt, 1971) and was not used to purify the actin shown in the figures presented here. This has now been incorporated as a routine step in the procedure.

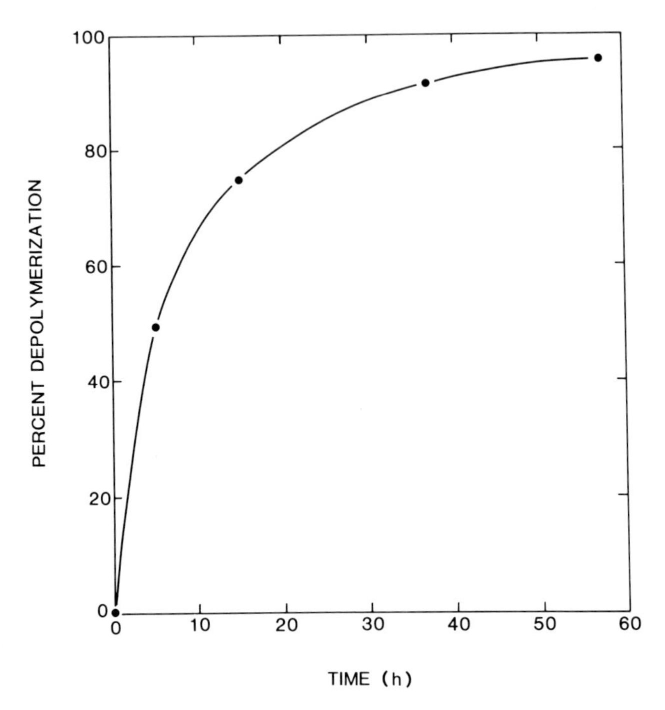

FIG. 4. Time course of depolymerization. Aliquots of dialysing actin from step 7 of the isolation procedure (Fig. 3) were centrifuged at 100,000 *g* (ave) for 2 hr at 4°C. Protein was determined in the supernatant (depolymerized actin), and % depolymerization was calculated. Dialysis was performed in 12-in. × ¼-in. dialysis bags mounted on a rapid dialyser in 1 liter of buffer A at 4°C. The buffer was changed at 5, 15, and 37 hr of dialysis. Total dialysis time was 57 hr.

It is not unusual to detect a residual viscosity in the dialysed actin. This viscosity is due to the presence of nondepolymerized actin which is complexed with myosin or other proteins. The quantity of myosin present can vary from preparation to preparation as a result of differing efficiencies of myosin removal during acetone powder preparation. Some actin losses are encountered with short dialysis times, but a myosin-free final product is obtained. Extensive dialysis results in eventual dissociation of actin–myosin complexes and appearance of myosin in the final product. Prolonged dialysis can also lead to observable actin proteolysis, and should therefore be avoided.

H. Clarification of G-Actin

The dialysed actin is centrifuged at 80,000 *g* (ave) for 3 hr. Shorter centrifugation times can be employed at greater *g* force; e.g., 150,000 *g* (ave) is now readily attainable in modern ultracentrifuges with the corresponding clearing time reduced to 1.5 hr. The supernatant fraction is saved. It is convenient to determine protein concentration at this point rather than on the subsequent viscous F-actin final product.

I. Polymerization

G-actin solutions, even in the presence of high concentrations of ATP (1 m*M*) and stored on ice, begin to lose polymerization activity after 2–3 days. Therefore, actin is stored as F-actin. To polymerize, add KCl to 50 m*M*, $MgCl_2$ to 1 m*M*, and ATP to 1 m*M* final concentrations. For storage, also add 0.02% NaN_3. The final product can be stored on ice as an F-actin solution or as pellets of F-actin. Pelleting provides additional stability. Actin should not be frozen or lyophilized.

The expected yield for this protocol is 20–30 mg actin per gram acetone powder. The resulting actin is generally highly purified (Fig. 5). However, depending upon individual technique, buffer purity, dialysis times employed, and so forth, the actin preparation can contain small amounts of contaminating protein, including proteolysed actin. Consequently, several techniques for further purification of the actin are discussed in Section VI.

IV. Analysis of Purity of Final Product

Contaminating proteins are most easily detected by SDS–polyacrylamide gel electrophoresis (Laemmli, 1970; Ames, 1974), preferably utilizing slab-type gels of 1.5–2.0 mm thickness. Our experience is that 10–12% acrylamide is an optimal gel concentration for detection of contaminants, since in 8% gels protein species of <25,000 daltons run with the ion front, whereas 15% gels do not allow discrimination of high-molecular-weight proteins. The limit of detection for Coomassie Blue stained protein bands on a 1.5-mm-thick slab gel system is about 0.05 μg/band. Consequently, visualization or densitometry of one 0.15% contaminant requires loading approximately 40 μg of the actin preparation (Fig. 6). Although this constitutes gross overloading of the actin band (actin band staining is linear only to 4 μg/band), minor contaminations of 0.15% can be detected and purity estimated. It is highly desirable to have this degree of sensitivity since very low levels of other components in actin preparations can significantly alter the properties of actin filaments (MacLean-Fletcher and Pollard, 1980). For example, a factor that alters the function of actin filaments by specifically binding to filament ends need only represent about 0.2% of the protein in a preparation of filaments about 1 μm long.

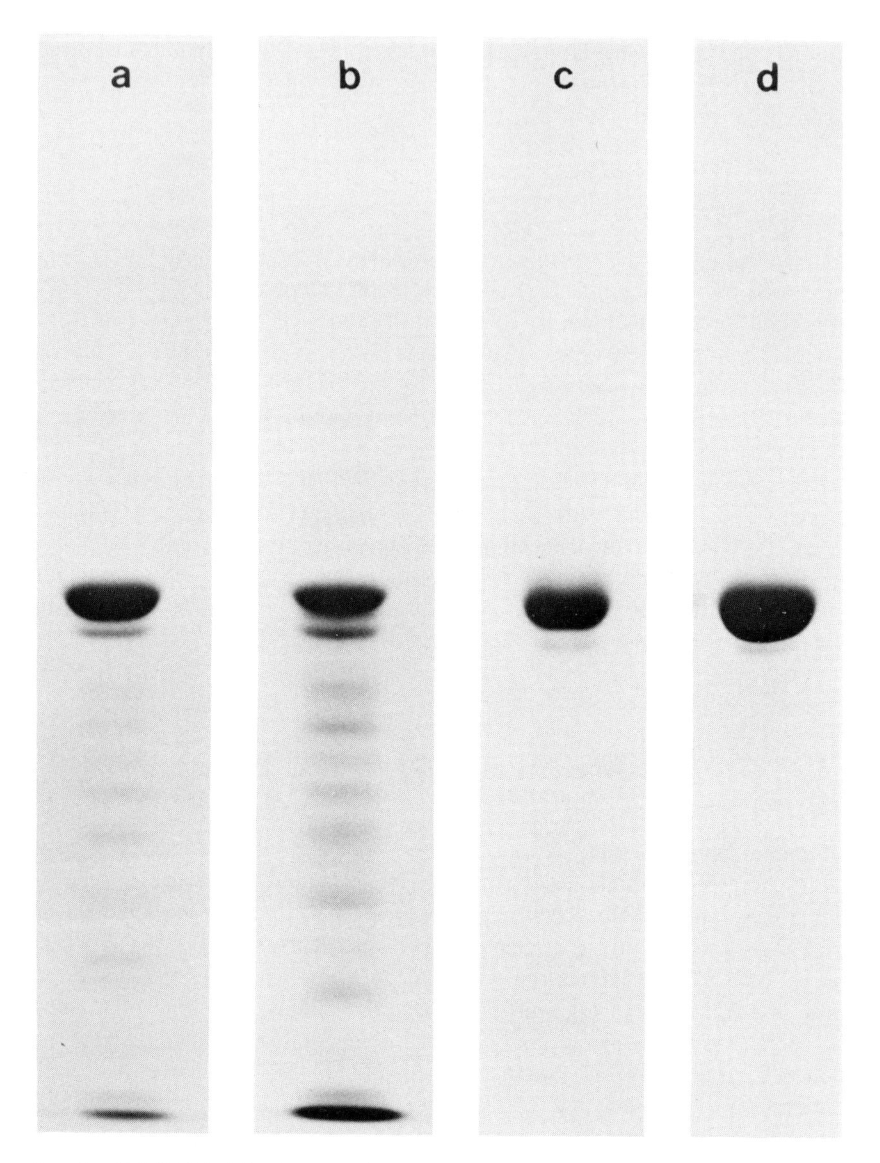

FIG. 5. SDS-PAGE of fractions at various stages in the actin purification. Samples taken during the isolation procedure were mixed 1:1 with a solution containing 2% SDS, 1% 2-mercaptoethanol, boiled for 3 min and applied to a 1.5-mm-thick slab gel containing 12% acrylamide, 0.2% methylene bisacrylamide, and 0.1% SDS in the buffer system of Laemmli (1970). Gels were stained overnight with 0.025% Coomassie Brilliant Blue G, 10% acetic acid, 25% isopropanol, and destained in 10% acetic acid by gentle shaking. Each gel lane contains about 6 μg of protein. (a) Muscle extract after 10,000 *g* clarification (Fig. 3, step 3). (b) Supernatant after incubation in 0.6 *M* KCl and centrifugation at 80,000 *g* (step 6). (c) Sedimented F-actin after incubation in 0.6 *M* KCl (step 6). (d) Final product (step 9).

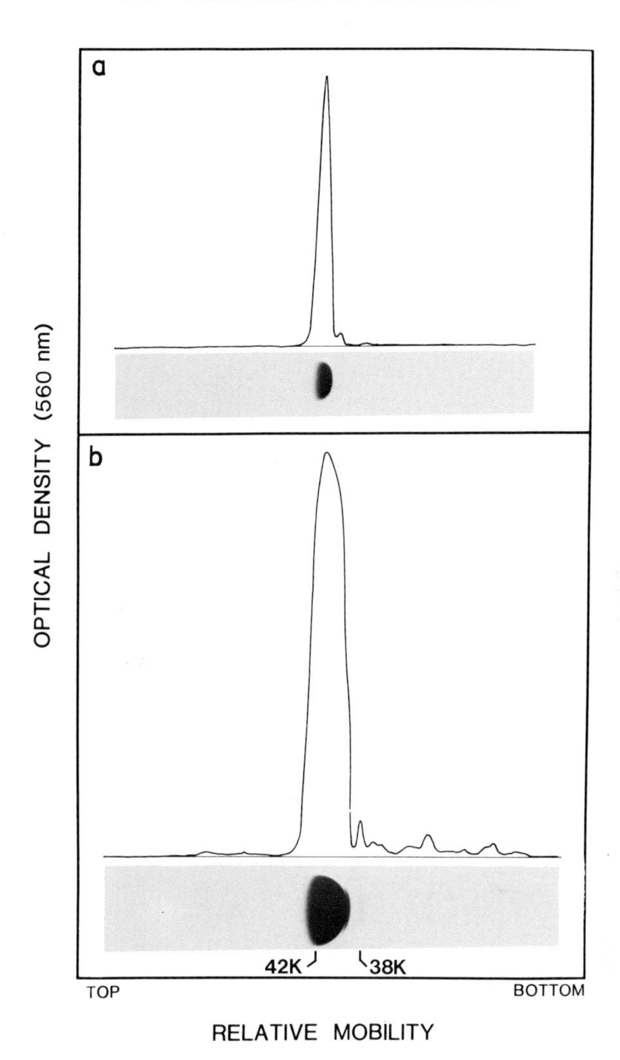

FIG. 6. Detection of actin contamination. Actin prepared as in Fig. 3 was electrophoresed in a 12% SDS-polyacrylamide gel. Stained gels were scanned at 560 nm on a RFT scanning densitometer (Transidyne). (a) 4 μg of isolated actin. Homogeneity was estimated at 98% by scan area. (b) 40 μg of isolated actin. Approximately eleven contaminants containing a total of about 2 μg of protein are detected by densitometry. Homogeneity $\cong$ 95%.

V. Sources of Contamination

Myosin contamination is sometimes observed and can be attributed to incomplete myosin extraction during acetone powder preparation. Proteolysis of actin is indicated by the appearance of a 38,000 dalton protease-resistant core (Jacobson and Rosenbusch, 1976) (Fig. 6b). Such proteolysis can arise either from bacterial contamination in extraction buffers or from proteolytic activity extracted with actin from some acetone powders (Fig. 7). While proteases are not prevalent in all acetone powder preparations, inspection of purified actin prepa-

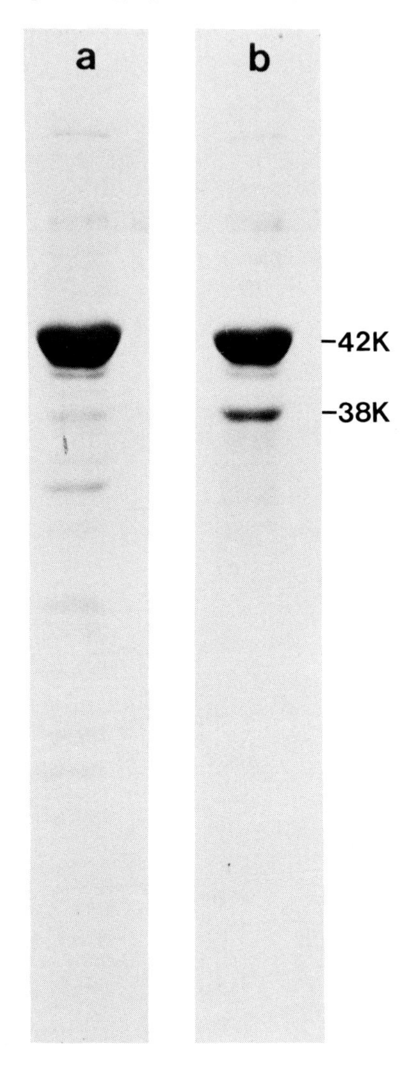

FIG. 7. Proteolytic activity in acetone powder extracts. (a) Acetone powder extract (10,000 *g* supernatant; Fig. 3, step 3) taken immediately after extraction. Note the presence of a small amount of 38,000-dalton proteolytic fragment. (b) The same acetone powder extract after 24-hr incubation at 20°C. Significant proteolysis of actin has occurred, resulting in a larger amount of the 38,000-dalton fragment. Not all of the acetone powders tested contained an active protease.

rations for protease activity is advisable. For those acetone powders that yield proteolytic activity, a slight modification of the actin preparation procedure is recommended. After sedimenting F-actin following the 0.6 *M* KCl treatment (see Fig. 3, step 6), the F-actin pellet is homogenized thoroughly into 100 ml of 4°C buffer A containing 50 m*M* KCl and 2 m*M* $MgCl_2$ and immediately recentrifuged at 150,000 *g* for 1.5 hr at 4°C. The soluble protease is fractionated away from F-actin in this step before depolymerization of filaments into protease-susceptible G-actin has been initiated. The washed pellet is then homogenized into buffer A and dialysed (step 7). An additional precaution when the acetone powder preparation contains proteolytic activity is to ensure that the pH of the extraction buffer is 8.0 at 25°C; high pH inhibits protease activity.

VI. Further Purification of Actin

Many current experiments in research on cell motility require actin completely free of trace contaminants such as myosin, tropomyosin, and other factors that are known to alter properties of actin assembly, disassembly, exchange, and ATPase activity. In addition, no specific steps in the purification shown in Fig. 3 are designed to efficiently remove ribonucleotides or polysaccharides, which are not detected by SDS-gel electrophoresis with Coomassie Blue staining. Consequently, we and others have designed additional steps to further purify the actin. The three following procedures can be considered alternatives or they can all be used.

A. Ion Exchange Chromatography

A highly recommended technique for obtaining highly purified muscle actin is ion exchange chromatography. This type of purification offers the considerable advantage of removing both protein and nonprotein contaminants from actin preparations. For further purification of muscle actin we use the following batch treatment of F-actin with DEAE cellulose.

1. DE-52 resin (Whatman) is prepared at 25°C in 50 m*M* triethanolamine buffer, pH 7.5, following the instructions provided with the resin. Two ml of settled resin are used to further purify 20 mg of actin isolated as shown in Fig. 3.
2. Two ml of settled resin are sedimented at 20,000 *g* for 20 min at 4°C, and resuspended with 200 ml of DEAE-buffer (10 m*M* imidazole, pH 8.0, at 4°, 0.1 *M* KCl, 0.1 m*M* $CaCl_2$, 1 m*M* ATP, 0.5 m*M* 2-mercaptoethanol, and 0.005% NaN_3). Equilibration is for 4 hr at 4°C with stirring.
3. Equilibrated resin is sedimented and resuspended with 20 ml of 1.0 mg/ml F-actin, which is prepared by diluting F-actin (~5 mg/ml) with cold DEAE-buffer and mixing for 30 min at 4°C.

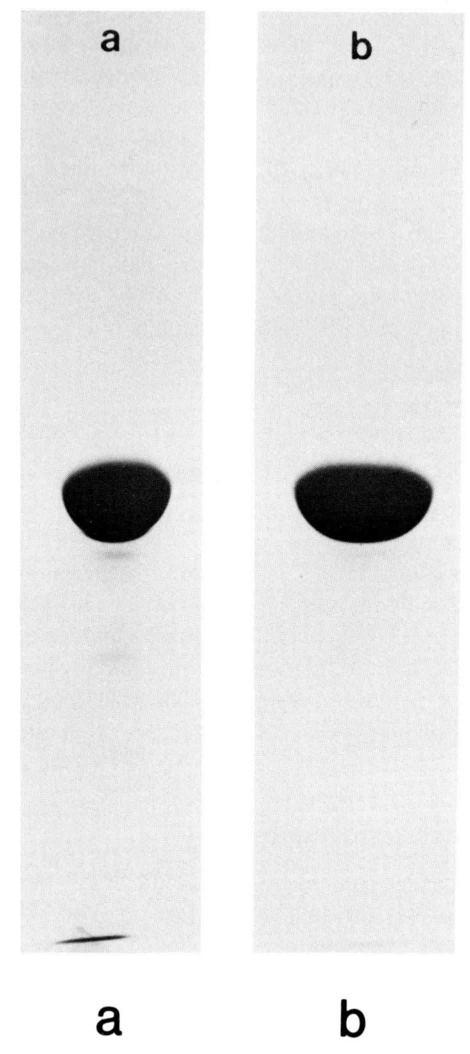

FIG. 8. DEAE purification of isolated actin. F-actin (20 mg) isolated as in Fig. 3 was treated with DE-52 by the batch method described in Section VI,A. (a) Isolated actin before DEAE treatment. Densitometry of contaminants indicates 1–2 μg of protein contamination per 40 μg of loaded protein, or about 95% actin homogeneity. (b) Isolated actin after DEAE purification. Densitometry of contaminants indicates <0.2 μg of contamination per 40 μg of loaded protein. Molecular weight homogeneity $>99\%$.

4. The F-actin, DE-52 mixture is stirred in a plastic beaker for 6 hr at 4°C. More than 95% of the actin is loaded onto the DE-52 in this step.

5. The resin is sedimented at 20,000 *g* for 20 min at 4°C and the supernatant is discarded.

6. The resin is resuspended by stirring in 20 ml of 10 m*M* imidazole, pH 6.4, at 4°C, 0.3 *M* KCl, 0.1 m*M* $CaCl_2$, 1 m*M* ATP, 0.5 m*M* 2-mercaptoethanol, and 0.005% azide. Actin is eluted for 2 hr at 4°C by mixing.

7. The resin is sedimented (20,000 *g*, 20 min, 4°C) and the supernatant (20 ml) is immediately dialysed against 2 liters of buffer A containing 50 m*M* KCl, 2 m*M* $MgCl_2$, and 1 m*M* ATP.

8. The dialysed F-actin is pelleted at 150,000 *g* for 1.5 hr at 4°C, homogenized into buffer A to a final actin concentration of 6–7 mg/ml, and depolymerized by overnight dialysis.

9. The resulting G-actin is clarified at 150,000 *g* for 1.5 hr at 4°C, assembled with 0.1 *M* KCl and 1 m*M* $MgCl_2$, and stored on ice in the presence of 0.005% sodium azide.

The recovery from this procedure is approximately 50% with a final product purity of greater than 99% (Fig. 8).

B. Depolymerization–Repolymerization

The actin can also be further purified by the following recycling protocol (Fig. 9). We often carry out this procedure just prior to using the actin in an experiment.

1. F-actin (stored as a viscous solution at 4–6 mg/ml) is diluted to 0.5 mg/ml and allowed to incubate at 4°C for 2 hr in the presence of 0.1 *M* KCl, 1 m*M* $MgCl_2$, and freshly added 1 m*M* ATP. Fresh ATP is always added to stored actin just prior to recycling. Incubated F-actin is sedimented at 150,000 *g* for 1.5 hr at 4°C, the supernatant decanted, and the tube and pellet rinsed carefully with buffer A.

2. The resulting F-actin pellet is rinsed carefully with buffer and gently homogenized into cold buffer A (see Section III,G) to a final actin concentration of 2–4 mg/ml and dialysed against 1 liter of buffer through a 10,000 dalton cutoff collodion bag (Schleicher and Schull) for 6 hr at 4°C with rapid stirring. Collodion bags are much more permeable to ATP in low-ionic-strength buffers than the traditionally used cellulose dialysis tubing (see Pardee and Spudich, 1982; Martonosi *et al.*, 1960). Consequently, actin depolymerization rates are enhanced and G-actin denaturation resulting from depletion of ATP within the dialysis bag is minimized. It is important to minimize dialysis time since pure actin depolymerizes quickly while actin associated with contaminants such as myosin and gelation factors depolymerizes more slowly; the basis of this purification step resides in these differential depolymerization rates.

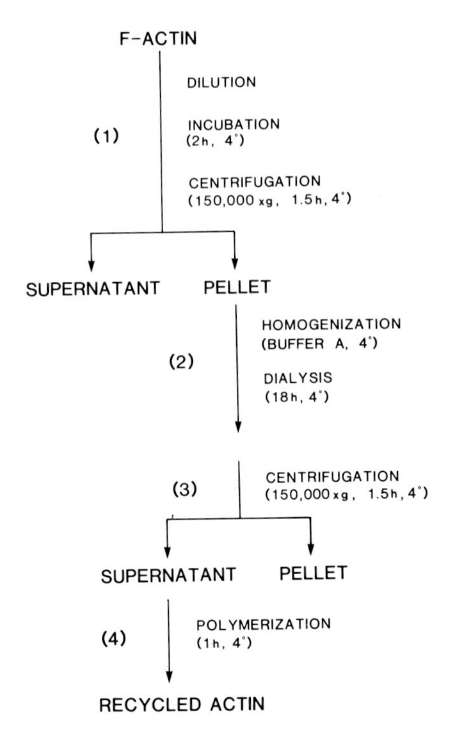

FIG. 9. Flow diagram for recycling the actin isolated as in Fig. 3. For complete explanation of the procedure, refer to the numbered steps in Section VI,B of the text. Dialysis is performed in 10,000 dalton cut-off collodion bags. Recovery is approximately 70% of starting material.

3. Dialysed actin (which still may retain a visible viscosity if actin associated contaminants are present in the starting material) is centrifuged at 150,000 *g* for 1.5 hr at 4°C, the supernatant is decanted, and any pelleted material is discarded.

4. The clarified G-actin is immediately polymerized by addition of ATP, KCl, and $MgCl_2$ to final concentrations of 1 m*M*, 0.1 *M*, and 1 m*M*, respectively. Full assembly is complete within 1 hr at 4°C. The F-actin solution at this stage should be quite clear and highly viscous. (The actin concentration is approximately 2 mg/ml). Purification may be complete at this stage, depending on the purity and type of contaminants originally present. However, a second recycling may be necessary when small amounts of proteolysed actin and low-molecular-weight contaminants persist.

The yield from this procedure depends on the level of contamination present in the starting actin material. For highly contaminated starting material (purity = 85–90%), approximately 50% recovery of highly purified actin is obtained. For actin that is 95% pure, recovery is approximately 75%. The value of recycling even apparently clean actin preparations is illustrated in Fig. 10. Although the

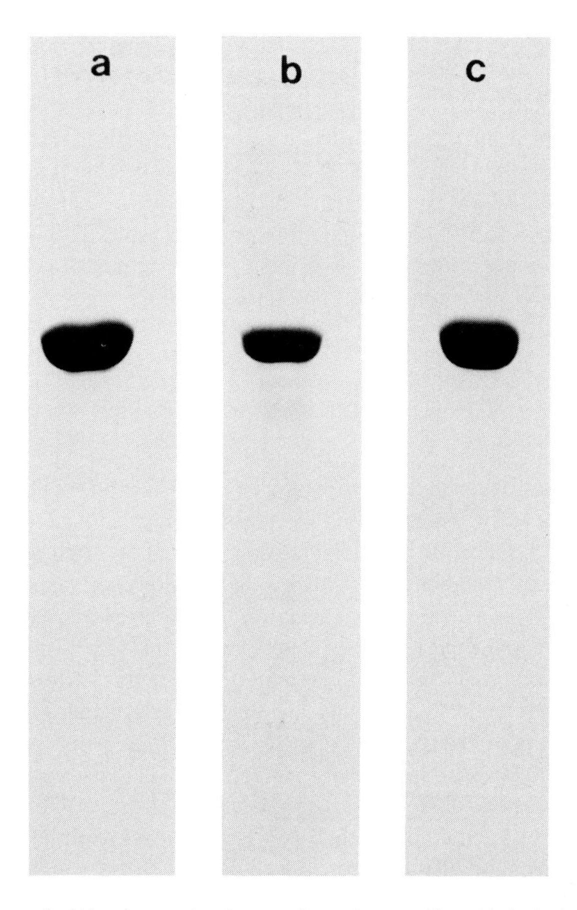

FIG. 10. Removal of F-actin-associated contaminants by recycling. (a) Actin from the isolation procedure shown in Fig. 3. (b) Sedimented material after depolymerization of the F-actin (Fig. 9, step 3). (c) Final product after recycling. Each lane was loaded with 20 μg of protein.

purity of the actin before recycling is high, the F-actin-associated impurities removed by a single recycling procedure are evident in the depolymerization pellet (Fig. 10b).

C. Sephadex Chromatography

G-150 Sephadex chromatography in depolymerization buffer can be used to provide additional purification as well as removal of actin oligomers from G-actin. Polymers are easily separated from the major peak of low-molecular-weight actin species, and if the trailing fractions of the eluted actin peak are pooled, a homogeneous population of monomeric actin can be obtained in addition to increasing the actin purity. The effects of purification by G-150 Sephadex on the properties of muscle actin have been studied recently by MacLean-Fletcher and Pollard (1980). Polymerization of column purified monomeric actin results in a significant increase in low-shear viscosity over that observed with actin before column purification. Moreover, some of the fractions from the Sephadex column show a significant viscosity-reducing activity when added back to the column-purified G-actin before polymerization. These results should be sufficient to warn of dramatic effects on the properties of actin by levels of contaminants heretofore considered negligible.

VII. Concluding Remarks

This review is devoted to the purification of muscle actin for two reasons. First, muscle actin has become a common laboratory reagent, and we therefore wished to discuss the details of the purification procedure of Spudich and Watt (1971) with special reference to possible problems that may be encountered. Second, the current level of sophistication in cell motility research places new demands on the level of acceptable purity of the actin preparation. Thus several additional steps are presented for further purification of muscle actin. Particularly useful is a simple batch treatment of F-actin with DEAE-cellulose, which gives good recovery of actin that is >99% pure.

In the last decade, actin from many cell types has become widely studied (for reviews, see Pollard and Weihing, 1974; Goldman *et al.*, 1976; Pollard, 1977; Weihing, 1976; Clarke and Spudich, 1977; Korn, 1978). Optimal purification of nonmuscle actins generally requires procedures specifically tailored to the cell type used (Yang and Perdue, 1972; Spudich, 1974; Hartwig and Stossel, 1975; Kane, 1975; Pollard *et al.*, 1976; Gordon *et al.*, 1976; Sheetz *et al.*, 1976; Hatano and Owaribe, 1977; Adelman, 1977; Uyemura *et al.*, 1978; Pardee and Bamburg, 1979; Weir and Frederiksen, 1980). An evaluation of various procedures for the purification of nonmuscle actins can be found in a recent review by Uyemura and Spudich (1980).

References

Adelman, M. R. (1977). *Biochemistry* **16,** 4862–4871.
Ames, G. F. C. (1974). *J. Biol. Chem.* **249,** 634–644.
Banga, I., and Szent-Györgyi, A. (1941). *Stud. Inst. Med. Chem., Univ. Szeged* **1,** 5.
Clarke, M., and Spudich, J. A. (1977). *Annu. Rev. Biochem.* **46,** 797–822.
Drabikowski, W., and Gergely, J. (1962). *J. Biol. Chem.* **237,** 3412–3417.
Ebashi, S., and Ebashi, F. (1965). *J. Biochem. (Tokyo)* **58,** 1–12.

Feuer, G., Molnar, F., Pettko, E., and Straub, F. B. (1948). *Hung. Acta Physiol.* **1,** 150–163.
Goldman, R., Pollard, T., and Rosenbaum, J., eds. (1976). "Cell Motility," Vol. 3. Cold Spring Harbor Lab., Cold Spring Harbor, New York.
Gordon, D. J., Eisenberg, E., and Korn, E. D. (1976). *J. Biol. Chem.* **251,** 4778–4786.
Hartwig, J. H., and Stossel, T. P. (1975). *J. Biol. Chem.* **250,** 5696–5705.
Hatano, S., and Owaribe, K. (1977). *J. Biochem.* (*Tokyo*) **82,** 201–205.
Jacobson, G. R., and Rosenbusch, J. P. (1976). *Proc. Natl. Acad. Sci. U.S.A.* **73,** 2742–2746.
Kane, R. E. (1975). *J. Cell Biol.* **66,** 305–315.
Kasai, M. (1969). *Biochim. Biophys. Acta* **180,** 399–409.
Korn, E. D. (1978). *Proc. Natl. Acad. Sci. U.S.A.* **75,** 588–599.
Laemmli, U. K. (1970). *Nature* (*London*) **227,** 680–685.
Laki, K., Bowen, W. J., and Clark, A. (1950). *J. Gen. Physiol.* **33,** 437–443.
Laki, K., Maruyama, K., and Kominz, D. R. (1962). *Arch. Biochem. Biophys.* **98,** 323–330.
MacLean-Fletcher, S., and Pollard, T. D. (1980). *Biochem. Biophys. Res. Commun.* **96,** 18–27.
Martonosi, A., Gouvea, M. A., and Gergely, J. (1960). *J. Biol. Chem.* **235,** 1700–1703.
Maruyama, K., Kaibara, M., and Fukada, E. (1974). *Biochim. Biophys. Acta* **371,** 20–29.
Mommaerts, W. F. H. M. (1951). *J. Biol. Chem.* **188,** 559–565.
Pardee, J. D., and Bamburg, J. R. (1979). *Biochemistry* **18,** 2245–2252.
Pardee, J. D., and Spudich, J. A. (1982). In preparation.
Pollard, T. D. (1977). *CRC Handb. Biochem. Mol. Biol.* Vol. **2,** 307–324.
Pollard, T. D., and Weihing, R. R. (1974). *CRC Crit. Rev. Biochem.* **2,** 1–65.
Pollard, T. D., Fujiwara, K., Niederman, R., and Maupin-Szamier, P. (1976). *In* "Cell Motility" (R. Goldman, T. Pollard, and J. Rosenbaum, eds.), pp. 689–724. Cold Spring Harbor Lab., Cold Spring Harbor, New York.
Sheetz, M. P., Painter, R. G., and Singer, S. J. (1976). *Biochemistry* **15,** 4486–4492.
Spudich, J. A. (1974). *J. Biol. Chem.* **249,** 6013–6020.
Spudich, J. A., and Watt, S. (1971). *J. Biol. Chem.* **246,** 4866–4871.
Straub, F. B. (1942). *Stud. Inst. Med. Chem., Univ. Szeged* **2,** 3.
Straub, F. B. (1943). *Stud. Inst. Med. Chem., Univ. Szeged* **3,** 23–37.
Straub, F. B., and Feuer, G. (1950). *Biochim. Biophys. Acta* **4,** 455–470.
Uyemura, D. G., and Spudich, J. A. (1980). *Biol. Regul. Dev.* **2,** 317–338.
Uyemura, D. G., Brown, S. S., and Spudich, J. A. (1978). *J. Biol. Chem.* **253,** 9088–9096.
Weihing, R. R. (1976). *In* "Cell Biology" (P. L. Altman and D. D. Katz, eds.), pp. 341–356. Fed. Am. Soc. Exp. Biol., Bethesda, Maryland.
Weir, J. P., and Frederiksen, D. W. (1980). *Arch. Biochem. Biophys.* **203,** 1–10.
Yang, Y. Z., and Perdue, J. F. (1972). *J. Biol. Chem.* **247,** 4503–4509.

Chapter 19

Directional Growth of Actin Off Polylysine-Coated Beads

SUSAN S. BROWN

Department of Structural Biology
Sherman Fairchild Center
Stanford School of Medicine
Stanford, California

I. Introduction—An Overview of the Actin Assembly Reaction

The intracellular distribution of actin in nonmuscle cells must change dramatically to function in such cellular processes as locomotion, endocytosis, and cell division. It has been suggested that redistribution may take place via actin assembly/disassembly reactions. These reactions can be studied in the test tube, using highly purified actin. Such studies have revealed that, upon addition of salt, G- (globular) actin monomers can come together to form F-actin filaments. Oosawa and Kasai (1971) proposed a two-step mechanism for assembly: The first step, nucleation, is slow, as several monomers must come together to make an actin filament end. In the subsequent step, elongation, monomers can add one by one to the growing filament end in a pseudo-first-order reaction. Several

ISBN 0-12-564124-9

laboratories have shown that there is a preferred end for elongation (Woodrum *et al.*, 1975; Hayashi and Ip, 1976; Kondo and Ishiwata, 1976; Spudich and Cooke, 1975).

The importance of these details of the assembly process to the functioning of actin *in vivo* is clear. There are likely to be sites within the cell where nucleation occurs, perhaps associated with membranes. The location of such sites could determine not only the location, but also the polarity of resulting filaments. This in turn would dictate the direction in which the filaments can move with respect to myosin (Huxley, 1963).

Recently, I have found that actin filaments can be assembled off polylysine-coated polystyrene beads (Brown and Spudich, 1979a). In this chapter, I will discuss how these beads can be used as a tool for studying actin assembly.

II. Preparation of Polylysine-Coated Polystyrene Beads

Polylysine adsorbs to the surface of polystyrene beads. This adsorption could be quantitated using [H^3]polylysine, prepared by the method of Means (1977). This approach is easy, because of the abundance of the epsilon amino groups on polylysine. First 1.8 m*M* formaldehyde is added slowly with stirring to 5 mg/ml polylysine (Sigma, type 1B) in 50 m*M* borate buffer, pH 9.32, on ice. Then 3.5 m*M* [H^3]$NaBH_4$ (Amersham) in methanol (10% final) is added, followed by ~0.5 mg/ml of unlabeled solid $NaBH_4$. The pH is then adjusted to about 5 with HCl, and the [H^3]polylysine dialyzed overnight versus H_2O. A specific activity of 7×10^4 cpm/μg is obtained. This corresponds to labeling 8% of the lysine groups.

Using this probe, we found that there were 1.3×10^7 lysine residues adsorbed per 1.1-μm bead. We also found that this polylysine slowly came off the bead surface after washing away free polylysine; the loss became significant after 24 hr. This indicated that it would be desirable to attach the polylysine covalently. The most convenient method is to first adsorb polylysine to the beads, then cross-link the polylysine molecules to one other. This approach is successful in eliminating desorption of polylysine; no losses were detected after a week, and these beads could be used for months without loss of activity.

Polylysine is covalently attached to polystyrene beads as follows (Brown and Spudich, 1981): 1 vol of 0.11-μm polystyrene beads (Sigma) or 10 vol of 1.1-μm polystyrene beads (Dow Chemical Co.) is added to 50 vol of 5 mg/ml polylysine (type IB Sigma) while vortexing. (The vortexing is important in avoiding bead "clumping," which can often be disrupted by sonication.) The mixture is stirred at 4°C for 3 hr to overnight. The beads are sedimented (10,000 rpm, 10 min, in an SS-34 rotor for 1.1-μm beads; 40,000 rpm, 20 min in a Ty 65 rotor for 0.11-μm beads) and resuspended by sonication in 2000 vol of 0.1 *M* sodium phosphate buffer, pH 6.2. While stirring in the hood, sodium cyanoborohydride (Sigma) is added to about 10^{-3} *M*, and glutaraldehyde to 1.6×10^{-4} *M* immediately thereafter. Beads are then sedimented and resuspended in 200 vol of methanol to which about 0.2 *M* sodium borohydride is then added to reduce any remaining aldehydes. Beads are well washed with water and stored at 4°C in 0.02% sodium azide. Beads are usually sonicated shortly before use to break up any clumps that formed upon standing.

Bead concentration is expressed as mg/ml polystyrene, which is measured spectrophotometrically after dissolving the beads in dioxane (Weisman and Korn, 1967). Aldrich dioxane was satisfactory; other brands absorbed in the UV. Number of beads/ml could be calculated from polystyrene concentration and was confirmed in the case of 1.1-μm beads by counting in a hemocytometer. I found that there were 10^9 beads/mg polystyrene for the 1.1-μm beads.

The 0.11-μm beads are too small to be counted in the light microscope, and were therefore counted by examining a 1:1 mixture with the 1.1-μm beads in the electron microscope. I obtained 8×10^{11} 0.11 micron beads/mg polystyrene.

III. Nature of the Actin–Bead Interaction

A. Adsorption of Actin to Beads

Binding of actin monomer to polylysine-beads (Brown and Spudich, 1981) was followed using *in vivo* labeled [^{35}S]actin (Simpson and Spudich, 1980). Polylysine-beads are mixed with various concentrations of monomeric actin. The beads can then be sedimented to determine binding. Saturable binding at 1 mmole actin/mg 1.1-μm beads is seen, and beads are half saturated at $\sim 10^{-6}$ *M* actin. The binding is not a function of bead diameter; smaller beads have the same amount of actin bound per square micron of bead surface. There is about one actin bound for every 50 lysine residues on the bead. This actin desorbs rather slowly when free actin is removed, and the loss is not significant after 8 hr.

The number of monomers bound per bead (3.6 and 5×10^5 per 1.1-μm bead in two determinations) is enough to coat the bead surface completely approximately twice if it is assumed that each monomer occupies 25 nm^2 of bead surface (5 nm = diameter of a monomer). The numbers should not be taken to indicate that there is precisely a bilayer of monomers around the bead, but rather that the bead is completely coated with closely packed monomer.

To explain the interaction of a possible second layer of monomer with polylysine, it is only necessary to point out that the polylysine I am using has a molecular weight of 70–230K, and a portion of the molecule could easily project out far enough from the bead to bind a second layer of monomers.

Polylysine is responsible for the actin–bead interaction, as judged by the fact that beads not treated with polylysine do not nucleate actin assembly. Also, free polylysine can nucleate actin assembly.

Since the reaction of coupling polylysine around the beads produces secondary and mostly tertiary amines, I wanted to know whether the degree of substitution affected the ability of the lysine residues to support actin assembly. Therefore, the glutaraldehyde concentration in the reaction was varied between 1.6×10^{-4} and 3.2×10^{-3} M (the concentration of lysine residues on the beads was 6.8×10^{-4} M). This had no effect on the ability of the beads to nucleate assembly. This may not be surprising if the negatively charged actin and positively charged polylysine are involved in a nonspecific electrostatic interaction, since the pK of the amino group is hardly affected by the substitutions.

B. Effect of Polylysine-Beads on Actin Assembly

All the binding studies in the previous section were done under buffer conditions where the actin is monomeric and binds directly to the bead. A second type of actin–bead interaction is seen (Brown and Spudich, 1979a) under assembly conditions. When the salt concentration is increased, the polymerization of the monomer that is not bound to beads is induced. The electron microscope reveals that when actin is assembled in the presence of polylysine-beads (Fig. 1), all

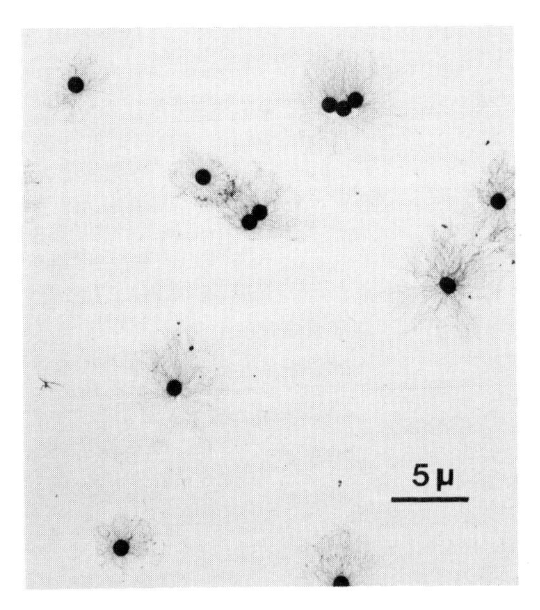

Fig. 1. Here 0.2 mg/ml actin is assembled in the presence of 0.4 mg/ml polylysine-coated 1.1 μm polystyrene beads. At this low magnification, the actin filaments are seen as "fuzz" radiating from the beads. Note that there are no free filaments.

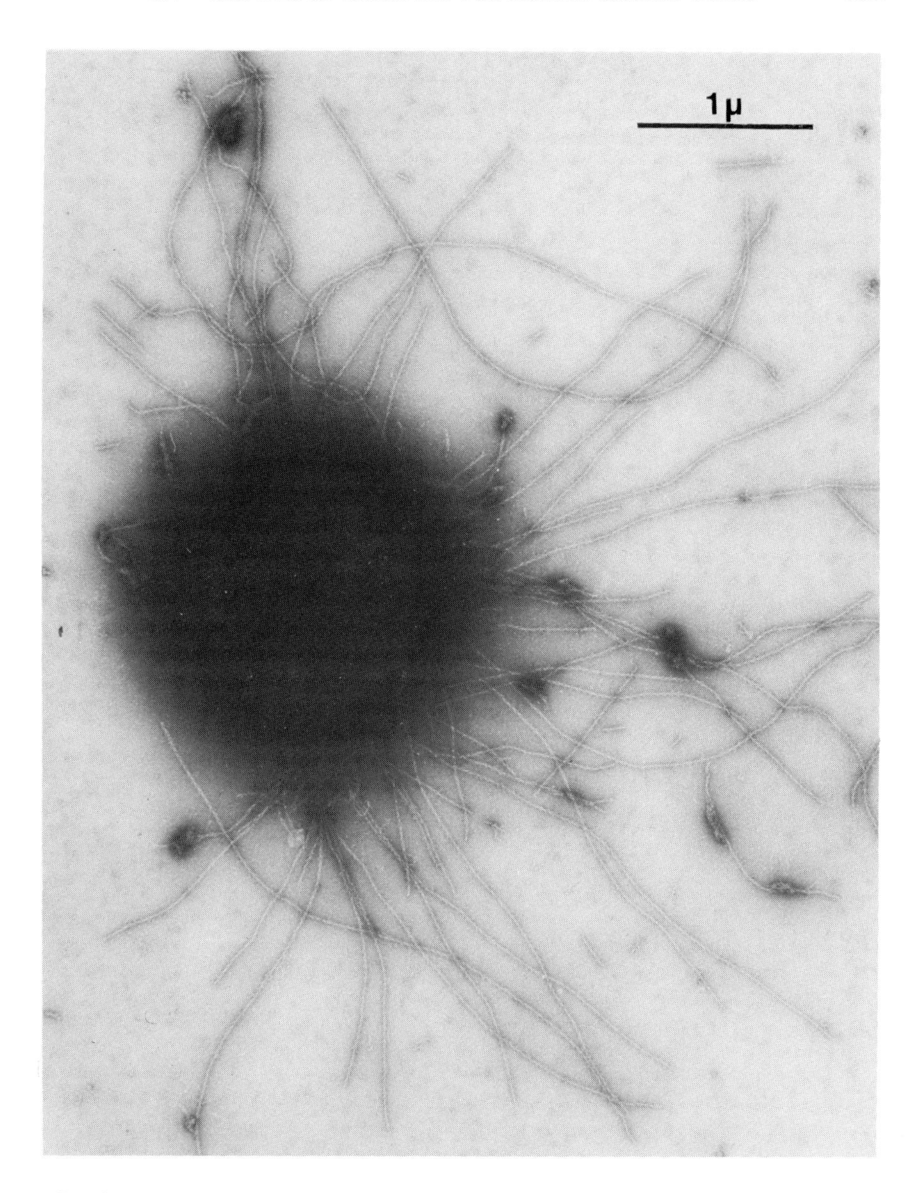

FIG. 2. Actin assembled in the presence of beads as in Fig. 1, and decorated with S_1 to indicate the polarity of the filaments. Note that the arrowhead decoration points toward the bead.

resulting filaments are found in association with the beads. The association of filaments with beads clearly occurs via the assembly reaction; a very different picture is obtained when actin that has already been assembled into filaments is mixed with the beads. Then there are many free filaments, and only a few seem attached to beads.

The filaments are associated with the beads in a very specific way. S_1 decoration (Cooke, 1972; Moore *et al.*, 1970) of the filaments radiating from the beads (Fig. 2) revealed that all the filaments have the same polarity; the arrowhead decoration points toward the bead on all filaments. In contrast, when F-actin is mixed with beads, the few bead-associated filaments are random in polarity.

The interpretation given to these findings is as follows (Fig. 3): The adsorption of actin to beads brings monomers close together (as shown in Section III,A) and thereby accelerates the normally slow nucleation reaction (see Introduction). Remaining free monomers can then assemble onto the nuclei that have formed at the bead surface. Since monomers add preferentially at one end of the filament (see Introduction), the filaments growing off the bead all have the same polarity. The arrowheads on decorated filaments point toward the bead because the preferential addition of monomer is at the barbed end of the filament, distal to the bead.

If the beads increase the number of nuclei for monomers to add onto, the overall rate of assembly should be faster in the presence of polylysine-beads. I

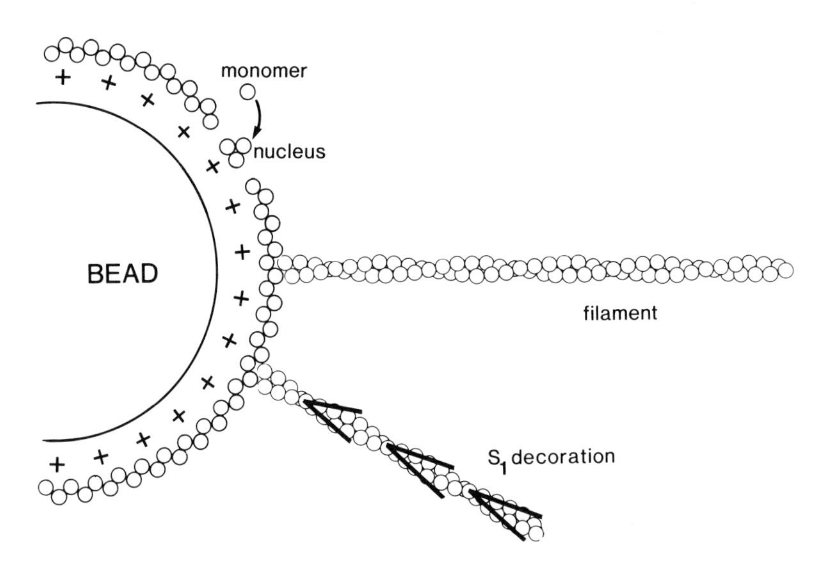

FIG. 3. Diagrammatic representation of the proposed mode of action of the polylysine beads: Actin monomer adsorbs to the polylysine coating, some of the bound monomers interact to form nuclei, more monomers add on to the nuclei to form filaments. S_1 decoration reveals that these filaments grow in one direction, i.e., have a uniform polarity.

have shown that this is in fact the case (Brown and Spudich, 1979a). Assembly rate can be followed in a number of ways: by viscometry or intrinsic ATPase activity (Oosawa and Kasai, 1971), change in absorption at OD_{232} (Cooke, 1975), DNase I inhibition (Blikstad *et al.*, 1978), and sedimentation. Beads do not interfere with any of these assays with the exception of absorbance change.

C. Ratio of Actin to Polylysine-Beads

The ratio of actin to beads is critical in obtaining the preceding results. It is, of course, necessary to use an actin concentration greater than the critical concentration for assembly (Oosawa and Kasai, 1971). If the ratio of actin to beads is too low (e.g., less than 0.05 mg actin/mg beads), there is not enough monomer to saturate the bead surface (Brown and Spudich, 1981). As a result, monomers are not brought into close apposition to make nuclei. If the ratio is too high (e.g., 2 mg actin/mg beads), many free filaments that are not attached to beads are seen, presumably because of increased numbers of endogenous nuclei (versus bead nuclei) at high actin concentrations.

Between these limits, varying the ratio provides a means of varying filament length and number. If actin is held constant and the number of beads increased, the number of filaments will also increase, since the number of filaments per bead can be seen by electron microscopy to stay the same. These filaments are shorter, since actin concentration is held constant.

D. Role of Nucleotide

One of the remaining mysteries in the actin assembly reaction is the role of nucleotide. Under normal conditions, actin monomer has a bound ATP that is required to prevent denaturation. This ATP is cleaved during assembly to give a filament with bound ADPs. It is clear, however, that ATP is not required for assembly and that assembly can occur in the complete absence of nucleotide (Barany *et al.*, 1966).

The beads allow me to look at effects of nucleotide on nucleation, which has not been possible before. Up until now, only the overall rate of assembly could be examined. The overall rate is slower in the absence of ATP (Hayashi and Rosenbluth, 1960); thus ATP may be affecting the rate of nucleation, the rate of elongation, or both.

Since I already suspect that the interaction between the net negatively charged actin and positively charged beads is electrostatic, and since actin has a bound negatively charged ATP, I was first interested in knowing whether ATP alone interacted with the polylysine-beads, and whether the binding of actin to beads was affected by the species of bound nucleotide.

The binding of ATP to beads was assayed by equilibrium dialysis. There was no high affinity binding, unlike the case for actin, but I estimated that there might be about 10^6 molecules of ATP per bead bound with low affinity under the conditions of assembling actin off beads (i.e., more than the amount of monomer bound).

The binding to beads of actin with bound ATP, ADP, or no nucleotide was assayed as before using [^{35}S]actin. Since monomeric actin denatures in the absence of ATP, special conditions must be employed: 50% sucrose (Kasai *et al.*, 1965) or 3 mM ADP (Brown and Spudich, 1979b) will stabilize the actin. I found that the same amount of actin was bound to the beads in all cases. The species of bound nucleotide did, however, have a dramatic effect on the ability of the beads to nucleate assembly. Beads were completely unable to nucleate if the actin had no or little bound nucleotide. Saturation of actin monomer with bound ADP allowed some nucleation, but less than in the presence of ATP. From these data, I cannot determine whether there is an effect of nucleotide on elongation rate. However, nucleotide clearly affects rate of nucleation; nucleation apparently requires binding of some nucleotide to actin and ATP is more effective than ADP. Preliminary evidence suggests that hydrolysis does not occur in the nucleation step.

E. Use of Cytochalasin in Conjunction with Beads

Recent evidence suggests very strongly that cytochalasin interferes with assembly by binding to the preferential assembly end of the filament (see Brown and Spudich, 1981). Thus, cytochalasin can be used to quantitate the ability of beads to make actin nuclei, which have the conformation of filament ends. This approach reveals that the actin directly bound to the bead can interact to form nuclei, even when the remaining free actin is washed away and no visible filaments are assembled off the bead. It can be used to demonstrate that the decreased bead nucleation seen in ADP (Section III,D) corresponds to a decrease in the number of nuclei or cytochalasin binding sites on the bead; the affinity of the sites for cytochalasin does not change. Finally, it can be demonstrated that if the actin–bead ratio is low, so that the bead surface is not saturated with monomer, an increase in bead concentration gives rise to a decrease in cytochalasin binding (Brown and Spudich, 1981). This fits with the idea that beads accelerate nucleation by bringing monomers into close apposition; as the close packing of monomers is lost, so is the ability to make nuclei.

IV. Summary of the Usefulness of Polylysine-Beads as a Tool for Studying Actin

I have concluded that polylysine-beads accelerate the nucleation step, but do not affect the elongation step, of actin assembly. This makes the beads a very powerful tool in dissecting out these two steps, so that they can be studied separately. This sort of dissection has not previously been possible.

Now one can ask how the nucleation step is affected by conditions of temperature, salt, pH, etc. Importantly, the beads provide an assay for nonmuscle factors that might specifically affect nucleation. Finally, information gained on the requirements of the nucleation reaction could lead to an assay for the physiological nucleation sites in the cell.

Let us look closely at what information can be obtained with the beads. A comparison of the relative rates of assembly ± beads gives a measure of the ability of the beads to make nuclei. Protein factors or changing buffer conditions may affect elongation as well as nucleation, but that effect can be subtracted out by virtue of the control experiment of assembly in the absence of beads. Note that this experiment can clearly assay nucleation, but may not provide information about whether elongation is also affected. For example, if nucleation is inhibited, one cannot tell whether the control rate in the absence of beads is slower simply because endogenous nucleation is affected, or because elongation has also been slowed down. A boundary condition for this experiment is that endogeneous nucleation should be a small fraction of bead nucleation (i.e., there should be a several-fold difference in rate of actin assembly ± beads).

Can the beads serve as a good model of an *in vivo* nucleation site? The details are almost certain to be different; for instance, an *in vivo* site should be more efficient than the beads, where only one nucleus is formed per 10^3 bead-bound monomers. However, the principle should be the same: Since monomers can apparently form nuclei on their own, but only very slowly, it seems likely that all that is required of an *in vivo* nucleation site is that it facilitate the endogenous reaction by bringing monomers together. I envision a nucleation site that binds several monomers in an orientation favorable for nucleus formation. (On a bead, orientation of monomer might be random, thus the low efficiency.)

The beads have already proved useful in devising a quick assay for cytochalasin binding and in examining the mechanism of action of cytochalasin (Brown and Spudich, 1981). In this case, beads were used to vary the number of filament ends in solution, a task that has proved hopeless by other means. For example, sonication fragments filaments, but only transitorily. Cytochalasin provides a complementary tool to the beads; whereas beads affect nucleation, cytochalasin effects elongation.

The beads can be used to obtain a population of oriented filaments of uniform polarity, which might prove useful in optical diffraction studies. If beads with filaments of the opposite polarity are desired, it is possible they could be produced by assembling actin off beads in the presence of cytochalasin (MacLean-Fletcher and Pollard, 1980). The two types of beads could possibly be used in studies of the effect of blocking one or the other end of the filament, or as labels for sites in the cell that bind one or the other filament end.

REFERENCES

Barany, M., Tucci, A. F., and Conover, T. W. (1966). *J. Mol. Biol.* **19,** 483–502.
Blikstad, I., Markey, F., Carlsson, L., Persson, T., and Lindberg, U. (1978). *Cell* **15,** 935–943.
Brown, S. S., and Spudich, J. A. (1979a). *J. Cell Biol.* **80,** 499–504.
Brown, S. S., and Spudich, J. A. (1979b). *J. Cell Biol.* **83,** 657–662.
Brown, S. S., and Spudich, J. A. (1981). *J. Cell Biol.* **88,** 487–491.
Cooke, R. (1972). *Biochem. Biophys. Res. Commun.* **49,** 1021–1028.
Cooke, R. (1975). *Biochemistry* **14,** 3250–3256.
Hayashi, R., and Ip, W. (1976). *J. Mechanochem. Cell Motil.* **3,** 162–169.
Hayashi, R., and Rosenbluth, R. (1960). *Biol. Bull.* (*Woods Hole, Mass.*) **119,** 290.
Huxley, H. E. (1963). *J. Mol. Biol.* **7,** 281–308.
Kasai, M., Nakano, E., and Oosawa, F. (1965). *Biochim. Biophys. Acta* **94,** 494–503.
Kondo, H., and Ishiwata, S. (1976). *J. Biochem.* (*Tokyo*). **79,** 159–171.
MacLean-Fletcher, S. D., and Pollard, T. D. (1980). *Cell* **20,** 329–341.
Means, G. E. (1977). *In* "Methods in Enzymology," (C. H. W. Hirs and S. Timasheff, eds.), Vol. 47, pp. 469–478. Academic Press, New York.
Moore, P. B., Huxley, H. E., and DeRosier, D. J. (1970). *J. Mol. Biol.* **50,** 279–295.
Oosawa, F., and Kasai, M. (1971). *In* "Subunits in Biological Systems" (S. N. Timasheff, ed.), Chap. 6. Dekker, New York.
Simpson, P. A., and Spudich, J. A. (1980). *Proc. Natl. Acad. Sci. U.S.A.* **77,** 4610–4613.
Spudich, J. A., and Cooke, R. (1975). *Fed. Proc., Fed. Am. Soc. Exp. Biol.* **34,** 540.
Weisman, R. A., and Korn, E. D. (1967). *Biochemistry* **6,** 485–497.
Woodrum, D. T., Rich, S. A., and Pollard, T. D. (1975). *J. Cell Biol.* **67,** 231–237.

Chapter 20

A Falling Ball Apparatus to Measure Filament Cross-linking

THOMAS D. POLLARD

Department of Cell Biology and Anatomy
Johns Hopkins University School of Medicine
Baltimore, Maryland

I. Introduction

This chapter describes a low shear falling ball viscometer that can be used to measure consistency changes in cellular extracts and the gelation of actin filament networks. It is particularly useful for routine assays of actin filament interactions during the purification of molecules that either promote or inhibit the network formation. Although this device cannot yet be used for rigorous quantitative analysis, the assay does have a number of practical advantages over other routine assays. A brief account of the falling ball device originally appeared in an article by MacLean-Fletcher and Pollard (1980a) in the *Journal of Cell Biology*.

METHODS IN CELL BIOLOGY, VOLUME 24

ISBN 0-12-564124-9

II. Theory

The falling ball apparatus consists of a capillary tube containing the sample and a steel ball (Fig. 1) that will fall through the sample, providing that the yield strength of the material is low enough. The force balance in the system and Stokes' equation yield an expression relating the observed velocity of the ball to the absolute viscosity of Newtonian samples and the apparent viscosity of non-Newtonian samples. For samples with high yield strengths, one can measure the

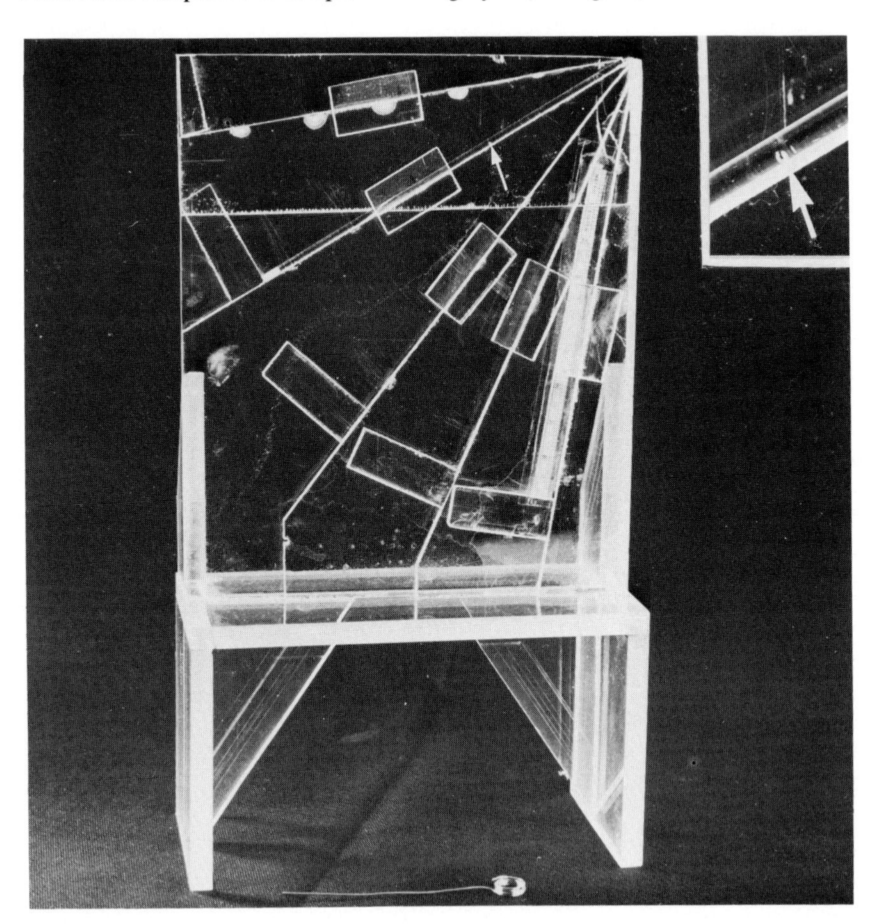

FIG. 1. A photograph of the plastic stand used to support the capillary tubes at various fixed angles (10°, 30°, 50°, 65°, 80° from horizontal) in the water bath. A rule affixed to the back of the apparatus is used to measure the velocity of the ball. The capillary is plugged at its lower end with "Seal Ease." The inset is a closeup of the inclined capillary containing the measuring ball (arrow). The wire used to push the ball through the meniscus is shown below.

yield strength from the maximum force per unit mass that can be applied to the system without movement of the ball.

The force balance for a ball immersed in a medium is

$$F_b = F_y + F_m + F_b \tag{1}$$

where F_b, the ball force, is the force acting on the medium. It is balanced by the forces acting on the ball, which are F_y, the static yield force, F_m, the buoyant force, and F_f, the frictional force. Equation (1) is valid when there is no acceleration in the system, i.e., when the ball is either stationary or moving at a constant velocity.

$$F_b = \tfrac{4}{3}\pi R^3 \rho_b \alpha \tag{2}$$

where R is the radius of the ball, ρ_b is the density of the ball, and α is the force per unit mass applied to the ball.

$$F_y = T_0 \pi R^2 C \tag{3}$$

The static yield strength T_0 is the maximum force per unit area that the medium will support without displacement and C is a geometrical constant to correct for the nonhomogeneous distribution of shear stress around the ball. For a sphere, C has been determined empirically to be about 1.75 (Johnson, 1970).

$$F_m = \tfrac{4}{3}\pi R^3 \rho_m \alpha \tag{4}$$

where ρ_m is the density of the medium and α is the force per unit mass applied to the medium.

$$F_f = fu \tag{5}$$

where f is the dynamic frictional coefficient and u is the velocity of the ball.

A. Yield Strength

In the case where the medium is strong enough to support the ball, there is no acceleration, the forces are balanced, $F_f = 0$ and $F_b = F_y + F_m$. Under these conditions, it is possible to estimate the static yield strength T_0 of the medium by determining α_o, the maximum force per unit mass that can be applied to the system without movement of the ball. As described in a later section, this is accomplished by centrifuging a capillary containing a ball immersed in the test medium. This applies to the system a force per unit mass of $\omega^2 r$, where ω is angular velocity in radians per second and r is the radius. Solving for T_0

$$T_o = \frac{\tfrac{4}{3}\pi R^3 \alpha_o(\rho_b - \rho_m)}{1.75\pi R^2} = 0.76 R\alpha_o(\rho_b - \rho_m) \tag{6}$$

B. Viscosity

If the medium is not strong enough to support the ball, the ball will accelerate until it reaches a constant "terminal" velocity. Under these conditions the forces are balanced and the absolute viscosity of a Newtonian fluid can be determined from the velocity using Stokes' equation

$$\eta = \frac{f}{6\pi R} \tag{7}$$

If F_y is negligible compared with the other forces in the system (i.e., $F_b - F_m - F_f$ is small), then $F_f = F_b - F_m$ and

$$f = \frac{m\frac{4}{3}\pi R^3 \alpha(\rho_b - \rho_m)}{u} \tag{8}$$

The term m is a calibration constant required when the ball and the medium are confined to a narrow tube and inclined at angle Θ. It corrects for drag on the ball resulting from wall effects (Van Wazer *et al.*, 1963) and friction at the point of contact between the ball and the wall (Geils and Keezer, 1977). The value of m can be determined empirically (see below) and is a function of the radius of the ball, the radius of the tube, and the angle of inclination. The force per unit mass (α) along the axis of a tube inclined at angle Θ from the horizontal is $g \sin \Theta$. Solving Eq. (7) and Eq. (8) for η

$$\eta = \frac{0.22 m R^2 g \sin \theta(\rho_b - \rho_m)}{u} \tag{9}$$

III. Design and Operation of the Falling Ball Apparatus

For our routine measurements a sample of about 170 μl is contained in a capillary tube, 1.3 mm i.d., 12.6-cm long (100-μl micropipette from VWR Scientific Inc., Univar Corp., San Francisco, CA). For small samples with high viscosities we use shorter tubes. For more accurate measurements of low viscosities, a longer tube is advisable. To avoid the wall effects discussed below, a wider tube can be used. The sample is introduced into the tube by capillary action or by using a Clay Adams Pipet Filler (Clay Adams, Div. Becton, Dickinson & Co., Parsippany, NJ). The tube is then sealed at one end with Clay Adams Seal Ease. The tube is held by a Plexiglas stand (Fig. 1) at a fixed angle of inclination in a temperature-controlled water bath with transparent walls, such as a $5\frac{1}{2}$-gal aquarium equipped with a thermocirculator. A stainless steel ball

(0.64-mm diameter, density 7.2 g/cm², grade 10, gauge deviation ±0.000064 mm, material 440C from the Microball Company, Peterborough, NH) is placed on the meniscus of the sample by hand. A convenient method is to use a pair of fine forceps with magnetized tips. Larger or smaller stainless steel balls can be purchased from the Microball Company or other suppliers of ball bearings such as Winsted Precision Ball Corp., Winsted, CT 06098. At the appropriate time the ball is pushed through the meniscus with a thin metal wire to initiate its fall. The velocity of the ball is measured by recording the time required for the ball to pass 2-cm intervals beginning about 1 cm below the meniscus. For samples with very high viscosity, a shorter interval, down to 2 mm, can be used, while for low-viscosity samples, a longer distance improves accuracy. For making sequential measurements on a single sample, a printing timer such as an "All Sport Timer," from the Chronomix Corp., Sunnyvale, CA, is helpful. Better yet, the measurements could be automated using photocells to detect the ball interrupting a beam of light.

For many non-Newtonian samples, such as delicately cross-linked actin filament gels, it is necessary to fill the tube with monomer and allow polymerization and cross-linking to take place within the tube while the sample is at rest. This prevents destruction of gelled samples and the orientation of fibers along the axis of the tube by flow.

If the sample has a yield strength >146 dyn/cm², the ball will fail to move at an angle of 80° under the influence of gravity alone. A measurement of yield strength can be made by increasing the ball force. This is accomplished by pushing the ball about 0.5 cm below the meniscus and centrifuging in a stepwise fashion for 30 sec at progressively higher speeds. For samples with T_0 <146 dyn/cm², a rough estimate of yield strength can be made by placing the ball in a horizontal tube and slowly increasing the angle of inclination until the ball moves. In this case $\alpha_0 = g \sin \Theta$.

IV. Testing and Calibration of the Falling Ball Apparatus

The viscometer was tested with glycerol/water mixtures, which are Newtonian solutions of known viscosity ("Handbook of Chemistry and Physics," 1958–59). The ball has a constant velocity throughout the measuring section of the capillary for any given sample over the entire viscosity range tested (1–12,000 cP). At any angle between 10 and 90°, u^{-1} is proportional to viscosity up to the point where the ball will not move (Fig. 2). With an experienced operator the measurements are highly reproducible with a standard deviation from the mean velocity of <2% over the entire range. The variation is larger at high viscosity than at low viscosity. Tilting the capillary at various angles Θ allows one to vary the useful range of the apparatus. Most of our measurements are made at 50 or 80° because our samples have high viscosities. An angle of 10° is more convenient for low-viscosity samples.

We determined the calibration constant m for the 0.32-mm radius stainless steel ball in tubes of various radii held at several angles. In tubes with radii $R_c > 16$ mm, $m = 1$. For R_c between 2.7 and 16 mm, m can be calculated from the empirical equation of Faxen (Van Wazer *et al.*, 1963). For $R_c < 2.7$ mm, m must be determined empirically and is a function of both R_c and Θ. For the capillaries we use with R_c of 0.65 mm, m is 0.13 at 80°, 0.09 at 50°, and 0.06 at 10°. These values do not vary with the viscosity of the sample (Fig. 2). The linearity of this relationship justifies, for these Newtonian fluids, our omission of the F_y term from the viscosity equation. (This result is expected from the magnitude of the forces in the system. At 80°, $F_b - F_m = 0.82$ dyn. For $F_y = 1\%$ of this force, $T_0 = 1.46$ dyn/cm^2. The standards used for calibration have yield strengths even less than this.) These tests establish that the viscometer can be used to measure

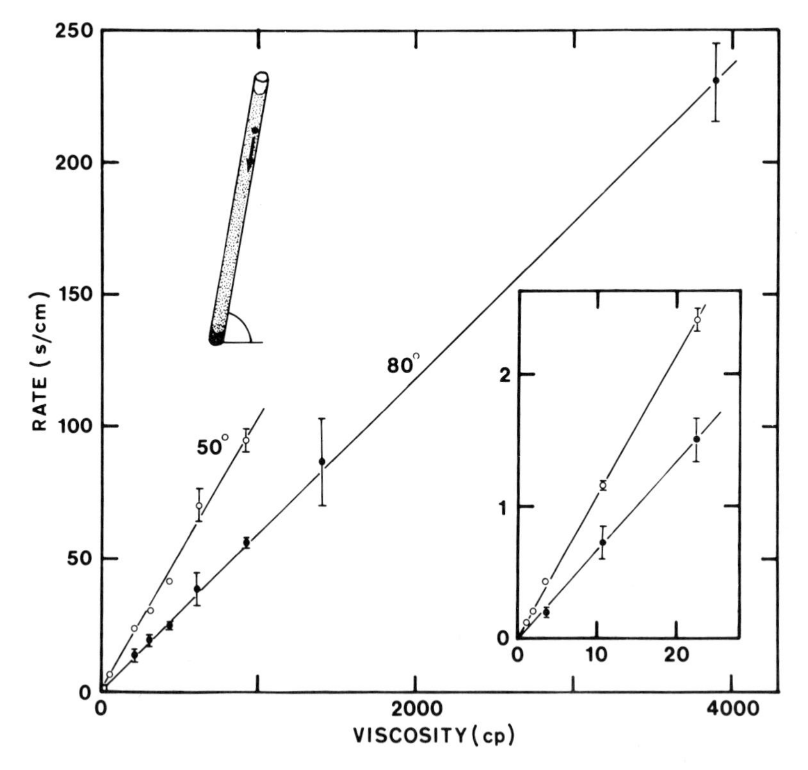

FIG. 2. Calibration of the falling ball viscometer with glycerol/water mixtures. Mean ±1 standard deviation is plotted versus viscosity. The tube, sample, ball, and plug are shown diagrammatically at the upper left.

the absolute viscosity of Newtonian solutions in the range of 1–12,000 cP with an error of $<2\%$.

The use of the apparatus for yield strength measurement was tested with solidifed gelatin. The precision of this measurement is limited by the crude method of applying centrifugal force in graded steps, but we obtained values close to those in the literature: 2×10^4 dyn/cm^2 for 4% gelatin and 1×10^4 dyn/cm^2 for 3% gelatin (Ferry, 1948). This confirms that the empirical geometrical factor [1.75 in Eq. (3)] is approximately correct for our apparatus. Using gelatin, we measured the same value for the yield strength in tubes with radii from 0.65 to 14 mm, demonstrating that there was no detectable wall effect on yield strength measurement.

V. Application of the Falling Ball Viscometer to Non-Newtonian Fluids

In a falling ball viscometer with acceleration provided by gravity, the velocity of the ball, and hence the maximum shear rate of a Newtonian fluid, $d = 1.5\ R/u$ (Van Wazer *et al.*, 1963), varies inversely with viscosity. This is of no consequence with a Newtonian fluid in which viscosity is independent of shear rate.

In contrast, the absolute viscosity of a non-Newtonian fluid varies with the shear rate, so that the observed ball velocity is used to calculate an "apparent viscosity" from Eq. (5). These apparent viscosities are expressed in centipoise for comparison with the absolute viscosities of Newtonian fluids through which the ball falls at the same velocity, with the full understanding that these apparent viscosities are not absolute, but depend on the shear rate around the ball. Absolute viscosities could be calculated directly from these apparent viscosities if the shear rate around the ball could be determined exactly and the dependence of the absolute viscosity upon the shear rate were established. However, to my knowledge there is no theoretical treatment of shear rates around balls falling through non-Newtonian fluids, nor have we or others established the dependence of the absolute viscosity of our samples on the shear rate.

A second note of caution is that the shear dependence of the viscosity of non-Newtonian samples such as actin filaments (Maruyama *et al.*, 1974) will amplify differences in the apparent viscosity. Take, for example, the case of actin filaments and other materials in which viscosity is inversely proportional to shear rate. If sample *A* has a higher viscosity than sample *B* when the two are measured at the same shear rate, the velocity of the ball falling under the influence of gravity will be less in *A* than in *B*. Thus the shear rate in *A* will be less than in *B*. As a result of the lower shear rate in *A*, the observed difference in the viscosity of *A* and *B* is greater than the difference measured at constant sheer rates. In contrast, with rheopexic materials the apparent difference will be less than the actual difference.

A final point is that the static yield strength of a sample will contribute to the apparent viscosity if the magnitude of the yield force approaches that of the other forces in the system. In our device at 80° inclination, F_y would have to be >0.02 dyn to have a detectable (>2%) effect on the observed viscosity. F_y of 0.02 dyn corresponds to T_0 of 3 dyn/cm^2, which is greater than the T_0 of low concentrations of actin filaments (Brotschi *et al.*, 1978), so we expect that F_y has little or no effect on the apparent viscosity of actin filament solutions. Even if the yield strength of a sample were 70 dyn/cm^2 (F_y = 0.41 dyn), it would increase the viscosity by only a factor of 2. On the other hand, when $F_y = F_b - F_m$, $u = 0$ and the apparent viscosity is, of course, infinite. These are maximum effects of yield strength on observed viscosity, because it is possible that the dynamic yield strength is much less than the static yield strength, in which case the F_y term is less consequential than stated here.

VI. Practical Applications of the Falling Ball Apparatus

To date, this falling ball apparatus has been used for a number of studies in our laboratory and by others. These include analysis of the interaction of actin filaments with microtubules and microtubule-associated proteins (Griffith and Pollard, 1978), with immunoglobulins (Fechheimer *et al.*, 1979), with *Acanthamoeba* gelatin factors (MacLean-Fletcher and Pollard, 1980a,b) and with erythrocyte spectrin and band 4.1 (Fowler and Taylor, 1980). In all of these studies the added proteins caused the actin filament samples to have a viscosity more than 10 times higher than the viscosity of actin alone, presumably due to the formation or stabilization of cross-links between the actin filaments. The apparatus was also used to study the kinetics of *Acanthamoeba* extract gelation (MacLean-Fletcher and Pollard, 1980a). Another type of application has been the detection of very weak interactions between actin filaments and the demonstration that these filament self-associations can be inhibited by cytochalasins (MacLean-Fletcher and Pollard, 1980b), by an unpurified protein factor that contaminates conventional muscle actin preparations (MacLean-Fletcher and Pollard, 1980c), and by a purified *Acanthamoeba* capping protein (Isenberg *et al.*, 1980).

This device also has many potential uses outside of the contractile protein field where small sample size, broad range, and high sensitivity are required. For example, I have received personal communications regarding its successful application in the analysis of complex carbohydrates.

VII. Conclusion: Evaluation and Comparison with Other Assays

The falling ball assay is simple, inexpensive, and highly reproducible. The sample size is small. The sample is stationary before interaction with the ball. The viscosity range is wide. Kinetic data are obtained with ease. It is possible to keep the maximum shear rate for Newtonian fluids below 10 sec^{-1} for samples with viscosities >6cP by choosing an appropriate angle of inclination. For comparison, Ostwald viscometers which we use have shear rates >5000 sec^{-1}.

It is the wide range and especially the low shear rate of the viscometer that enable one to study actin filament associations. The broad range is important because the samples can vary in apparent viscosity by more than three orders of magnitude. The low shear rate is essential, because the networks are destroyed by shearing. As a consequence, neither actin filament cross-linking nor inhibition of actin filament self-association is readily measured at high shear rates.

The ball centrifugation assay is a simple approach to measuring yield strength. Its major limitations at the present time are some uncertainty about the value of the geometrical constant in Eq. (3) and the crude, stepwise application of force by centrifugation. A better method would be to observe the ball continuously as the centrifugal force is gradually increased.

Although the falling ball method has many advantages, it has three main limitations in studies of gelation. First, only two structural parameters (apparent viscosity and yield strength) can be measured. Second, for a given sample the shear rate can only be varied by a factor of ~12 when using gravitational acceleration and angles of inclination between 10 and 80°. Finally, even at the relatively low shear rates in the falling ball device, the stress applied to these delicate samples by the falling ball is destructive.

These factors are limitations because, in our studies, the actin networks are complex materials, possessing the properties of both liquids (above their yield points) and solids (below their yield points). Moreover, microscopic observations (Condeelis and Taylor, 1977; Taylor, 1977) suggest that they are elastic as well. It is also likely that their physical properties change with time under stress. Eventually it will be necessary to make a thorough analysis of the physical properties of these gels, including a complete stress/strain curve versus time for the viscous component and a restoring force versus displacement curve for the elastic component. No simple assay can provide all of this information, but it can be obtained with a number of different complex devices such as the micromagneto-rheometer developed by Litt and colleagues (Lutz *et al.*, 1973).

A number of other assays have been used to study gelation of cytoplasmic extracts and systems reconstituted from purified actin and various cross-linking proteins. In most cases, the assay has been the inversion of test tubes to demonstrate sample consistency (Bryan and Kane, 1978; Condeelis and Taylor, 1977;

Kane, 1975; Pollard, 1976a,b; Stossel and Hartwig, 1976; Wang and Singer, 1977). This assay is both subjective and impossible to quantitate. Other inadequate assays include capillary viscometry, strain birefringence, pelleting, and turbidity. Capillary viscometry has been used to show that the viscosity increases during gelation, but the apparent viscosities obtained are unreliable because the gel is fragmented during flow through the capillary (Pollard and Ito, 1970). Moreover, stationary samples allowed to gel in such viscometers will not flow because their viscosity exceeds the range of the viscometer, an observation no more quantitative than test-tube inversion. Strain birefringence measurements (Condeelis and Taylor, 1977) could be made quantitative if the stress were applied quantitatively, but this has not been done. Because cross-linked actin filaments will sediment at low centrifugal forces, it has been possible to use a pelleting assay for gelation (Brotschi *et al.,* 1978; Ishiura and Okada, 1979). This assay may be misleading under some conditions because aggregated proteins also pellet. Turbidity changes have been used to follow the progress of the gelation reaction (Condeelis and Taylor, 1977; Pollard, 1976a,b; Stossel and Hartwig, 1976). Although the method is ideal for kinetic analysis, turbidity changes resulting from other reactions, including precipitation, complicate this approach.

Perhaps the only other satisfactory assay for gelation is a measurement of yield point in a device developed by Brotschi *et al.* (1978) called a gelmeter. This apparatus has been used successfully for the analysis of actin filament cross-linking (Brotschi *et al.,* 1978) and of the influence of cytochalasin B on actin filament network formation (Hartwig and Stossel, 1979). Because this device is rather cumbersome, it is difficult to use for kinetic studies or routine assays.

Acknowledgments

Development of this assay was supported by NIH Research Grants GM-19654 and GM-26338. Harvey Kaufman, Linda Griffith, Susan MacLean-Fletcher, and Janelle Levy all made important contributions to the design, testing, and application of the apparatus.

References

Brotschi, E. A., Hartwig, J. H., and Stossel, T. P. (1978). *J. Biol. Chem.* **253,** 8988–8993.
Bryan, J., and Kane, R. E. (1978). *J. Mol. Biol.* **125,** 207–224.
Condeelis, J. S., and Taylor, D. L. (1977). *J. Cell Biol.* **74,** 901–927.
Fechheimer, M., Daiss, J., and Cebra, J. J. (1979). *Mol. Immunol.* **16,** 881–888.
Ferry, J. D. (1948). *Adv. Protein Chem.* **4,** 1–78.
Fowler V., and Taylor. D. L. (1980). *J. Cell Biol.* **85,** 361–376.
Geils, R. H., and Keezer, R. C. (1977). *Rev. Sci. Instrum.* **48,** 783–785.
Griffith, L., and Pollard, T. D. (1978). *J. Cell Biol.* **78,** 958–965.

''Handbook of Chemistry and Physics'' (1958–59). (C. D. Hodgman, ed.). Chem. Rubber Publ. Co., Cleveland, Ohio.

Hartwig, J. H., and Stossel, T. P. (1979). *J. Mol. Biol.* **134,** 539–553.

Isenberg, G. H., Aebi, U., and Pollard, T. D. (1980). *Nature* **288,** 455–459.

Ishiura, M., and Okada, Y. (1979). *J. Cell Biol.* **80,** 465–480.

Johnson, A. M. (1970). ''Physical Processes in Geology,'' pp. 481–491. Freeman, San Francisco, California.

Kane, R. E. (1975). *J. Cell Biol.* **66,** 305–315.

Lutz, R. J., Litt, M., and Chakrin, L. W. (1973). *In* ''Rheology of Biological Systems'' (H. L. Gabelnick and M. Litt, eds.), pp. 119–157. Thomas, Springfield, Illinois.

MacLean-Fletcher, S., and Pollard, T. D. (1980a). *J. Cell Biol.* **85,** 414–428.

MacLean-Fletcher, S., and Pollard, T. D. (1980b). *Cell* **20,** 329–341.

MacLean-Fletcher, S., and Pollard, T. D. (1980c). *Biochem. Biophys. Res. Commun.* **96,** 18–27.

Maruyama, K., Kaibara, M., and Fukada, E. (1974). *Biochim. Biophys. Acta* **371,** 20–29.

Pollard, T. D. (1976a). *J. Cell Biol.* **68,** 579–601.

Pollard, T. D. (1976b). *J. Supramol. Struct.* **5,** 317–334.

Pollard, T. D., and Ito, S. (1970). *J. Cell Biol.* **46,** 267–289.

Stossel, T. P., and Hartwig, J. H. (1976). *J. Cell Biol.* **68,** 602–619.

Taylor, D. L. (1977). *Exp. Cell Res.* **105,** 413–426.

Van Wazer, J. R., Lyons, J. W., Kim, K. Y., and Colwell, R. E. (1963). ''A Laboratory Handbook of Rheology.'' Wiley (Interscience), New York.

Wang, K., and Singer, S. J. (1977). *Proc. Natl. Acad. Sci. U.S.A.* **74,** 2021–2025.

Chapter 21

Antibody Production and Immunofluorescent Characterization of Actin and Contractile Proteins

ELIAS LAZARIDES

Division of Biology
California Institute of Technology
Pasadena, California

I. Introduction

The successful generation of experimentally defined antibodies to vertebrate actin and their use in immunofluorescence microscopy to study the intracellular distribution of actin, opened a new methodological approach to our understanding of cell structure and cell motility (Lazarides and Weber, 1974; Lazarides, 1975b). The technique of immunofluorescence was soon extended to determine the intracellular distribution of a number of other actin-associated contractile

ISBN 0-12-564124-9

proteins as well as to determine the distribution of the other two major cytoplasmic filament systems, microtubules and intermediate (10-nm) filaments (for reviews, see Brinkley *et al.*, 1980; Lazarides and Revel, 1979; Lazarides, 1980a,b).

The highly conserved primary structure of actin has long been thought to be the prohibiting factor in inducing experimentally defined antibodies against this molecule. Indeed, a number of attempts to induce antibodies against actin resulted in the production of antibodies against protein contaminants in the original antigen preparation. However, we observed (Lazarides and Weber, 1974) that the poor antigenicity of actin can be overcome by denaturation of the molecule with sodium dodecyl sulfate (SDS) and reduction with a reducing agent prior to immunization. The possibility of obtaining secondary antibodies to proteins contaminating the original antigen preparation can be minimized by purification of the antigen protein by SDS-polyacrylamide gel electrophoresis prior to immunization. Since proteins migrate on such a gel system primarily according to their molecular weight (Shapiro *et al.*, 1967; Weber and Osborn, 1969), the possibility of obtaining antibodies with a molecular weight different from that of the desired antigen is minimized. Using this approach we have prepared antibodies against actin purified from a clone of SV40-transformed 3T3 cells (Lazarides and Weber, 1974), calf thymus actin (Lazarides and Lindberg, 1974), and chicken gizzard actin (Lazarides, 1975a,b). These antibodies can then be used in direct or indirect immunofluorescence to determine the cytoplasmic distribution of actin. The immunofluorescence technique used to determine the localization of actin can be used, unaltered or with minor modifications, to determine the distribution of a number of other cytoplasmic structural proteins, as well as microtubules and intermediate filaments. It is the intention of this review to provide readers with a few well-documented ways of obtaining antibodies to actin and other structural proteins and their use in immunofluorescence microscopy to determine their cytoplasmic distribution.

II. Methods

A. Antigen Purification

Since actin is such a highly conserved protein, a number of higher vertebrate cell types can be used as a source for the purification of this molecule. In general, antibodies produced against SDS-denatured higher vertebrate actin show a wide cross-reactivity with actin from a variety of cell types and species. From muscle cells, actin can be purified from an acetone powder by cycles of polymerization–depolymerization (Spudich and Watt, 1971; Jockusch *et al.*,

1978). The same approach can be used for actins from nonmuscle cells. If the actin is to be purified by preparative SDS-gel electrophoresis prior to immunization, then the initial protein preparation does not need to be homogeneous. From most cell types grown in tissue culture, actin can be partially purified from an acetone powder before final purification by preparative SDS-gel electrophoresis. Acetone powders from cells grown in tissue culture are prepared as follows: The cells are grown to confluency, the medium is removed, and the cells are washed with 0.15 *M* NaCl, 0.01 *M* sodium phosphate pH 7.2 (PBS). The cells are scraped off the plastic dishes with PBS and collected by centrifugation. They are then suspended in a minimal volume of PBS and diluted with absolute acetone precooled to 0°C (10 vol). After 10 min on ice, the cells are collected again by centrifugation, washed once more in acetone, collected, and washed rapidly in ether and the pelleted protein material is allowed to air-dry overnight. The air-dried material is then extracted on ice for 2–3 hr with 1 m*M* Tris HCl, pH 7.5, 1 m*M* ATP, 5 mM β-mercaptoethanol. The supernatant is brought to 30% saturation with ammonium sulfate, and after 30 min on ice the pellets (10,000 *g*, 10 min) are dialyzed against the same buffer. This material can be used for preparative SDS-gel electrophoresis for the purification of actin. Before further purification, these partially purified actin preparations are analyzed by one-dimensional SDS–polyacrylamide gel electrophoresis and by two-dimensional isoelectric focusing (IEF) SDS-gel electrophoresis to ensure that the actin is not contaminated by other molecules of very close molecular weight or molecules of the same molecular weight but different isoelectric point.

The most convenient reliable source of actin to be used as an antigen is the chicken gizzard. Actin can be purified from this cell type using polymerization-depolymerization of actin extracted from an acetone powder (Spudich and Watt, 1971; Herman and Pollard, 1979). Alternatively, the actin is purified from actomyosin. The gizzard is freed from connective tissue and the muscle cells are blended in 4 m*M* EGTA, pH 7.0. The disrupted cells are extracted three times in EGTA at 4°C each time for 2–4 hr. The muscle is then extracted with 0.6 *M* KCl, 0.01 *M* sodium phosphate, pH 7.2, 5 and m*M* β-mercaptoethanol for 1 hr, and the procedure of high-salt extraction is repeated twice, each subsequent time for 4 hr. Actomyosin is precipitated from the second or third extract by dialysis against 0.01 *M* sodium phosphate, pH 7.2, and 5 m*M* β-mercaptoethanol overnight. The material can then be used for preparative SDS-gel electrophoresis or actin and myosin can be purified away from each other by chromatography on a column of Sepharose 4B (or Biogel P150) equilibrated in 0.6 *M* KCl, 0.01 *M* sodium phosphate, pH 7.2, 0.01 *M* sodium pyrophosphate, and 5 m*M* β-mercaptoethanol. The actin or the myosin peak can be used for the preparative purification of actin or myosin respectively by SDS-gel electrophoresis. Gizzard actin or myosin, purified by preparative SDS-gel electrophoresis prior to immunization, consistently result in the production of antibodies against them when injected in rabbits. However, it should be noted that any of a number of methods published for the purification of actin or myosin can be used and the proteins can be finally purified prior to immunization by preparative SDS-gel electrophoresis.

Final purification of actin is achieved by preparative SDS-gel electrophoresis using the Tris-glycine system of Laemmli (1970). The partially or fully purified protein preparation (1–2 mg/ml) is solubilized for 5 min in a boiling H_2O bath in sample buffer: 50–100 m*M* β-mercaptoethanol, 10% glycerol, 2.0% SDS, 0.08 *M* Tris-HCl pH 6.8, Bromphenol blue. The protein solution is then analyzed on a preparative SDS-polyacrylamide slab gel (2–4 mm thick). At the end of the run the slab gel is stained for 15–25 min in 0.25% Coomassie Brilliant Blue R, 47.5% ethanol, and 10% acetic acid and destained in 20% ethanol, 7.5% acetic acid. The protein bands of interest (actin) are excised and neutralized exhaustively in 0.1 *M* Tris-HCl, pH 7.5, and finally in 0.1 *M* Tris-HCl, pH 6.8. The protein-polyacrylamide bands are then sliced into small pieces and the protein is eluted by one of two methods: (a) by incubation into 2–3 changes of sample buffer at 37°C each for 12 hr or (b) by electrophoretic elution. This technique is a modification (by Dr. B. L. Granger) of the previously published elution procedure (Lazarides, 1976b). Elution columns are prepared from the lower 6-ml portion of glass or plastic disposable 10-ml pipettes. About one-fourth of an inch is removed from the tip end of the pipettes to widen the hole and facilitate current flow. The tips of the pipettes are plugged with a piece of parafilm and 3 ml of 1% agarose in 0.08 *M* Tris-HCl, pH 6.8, 0.1% SDS is poured into the columns and allowed to polymerize. The neutralized protein-polyacrylamide bands are equilibrated in sample buffer at room temperature for 30 min, sliced into small pieces, and placed on top of the stacking gel. The parafilm is removed from the bottom of the stacking gel and a small piece of dialysis tubing ($\frac{1}{4}$ in. in diameter) is affixed around the end of the column. Before attaching to the column, one end of the dialysis membrane (2 in. long) is clamped with a "Spectrum" clamp (Spectrum Medical Industries) and filled with running buffer (per liter: 14.4 g glycine, 3 g Tris base and 1 g SDS, pH 8.3). The dialysis membranes are immersed halfway in running buffer (lower chamber anode) and the upper chamber is filled with running buffer (cathode). Elution is accomplished at 120–140 V applied for approximately 18 hr. After elution the spectrum clamp is removed and the dialysis membrane buffer collected, dialyzed against 0.05 *M* NH_4HCO_3, and lyophilized. The protein is dissolved in a small volume of 0.15 *M* NaCl, 0.01 *M* sodium phosphate, pH 7.2, and precipitated with the addition of $AlCl_3$ (or KAl $(SO_4)_2$) to 1%. Upon addition of $AlCl_3$ the pH drops and it is readjusted to pH 7.0–7.3 with 1 *M* NaOH. The final product is a heavy precipitate. In general, actin dissolved in SDS directly is not a good immunogen and should be precipitated prior to immunization. Alternatively the eluted and lyophilized actin is resuspended in PBS and incubated for 2–10 hr at room temperature with 0.1% glutaraldehyde prior to $AlCl_3$ precipitation. The $AlCl_3$

precipitated actin or the glutaraldehyde cross-linked and precipitated actin can now be used for immunization. If actin can be purifed to homogeneity by conventional chromatographic techniques or by polymerization-depolymerization, a couple of different techniques can be used for the preparation of the antigen prior to immunization.

1. Actin is dissolved in sample buffer and incubated for 3 min at 100°C. The protein solution (1 mg/ml) can now be directly precipitated with 1% $AlCl_3$ and neutralized with 1 *N* NaOH; alternatively it can first be cross-linked with 0.1% glutaraldehyde and subsequently precipitated with 1% $AlCl_3$. The precipitated antigen is very stable and can be stored at 4°C or frozen at −20%C. This technique for the purification of actin by preparative SDS–polyacrylamide gel electrophoresis has been used successfully for the preparation of antibodies to chicken gizzard actin, chicken skeletal muscle actin, mouse fibroblast (3T3) actin, and calf thymus actin (Lazarides and Weber, 1974; Lazarides and Lindberg, 1974; Lazarides, 1975a,b; Jockusch *et al.*, 1978; Lessard *et al.*, 1979).

2. An alternative method for the preparation of actin as an antigen for immunization has been introduced recently by Herman and Pollard (1979). Actin is purified from a chicken gizzard acetone powder by cycles of polymerization-depolymerization and finally over a column of Sephadex G150. Purified actin is induced to polymerize by the addition of KCl to 100 m*M* and $MgCl_2$ to 2 m*M*. The polymerized actin is then treated with 0.1% glutaraldehyde at 0°C for 1 hr and dialyzed against several changes of PBS. The actin is then used for immunization.

B. Immunization

A number of immunization schemes have been used successfully for the generation of antibodies to actin purified and prepared for immunization as described above. In general, there is no fixed immunization schedule and the reader may find different schemes to work equally as well. I will describe a few standard immunization schemes that have proved reproducibly successful.

On day 0, the rabbits are prebled for preimmune sera. This is especially important since different rabbits carry autoantibodies against a variety of antigens including microtubules, intermediate filaments, Z lines, M lines, and H zones of myofibrils (Karsenti *et al.*, 1977; Osborn *et al.*, 1977; Gordon *et al.*, 1978; E. Lazarides, unpublished observations). Each preserum should be tested by immunofluorescence of the cell type to be used to determine if it is suitable for immunization.

(a) On day 1, 0.3 mg of the $AlCl_3$ precipitated protein is emulsified with an equal volume of Freund's Adjuvant Complete (FAC) and injected in the animal's footpads, or subcutaneously at multiple sites on the back, or both.

(b) On day 14 or day 21 or day 28, 0.3 mg of $AlCl_3$ precipitated protein is emulsified with an equal volume of Freund's Adjuvant Incomplete (FAI) and injected at multiple sites intramuscularly and subcutaneously. If some of the footpads have not been injected the first time, they can be used as well this time.

(c) On day 21 or day 28 or day 35, the rabbits are bled from the marginal ear vein (approximately 20–30 ml of blood). In general, the antisera are tested for antiactin antibodies by immunofluorescence. This assay is much more sensitive than classical immunodiffusion analysis and can detect nonprecipitating antisera as well. A positive antiserum begins to show staining of the actin filament bundles at a dilution of 1:5–1:10 or higher. If the antigenic response is weak, it can be maximized by following an intravenous injection schedule.

(d) Three to five days after bleeding (or seven days after the last injection), the animals are injected intravenously every two days, six times, with 100 μg of the $AlCl_3$ precipitated protein. Three days after the sixth injection the animals are injected with 0.3 mg of the $AlCl_3$ precipitated protein and three days later with 0.5 mg of protein. The animals are bled five days later. If the titers are high, then the animals are bled every 5 days until the titers drop. Then they can be reboosted intravenously with 0.5 mg of protein. If the animals have not responded by this time, then the immunization is usually terminated.

Blood is allowed to clot at room temperature for 3 hr in a plastic centrifuge tube, is rimmed with a wooden applicator stick, or a spatula, and is allowed to retract at 4°C overnight. Serum is collected by centrifugation at 10,000 rpm for 10 min. After the serum is collected, ϵ-amino caproic acid is added to 1–2 mM to inhibit plasminogen activation. The serum is partially purified by the addition of an equal volume of saturated ammonium sulfate on ice for 15 min. Partially purified globulins are collected at 10,000 rpm for 10 min and dialyzed against 0.15 M NaCl, 0.02 M Tris-HCl pH 7.5, 1 mM ϵ-amino caproic acid. A small amount of the globulin preparation is diluted for immunofluorescence in PBS containing 5 mM NaN_3 and kept at 4°C. The rest of the globulin is stored at −70°C.

C. Antibody Purification and Test of Specificity

The partially purified globulins are tested for the presence of antiactin antibodies by immunofluorescence as described further below. In general, different rabbits carry autoimmune antibodies to a variety of different antigens and most notably to structural proteins. As a first test the preimmune partially purified globulins should be tested in immunofluorescence at the same protein concentration as the partially purified immune globulins. As a further test of specificity, actin specific immunoglobulins are purified over an actin affinity column.

1. Affinity Purification of Immunoglobulins*

Biogel A15 (25 g; Biorad) is washed with distilled water and resuspended in 60 ml of water. Cyanogen bromide (4.5 g) is dissolved in 135 ml of water and added to the suspended agarose beads. The mixture is stirred with dropwise addition of 2 *M* NaOH to keep the pH at 11.0–11.5. After completion of the reaction (20–25 min), the resin is washed on a Buchner funnel with 0.1 *M* $NaHCO_3$ and resuspended in 25 ml of 0.1 *M* $NaHCO_3$. Actin, 20–30 mg dissolved in 2 ml of 0.1 *M* $NaHCO_3$, 0.1 m*M* $CaCl_2$, 0.2 *M* KCl, pH 8.5, is mixed immediately after dissolving with the resin and stirred for 4–8 hr at room temperature. Occasionally some preparations of actin are turbid when suspended in 0.1 *M* $NaHCO_3$, 0.1 m*M* $CaCl_2$. If this is the case, the protein solution is adjusted to 0.1% SDS and the addition of KCl is omitted. The presence of SDS does not interfere with coupling. Under these conditions more than 90% of the actin is coupled. Unbound protein is washed with 0.1 *M* $NaHCO_3$, 0.1 m*M* $CaCl_2$, 0.05% SDS. Finally, the affinity resin is incubated with 1 *M* ethanolamine, pH 8.0, for 1–2 hr. The resin is then washed with 0.1 *M* $NaHCO_3$, 0.1 m*M* $CaCl_2$, 0.05% SDS, and finally with 0.15 *M* NaCl, 0.01 *M* sodium phosphate, pH 7.2 (PBS). Partially purified immune globulins (20 mg/ml) are passed slowly through the column (1–5 ml of original serum per milliliter of affinity resin). The effluent is collected and the column is washed further with PBS. The column is then washed with 0.6 *M* KCl, 0.01 *M* sodium phosphate, pH 7.2, and the effluent collected. Specific actin globulins are eluted with 4 *M* $MgCl_2$ or with 0.2 *M* glycine HCl, pH 2.5. However, this method of elution does not elute quantitatively the bound IgG. Complete recovery can be achieved by using 6–8 *M* urea in 0.01 *M* sodium phosphate, pH 7.2. The eluted globulins are rapidly dialyzed against PBS containing 5 m*M* NaN_3 and 1 m*M* ϵ-amino caproic acid. The urea-eluted globulins are less stable and are usually used within 2–3 weeks. The columns can be regenerated with PBS and used again for up to 1 month (three to five times). This method of purification of actin-specific immunoglobulins has been used successfully both by us and, with certain modifications, by other investigators (Jockusch *et al.*, 1978; Herman and Pollard, 1979). All eluted column fractions are then tested by immunofluorescence. In general, all the immune activity elutes with urea, $MgCl_2$, or pH 2.5, but occasionally a small amount of fluorescence is detected in the 0.6 *M* KCl wash. If the immune sera can be proved specific by this approach, then they are used routinely without affinity purification, provided the immunofluorescent patterns observed in different cell types are typical. Any atypical fluorescence patterns should always be verified with affinity-purified material. In different animals responding to actin,

*Lazarides and Lindberg, 1974; Jockusch *et al.*, 1978; Herman and Pollard, 1979.

the specific actin IgG is usually between 0.1 and 2.0 mg per 10 ml of serum applied to an affinity column. The same range of specific immunoglobulins is also obtained using tubulin as an antigen (Brinkley *et al.*, 1980). Classical double immunodiffusion analysis and immunoelectrophoresis can also be utilized to assess the specificity of the antiserum. However, some of the antiactin antisera are only weakly precipitating and these two tests are not sufficiently sensitive to assess the specificity of the antibodies for actin.

2. Assessing the Specificity of Actin Antibodies by Immunoautoradiography

Even if the actin-specific antibodies are purified by affinity chromatography and are positive by immunofluoresence, it is generally desirable to use a technique that can unequivocally demonstrate the reaction of antiactin antibodies with only actin in a whole-cell extract. This is particularly desirable since the actin-specific antibodies may carry a weak cross-reactivity with antigens unrelated to actin. For this purpose the technique of immunoautoradiography has been used successfully on one-dimensional SDS-polyacrylamide gels to show that actin antibodies react only with actin in cell extracts (Burridge, 1976). We have used immunoautoradiography on two-dimensional IEF/SDS-polyacrylamide gels not only as a means of demonstrating the specificity of a given antibody, but also, in the case of actin, to show whether all three isoelectric variants of actin α, β, and γ react with equal efficiency. This technique is lengthy (2 weeks) but offers extreme sensitivity (detection of less than 50 ng of protein on the gel) and reliability and provides convincing evidence of the specificity of the antisera (Granger and Lazarides, 1979). Since it is applicable to a variety of antigens, it is described here in detail and it is recommended as an ultimate test for the specificity of any of the structural antigens.

For testing the specificity of antibodies to actin, to the intermediate filament subunits desmin and vimentin, to tropomyosin, to tubulin, and to α-actinin a variety of cell extracts from embryonic tissues, cells grown in tissue culture, or isolated myofibrils can be used. To obtain the maximum number of cytoskeletal proteins on the gel we routinely use embryonic muscle extracts. Leg and breast muscle of a 14-day chick embryo is disrupted in a Dounce homogenizer in 4 ml of 20 m*M* Tris-HCl (pH 8.0), 5 mM EGTA, 1 m*M* *o*-phenanthroline, and 0.5 m*M* phenylmethyl sulforyl fluoride (PMSF) and spun for 1 hr at 150,000 *g*. The supernatant is made 10 *M* in urea, 2% in NP-40, and 0.05% in 2-mercaptoethanol and 75–100 μg of protein are loaded on two-dimensional gels. Two-dimensional IEF/SDS-polyacrylamide gel electrophoresis is carried out as described by O'Farrell (1975) with modifications to maximize the resolution of the various isoelectric variants of actin (Hubbard and Lazarides, 1979).

These extracts contain ample quantities of tubulin, α-actinin, myosin, the various actin variants, the intermediate filament subunits of desmin and vimentin, the light chains of myosin, the various troponin subunits, the tropomyosin subunits, etc. (Granger and Lazarides, 1979).

After electrophoresis the gels are fixed for 6 hr in 50% ethanol, 10% acetic acid (all incubations are carried out at room temperature in rocking Pyrex baking dishes) and then neutralized for 1 day in several changes of 50 m*M* Tris-HCl, pH 7.5, 0.1 *M* NaCl, 10 m*M* NaN_3. After another day of washing in buffer A (0.01 *M* Tris-HCl, pH 7.5, 0.140 *M* NaCl, 0.01 *M* NaN_3, 0.1% gelatin (Difco; dissolved by heating the buffer to 55°C), the gels are incubated for 1 day in 100 ml of buffer to which 100 μl of antiserum has been added (1:1000 dilution). The actual final dilution of the antiserum depends on the titer of the antibodies. Good titer antisera can be diluted 1:500–1:1000 but lower or higher dilutions need to be tested. If too low a dilution is used, nonspecific staining of a number of proteins results. At too high an antibody dilution no staining is obtained.

After incubation with the antibody solution, the gels are washed for 3 days in several changes of buffer and they are then incubated for 1 day in 100 ml of buffer A containing [^{125}I]protein A (see below). Two days of washing in buffer A and 1 day in buffer A *without* the gelatin are followed by staining in 0.1% Coomassie Brilliant Blue R, 47.5% ethanol, 10% acetic acid and destaining in 12.5% ethanol, 5% acetic acid. If the gels are not washed in buffer A without gelatin prior to staining, background staining is very high. The destained gels are dried, photographed using a Polaroid setup and P/N 55 film, and autoradiographed for 12–25 hr at −70°C on Kodak X-Omat R XR5 film with a DuPont Cronex Lightning-Plus intensifying screen. Autoradiograms are developed in Kodak X-ray developer after 1 day (short) or 3–4 days (long) exposure, or longer if necessary.

Protein A (from Pharmacia) is iodinated by the chloramine T method as described by Greenwood *et al.* (1963) with the exception that the reaction is terminated by adding an excess of tyrosine as follows: 100 μl of 0.5 *M* potassium phosphate (pH 7.5) is added to 1 mCi of [^{125}I]Na (ICN; in 2 μl of NaOH, final pH 8.8); 20 μl of protein A (5 mg/ml) and 20 μl of chloramine T (2.5 mg/ml) are added next. After 2 min, 150 μl of tyrosine (0.4 mg/ml) are added. The mixture is passed centrifugally through a 3-ml bed of Sephadex G-25 and the void fraction is used for labeling. Specific activities on the order of 0.1 μCi/μg protein are obtained and each gel is incubated with approximately 1 μCi. The iodinated protein A is used within a week.

The entire procedure requires approximately 14 days to complete (including exposure of the gels on film). Using this technique a number of proteins that react with the antibodies and which do not stain with Coomassie Blue can be detected. We have estimated that proteins present in quantities less than 50 ng can be detected.

D. Immunofluorescent Localization of Actin

1. Cell Cultures

Immunofluorescence studies can be carried out on cell monolayers, cell suspensions, or tissue sections. Since actin filaments are long structures, frequently aggregated laterally to supramolecular assemblies in most cells, studies of sectioned tissues are not likely to reveal much information about actin filament distribution in individual cells except to demonstrate the presence (or absence) of the antigen. In this report I will restrict my description to the method as applied to cells grown in tissue culture in monolayers. In general, it is best if cells can be grown as monolayers. Cells that exhibit a high degree of adhesion to a substrate and can assume diameters of many tens of microns are ideal for obtaining detailed information about the structure and distribution of actin filaments and their associated proteins. Cells are usually plated on No. 1 coverslips, round or square, and used before they reach confluency (confluent cells underlap or overlap each other and tend to be rounder and thicker, thus making actin filament detection difficult).

2. Fixation and Staining

A variety of fixation protocols can be used for the localization of actin and other structural proteins in tissue culture cells. I will summarize a few methods that we use routinely in the laboratory for the localization of actin and other structural proteins.

a. Formaldehyde-Ethanol (or Acetone) Fixation (Lazarides and Weber, 1974; Lazarides, 1975a, 1976a,b). Cells are removed from the petri dishes, dipped once in PBS (0.15 *M* NaCl, 0.01 *M* sodium phosphate, pH 7.2) at 37°C, drained rapidly by touching their edge on a paper towel, and immersed in 3.7% formaldehyde (Mallinckrodt) in PBS at 37°C. Alternatively, 4% paraformaldehyde in PBS can be used. Fixation proceeds for 10–30 min, and the coverslips are then immersed in 47.5% ethanol (or 50% acetone) at room temperature for 15 min. In general, ethanol is preferable to acetone. We have used variably 95% ETOH or 100% acetone, but the brightest fluorescence is obtained using 47.5% ethanol. The coverslips are then transferred back to PBS for 5 min. They are then placed in a moist chamber and covered with a drop of PBS. During the whole immunofluorescent staining technique the cells are not allowed to air-dry. Air drying of the coverslips, especially after the application of the antibodies, results in antibody precipitation and a very high nonspecific fluorescent background. A few microliters of the diluted antiactin antibody (20–30 μl of a 1:20 dilution in PBS, depending on the titer) are added to each coverslip and incubation proceeds for 45 min at 37°C in a humidified atmosphere (tissue culture incubators). The coverslips are then drained on a paper towel, washed in PBS for 15 min, and incubated in the second antibody solution (the indirect technique).

The second antibody can be coupled either with fluorescein or with rhodamine. We generally use fluorescein, since, in our experience, it gives visually a sharper image and is better for low titer antisera. However, quenching is a serious problem with fluorescein when using an ultraviolet mercury arc lamp, whereas rhodamine does not fade as rapidly under the beam or in storage. The second antibody can be purchased commercially from a number of outfits. In general, we use goat anti-rabbit IgG purchased from Miles or Meloy; however, we have used antibody purchased from Antibodies Incorporated with equal success. The actual dilution of this antibody depends on the protein concentration and the dye-to-protein ratio but in general we dilute the antibodies 1:50 in PBS. Antibodies coupled to fluorescein are not stable to freezing and thawing and should not be frozen and thawed more than twice. The working dilution of these antibodies can be kept at 4°C for up to 1 month. Before use, the secondary antibody is centrifuged at top speed in a clinical centrifuge for 5 min to remove insoluble or aggregated material. This step is essential in order to minimize background fluorescence. The coverslips are incubated with 20–30 μl of the secondary antibody at 37°C for 30–45 min. For each batch of fluorescein-labeled goat anti-rabbit IgG, control coverslips are used to ensure that the secondary antibody does not give any artificial background. The coverslips finally are washed in PBS for 15 min and briefly dipped in distilled H_2O before mounting.

Coverslips can be mounted with a drop of 25% Elvanol (DuPont) in water, which results in an ultraviolet inert permanent mount. However, Elvanol reduces the fluorescence. For this reason coverslips are usually mounted in 90% glycerol made up in PBS and sealed with nail polish. For best results and the sharpest fluorescence images, the coverslips are photographed within 1 hr. They can be stored at room temperature in light-tight boxes for several days, but usually will not yield good fluorescent micrographs.

b. Triton X-100-Formaldehyde-Ethanol. Considerable improvement in the visualization of actin filaments can be achieved in some cells by treatments with mild detergents prior to staining. The cells can be made permeable to antibodies by immersion in a detergent solution. This method extracts some of the cytoplasmic proteins and yields even crisper and sharper fluorescent images of actin. After dipping briefly in PBS at 37°C, the cells are immersed for 5 min in a solution containing 0.5% Triton X-100 (NP-40), 0.37% formaldehyde in PBS at 37°C. They are then postfixed in 3.7% formaldehyde in PBS and in ethanol as described in the preceding method. The ethanol step can be omitted in this case with no change in the results. The rest of the procedure is the same except that 0.5% Triton X-100 is included in all the PBS washes.

c. Triton-SDS-Formaldehyde. This last method provides considerable improvement in the visualization of actin, intermediate filaments, and a number of other structural proteins (developed by B. L. Granger in my laboratory). This fixation method is especially useful with cells that are not fully spread out and in particular with myotubes. Coverslips are dipped in PBS at 37°C, drained, and immersed in 3.7% formaldehyde, 0.15 *M* NaCl, 0.01 *M* sodium phosphate, pH 7.2, for 10 min at 37°C. The coverslips are then immersed in PBS at room temperature for 5 min and transferred into a buffer containing 0.5% Triton X-100, 0.1% SDS, 0.01 *M* sodium phosphate, pH 7.5, 0.15 *M* NaCl, and 0.1% gelatin, and incubated for 10 min at room temperature. After the incubation each coverslip is covered with 1–2 drops of the Triton-SDS buffer and 20–30 μl of the antibody solution (in PBS) is added directly to the coverslips. The same is done for the second antibody. The rest of the method is as described above except that washings are done in the Triton-SDS buffer for 15 min.

3. Controls

With actin, and actin-associated proteins, different cell types require different controls unless typical patterns are obtained. However, certain standard controls should always be applied: (a) the ammonium sulfate purified globulins and the affinity purified globulins should stain the same structures; (b) the preimmune sera at the same or higher IgG concentration should be uniformly negative (this is the major serious problem with immunofluorescence since different rabbits carry different autoantibodies for different tissues); (c) the secondary antibody alone should not provide any staining; (d) in the presence of an excess of the purified antigen, the staining should be abolished.

In general, we test the presence of antibodies to actin by indirect immunofluorescence in cells that are very flat and contain many phase-contrast fibers (actin filament bundles). This is our first indication that the rabbits are responding to the antigen, but is not the final. As is the case with many cytochemical techniques, not all cells respond similarly to a given fixation protocol. Thus, if a particular cell type is negative for a given antigen (e.g., actin) the antiserum should not be discarded as being negative. Different fixation protocols should be followed at different fixative concentrations and alternate tests should be sought to ascertain the presence (or absence) of antibodies in a given serum preparation.

E. Double Immunofluorescent Localization of Actin and Other Structural Proteins

One of the distinctive advantages of immunofluorescence is that it allows the localization of two antigens simultaneously using two fluorochromes, fluorescein and rhodamine. This can be achieved in one of the following three ways: (a) The antibodies to the two primary antigens are each made in different animals. In this case each antigen can be localized using indirect immunofluorescence with rhodamine or fluorescein conjugated secondary antisera, which can be purchased commercially (e.g., goat anti-rabbit IgG fluorescein labeled and sheep anti-guinea pig or goat rhodamine labeled). (b) The antibodies, made in the same or different animals, are coupled directly with fluorescein or rhodamine. Details of this approach with myosin and tubulin as examples have been described by Fujiwara and Pollard (1978). (c) The two antisera can be produced in the same animal and only one of them is coupled directly with a fluorochrome, e.g., rhodamine. In this case we have followed the technique of Hynes and Destree (1978) for the successful localization of actin and fibronectin. This indirect/direct staining method is convenient and is described in more detail here. Cells are fixed for immunofluorescence as described earlier. Coverslips containing fixed cells are first incubated with one of the antibodies made in rabbits, washed, and incubated with goat anti-rabbit IgG. Unreacted anti-rabbit IgG is blocked for 1–3 hr with a 1:5 dilution of normal rabbit serum or ammonium sulfate purified globulin. This treatment is sufficient to eliminate completely any binding of the directly labeled antibodies for the preexisting rabbit IgG–goat anti-rabbit IgG complex. Cells are then washed and incubated with the directly conjugated rhodamine antisera. For color photography we have used Kodak Ektachrome 200 color slide film, developed commercially. Conjugation of rhodamine or fluorescein isothiocyanate with antibodies and their subsequent purification is carried out according to the method of Cebra and Goldstein (1965). Double immunofluorescent localization of actin can be carried out successfully also by two nonimmunological techniques that involve the specific binding of the tryptic fragment of myosin, heavy meromyosin (HMM), to actin. HMM either is labeled directly with fluorescein isothiocyanate or is coupled to biotin and biotin is localized with fluorescently labeled avidin (Heggeness and Ash, 1977). The second antigen can be localized with antibodies labeled directly with rhodamine or using the indirect technique.

F. Microscopy and Photography

There are several excellent ultraviolet microscopes available for immunfluorescence studies. Here I will describe only those instruments that we have used routinely in our work. With certain modifications these microscopes can be used to obtain stereoimmunofluorescent images (Osborn *et al.*, 1978a). Our best illumination is obtained using an epifluorescence system. Either a Leitz or a Zeiss microscope is superb for a combination of phase-epifluorescence microscopy and each instrument has a set of superb lenses for combination phase-fluorescence or for fluorescence alone. We use a Leitz Orthoplan microscope equipped with the following objectives: a 16× (NA = 0.45) phase-fluorescence, a 40× (NA = 1.30, oil) fluorescence, 40× (NA = 1.0, oil) phase-fluorescence, 54× (NA = 0.95, oil), 50× (NA = 1.0, oil), a Zeiss 63× (NA = 1.4, oil), and a 100× (NA = 1.32, oil). The 63× Zeiss lens can be used with the Orthoplan without any adjustments in the tube length for fluorescence. For phase, the No. 4 phase ring setting of the condenser should be used.

For most of our work we use the 63×, 54×, and 50× lenses. For studies on skeletal myofibrils we invariably use the 100× objective. The microscope is equipped with epi-illumination and a Leitz Ploem Illuminator (Ploemopak 2.1). The illumination source is a 100-W high-pressure mercury arc lamp used in combination with the following Leitz filter modules: for fluorescein, H wide band blue high intensity, and K extremely narrow band blue, peak at 495 nm to eliminate autofluorescence and for rhodamine, N.2 narrow band green fluorescein excitation excluded. With good staining, exposure times vary from 1 sec up to 30 sec (using Tri-X film at a din setting of 31). When Plus X film is used (din 28) exposures are usually 20–40 sec. Films are developed in Diafine (Acufine Corp.) which enhances the ASA of the film approximately threefold.

G. Antibody Generation to Other Major Cytoplasmic Structural Proteins

In general, most contractile proteins other than actin, which can be purified by conventional biochemical techniques, are quite antigenic. However, we have used SDS–polyacrylamide slab gel electrophoresis for the purification of a number of structural proteins destined for immunization. Some of these proteins can be purified by conventional biochemical techniques and others are more difficult to purify. We have succeeded in reproducibly generating antibodies against the heavy chains of smooth and skeletal muscle myosin, platelet myosin heavy chain, skeletal muscle tropomyosin, skeletal muscle troponin, skeletal and smooth muscle α-actinin, filamin (Wang *et al.*, 1975), the intermediate filament subunits desmin and vimentin (Lazarides and Hubbard, 1976; Granger and Lazarides, 1979), and tubulin (Lazarides, 1976b). For all of these, either native or purified by SDS-gel electrophoresis, the immunization schemes and subsequent methods are the same as described for actin. Quite frequently the procedure of antigen elution from the gels can be shortened appreciably by injecting the whole acrylamide-protein band in the animal. In this case the polyacrylamide-protein bands, after staining–destaining and neutralization, are cut into small pieces and homogenized in a small volume of PBS with a motor-driven Teflon pestle. The homogenized acrylamide-protein material is then adjusted to 1% $AlCl_3$ and neutralized to pH 7.0 with 1*N* NaOH. The whole highly viscous precipitate is then emulsified with Freund's Adjuvant and injected in the footpads as described earlier, subcutaneously or intramuscularly. With homogenized acrylamide material, the animals are boosted intramuscularly or subcutaneously weekly or biweekly. Blood is collected seven days after each boosting. This approach is much faster and more convenient than elution and has worked well with all the antigens described earlier (for an example, with desmin and vimentin, see Lazarides and Hubbard, 1976; Granger and Lazarides, 1979). In all of these cases we have found that antibodies to the SDS-gel-purified proteins cross-react with the native molecules. Similarly, antibodies raised to SDS-denatured *Limulus* paramyosin and myosin cross-react with the native molecule (Elfvin *et al.*, 1979). In the case of smooth or skeletal muscle α-actinin, skeletal muscle tropomyosin, smooth muscle myosin, platelet myosin, filamin, and the intermediate filament subunit vimentin, the antibodies to the SDS-denatured molecules exhibit a wide cross-reactivity in both muscle and nonmuscle cells (Lazarides, 1975a,b; Fujiwara and Pollard, 1976; Franke *et al.*, 1978; Bennett *et al.*, 1978; Granger and Lazarides, 1979). However, the tissue specificity exhibited by antibodies to native smooth and skeletal muscle myosin (Fujiwara and Pollard, 1976) is not overcome by antibodies prepared against SDS-gel-purified smooth muscle myosin heavy chain (Lazarides, unpublished observations). Antibodies to the smooth muscle myosin heavy chain react with smooth muscle and nonmuscle myosin but they do not cross-react with skeletal muscle myosin and vice versa for skeletal muscle myosin; it does not cross-react with

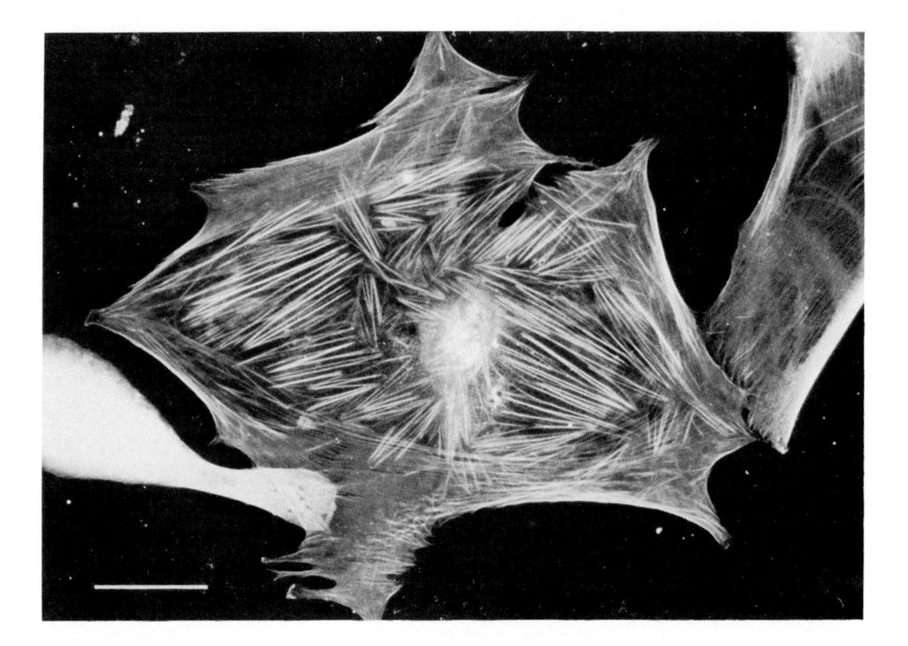

FIG. 1. The distribution of actin in a human skin fibroblast using antibodies to chicken gizzard actin and indirect immunofluorescence. The antigen was purified by preparative SDS-slab gel electrophoresis prior to immunization. This immunofluorescent picture depicts a typical distribution of the actin filament bundles in fully spread-out cells. Bar = 50 μm.

smooth or nonmuscle myosin. The immunofluorescent localization of these molecules has been adequately described in the literature, in reviews, and in other sections of this volume and will not be described here. An example of the immunofluorescent localization of actin is given further in Fig. 1. For examples of the immunofluorescent localizations of the various contractile and structural proteins described here the reader is referred to the following articles:

Actin: Lazarides and Weber (1974); Lazarides (1975a,b, 1976a,b); Lazarides and Revel (1979);

Tropomyosin: Lazarides (1975a,b, 1976a,b);

α-Actinin: Lazarides and Burridge (1975); Lazarides (1976a);

Myosin: Weber and Groeschel-Stewart (1974); Fujiwara and Pollard (1976);

Filamin: Wang *et al.* (1975);

Tubulin: Fuller *et al.* (1975); Osborn and Weber (1977); Osborn *et al.* (1978b) and for a review of antibody production to tubulin and its immunofluorescent localization, see Brinkley *et al.* (1980);

Intermediate filaments: for a review and full references, see Lazarides (1980).

III. Immunological Characterization of Antiactin Antibodies

In general, higher vertebrate (avian and mammalian) native actin has been found to be a poor immunogen. However, in some rather rare cases chemically unmodified native actin elicits an immunogenic response in rabbits. Interestingly enough, these antibodies (made against native chicken gizzard actin) show restricted cell but unrestricted species cross-reactivity; they cross-react with avian or mammalian smooth, skeletal, and cardiac muscle actin specifically and they do not cross-react with nonmuscle actins (Gröschel-Stewart *et al.*, 1977; Chamley *et al.*, 1977). Similarly, antibodies can be induced against native actin from the plasmodium of the myxomycete, *Physarum polycephalum*. However, in this case the antibodies exhibit a wide cell and species cross-reactivity (Owaribe and Hatano, 1975; Owaribe *et al.*, 1979). With the exception of one case discussed above, one of the most general characteristics of the antiactins prepared from a variety of cell types and species so far is their wide phylogenetic cross-reactivity. This has been the case with antibodies to avian actins (Lazarides, 1975b, 1976a,b; Jockusch *et al.*, 1978; Herman and Pollard, 1979), mammalian actins (Lazarides and Weber, 1974; Lazarides and Lindberg, 1974; Edelman and Yahara, 1976), invertebrate actins (Kleve *et al.*, 1979), and *Physarum* actin (Owaribe and Hatano, 1975; Owaribe *et al.*, 1979). This indicates that all these actins share at least one common antigenic determinant (Eckert and Lazarides, 1978). This observation is substantiated with the demonstration that the primary structures of actins as phylogenetically apart as *Physarum polycephalum* and mammalian muscle exhibit a striking similarity (Vanderkerckhove and Weber, 1978).

IV. Autoimmune Antibodies to Actin

Human patients with chronic hepatitis have been found to carry in their sera autoantibodies to actin. Over the past several years these antibodies from a number of patients have been characterized by a variety of immunological techniques and have been shown to be specific for actin (Trenchev and Holborow, 1976; Anderson *et al.*, 1976; Gabbiani *et al.*, 1976; Norberg *et al.*, 1975; Lidman *et al.*, 1976). As is the case with the experimentally induced actin antibodies, the chronic hepatitis antiactin antibodies (originally called smooth muscle autoantibodies) are species and organ nonspecific. These antibodies have been used considerably in the immunofluorescent localization of actin in tissue sections and in cells grown in tissue culture (see preceding list of references). In general, they exhibit the same immunofluorescent characteristic distribution of actin as seen with the experimentally induced actin antibodies described above. Since, however, these antibodies are not experimentally induced, each new batch of serum from a different patient should be tested for its specificity by any of the criteria of specificity for actin discussed above. If these criteria are met, then these antibodies can be used in place of the experimentally induced antibodies to actin.

V. Immunofluorescent Distribution of Actin in Cells Grown in Tissue Culture

A full description of the immunofluorescent localization of actin is beyond the scope of this article and the reader is referred to the numerous immunofluorescent papers that describe it (see preceding list of references). Figure 1 presents a typical example of the distribution of actin filament bundles (stress fibers) in a cell grown in tissue culture using rabbit antibodies against chicken gizzard actin, purified by preparative SDS–polyacrylamide slab gel electrophoresis prior to immunization. The indirect immunofluorescence technique used here is that presented in Section II,D,2,a.

Acknowledgments

I thank Janet Sauer for her help in the preparation of the manuscript. This work was supported by grants from the National Institutes of Health (PHS-GM-06965), the Muscular Dystrophy Association of America, and the National Science Foundation. The author is a recipient of a Research Career Development Award from the National Institutes of Health.

References

Anderson, P., Small, J. V., and Sobieszek, A. (1976). *Clin. Exp. Immunol.* **26,** 57–66.
Bennett, G. S., Fellini, S. A., Croop, J. M., Otto, J. J., Bryan, J., and Holtzer, H. (1978). *Proc. Natl. Acad. Sci. U.S.A.* **75,** 4364–4368.
Brinkley, B. R., Fistel, S., Marcum, J. M., and Pardue, R. L. (1980). *Int. Rev. Cytol.* **63,** 59–95.
Burridge, K. (1976). *Proc. Natl. Acad. Sci. U.S.A.* **73,** 4457–4461.
Cebra, J. J., and Goldstein, G. (1965). *J. Immunol.* **95,** 230–245.
Chamley, J. H., Gröschel-Stewart, U., Campbell, G. R., and Burnstock, G. (1977). *Cell Tissue Res.* **177,** 445–457.
Eckert, B. S., and Lazarides, E. (1978). *J. Cell Biol.* **77,** 714–721.
Edelman, G. M., and Yahara, I. (1976). *Proc. Natl. Acad. Sci. U.S.A.* **73,** 2047–2051.
Elfvin, M. J., Levine, R. J. C., Pepe, F. A., and Dewey, M. M. (1979). *J. Histochem. Cytochem.* **27,** 1478–1482.
Franke, W. W., Schmid, E., Osborn, M., and Weber, K. (1978). *Proc. Natl. Acad. Sci. U.S.A.* **75,** 5034–5038.
Fujiwara, K., and Pollard, T. D. (1976). *J. Cell Biol.* **71,** 847–875.
Fujiwara, K., and Pollard, T. D. (1978). *J. Cell Biol.* **77,** 182–195.
Fuller, G. M., Brinkley, B. R., and Boughter, J. M. (1975). *Science* **187,** 948–950.
Gabbiani, G., Csank-Brassert, J., Schneeberger, J.-C., Kapanci, Y., Trenchev, P., and Holborow, E. J. (1976). *Am. J. Pathol.* **83,** 457–474.
Gordon, W. E., Bushnell, A., and Burridge, K. (1978). *Cell* **13,** 249–261.
Granger, B. L., and Lazarides, E. (1979). *Cell* **18,** 1053–1063.
Greenwood, F. C., Hunter, W. M., and Glover, J. S. (1963). *Biochem. J.* **89,** 114–123.
Gröschel-Stewart, U., Ceurremans, S., Lehr, I., Mahlmeister, C., and Paar, E. (1977). *Histochemistry* **50,** 271–279.
Heggeness, M. H., and Ash, J. F. (1977). *J. Cell Biol.* **73,** 783–788.
Herman, I. M., and Pollard, T. D. (1979). *J. Cell Biol.* **80,** 509–520.
Hubbard, B. D., and Lazarides, E. (1979). *J. Cell Biol.* **80,** 166–182.
Hynes, R. O., and Destree, A. T. (1978). *Cell* **15,** 875–886.
Jockusch, B. M., Kelly, K. H., Meyer, R. K., and Burger, M. M. (1978). *Histochemistry* **55,** 177–184.
Karsenti, E., Guilbert, B., Bornens, M., and Avrameas, S. (1977). *Proc. Natl. Acad. Sci. U.S.A.* **74,** 3997–4001.
Kleve, M. G., Fuseler, J. W., and Clark, W. H., Jr. (1979). *J. Exp. Zool.* **209,** 21–32.
Laemmli, U. K. (1970). *Nature (London)* **227,** 680–685.
Lazarides, E. (1975a). *J. Cell Biol.* **65,** 549–561.
Lazarides, E. (1975b). *J. Histochem. Cytochem.* **23,** 507–528.
Lazarides, E. (1976a). *J. Cell Biol.* **68,** 202–219.
Lazarides, E. (1976b). *J. Supramol. Struct.* **5,** 531–563.
Lazarides, E. (1980). *Nature (London)* **283,** 249–256.

Lazarides, E., and Burridge, K. (1975). *Cell* **6,** 289–298.
Lazarides, E., and Hubbard, B. D. (1976). *Proc. Natl. Acad. Sci. U.S.A.* **73,** 4344–4348.
Lazarides, E., and Lindberg, U. (1974). *Proc. Natl. Acad. Sci. U.S.A.* **71,** 4742–4746.
Lazarides, E., and Revel, J. P. (1979). *Sci. Am.* **240,** 100–113.
Lazarides, E., and Weber, K. (1974). *Proc. Natl. Acad. Sci. U.S.A.* **71,** 2268–2272.
Lessard, J. L., Carlton, D., Rein, D. C., and Akeson, R. (1979). *Anal. Biochem.* **94,** 140–149.
Lidman, K., Biberfeld, G., Fagraeus, A., Norberg, R., Torstensson, R., Utter, G., Carlsson, L., Luca, J., and Lindberg, U. (1976). *Clin. Exp. Immunol.* **24,** 266–272.
Norberg, R., Lidman, K., and Fagraeus, A. (1975). *Cell* **6,** 507–512.
O'Farrell, P. H. (1975). *J. Biol. Chem.* **250,** 4007–4021.
Osborn, M., and Weber, G. (1977). *Exp. Cell Res.* **106,** 339–349.
Osborn, M., Franke, W. W., and Weber, G. (1977). *Proc. Natl. Acad. Sci. U.S.A.* **74,** 2490–2494.
Osborn, M., Born, T., Koitsch, H.-J., and Weber, K. (1978a). *Cell* **14,** 477–488.
Osborn, M., Webster, R. E., and Weber, K. (1978b). *J. Cell Biol.* **77,** R27–R34.
Owaribe, K., and Hatano, S. (1975). *Biochemistry* **14,** 3024–3029.
Owaribe, K., Izutsu, K., and Hatano, S. (1979). *Cell Struct. Funct.* **4,** 117–126.
Shapiro, A. L., Viñuela, E., and Maizel, J. V. (1967). *Biochem. Biophys. Res. Commun.* **28,** 815–820.
Spudich, J. A., and Watt, S. (1971). *J. Biol. Chem.* **246,** 4866–4871.
Trenchev, P., and Holborow, E. J. (1976). *Immunology* **31,** 509–517.
Vanderkerckhove, J., and Weber, K. (1978). *Nature (London)* **276,** 720–721.
Wang, K., Ash, J. F., and Singer, S. J. (1975). *Proc. Natl. Acad. Sci. U.S.A.* **72,** 4483–4486.
Weber, K., and Groeschel-Stewart, U. (1974). *Proc. Natl. Acad. Sci. U.S.A.* **71,** 4561–4564.
Weber, K., and Osborn, M. (1969). *J. Biol. Chem.* **244,** 4406–4412.

Chapter 22

Myosin Purification and Characterization

THOMAS D. POLLARD

Department of Cell Biology and Anatomy
Johns Hopkins University School of Medicine
Baltimore, Maryland

I. Introduction

Myosin is the energy-transducing enzyme responsible for muscle contraction. During its interaction with actin, myosin converts the chemical energy stored in

METHODS IN CELL BIOLOGY, VOLUME 24

ISBN 0-12-564124-9

ATP into mechanical force. Myosin is also the only energy-transducing enzyme known to participate in cytoplasmic streaming and contractility. (Dynein is, of course, the well-established energy-transducing enzyme responsible for ciliary, flagellar, and axostylar motion, but it has never been proved to power any other form of cell motility.) Consequently, a full understanding of both muscle contraction and cell motility depends on the detailed characterization of myosin molecules. Nature has made this enterprise challenging by evolving a wide variety of distinct myosin molecules to meet the special needs of different cell types.

The objective of this chapter is to describe and evaluate the methods available for purifying and characterizing myosin from muscle and nonmuscle cells. The approach will be first to describe the various procedures and then to give a few examples of how they can be combined to purify myosin. The rationale for this approach is that myosins from various cells differ sufficiently that no single procedure is generally useful for myosin purification. However, a suitable combination of the procedures now available should make it possible to purify any myosin type.

II. Myosin Assays

The first prerequisite in the purification of any protein is reliable, simple assays. Fortunately, the distinctive ATPase activities of the myosins and, in most cases, the unique high molecular weight of the myosin heavy chains make this a simple matter.

A. ATPase Assays

In most cases myosins have the unique property of having low ATPase activity in Mg^{2+} and high activity in Ca^{2+} and in EDTA, providing high concentrations of K^+ are present. The high ionic strength also prevents any actin in crude fractions from activating the Mg^+ ATPase. Most other ATPases are more active in Mg^{2+} than Ca^{2+} and are inactive in EDTA. Consequently, one can usually follow myosin purification from the crude extract stage onward with a K^+-EDTA ATPase assay. However in a few cases a Ca^{2+} ATPase assay is useful or essential. For example, *Acanthamoeba* myosin-II has a higher Ca^{2+} ATPase than K^+-EDTA ATPase activity (Maruta and Korn, 1977a; Pollard *et al.*, 1978) and *Physarum* myosin has Ca^{2+} ATPase activity but no K^+-EDTA ATPase activity (Adelman and Taylor, 1969).

Procedures for ATPase assay:

1. Determination of inorganic phosphate by Pollard and Korn's (1973a) modification of the method of Martin and Doty (1949).

Stock solutions

A. Isobutanol/benzene, 1:1. Store at room temperature.

B. Silicotungstic-sulfuric acid. Mix 2 parts of 10 *N* sulfuric acid (made by adding 150 ml of concentrated sulfuric acid to 350 ml of water) with 5 parts of 6% aqueous silicotungstic acid (Fisher #A-289). Store at room temperature.

C. Ammonium molybdate. Dissolve 10 g of ammonium molybdate (Baker #0716) in water to give a final volume of 100 ml. Store at room temperature. Discard when a precipitate forms.

D. Acid alcohol. Mix 1 pint of cold 100% ethanol with 15 ml of concentrated sulfuric acid.

E. Stannous chloride. Dissolve 1 g of fresh stannous chloride Baker (#3980) in concentrated HCl to give a final volume of 10 ml. Store in a light-tight container in a refrigerator, where it will keep for several months. Also store dry stannous chloride in the refrigerator. *If this assay fails to give a linear standard curve, stannous chloride is at fault until proved otherwise.*

Working solutions (to be made up fresh daily)

F. Stop mixtures. Mix 1 ml of A with 0.25 ml of B in small test tubes. Two phases will be present. Alternatively these two components can be pipeted into the assay tubes directly.

G. Diluted stannous chloride. Dilute 1 part stannous chloride stock solution (E) with 25 parts of 1 *N* sulfuric acid and store in an ice bath on day of use.

Assay

A. Pour one tube of stop mixture into each 0.5-ml sample containing 2–300 nmoles of inorganic phosphate and immediately vortex for 2 sec. If this is an ATPase assay, the acid will stop the reaction and precipitate the protein.

B. Immediately add 0.1 ml of ammonium molybdate (solution C) and vortex again for about 10 sec.

C. Let phases separate. The yellow phosphomolybdate complex will be extracted quantitatively into the upper (organic) phase, while organic phosphates such as ATP or ADP will remain in the lower aqueous phase. If the phosphate is radioactive, an aliquot of the organic phase can be used for scintillation or Cherenkov counting. If there is a very large protein precipitate, low-speed centrifugation may be necessary to layer it at the interface between the two phases.

D. Using a 1-ml pipette with a propipette, transfer 0.5 ml of the upper phase into a clean test tube and, using the same pipette, add 1.0 ml of acid alcohol (solution D). This solution is stable.

E. Develop the peacock blue color by reducing the phosphomolybdate complex with 0.05 ml of diluted stannous chloride (solution G) and vortexing to mix. If the color is navy blue, the stannous chloride is bad.

F. Read the absorbance at 720 nm against a blank without phosphate.

G. Construct a standard curve with 0.1, 0.2, and 0.3 ml of 1 m*M* phosphate (i.e., 100, 200, and 300 nmoles) in 0.5 ml total volume. The extinction coefficient is about 0.007 OD per nmole.

2. Myosin ATPase by the method of Pollard and Korn (1973a).

Stock solutions

A. Buffer mixture to give final concentrations of 0.5 *M* KCl, 10 m*M* imidazole (pH 7) in the assay [25 ml of 2 *M* KCl, 1 ml of 1 *M* imidazole (pH 7), 44 ml of water].

B. 100 m*M* $CaCl_2$.

C. 100 m*M* $MgCl_2$.

D. 20 m*M* EDTA neutralized with KOH.

E. 20 m*M* ATP made by diluting a 100 m*M* stock solution. This stock is made from disodium ATP neutralized with KOH immediately after dissolving in water. Store frozen.

Assay

A. Set up a stop tube for each assay as described under inorganic phosphate determination.

B. Set up assay tubes by adding 0.35 ml of ATPase mix, 50 μl of 20 m*M* ATP, and 50 μl of either EDTA, $CaCl_2$, or $MgCl_2$.

C. Start the reaction by adding 50 μl of enzyme to an appropriate tube, shaking gently and placing in a water bath set at an appropriate temperature. Tubes are conveniently started at 30-sec intervals.

D. After a suitable time interval (which will be determined by the activity of the sample) stop the reaction by adding the stop solution and vortexing. Alternatively, the silicotungstic-sulfuric acid and the isobutanol benzene can be pipetted separately into the assay tube, but this is slower.

E. During the 30 sec before the next tube is to be stopped, add 0.1 ml of 10% ammonium molybdate and vortex 10 sec.

F. Measure the released inorganic phosphate when all the assays have been stopped and extracted.

Another absolutely unique feature of myosin is that actin filaments stimulate the Mg^{2+} ATPase activity, providing the ionic strength is low. Nevertheless this assay is not generally useful for following myosin purification for two reasons. First, the actin-Mg^{2+} ATPase activity of myosin is only a small fraction of the total cellular Mg^{2+} ATPase activity in nonmuscle cells. Second, many myosins require activating enzymes, such as the *Acanthamoeba* cofactor protein (heavy-chain kinase) (Pollard and Korn, 1973a; Maruta and Korn, 1977b) or the cytoplasmic and smooth muscle myosin light-chain kinases (Adelstein and Conti, 1975; Dabrowska *et al.*, 1977), for actin-activated ATPase activity. When these activating enzymes are separated from myosin during purification, the myosin loses its actin-Mg^{2+} ATPase activity. Thus, although actin-Mg^{2+} ATPase of myosin cannot be used for quantitating purification, it is an essential feature of myosin and should be studied as part of myosin characterization.

It is difficult to specify exact conditions because they vary considerably for different myosins. For starters I recommend trying the following conditions: 10–30 m*M* KCl, 2 m*M* $MgCl_2$, 1 m*M* ATP, 0.1 m*M* $CaCl_2$, 10 m*M* imidazole (pH 7), 0.5 mg/ml actin. The rationale for these conditions is that all actomyosin ATPases are inhibited by high concentrations of salt and ATP, all require Mg^{2+}, and many are activated directly or indirectly by Ca^{2+} (see Moos, 1973).

An important factor to be tested is the dependence of the myosin ATPase activity on actin concentration. If the myosin is sufficiently soluble, the plot of ATPase versus actin concentration is hyperbolic and a double reciprocal plot (1/rate versus 1/actin) will be linear. The Y intercept is the V_{max} and the X intercept gives the K_a (the apparent affinity of myosin for actin). Thus it is possible to test whether any particular parameter affects the activity by altering V_{max}, K_a, or both (see Moos, 1973).

B. Gel Electrophoresis

Myosin polypeptides have been analyzed by polyacrylamide gel electrophoresis using any number of buffer systems, but I have found one simple tube gel formulation and one slab gel system particularly useful because *both* the heavy and light chains are well resolved without using a polyacrylamide gradient. The tube gel method is a variation by Stephens (1975) of a system originally used by J. Bryan.

Procedure for tube gel electrophoresis:

1. *Stock solutions*

A. Acrylamide-bis: 30 g acrylamide (Eastman #5521) and 0.8 gm *N,N'*bis-methylene acrylamide (Eastman #8383) to 100 ml with water. Filter and store at 4°C.

B. SDS: 10 g sodium dodecyl sulfate (Sigma #L5750) to 100 ml with water. Store at room temperature.

C. Tris-glycine, "250 m*M*": 30.3 g Tris base (Sigma #T-1503) and 144 g glycine (Sigma #G-7126) to 1 liter with water. The pH is about 8.6; do not adjust. Store at room temperature.

D. Ammonium persulfate (Bio-Rad): 0.1 g per ml water. Store at −20°C.

E. Electrode chamber buffer: 100 ml Tris-glycine stock and 10 ml SDS stock per liter. Store at room temperature.

F. 2× sample buffer: 20 ml SDS stock, 2 ml Tris-glycine stock, 2 ml β-mercaptoethanol, 20 ml glycerol, 10 mg bromophenol blue to 100 ml with water. Store at room temperature.

G. TEMED: *N,N,N',N'*-tetramethylethylene-diamine (Bio-Rad).

2. *Formulations for 12 0.5 × 7 cm tube gels*

	5%	7.5%	10% acrylamide
Water	14.25 ml	12.7	11.0
SDS	0.2	0.2	0.2
Tris-glycine	2.0	2.0	2.0
Acrylamide	3.33	5.0	6.67
TEMED	0.01	0.01	0.01
Persulfate	0.20	0.15	0.10

Pour the solution into 0.5 (ID) × 10-cm tubes sealed at the bottom with parafilm. Overlayer with water-saturated butanol during polymerization. The 7.5% gel is particularly suitable for resolution of both myosin heavy chains and light chains.

3. *Samples:* Dialyze samples containing 0.2 to 1.0 mg/ml protein for 2 hr versus water to remove salts. Mix samples with an equal volume of 2× sample buffer and *immediately* heat to 100°C in a boiling water bath for 2 min. Apply 10–50 μl samples to the gels and run at 100–150 V. The tracking dye will reach the bottom of the gel in 60–80 min.

4. *Stain*: Use the method of Fairbanks *et al.* (1971).

The procedure for slab gel electrophoresis is a modification by Gibson (1974) of the Laemmli (1970).

Procedure for slab gel electrophoresis:

1. *Stock solutions*

A. 20% SDS: 20 g sodium dodecyl sulfate (BDH #44244) to 100 ml with water. Store at room temperature.

B. 3 *M* Tris-HCl (pH 8.8) 36.3 g Tris base plus HCl to pH 8.8 in 100 ml. Store at 4°C.

C. 1 *M* Tris-HCl (pH 7): 12.1 g Tris base plus HCl to pH 7.0 in 100 ml. Store at 4°C.

D. Resolving gel acrylamide: 28 g acrylamide (Eastman #5521) 1.09 g diallyltartardiamide (Eastman #11444) plus water to 100 ml. Store at 4°C.

E. Stacking gel acrylamide: 21.5 g acrylamide, 3.75 g diallyltartardiamide plus water to 100 ml. Store at 4°C.

F. Stacking gel buffer: 19.2 ml 1 *M* tris (pH 7) plus 0.8 ml 20% SDS. Store at 4°C.

G. Ammonium persulfate: 0.14 g in 5 ml. Store at −20°C.

H. Electrode chamber buffer: 28.8 g glycine, 6 g Tris base, 2 g SDS plus water to 2 liter. Store at room temperature.

I. 2× sample buffer: 10 ml 20% SDS, 10 ml β-mercaptoethanol, 5 ml glycerol, 5 ml 1 *M* Tris (pH 7), 5 mg bromophenol blue, 20 ml water. Store at room temperature.

2. Formulation for a 1-mm × 10-cm × 12-cm slab gel with 14% acrylamide.
   Mix
   Tris-HCl (pH 8.8), 1.65 ml
   Resolving gel acrylamide, 6.55 ml
   20% SDS, 0.065 ml
   Water, 4.45 ml
   Then degas in a sidearm flask for 2 min using a water pump. Then add
   Persulfate, 0.33 ml
   TEMED, 3 μl
   Mix gently and pour between plates. Overlayer with 1 ml 0.1 SDS and allow to polymerize for 1 hr only.

3. Formulation for the stacking gel. Mix
   1 *M* Tris-HCl (pH 7), 1 ml
   20% SDS, 40 μl
   Stacking gel acrylamide, 1.6 ml
   Water, 5.3 ml
   Degas for 2 min and then add
   Persulfate, 0.4 ml
   TEMED, 2.5 μl
   Mix gently, drain the 0.1% SDS from surface of the resolving gel, pour the stacking gel and insert the sample well comb and polymerize for 60 min only.
4. Prepare the samples as described for the tube gels using the 2× slab gel sample buffer. Apply 10–30 μl samples in the sample wells underneath the electrode buffer and run at 150 V. The tracking dye will reach the bottom of the gel in about 4 hr.

C. Negative Staining for Electron Microscopy

During the characterization of a new myosin it is usually essential to examine the structure of myosin filaments and of the complex of myosin with actin filaments. Although the preparation of negatively stained specimens is simple, at least in principle, marginal micrographs are occasionally published, because some investigators are not familiar with the technical details that improve the chances of making high-quality specimens.

Procedure for negative staining:

1. Make grids coated with formvar or colloidin as described in electron microscopy handbooks and coat with carbon.

2. Just prior to specimen application, make the grid surface hydrophilic by glow discharge in a vacuum evaporator. If the evaporator is not equipped with such a device, an adequate substitute is to use an inexpensive Tesla coil, sold by scientific supply houses as a "gas leak detector." Place the grids on a grounded metal plate inside the evaporator bell jar, evacuate to about 0.1 torr, and then create a discharge inside the bell jar by touching the tip of the active Tesla coil to the external end of the one of the metal lead-throughs into the bell jar. If successful, a purple discharge will emanate from the internal tip of the lead-through. Usually 30 sec of discharge will make the grids hydrophilic, although some experimentation will be necessary with an improvised device.

3. Make myosin filaments by dilution or dialysis into a suitable buffer. Start with 50 m*M* KCl, 2 m*M* $MgCl_2$, 10 m*M* imidazole (pH 7), and a myosin concentration of 0.1 to 1 mg/ml. Just before application to the grid, dilute to about 20 to 80 μg/ml in the buffer.

4. While holding the grid with fine forceps, place a small drop of diluted

myosin filaments on the grid for about 30 sec. Then withdraw excess sample by touching the edge of the grid to filter paper. If the grid is hydrophilic, a thin film of sample will cover the surface uniformly and resist removal by the filter paper.

5. The grid can then be stained directly by applying a drop of aqueous 1% uranyl acetate for 15–30 sec. All but a thin film of stain is then removed with filter paper and the specimen is air-dried. Alternatively, the background on the grid can be cleaned up and the staining made more uniform by the following steps recommended by Dr. U. Aebi (Johns Hopkins Medical School, Department of Cell Biology and Anatomy). Transfer the grid with absorbed specimen through 3 drops of buffer on Parafilm for 5 sec each. Merely touch the specimen side of the grid to each drop; do not submerge the grid. Then transfer through 3 drops of water for 5 sec each, withdraw the water by touching the edge of the grid to filter paper, transfer through 3 drops of 1% uranyl acetate or 0.75% uranyl formate for 5 sec each, and finally withdraw excess stain by touching the edge of the grid to filter paper. It is also helpful to use a glass capillary connected to a vacuum to clean off any droplets of stain from the edge of the grid to aid in the uniform drying of the stain.

It is difficult to obtain uniformly excellent grids, so it is necessary to make multiple specimens and to search each grid for the best staining. If the staining is uniformly poor, first check that the grids are hydrophilic and then *dilute* the specimen.

Actin filaments can be decorated with myosin either in solution or on the grid itself. The most common cause of failure is the presence of ATP in the actin buffer. Therefore wait long enough after mixing the proteins for the myosin to hydrolyze the ATP or first apply about 50 μg/ml of actin filaments to the grid, wash through 3 or 4 drops of buffer without ATP, and then apply the myosin.

III. Myosin Extraction from Cells

Extraction of myosin from the cell requires lysis of the cell into a buffer that solubilizes myosin. Three general types of extracting solutions have been used to solubilize myosin: high ionic strength, sucrose, or pyrophosphate buffers (Table I). Both the high ionic strength and pyrophosphate buffers solubilize myosin by dissociating myosin filaments into myosin monomers. To my knowledge, no one has investigated the form of myosin in the sucrose buffers, but it is clear that a number of cytoplasmic myosins (macrophage, *Dictyostelium, Acanthamoeba*) are soluble in sucrose. There are no absolute reasons why one or the other of these general buffer types should be superior for extracting myosin from a given cell and, with the exception of some early work (Adelman and Taylor, 1969;

TABLE I

MYOSIN EXTRACTION SOLUTIONS

High ionic strength	Cell types	Reference
0.5 *M* KCl 0.1 *M* $KHPO_4$	Skeletal muscle	Kielley and Bradley (1956)
0.6 *M* KCl 15 m*M* Tris (pH 7.5) 10 m*M* Dithiothreitol 3% Butanol	Platelets Vertebrate tissue culture Brain	Adelstein *et al.* (1971) Adelstein *et al.* (1972) Burridge and Bray (1975)
0.9 *M* KCl 15 m*M* Sodium pyrophosphate (pH 7.0) 5 m*M* $MgCl_2$ 3 m*M* Dithiothreitol 30 m*M* Imidazole (pH 7.0)	Platelets	Pollard *et al.* (1974)
0.6 *M* Ammonium acetate (pH 6.5) 2 m*M* Magnesium acetate 2 m*M* Sodium pyrophosphate 0.5 m*M* Dithiothreitol	Platelets Skeletal muscle	Trayer and Trayer (1975)
0.3 *M* Ammonium acetate 10 m*M* Sodium bicarbonate (pH 7.2) 2 m*M* $CaCl_2$	Thyroid	Kobayashi *et al.* (1977)
Pyrophosphate		
50 m*M* Sodium pyrophosphate 10 m*M* Tris-maleate (pH 6.8)	*Physarum*	Adelman and Taylor (1969)
Sucrose		
0.34 *M* Sucrose 2 m*M* ATP 10 m*M* Dithiothreitol 10 m*M* Tris-maleate (pH 7)	Macrophages Neutrophils	Stossel and Pollard (1973)
0.34 *M* Sucrose 1 m*M* ATP 1 m*M* Dithiothreitol 10 m*M* Imidazole (pH 7.0)	*Acanthamoeba*	Pollard *et al.* (1978)
0.87 *M* Sucrose 1 m*M* EDTA 0.1 m*M* Dithiothreitol 10 m*M* Tris (pH 7.5)	*Dictyostelium*	Clarke and Spudich (1974)

Pollard and Korn, 1973a), few comparative studies have been made with different extracting solutions. The sucrose buffers seem to have one distinct advantage for nonmuscle cells: they minimize rupture of lysosomes and make it possible to isolate intact myosin from cells, such as macrophages, that are rich in lysosomes (Stossel and Pollard, 1973). (In concentrated KCl buffers enough lysosomal enzymes are released that most of the macrophage myosin heavy chains are cleaved by proteases.) A further advantage of sucrose extraction is that the extract has a low ionic strength and is therefore suitable for ion exchange chromatography.

Besides the high salt, pyrophosphate, or sucrose to solubilize the myosin, the extracting solutions usually contain a pH buffer and a sulfhydryl reducing agent. Some also contain a low concentration ATP or pyrophosphate to dissociate actin from myosin and a detergent or organic solvent to solubilize membranes.

Cell lysis techniques vary considerably. Striated muscle cells are large enough to be broken with a meat grinder. Most cells can be broken gently in Potter Teflon–glass or Dounce glass homogenizers or by nitrogen cavitation in a Parr bomb. Platelets are too small to break mechanically, so they are lysed by freeze-thawing or by membrane dissolution with butanol or Triton X-100. When the specimen is very tough, such as smooth muscle, it may be necessary to resort to more violent means of cell lysis such as a blade-type homogenizer. However, it is wise to be as gentle as possible and to avoid bubbles during homogenization and all subsequent steps, because myosin is particularly sensitive to surface denaturation at air–water interfaces.

Various investigators have used extraction times from a few minutes to many hours. In those cases where the extent of extraction has been measured as a function of time (e.g., Pollard *et al.,* 1974), extraction was complete in less than 30 min, so there is probably no need for prolonged extraction.

Once the myosin is solubilized, the insoluble cell remnants are invariably removed by centrifugation. High-speed (100,000 *g*), long (60-min) centrifugation has the advantage of pelleting microsomes as well as larger membrane fragments and yields an extract of soluble components. Low-speed centrifugation (25,000 *g*) for 15 min is sufficient for muscle and is commonly used for other cells as well. If a pellicle of lipid floats to the top of the centrifuge tube, it can be removed with a Pasteur pipette connected to a vacuum via a trap. Large chunks of fat can be filtered out with glass wool.

A completely different approach is to solubilize as much of the cell as possible except for the molecules of interest. This has been done for years with skeletal muscle where insoluble myofibrils can be separated from the cytosol by cell lysis and extraction under conditions that stabilize the actin and myosin flaments (see, e.g., Potter, 1974). Myofibrils have also been isolated from cardiac (Solaro *et al.,* 1971) and smooth (Sobieszek and Bremel, 1975) muscles. To purify the myosin, the myofibrils are extracted by one of the methods used to solubilize myosin.

IV. Fractional Precipitation Methods

A. Actomyosin Precipitation

Precipitation of the specific combination of myosin filaments with actin filaments is one of the original methods for purifying myosin (Szent-Györgyi, 1951) and remains one of the most useful. This precipitation requires low ionic strength for myosin filament formation and the absence of ATP for the combination of myosin with actin filaments. This is usually accomplished by dilution or dialysis. Dilution is faster but dialysis may give a higher yield in some cases where the myosin and/or actin concentrations are low. The final ionic strength required to precipitate actomyosin is in the range of 50–100 m*M* depending on the properties of the myosin. The pH is often an important consideration. Usually it must be slightly acidic. For example, platelet actomyosin fails to precipitate above pH 7 and precipitates quantitatively at pH 6.4 (Pollard *et al.*, 1974). ATP in the extract can be removed enzymatically with hexokinase and glucose (Pollard *et al.*, 1978) if the myosin ATPase activity is too low to hydrolyze the ATP quickly.

B. Ammonium Sulfate Precipitation

Ammonium sulfate precipitation has been used successfully for years in muscle myosin purification (Kielley and Bralley, 1956) and has also been widely used for nonmuscle myosin purification. Solid ammonium sulfate can be added directly to the sample, but the evolution of gas bubbles may lead to some myosin denaturation. Consequently, I usually use a saturated solution of ammonium sulfate containing EDTA to bind minor heavy metal contaminants.

Procedure for making saturated ammonium sulfate (from Wayne Kielley):

1. Add 780 g ultrapure ammonium sulfate (Schwartz-Mann #1946) to 1 liter of water and heat until dissolved.
2. Add 0.2 *M* EDTA (pH 7) to give a concentration of 10 m*M*.
3. Cool to 4°C overnight to crystallize excess ammonium sulfate.
4. Neutralize by adding ammonium hydroxide to give an apparent pH of 8.2. To confirm that the actual pH is 7.0, dilute a small sample 1:10 with water and read the pH.

The ammonium sulfate solution can be used to precipitate myosin in two different ways: Usually the desired volume of ammonium sulfate solution is added slowly to the protein solution with gentle stirring. After about 15 min, the precipitate is pelleted at 25,000 *g* for 15 min. This is satisfactory when the protein concentration is relatively high. When the protein concentration is low, the sample may be dialyzed against the ammonium sulfate. In this case, use the same volume of saturated ammonium sulfate that one would add to the protein solution to give the desired salt concentration. For example, for a final concentration of 1.9 *M* ammonium sulfate (50% saturation) either add 1 vol of saturated ammonium sulfate solution to 1 vol of protein or dialyze 1 vol of protein versus 1 vol of saturated salt solution. The time required for completion of dialysis will depend on the diameter of the dialysis bag. About 4 hr is sufficient for a 2-cm-diameter bag.

Usually the concentration of ammonium sulfate is raised in steps and the precipitated protein is pelleted after each step. The exact concentration required to precipitate myosins varies, but most precipitate between 1 and 2 *M*. Pilot experiments using steps of 0.2 *M* are recommended to determine the optimal conditions.

V. Chromatography of Myosin

Column chromatography has been used as a final step to remove minor adherent contaminants such as C-protein from conventional muscle myosin preparations (Offer *et al.*, 1973) and is an absolutely essential step in the purification of all cytoplasmic myosins. The theory of each type of column is well covered in other ''Methods'' series, so I will emphasize the specific advantages and disadvantages of each for myosin purification.

A. Gel Filtration Chromatography

Gel filtration has been used very successfully in myosin purification because few other cellular components share the large Stokes' radius of myosin. Oligomeric actin is one of the only cellular proteins that is not separated from myosin on these columns, and it can be eliminated by complete depolymerization of the actin in KI (Puszkin and Berl, 1972; Nachmias, 1974; Pollard *et al.*, 1974).

Beaded 4% agarose gels are usually selected for myosin gel filtration. As in all other forms of chromatography, higher resolution is achieved if the gel particles are small and uniformly sized. For this reason the resolution of 200–400 mesh 4% agarose beads from Bio-Rad Laboratories (A-15m) is much better than the larger (100–200 mesh) A-15m beads from Bio-Rad or the widely used 4% agarose beads from Pharmacia Fine (Sepharose 4B). Although the peaks are sharper, there is a penalty for this high resolution; the flow rates are lower with the smaller beads.

These gel filtration columns are usually equilibrated with high concentration of KCl plus a pH buffer and a sulfhydryl reducing agent. When KI is used to depolymerize actin in the column sample, one can equilibrate the whole column with KI (Stossel and Pollard, 1973), but the prolonged exposure to KI usually denatures the myosin. Consequently many laboratories now use a two-buffer system in which the myosin passes through a KI zone into KCl while the much smaller actin monomers move more slowly down the column and remain in the zone of KI applied ahead of the sample (Pollard *et al.*, 1974).

Procedure (see Fig. 1):

1. A 2.6 × 90-cm column of Bio-Rad A-15m (200–400 mesh) is equilibrated with 600 ml 0.6 *M* KCl, 1 m*M* dithiothreitol, 10 m*M* imidazole (pH 7). Using 55 cm of hydrostatic pressure, the flow rate at 4°C is about 15–20 ml/h.

2. Before preparing the sample, the KCl buffer is removed from the top of the column and 50–60 ml (~14% of column volume) of 0.6 *M* KI, 5 m*M* ATP, 5 m*M* dithiothreitol, 1 m*M* $MgCl_2$, 20 m*M* imidazole (pH 7) is allowed to run into the column bed.

3. An actomyosin sample containing about 40–120 mg of protein is dissolved by gentle Dounce homogenization in 10–15 ml of 0.6 *M* KI, 5 m*M* ATP, 5 m*M*

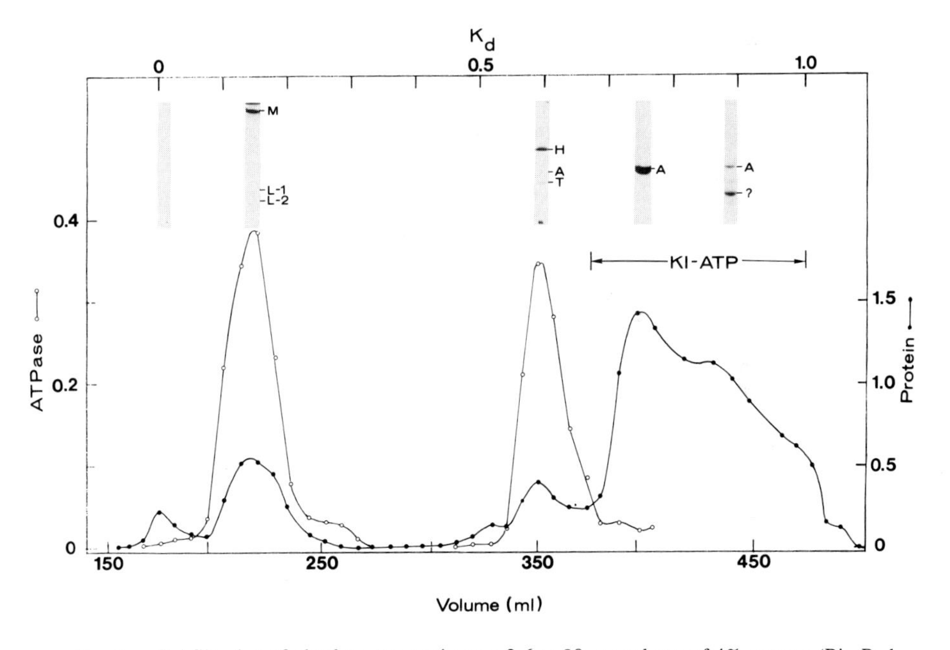

FIG. 1. Gel filtration of platelet actomyosin on a 2.6 × 90-cm column of 4% agarose (Bio-Rad A-15m, 200–400 mesh) according to Pollard *et al.* (1974). The zone of KI/ATP buffer is indicated. K^+-EDTA ATPase activity in μmoles min^{-1} ml^{-1} (○ -○). Protein concentration in mg ml^{-1} (● - ●). The protein composition of selected fractions is shown by the tube gel electrophoresis in SDS. Myosin heavy (M) and light chains (L-1 and L-2). Head fragment of myosin (H). Actin (A). Tropomyosin (T). (From Pollard *et al.*, 1974.)

dithiothreitol, 1 m*M* $MgCl_2$, 20 m*M* imidazole (pH 7) and is immediately clarified by centrifugation at 100,000 *g* for 20–30 min.

4. The clarified sample is applied to the column immediately behind the KI zone.

5. KCl buffer is layered over the sample and used to elute the column.

6. Myosin elutes with a partition coefficient of about 0.25 (depending on the particular batch of agarose) and can be detected by absorbance at 280 nm, protein assay, and ATPase assay. Actin elutes with a partition coefficient of about 0.75. Since it is, at least in part, contained in the KI zone, the actin will be denatured unless the KI is removed by dialysis.

7. The column is stored in 0.6 *M* KCl, 0.05% sodium azide, 2 m*M* EDTA.

There are also successful procedures for myosin gel filtration without KI. Adelstein *et al.* (1971, 1972) and Ostlund and Pastan (1976) have purified myosin from platelets and from cultured cells using agarose columns equilibrated with 0.6 *M* KCl, 1 m*M* ATP, 1 m*M* $MgCl_2$, 5 m*M* dithiothreitol, 15 m*M* Tris (pH 7.5). I have had more success separating myosin from actin using the two-buffer system.

B. Anion Exchange Chromatography

Chromatography on DEAE columns is one of the better ways to remove minor contaminants such as C-protein from muscle myosin (Offer *et al.*, 1973) and is an important step in the purification of *Acanthamoeba* myosins (Pollard and Korn, 1973a; Maruta and Korn, 1977a,b; Pollard *et al.*, 1978). It has also been used to separate muscle myosin subfragment-1 isozymes (Weeds and Taylor, 1975). My personal experience has been that DEAE-cellulose is easy to work with and gives better resolution of myosins than DEAE-Sephadex or DEAE-agarose.

A limitation of ion exchange chromatography is the insolubility of most myosins at low ionic strength where they bind to the column. This can be overcome by using pyrophosphate (Richards *et al.* 1967) or phosphate buffers (Offer *et al.*, 1973), which dissociate myosin filaments at relatively low ionic strength. For cytoplasmic myosins, sucrose-extracting solutions can be used to solubilize the myosin at low ionic strength.

Procedure (see Fig. 2):

1. Wash fresh or previously used DEAE-cellulose according to the manufacturers instructions. For Whatman DE-52, DE-53 or DE-32, this involves the following steps: (a) Stir 30 min in 0.5 *M* NaOH to release tightly bound materials; (b) wash with 20 vol of water on a Buchner funnel; (c) stir 30 min under vacuum in 0.5 *M* HCl to release bound bicarbonate; (d) wash with 20 vol of water on a Buchner funnel; and (4) titrate to desired pH using the free base used

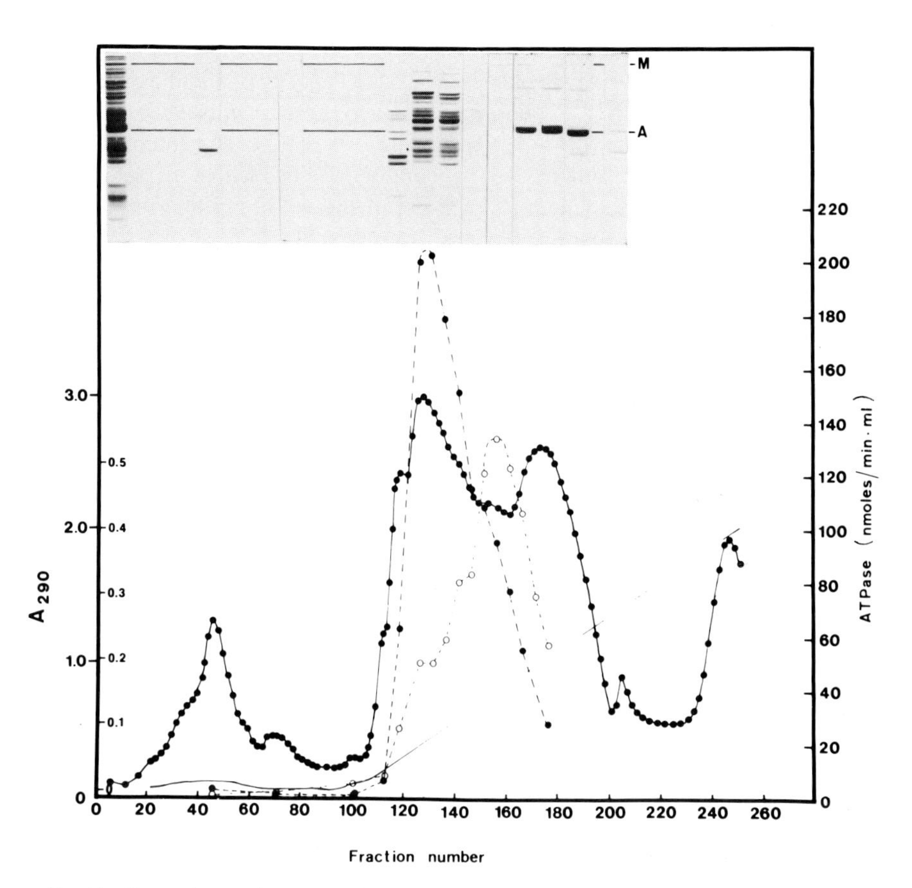

FIG. 2. Ion exchange chromatography of 230 ml of *Acanthamoeba* sucrose extract on a column (4 × 16 cm) of DEAE-cellulose as described (Pollard *et al.*, 1978). A_{290} (●——●). K^+-EDTA-ATPase (●- - -●). Ca^{2+} ATPase (○- - -○). KCl gradient in moles per liter on the scale on the left (——). Gel electrophoretic analyses of selected fractions are shown at the top starting with the sample of extract applied to the column on the far left. The photograph of each gel is centered over the fraction analyzed. M marks the mobility of the *Acanthamoeba* myosin-II heavy chain and A marks the mobility of actin. (From Pollard *et al.*, 1978.)

in the buffer (e.g., 1 *M* Tris base or imidazole) and monitoring the pH by filtering small aliquots and returning the filtered DEAE-cellulose to the sample.

2. If the column is equilibrated with ATP (e.g., Gordon *et al.*, 1976), add it at this point.

3. Pour the column using a high flow rate (achieved by 1 to 2 m hydrostatic pressure, by pumping, or by N_2 pressure) to pack the bed firmly (see Peterson, 1970, for details). Once packed the column is run at lower flow rates on the order of 1 column volume per hour or less. Choose a bed volume appropriate for the

sample, for example, 1 ml of DE-52 per 20 mg of *Acanthamoeba* extract protein (Pollard *et al.*, 1978). DEAE-Sephadex A50 has a capacity of about 5 mg muscle myosin per milliliter in 0.15 *M* phosphate (pH 7.5) (Offer *et al.*, 1973) or in 0.04 *M* sodium pyrophosphate (pH 7.5) (Richards *et al.*, 1967). The column dimensions will depend on the sample size. I generally use a bed with a diameter about one-sixth to one-tenth the length (e.g., 5-cm diameter, 30 cm long) to achieve high flow rates.

4. Equilibrate the column with three or more column volumes of the desired buffer (e.g., 0.04 m*M* sodium pyrophosphate (pH 7.5) as used by Richards *et al.* (1967)) or 1 m*M* ATP, 0.2 m*M* $CaCl_2$, 0.5 m*M* dithiothreitol, 20 m*M* imidazole (pH 7.5) as used by Gordon *et al.* (1976) Equilibration can be carried out much faster on a Buchner funnel than in a chromatographic column. Check the pH and conductivity of the eluate to confirm that equilibration has been successful.

5. Apply the sample. If it is very large, it may cake up the top of the column and lead to fissuring and channeling. This will destroy the performance of the column. The problem can be avoided by dividing the DEAE-cellulose into two parts. About one-third to one-half of the total is poured into the column and packed as usual. The remainder is mixed with the sample for 15–30 min and then carefully layered and packed on top of the DEAE-cellulose in the column.

6. Wash the column with 1–3 column volumes of starting buffer.

7. Elute the bound myosin by raising the salt concentration of the buffer. Most often this is done with a gradient of KCl or NaCl, but steps may be successful in some cases. I generally use a concave gradient because more proteins elute at low than high salt concentrations and they will be spread out more evenly among the fractions by a concave gradient. Such a gradient can be made using three identical chambers: one (a) containing the final salt concentration in the starting buffer and two (b and c) containing the starting buffer. Flow is from (a) to (b) to (c) to the column. Chambers (b) and (c) are stirred. In every case I know of myosin elutes between 0 and 300 m*M* KCl, so these would be appropriate limits for the gradient.

8. Assay fractions by measuring conductivity, protein concentration, and ATPase activity.

9. The DEAE-cellulose can be reused many times providing it is washed carefully and bacterial growth is inhibited during storage by azide or toluene.

C. Hydroxylapatite Chromatography

Hydroxylapatite (an insoluble crystalline form of calcium phosphate) is one of the most useful chromatographic media for myosin, because myosin binds to hydroxylapatite even in high concentrations of KCl or NaCl. For example, *Acanthamoeba* myosin-I is purified on hydroxylapatide in 0.5 *M* KCl (Pollard and Korn, 1973a; Maruta and Korn, 1977b). Proteins binding to the column are eluted with phosphate buffer. The mechanism by which hydroxylapatite binds proteins is still somewhat mysterious (see Bernardi, 1971).

In my experience, homemade hydroxylapatite (Bernardi, 1971) is superior to commercial preparations, but Bio-Rad Biogel HT is satisfactory for many applications.

Procedure (see Fig. 3):

1. Wash about 1 ml hydroxylapatide per 2 mg myosin in 5 vol of starting buffer (e.g., 0.5 *M* KCl, 1 m*M* DTT, 10 m*M* imidazole (pH 7.5)) by suspension and settling at 1 *g*. Repeat three times, removing any fines from the supernate each time. Handle the hydroxylapatite with great care! No magnetic stirring bars. No scraping between two hard objects. The crystals are easily broken and the resulting fines will greatly reduce the flow rate.

2. Pour a 1:1 slurry of hydroxylapatite/buffer into a column and equilibrate with 3 column volumes of starting buffer. To achieve an adequate flow rate use a

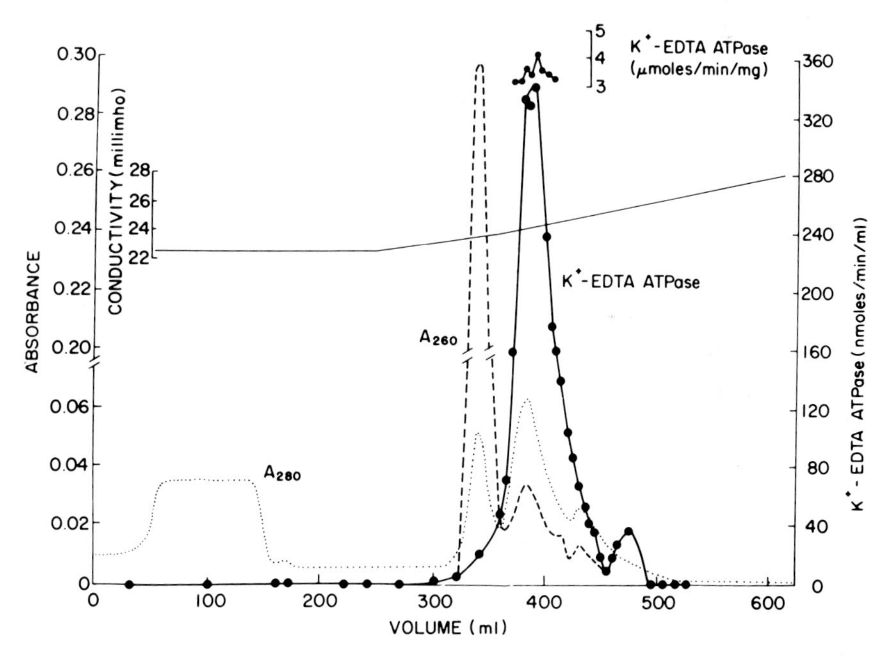

FIG. 3. Hydroxylapatite chromatography of partially purified *Acanthamoeba* myosin-I (Pollard and Korn, 1973a). A 100-ml sample containing about 4 mg of myosin-I was applied to a 1.5 × 26-cm column equilibrated with 0.5 *M* KCl, 1 m*M* dithiothreitol, 1 m*M* phosphate, and 5 m*M* imidazole (pH 7) and eluted at a flow rate of 30 ml/hr with a 600-ml concave 1–300 m*M* potassium phosphate gradient in the column buffer. A_{280} (.....). A_{260} (---). K^+-EDTA ATPase activity (● - ●). The specific activity of peak fraction is shown at the top. (From Pollard and Korn, 1973a.)

relatively short (e.g., 1.5 × 10-cm) column and 1–2 m of hydrostatic pressure or a peristaltic pump.

3. Apply the sample and wash with 1–2 column volumes of starting buffer.

4. Elute the bound proteins with a phosphate gradient. Potassium phosphate is more soluble than sodium phosphate, so it has advantages in the cold room. As in the case of DEAE, concave gradients are probably the best. In every case examined myosin elutes between 0–200 m*M* phosphate.

5. Assay fractions for conductivity, protein concentration, and ATPase activity (using [γ^{32}P]ATP). Simply count an aliquot of the organic phase in the ATPase assay (Section II,A). The high concentration of phosphate in the fractions will inhibit the extraction of $^{32}P_i$ in the ATPase assay, but the position of the peak can, at least, be determined before dialysis or precipitation of myosin to remove the phosphate.

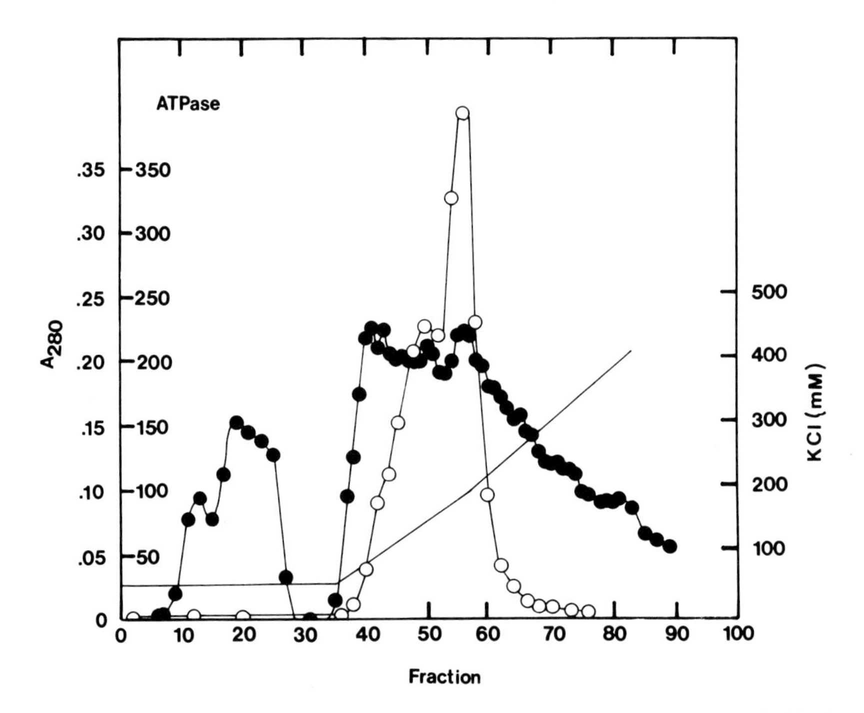

FIG. 4. Agarose adsorption chromatography of partially purified *Acanthamoeba* myosin-I by the method of Maruta *et al.* (1979). The sample was obtained by DEAE-cellulose chromatography, ammonium sulfate precipitation, and ADP-agarose chromatography and applied to a 2.5 × 27-cm column of Bio-Rad A-1.5m (200–400 mesh) equilibrated with 50 m*M* KCl, 1 m*M* dithiothreitol, 15 m*M* imidazole (pH 7.5). The myosin was eluted with a 400-ml 50–500 m*M* KCl gradient (—). A_{280} (● — ●). K^+-EDTA ATPase activity in nmoles min^{-1} ml^{-1} (○ - ○). Fraction volume was 6 ml. (From T. D. Pollard, unpublished observations.)

D. Agarose Adsorption Chromatography

This is not a common chromatographic procedure, but Pollard and Korn (1973a,b) found that *Acanthamoeba* myosin-I binds to Bio-Rad A-1.5m agarose beads. Since few of the contaminating proteins bound, this proved to be the most important step in the purification procedure. Later, another batch of A-1.5m was found to separate two myosin-I isozymes (Maruta *et al.*, 1979). In recent experiments with another batch of A-1.5m, I have also been able to separate these two isozymes, but both elute at much higher concentrations of KCl than in the earlier work (Fig. 4). The mechanism is not known, but binding is probably due to traces of carboxyl groups and/or sulfate groups on the agarose. Different batches of A-1.5m can differ considerably in their affinity for myosin-I. The manufacturer attributes this to variability in the crude agar received from different suppliers. Consequently it is impossible to specify exact conditions for chromatography. The original method employed a buffer consisting of 0.2 m*M* ATP, 1 m*M* dithiothreitol, 2 m*M* Tris (pH 7.6). The myosin-I bound to the column and was eluted with a gradient of KCl (Pollard and Korn, 1973a). A later modification by Maruta *et al.* (1979) employes a column equilibrated with 42 m*M* KCl, 1 m*M* dithiothreitol, 10 m*M* imidazole (pH 7.5) and the sample dissolved in the same buffer with 0.34 *M* sucrose. The bound proteins are eluted with a gradient of KCl (Fig. 4).

E. Affinity Chromatography

Affinity chromatography of myosins is still in its infancy, but already a number of ligands immobilized on agarose beads are known to bind myosin and myosin fragments (Table II). These affinity columns are useful for (a) separation of myosin from other cellular proteins; (b) separation of active from inactive myosin; (c) separation of myosin isozymes; and (d) concentration of dilute myosin solutions. As more experience is gained, I expect that these techniques will gain in importance for the purification of myosins.

The immobilized ligands used for myosin include Cibachron Blue, adenine nucleotides, pyrophosphate, actin, and antibodies (Table II). Of these ligands actin and antibodies are the most specific for myosin, but they have been little used for preparative purposes. In contrast, although many enzymes bind to adenine nucleotides, pyrophosphate, and Cibachron Blue, they each have been used successfully for myosin purification. There are many different immobilized myosin ligands available, so I will select one procedure from each class to illustrate the principles of these methods and, where appropriate, will make brief comments about the relative merits of the alternatives.

In the development of these affinity columns several general principles were important. It is important to keep these same principles in mind when these

TABLE II

AFFINITY CHROMATOGRAPHY MATERIALS FOR MYOSIN

	Reference
1. Cibachron Blue	
Blue dextran–Sepharose	Kobayashi *et al.* (1977)
Cibachron Blue–Sepharose	Toste and Cooke (1978)
2. Nucleotides	
Sepharose–adipic acid hydrazide–ATP	Lamed *et al.* (1973)
Sepharose–sebacic acid hydrazide–ATP	Lamed and Oplatka (1974)
Sepharose–adipic acid hydrazide–ADP	Wagner (1977)
Sepharose-N^6-(6-aminohexyl)-ADP	Trayer *et al.* (1974)
Sepharose-8-(6-aminohexyl)-ADP	Trayer *et al.* (1974)
3. Sepharose–aminohexyl pyrophosphate	Trayer *et al.* (1974)
4. Actin	
Sepharose-G–actin	Bottomley and Trayer (1975)
Sepharose-F–actin	Bottomley and Trayer (1975)
Sepharose–cross-linked F-actin	Winstanley *et al.* (1977)
Sepharose–cross-linked F-actin–tropomyosin	Winstanley *et al.* (1977)
Sepharose–phalloidin–stabilized F-actin	Winstanely *et al.* (1979)
Phalloidin F-actin trapped in Sepharose	Grantmont-Leblanc and Gruda (1977)
5. Antibodies	
Sepharose–anti-*Acanthamoeba* myosin-I	T. D. Pollard (unpublished observations)
Sepharose–anti-*Acanthamoeba* myosin-II	T. D. Pollard (unpublished observations)

columns are used for new applications, because it is by no means certain that the conditions used for the binding and elution of one myosin will be optimal (or even successful) for another myosin. The principles include the following:

1. The ligand must be attached to the insoluble matrix (usually agarose beads) in such a way that myosin can bind to the ligand. This may require a hydrocarbon spacer between the agarose and the ligand. Small differences in the length of this spacer may determine whether the myosin will bind.

2. The conditions for myosin binding and elution may not conform to one's preconceived notions about myosin–ligand interaction. Therefore it may be useful to experiment with unusual conditions. The following variables should be tested; the ionic strength; the concentration and type of anion; the concentration and type of divalent cation; the concentration and type of nucleotide (or pyrophosphate); and the temperature. Every one of these variables has been used in the selective binding and elution of myosin in affinity chromatography.

1. CIBACHRON BLUE

The blue dye Cibachron Blue F3GA is a structural analogue of the adenine nucleotides and binds to a number of enzymes with binding sites for NADH,

ADP, or ATP. The dye binds with high affinity ($K_i \sim 0.2\ \mu M$) and inhibits myosin ATPase activity (Toste and Cooke, 1978). This dye can be attached directly to agarose beads or first conjugated to dextran, which is then attached to cyanogen bromide activated agarose beads (Kobayashi *et al.*, 1977). Both types of immobilized Cibachron Blue have been used successfully for affinity chromatography of myosin. "Blue Sepharose" can be purchased from Pierce Biochemicals, Bio-Rad, or Pharmacia Fine Chemicals.

Blue-Sepharose binds muscle myosin subfragment-1 and heavy meromyosin at low ionic strength (0.5 m*M* dithiothreitol, 3.2 m*M* sodium azide, 20 m*M* TES, pH 7) with a capacity of about 3.5 mg/ml (Toste and Cooke, 1978). No conditions were found where muscle myosin was soluble and would bind quantitatively to Blue-Sepharose. Subfragment-1 and heavy meromyosin are eluted by KCl in the range 10–300 m*M*, by 20 m*M* TES (pH 9), and by Cibron Blue at 1 m*M*, but are not eluted by 1 m*M* NADH, Mg ATP, Ca ATP, Mg pyrophosphate, or EDTA. Actin does not bind, so application of a mixture of subfragment-1 and actin to Blue-Sepharose in 1 m*M* Mg ATP results in a clean separation of the two proteins. Blue-Sepharose also can be used to fractionate isolated muscle myosin light chains. The A-1 light chain does not bind; the A-2 light chain binds very weakly; and the "DTNB" light chain binds tightly.

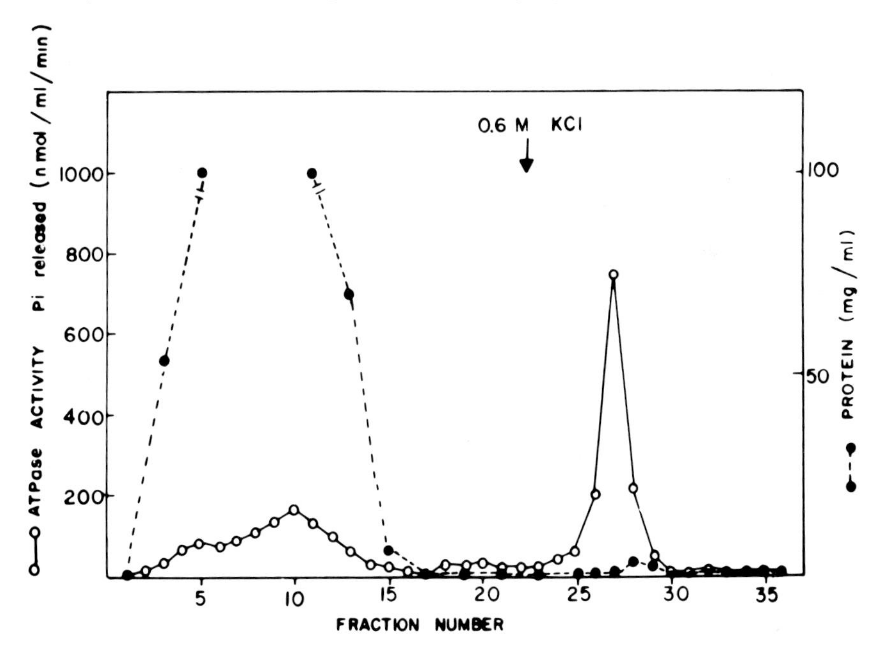

FIG. 5. Blue dextran–Sepharose affinity chromatography of 20 ml of a crude extract of thyroid gland on a 1 × 12-cm column by the method of Kobayashi *et al.* (1977). The bound myosin was eluted with 0.6 *M* KCl. Protein concentration in mg ml^{-1} (● — ●). K^{+}-EDTA ATPase activity (○ — ○). Fraction volume 3 ml. (From Kobayashi *et al.*, 1977.)

Given these properties, Blue-Sepharose would appear to be ideal for purification of some cytoplasmic myosins that are more soluble at low ionic strength than muscle myosin. The ability to separate actin and myosin cleanly would appear particularly valuable. A limitation will be that other proteins (dehydrogenases and kinases) also bind to Cibachron Blue, but selective elution of some of these enzymes can be achieved with their nucleotide substrates (see Toste and Cooke, 1978).

The partial purification of thyroid myosin on Blue dextran–Sepharose by Kobayshi *et al.* (1977) is a good example of how these columns can be used preparatively.

Procedure (see Fig. 5):

1. A crude extract is prepared by homogenizing the tissue in 3 vol of ice-cold 0.3 *M* ammonium acetate, 2 m*M* $CaCl_2$, 10 m*M* sodium bicarbonate, (pH 7.2) and centrifuging at 22,000 *g* for 1 hr.
2. 10–21 ml of extract supernatant is applied to a 0.9 × 12 cm column of Blue dextran–Sepharose equilibrated with the 0.3 *M* ammonium acetate buffer.
3. After washing the column with 30 ml of buffer, the myosin is eluted with 0.6 *M* KCl. In the case of thyroid myosin, the specific ATPase activity increases about 50-fold by this procedure. The recovery is about 60%.
4. Final purification is achieved by gel filtration with the discontinuous buffer system of Pollard *et al.* (1974).

2. Nucleotides

Both immobilized ATP and ADP have been used successfully for myosin affinity chromatography. The most thoroughly tested immobilized nucleotide for myosin chromatography is ADP linked by nitrogen-6 of the adenine via a 6 carbon spacer to Sepharose 4B (Trayer *et al.*, 1974). Another successful immobilized nucleotide is ADP linked by the ribose via a 6 carbon chain to Sepharose by the method of Lamed *et al.* (1973). Both are commercially available from PL Biochemicals and have been used for chromatography of muscle myosin and its proteolytic fragments (Trayer *et al.*, 1974; Lamed and Oplatka, 1974) and of cytoplasmic myosins from platelets (Trayer and Trayer, 1975) and *Acanthamoeba* (Maruta *et al.*, 1979).

Procedure (Trayer and Trayer, 1975):

1. A small column (0.8 × 8-cm) of Sepharose-N^6-ADP is equilibrated at 4°C with 0.6 *M* ammonium acetate, 5 m*M* EDTA, 2 m*M* sodium pyrophosphate, 0.25 m*M* DTT (pH 7.5).
2. Myosin is extracted from 1 g of minced muscle for 30 min at 4°C with 5 vol of 0.6 *M* ammonium acetate, 2 m*M* $MgCl_2$, 2 m*M* sodium pyrophosphate, 0.5 m*M* DTT (pH 6.5) and insoluble material is removed by centrifugation at

2000 *g* for 15 min. After dilution with 5 vol of column equilibration buffer the crude extract is applied to the column at a flow rate of 15–20 ml/hr.

3. The column is washed with equilibration buffer until A_{280} reaches zero. The unbound fraction contains virtually all the proteins in the extract except myosin.

4. The bound myosin is eluted either by replacing the ammonium acetate with ammonium chloride or by adding 50 m*M* ATP to the equilibration buffer. The eluted material is essentially pure myosin and has a higher specific activity than myosin prepared by conventional means. The yield is 10–20 mg/g of tissue.

5. The column is washed with 6 *M* urea, 2 *M* KCl. I store my columns of ADP-agarose for months at −20°C after first equilibrating them with 50% glycerol and 0.05% sodium azide.

Note: The use of EDTA in the buffer has the theoretical advantage that most other ADP binding proteins are thought to require divalent cations; however, Trayer and Trayer (1975) found similar results with muscle extracts with 5 m*M* EDTA or 5 m*M* $MgCl_2$.

The same procedure can be used for purification of platelet myosin with the following modifications (Trayer and Trayer, 1975): Extract frozen/thawed platelets with 10 vol of extracting solution, centrifuge at 80,000 *g* for 15 min, and dilute the supernate with 1 vol of water and 3 vol of column equilibration buffer containing 0.3 *M* ammonium acetate. The bound fraction is highly enriched in myosin but contaminated with actin. The actin can be removed by gel filtration with the two buffer system described above in Section V, A.

The Sepharose-N^6-ADP column can also be used to separate myosin subfragment-1 isozymes (Winstanley *et al.*, 1979). When the crude subfragment-1 is applied to the column in 5 m*M* triethanolamine (pH 7.5) denatured material fails to bind and the isozymes with the A1 and A2 light chains can be eluted separately with a gradient of 0–10 m*M* ADP.

Two other forms of immobilized ADP have been prepared: one linked from adenine C-8 to Sepharose via a 6 carbon hexane chain (Trayer *et al.*, 1974) and the other from the ribose to Sepharose by a 6 carbon adipic acid dihydrazide (Wagner, 1977). Using the latter column, Wagner was able to separate the muscle myosin subfragment-1 isozymes with a pyrophosphate gradient. Another application of the ribose-linked ADP-Sepharose is in the purification of *Acanthamoeba* myosin-I (Maruta *et al.*, 1979). This myosin is extremely difficult to purify and the KCl elution from immobilized ADP provides an essential 4–10-fold purification at an intermediate stage of purity (Fig. 6).

Immobilized ATP was actually the first nucleotide used for affinity chromatography of myosin (Lamed *et al.*, 1973; Lamed and Oplatka, 1974). It was found that muscle myosin, heavy meromyosin, and subfragment-1 all bound to ATP attached from the ribose to Sepharose with a 10 carbon spacer, but that only heavy meromyosin and subfragment-1 bound to ATP attached to Sepharose with a 6 carbon spacer. Binding was enhanced by divalent cations and in their

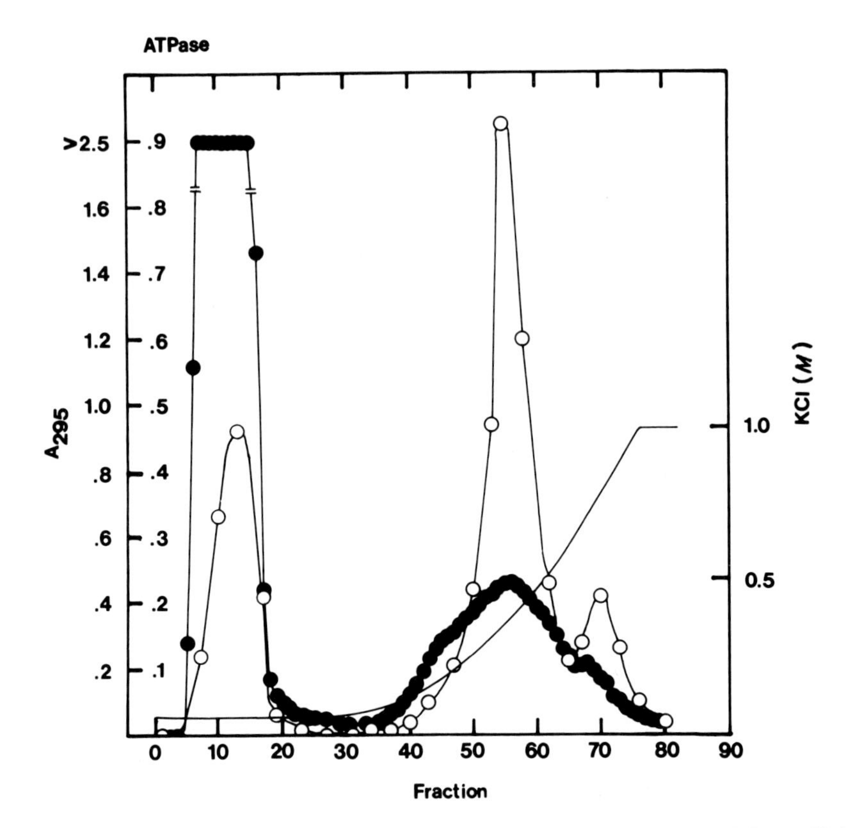

FIG. 6. Sepharose-adipic acid hydrazide-ADP affinity chromatography of partially purified *Acanthamoeba* myosin-I by the method of Maruta *et al.* (1979). The sample was obtained by DEAE-cellulose chromatography and ammonium sulfate precipitation and applied to a 1.5 × 15-cm column of Sepharose-ADP (PL Biochemicals #5568) equilibrated with 50 m*M* KCl, 1 m*M* dithiothreitol, 15 m*M* imidazole (pH 7.5). Myosin was eluted with a 240-ml concave 50–1000 m*M* KCl gradient (——). A_{295} (●—●). K^+-EDTA ATPase activity in μmoles min^{-1} ml^{-1} (○—○). Fraction volume was 3 ml. (From T. D. Pollard, unpublished observations.)

presence the gamma phosphate of ATP was hydrolyzed. Consequently the ATP columns could only be used one or two times before their properties changed. The bound myosin or its fragments could be eluted with gradients of salt or ATP. The columns successfully separated active from denatured myosin and subfragment-1 from heavy meromyosin. Myosin and single-headed myosin were partially separated by elution with a KCl gradient.

3. SEPHAROSE-AMINOHEXYL-PYROPHOSPHATE

Pyrophosphate can be attached to Sepharose at the end of a 6 carbon spacer (Trayer *et al.,* 1974). The resulting material is excellent for affinity chromatog-

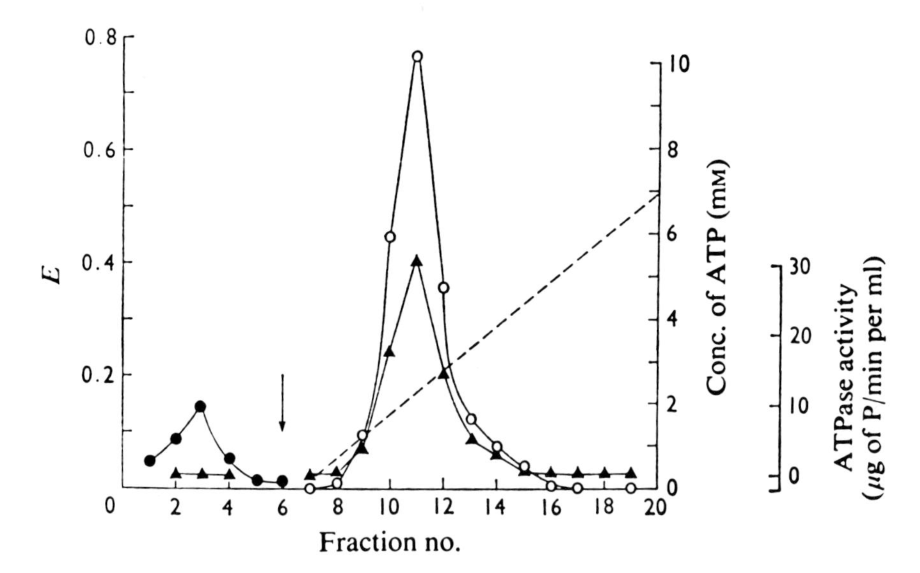

FIG. 7. Sepharose-6-aminohexylpyrophosphate affinity chromatography of crude heavy meromyosin according to Trayer *et al.* (1974). About 4 mg of protein was applied to an 0.8 × 10-cm column equilibrated with 2 m*M* $MgCl_2$, 2 m*M* β-mercaptoethanol, 5 m*M* Tris (pH 7.6) and eluted at 20 ml/hr with a linear gradient of 0–10 m*M* ATP (-). Protein concentration by A_{280} (● — ● (or turbidity at 360 nm after precipitation with 4% trichloroacetic acid (○ — ○). Ca^{++} ATPase affinity (▲ - ▲). Fraction volume was 2 ml. (From Trayer *et al.*, 1974.)

raphy of muscle myosin and its proteolytic fragments (Fig. 7). To my knowledge, it has not been used for cytoplasmic myosin purification, but it is used by Trayer's laboratory (in deference to Sepharose-ADP or actin) for the single-step purification of subfragment-1 from myosin digests. Because of its stability and modest expense, it would seem like an ideal addition to or substitute for the nucleotide affinity columns for cytoplasmic myosin purification. It is not commercially available, but the key reactant, aminoheyxlpyrophosphate can be synthesized relatively easily from 6-aminohexanol and pyrophosphoric acid and then conjugated with cyanogen bromide activated agarose beads.

4. ACTIN-AGAROSE

Two strikingly different approaches have been taken for preparing actin-agarose for affinity chromatography. On one hand, actin monomers or polymers can be coupled directly to cyanogen bromide activated agarose beads. On the other, actin monomers can be equilibrated with agarose beads and then polymerized inside the beads. Providing the actin-filament-stabilizing alkaloid phalloidin is present during polymerization, the filaments become permanently trapped inside the pores of the beads, even during prolonged washes with actin-free buffers. Both types of actin-agarose beads can be used for myosin affinity chromatography.

Actin is chemically coupled to agarose beads according to Bottomley and Trayer (1975) as follows: Sepharose 4B is activated with 100 mg of cyanogen bromide per gram of wet gel and mixed with 2–4 mg/ml of actin in an appropriate buffer. For actin monomers 5 m*M* triethanolamine, pH 8.5, is used. For actin filaments 50 m*M* KCl and 2.5 m*M* $MgCl_2$ are included. After reacting overnight at 4°C, the beads are washed in a Buchner funnel with 2 *M* KCl and then the coupling buffer. The coupling efficiency is about 70% and 1–2 mg of actin are bound per milliliter of beads. Actin filaments stabilized by glutaraldehyde cross-linking (Winstanley *et al.*, 1977) or by Phalloidin (Winstanley *et al.*, 1979) and glutaraldehyde cross-linked actin-tropomyosin filaments (Winstanley *et al.*, 1977) have also been coupled to cyanogen bromide activated agarose beads. Except for the actin filament–agarose conjugate, these various forms of immobilized actin are stable and useful for affinity chromatography.

Procedure (see Fig. 8):

1. Equilibrate a 1.2 × 10-cm column of actin monomer–Sepharose with 5m*M* triethanolamine, 2 m*M* $MgCl_2$ (pH 7.5) at 20°C and a flow rate of 20 ml/hour.
2. Apply 10 ml of 1.5 mg/ml of myosin digested with Sepharose-papain and then dialyzed against column buffer.
3. Wash with column buffer until the absorbance of the eluate is zero.
4. Elute the bound subfragment-1 with 0.25 *M* KCl in column buffer. Alternatively, subfragment-1 can be eluted with 2 m*M* Mg pyrophosphate, 3 m*M* Mg

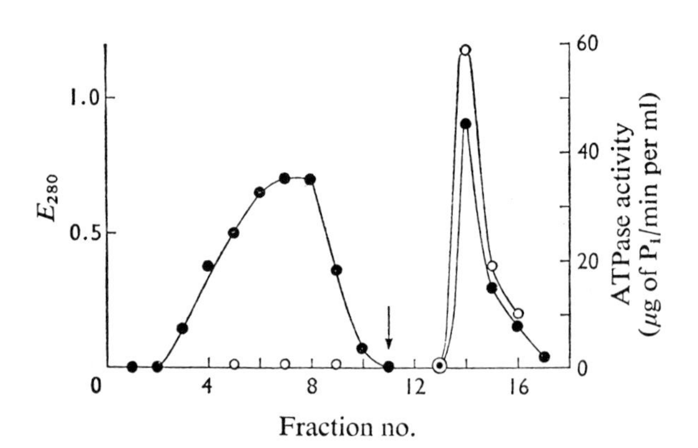

FIG. 8. Sepharose–G-actin affinity chromatography of crude muscle myosin subfragment-1 according to Bottomley and Trayer (1975). About 10 mg of protein in 2 m*M* $MgCl_2$, 5 m*M* triethanol amine (pH 7.5) was applied to a 1.2 × 10-cm column equilibrated with the same buffer at 20°C with a flow rate of 20 ml hr^{-1}. The bound subfragment-1 was eluted with 0.25 *M* KCl, starting at the arrow. A_{280} (● — ●). Ca^{2+}-ATPase activity (○). Fraction volume was 2 ml. (From Bottomley and Trayer, 1975.)

ATP or 4 m*M* Mg ADP. With the nucleotides and pyrophosphate the elution is "more efficient" if the column is cooled to 4°C and the eluted peaks are sharper when $MgCl_2$ omitted from the buffer.

5. The column is washed with 1 *M* KCl and stored in the column buffer. I would include 0.04% sodium azide to retard bacterial growth. The actin bound to the column is stable through repeated uses over a period of months.

These actin monomer–Sepharose columns also bind heavy meromyosin, but improved resolution of heavy meromyosin isozymes is possible on cross-linked actin filament–tropomyosin Sepharose. Using a KCl gradient HMM-A-1 and HMM-A-2 homodimers can be separated from heterodimers (Trayer *et al.*, 1977). After digestion of muscle myosin with chymotrypsin, subfragment-1 isozymes with either A-1 or A-2 light chains can be separated very cleanly on the actin-tropomyosin columns using the 5 m*M* triethanolamine, 2 m*M* $MgCl_2$ buffer described earlier. Only subfragment-1 with the A-1 light chain binds to the column (Winstanley *et al.*, 1977).

It has not been possible to find a concentration of KCl or ammonium acetate where muscle myosin is soluble and will bind to immobilized actin columns (Bottomley and Trayer, 1975). For cytoplasmic myosins, it would be worth trying sucrose buffers to solubilize the protein at a low ionic strength where it might bind to an immoblized actin column.

Actin filaments stabilized with phalloidin can be trapped in agarose beads for use in affinity chromatography by the method of Grantmont-Leblanc and Gruda (1977):

Procedure:

1. Wash Sepharose 4B with 0.2 m*M* ATP, 0.5 m*M* β-mercaptoethanol, 0.2 m*M* $CaCl_2$, 2 m*M* Tris (pH 8).

2. Mix 3–4 mg of actin monomer in the preceding buffer with each milliliter of packed Sepharose. (Although not specified in the original article, I would use a solution of actin monomer with a high concentration, 8–10 mg/ml.)

3. Add KCl, $MgCl_2$, and phalloidin to bring the concentrations to 0.1 *M* KCl, 1 m*M* $MgCl_2$, and 100 μM phalloidin and allow the actin to polymerize for 90 min at room temperature.

4. Wash the Sepharose with enough 0.1 *M* KCl, 1 m*M* $MgCl_2$, 0.5 m*M* β-mercaptoethanol, 20 m*M* Tris (pH 8) to bring the A_{280} of the filtrate of zero. The concentration of trapped actin is determined by pronase digestion and protein determination. The content of trapped actin is 0.1–0.3 mg/ml packed gel and is capable of binding 1.3–2.2 mg of muscle myosin per milligram of actin. The myosin binding capacity declines by about 50% in 7 weeks.

The trapped actin filament columns can be used for affinity chromatography of muscle myosin, heavy meromyosin, and subfragment-1. The method for myosin from a crude extract follows:

Procedure:

1. Equilibrate a 20-ml trapped actin column with 0.5 *M* KCl, 1 m*M* $MgCl_2$, 0.5 m*M* β-mercaptoethanol, 20 m*M* Tris (pH 7.6) at 4°C.
2. Apply a sample consisting of about 20 mg of protein extracted from muscle with a buffer similar to the column buffer.
3. Wash with column buffer until A_{280} is <0.05. Essentially all the extract proteins except myosin fail to bind to the column.
4. Elute the bound myosin with 1 m*M* sodium pyrophosphate in column buffer. The resulting myosin is very clean and has a high ATPase activity.

A similar procedure can be used for the purification of crude muscle myosin subfragment-1. This remarkable affinity column has an advantage over the other types of immobilized actin. It alone seems to be capable of binding muscle myosin in a high concentration of KCl. Consequently, it should be quite useful for purification of cytoplasmic myosins as well. A disadvantage is its low capacity.

5. Antibody Columns

In most cases conventional methods can be used to purify myosin, but there are difficult cases such as *Acanthamoeba* myosin-I where immunological techniques may be useful. For example, antibodies specific for either *Acanthamoeba* myosin-I or myosin-II can be coupled to agarose beads and used for affinity chromatography (T. D. Pollard, unpublished observations).

Procedure:

1. Homogenize *Acanthamoeba* in 4 vol of ice-cold 0.2 *M* NaCl, 20 m*M* sodium pyrophosphate (pH 7), 1 m*M* ATP, 10 μg/ml benzamidine (a protease inhibitor) and clarify at 120,000 *g* for 60 min at 4°C.
2. Mix 4 ml of antibody-Sepharose equilibrated with extracting buffer with 2 ml of the extract supernatant overnight at 4°C.
3. Pour into a 1.5-cm-diameter column and wash with 10 ml extracting buffer at about 1 ml/min.
4. Elute with 8 ml 1 *M* NaCl.
5. Elute with 8 ml 2.5 *M* $MgCl_2$.
6. Elute with 8 ml 4 *M* $MgCl_2$.
7. Elute with 8 ml 7 *M* urea.
8. Wash the column with extracting buffer.

In the case of the anti-myosin-II column, the myosin-II eluted in both the 2.5 and 4 *M* $MgCl_2$ fractions and none was found in the final urea wash (Fig. 9). With the anti-myosin-I column, a polypeptide with the electrophoretic mobility of myosin-IB eluted with 2.5 *M* $MgCl_2$ and another with the mobility of myosin-

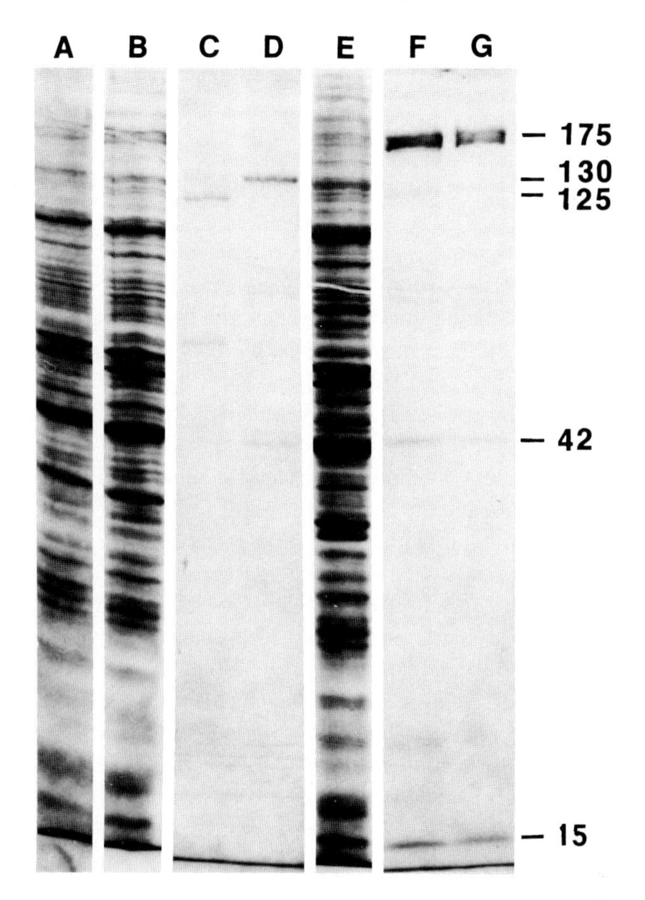

FIG. 9. Antimyosin affinity chromatography of *Acanthamoeba* myosin-I and myosin-II as described in the text. The polypeptide composition of various samples is shown by polyacrylamide slab gel electrophoresis by the method described in the text. (A) Crude extract. (B) Polypeptides that did not bind to an anti-myosin-I column. (C) Polypeptides eluted from the anti-myosin-I column with 2.5 M $MgCl_2$. (D) Polypeptides eluted from the anti-myosin-I column with 4 M $MgCl_2$. (E) Polypeptides that did not bind to an anti-myosin-II column. (F) Polypeptides eluted from the anti-myosin-II column with 2.5 M $MgCl_2$. (G) Polypeptides eluted from the anti-myosin-II column with 4 M $MgCl_2$. The molecular weight of the major polypeptides is indicated on the right in thousands.

IA eluted with 4 M $MgCl_2$ (Fig. 9). Thus it would appear that >50 purification of the three major myosins from *Acanthamoeba* is now possible with a simple, single-step procedure. The major limitation is the small capacity of these columns, but this will be overcome by using monoclonal antibodies.

Immobilized antibodies to myosin light chains have also been used by S. Lowey and her colleagues (personal communication) for the purification of a muscle myosin subfragment-1 and heavy meromyosin isozymes.

VI. Concentration

There are a variety of mild methods to concentrate myosin solutions. The two I favor are precipitation with ammonium sulfate and pelleting of myosin filaments, because they are mild and the recovery is excellent. Dialysis against 1.5 vol of ammonium sulfate with EDTA will remove some water from the sample and bring the final concentration of ammonium sulfate to 2.3 *M* (60% saturation), a concentration sufficient to precipitate all myosins that I know of. The precipitate is collected by centrifugation and resuspended gently in an appropriate buffer. Filaments can be formed from most myosins by dialysis against 50 m*M* KCl, 2 m*M* $MgCl_2$, 10 m*M* imidazole (pH 6.5) and then collected by centrifugation. Since some cytoplasmic myosin filaments are small, ultracentrifugation may be necessary.

Another mild, efficient method of concentration is binding to one of the affinity columns described in the previous section, followed by step elution. This procedure has the additional feature that any denatured myosin will not bind to the column and will be separated from the concentrated, active myosin.

Myosin can also be concentrated by various membrane filtration methods. I have had better success with osmotic pressure than with hydrostatic pressure methods. For example, excellent recovery is possible with "dialysis" against dry Sephadex G-150 or polyethylene glycol (Aquacide, Cal-Biochem). Dry sucrose (Maruta and Korn, 1977a) or 60% sucrose solution (Collins and Korn, 1980) can also be recommended for osmotic pressure dialysis. In contrast, I have had more difficulty recovering myosin concentrated by vacuum dialysis in cellulose dialysis sacks or by pressure ultrafiltration on several different Amicon filters. It is possible that the new generation of Amicon filters or similar products from other manufacturers will give better results.

Finally, it is possible to lyophilize muscle myosin subfragment-1 and heavy meromyosin after freezing with an equal quantity of sucrose (Yount and Koshland, 1963). Upon rehydration in a smaller volume of water, the protein will be concentrated. I am not aware that any intact myosin has been successfully lypohilized and rehydrated, but my experience in this matter is limited.

VII. Storage

Each myosin has different stability, but many (including muscle myosin, platelet myosin, and *Acanthamoeba* myosin-II) can be stored at concentrations of >1 mg/ml in 0.6 *M* KCl, 1 m*M* DTT, 10 m*M* imidazole (pH 7) for 5–10 days at 0°C with only slight loss of activity. The inclusion of 0.02% sodium azide in the buffer will prevent bacterial growth and does not, as far as I know, alter the myosin.

For longer-term storage, any of the following methods can be used:

1. Precipitate with 60% ammonium sulfate, 10 mM EDTA (see Section IV,B) and store the pellets under the ammonium sulfate supernate at 4°C for periods of months.

2. Mix the myosin solution with an equal volume of glycerol and store the solution at −20°C for months (Szent-Györgyi, 1951).

3. Dialyze the myosin against 60% sucrose and store at −20°C.

I would not recommend such long-term storage for myosin to be used for the most critical enzymatic analysis, but for routine applications these methods preserve the activity of myosins remarkably well.

If it is possible to lyophilize a particular myosin, this is the ideal way to preserve the protein.

VIII. Detailed Methods to Purify Specific Myosins

In this section I will give step-by-step procedures for purifying myosin from skeletal muscle and one nonmuscle source, human platelets. I have also included procedures for making some of the proteolytic fragments of muscle myosin. These protocols will illustrate how the methods reviewed in the preceding sections can be put together in specific cases. None of these methods is very difficult and anyone with some biochemical experience should be able to carry them out.

A. Skeletal Muscle Myosin*

1. Exsanguinate one rabbit by cutting the jugular vein (or two chickens by decapitation).

2. Quickly skin the rabbit and dissect out the back and hind leg muscles, placing them on ice. With the chicken dissect out the breast muscles. All subsequent operations are done at 0–4°C.

3. Grind the chilled muscle twice with a cold meat grinder and weigh. The example given here is for 300 g of muscle.

4. Mix the ground muscle with 3 vol of cold 0.5 M KCl, 0.1 M K_2HPO_4 and stir at 0–4°C for exactly 10 min.

5. Pellet the muscle residue in a Sorvall GSA or Beckman JA-10 rotor at top speed (~17,000 g) for 15 min. The supernatant contains the myosin. The pellets can be used for making an actin acetone powder.

*Modified from Kielley and Bralley, 1956.

6. Adjust the pH of the supernatant to 6.6 with 0.5 *M* acetic acid. This takes about 40 ml and must be done *very carefully* with adequate stirring to avoid local acid precipitation.

7. Dilute the supernatant slowly, with stirring, with 10 vol of cold distilled water to precipitate myosin along with any extracted actin. Here and everywhere else in this procedure be very careful to avoid bubbles, because myosin will precipitate at air–water interfaces. Check the pH and readjust to 6.6 if necessary.

8. Spin down the precipitate in a large-capacity centrifuge rotor (such as a JA-10) at 9000 rpm for 15 min. Discard the supernatant.

9. Carefully resuspend the pellet in 40 ml of 2 *M* KCl using a glass stirring rod. Transfer to a 1-liter graduated cylinder, measure the total volume, and calculate the volume of the pellet. Add enough 2 *M* KCl to make the solution 0.5 *M* KCl. The pellet is about 0.045 *M* KCl, so the volume of KCl to be added is 0.303 × volume of the pellet.

10. Check the pH of the solution and adjust to 6.7–6.8 if necessary using 1 *M* $NaHCO_3$.

11. *Carefully* dilute the solution to exactly 0.28 *M* KCl with cold distilled water using constant stirring. This will precipitate any actomyosin present.

12. Centrifuge down any precipitate in a JA-10 rotor at 9000 rpm for 15 min and save the supernatant. The pellet of ''actomyosin'' may actually be reasonably pure myosin and suitable for some purposes.

13. Measure the volume of the supernatant and dilute carefully to 0.04 *M* KCl using cold distilled water and constant stirring to precipitate the myosin.

14. Centrifuge down the myosin precipitate and discard the supernatant.

15. Dissolve the myosin pellet in 2 *M* KCl to give a concentration of approximately 0.5 *M* KCl as earlier and dilute to 450 ml with 0.5 *M* KCl.

16. Slowly add saturated neutralized ammonium sulfate with 10 m*M* EDTA (described earlier) to 40% saturation with constant stirring to get adequate mixing. The volume of ammonium sulfate added will be two-thirds the volume of the myosin solution.

17. After 15 min centrifuge down and save the supernatant. The precipitate is usually discarded, but as in step 12 this myosin may be useful for some purposes.

18. Add ammonium sulfate to the supernatant with constant stirring to bring the concentration to 50% saturation. The volume of ammonium sulfate to be added will be one-fifth the volume of the myosin solution.

19. After 15 min centrifuge down the precipitated myosin. This pellet may be stored at 4°C for months.

20. The precipitated myosin can be solubilized by placing the pellet in a small amount of 0.5 *M* KCl, 10 m*M* imidazole, (pH 7) and gently teasing it with a glass rod. Do not rush or traumatize the pellet. It will dissolve nicely if given time.

21. Dialyze the myosin solution against a suitable buffer (0.5 *M* KCl, 10 m*M*

EDTA, 10 m*M* imidazole pH 7) and clarify by centrifuging at 20,000 *g* for 30 min.

This procedure yields 1–2 g of myosin of high purity. If it is necessary to remove minor contaminants such as C-protein, ion exchange chromatography on DEAE-Sephadex A50 (Offer *et al.*, 1973) can be used. Although not routinely employed, one of the affinity chromatography methods (such as an ADP-agarose column) could be used to remove any denatured myosin.

B. Proteolytic Fragments of Muscle Myosin

1. Heavy Meromyosin

Heavy meromyosin is a two-headed fragment of myosin. It has a short tail but is soluble at low ionic strength.

Procedure of Lowey and Cohen (1962):

a. Dialyze muscle myosin at a concentration of 14 mg/ml against 0.5 *M* KCl, 0.1 *M* phosphate buffer (pH 7) at 4°C. Just before digestion, warm to room temperature.

b. Mix 10 vol of myosin with 1 vol of trypsin (0.5 mg/ml in 1 m*M* HCl, Worthington Biochemicals, twice recrystallized) and incubate with stirring for 4–5 min at room temperature. The digestion time can be varied, to increase the yield at the expense internal of cleavages of the myosin polypeptides.

c. Terminate the digestion with 1 vol of soybean trypsin inhibitor (1 mg/ml in water adjusted to pH 7.4 with NaOH, Worthington Biochemicals, crystallized from ethanol).

d. Precipitate light meromyosin (the distal part of the tail) and undigested myosin by dialyzing the reaction mixture against 7 m*M* phosphate or imidazole buffer (pH 7) at 4°C for several hours.

e. Pellet the precipitate by centrifugation at 100,000 *g* for 1 hr. The supernatant contains mostly heavy meromyosin.

f. For further purification, collect the 1.5–2.1 *M* ammonium sulfate fraction (40–55% saturated).

g. Dialyze the pelleted heavy meromyosin against 10 m*M* imidazole, 1 m*M* DTT (pH 7).

This conventional heavy meromyosin can be purified further in several ways. Perhaps the best is affinity chromatography on Sepharose-aminohexylpyrophosphate or Sepharose-ADP. Both remove any contaminating proteins as well as any denatured heavy meromyosin (see Section V, E). Other options are gel filtration (Section V, A) and anion exchange chromatography (Section V, B).

By using a low concentration of trypsin treated with L-(1-tosylamido-2-phenyl)-

ethyl chloromethyl ketone, it is possible to make heavy meromyosin with few or no proteolytic nicks in the heavy chain (Woodrum *et al.*, 1975).

2. Subfragment-1

Subfragment-1 is the single-headed fragment of myosin that possesses the catalytic site for ATP hydrolysis and the actin binding site. There are three commonly used methods to prepare muscle myosin subfragment-1: digestion with papain in the presence of $MgCl_2$ or EDTA or with α-chymotrypsin in EDTA. Each method yields slightly different subfragment-1. Papain-Mg subfragment-1 has all of its light chains. Papain-EDTA subfragment-1 is missing part of its nonessential DTNB light chain. α-Chymotryptic subfragment-1 has no DTNB light chain. There are also differences in the number of nicks in the heavy chains. Remarkably, all of these preparations have similar ATPase activities (Margossian and Lowey, 1978).

Procedure for papain subfragment-1 of Margossian and Lowey (1978):

1. Dialyze myosin at a concentration of 20 mg/ml against 0.2 *M* ammonium acetate (pH 7.2) with either 2 m*M* $MgCl_2$ or 2 m*M* EDTA at 4°C to precipitate the myosin in the form of thick filaments.
2. Warm the myosin filament suspension to room temperature and add papain (Worthington Biochemicals) to a concentration 30 μg/ml.
3. Stir for 7 min at room temperature.
4. Add 0.01 ml of 100 m*M* iodoacetic acid per milliliter to stop the reaction, and cool the protein to 0–4°C.
5. Centrifuge down the insoluble material, leaving subfragment-1 at about 10 mg/ml in the supernatant.
6. If further purification is desired, collect the protein precipitating between 1.6 and 2.2 *M* ammonium sulfate and dissolve the pellet in 0.1 *M* KCl, 1 m*M* dithiothreitol, 10 m*M* imidazole (pH 7). Dialyze the subfragment-1 against the same buffer to remove the ammonium sulfate.

Procedure for chymotrypic subfragment-1 of Weeds and Taylor (1975):

1. Dialyze myosin at a concentration of 15–20 mg/ml against 0.12 *M* NaCl, 1 m*M* EDTA, 20 m*M* sodium phosphate (pH 7) at 4°C to form myosin filaments.
2. Warm to room temperature and add α-chymotrypsin (Worthington Biochemicals) to a concentration of 50 μg/ml.
3. Stir for 7 min at room temperature.
4. Add 1 μl of fresh 100 m*M* phenylmethanesulfonyl fluoride (Sigma #P7626) in ethanol per milliliter of myosin to stop the reaction and cool the protein to 0°C. (A higher final concentration of phenylmethanesulfonyl fluoride, for example, 1 m*M* might be advisable because I. Herman has found traces of

active α-chymotrypsin in these preparations even after chromatography on DEAE-cellulose.)

5. Clarify and ammonium sulfate precipitate as earlier.

Subfragment-1 in the digestion supernatant can also be purified by chromatographic methods. Active papain subfragment-1 can be separated from denatured subfragment-1 on Sepharose-aminohexylpyrophosphate (Trayer *et al.*, 1974), ADP-agarose (Trayer *et al.*, 1974), or actin monomer–agarose (Bottomley and Trayer, 1975). Chymotryptic subfragment-1 with A-1 or A-2 light chains can be separated on DEAE (Weeds and Taylor, 1975), on ADP-agarose (Wagner, 1977; Winstanley *et al.*, 1979), or on immobilized actin (Winstanley *et al.*, 1979). Aggregates and the proteases can be removed by gel filtration on Sephacryl S-200 (Pharmacia Fine) in 0.5 *M* KCl, 10 m*M* EDTA, 10 m*M* imidazole (pH 7). These additional purification steps are necessary for the analysis of light chain function and are a wise precaution if traces of active protease in the subfragment-1 might influence the results of a particular experiment.

C. Human Platelet Myosin

This procedure with modifications has been used for purifying myosin from many nonmuscle cells.

Procedure of Pollard et al. *(1974):*

1. Mix washed frozen platelets with 2 ml of extracting solution (0.9 *M* KCl, 15 m*M* pyrophosphate (pH 7), 30 m*M* imidazole (pH 7), 5 m*M* $MgCl_2$, 3 m*M* dithiothreitol) per gram of platelets and stir on ice for 30 min. The platelets will lyse during thawing.
2. Centrifuge in a Beckman JA-20 or Sorvall SS-35 rotor at 16,000 rpm for 30 min at 4°C. Save the supernatant.
3. The pellet can be reextracted with 2 vol of extracting solution (diluted with 0.5 part water) to increase the yield.
4. Take the supernatant(s), measure the volume, and carefully add 3 vol of ice-cold 2 m*M* $MgCl_2$ with stirring. Then adjust the pH to 6.4 with 0.5 *M* acetic acid being careful not to overshoot (the pH meter is sluggish in the cold). Stir for 15 min.
5. Pellet the actomyosin precipitate by centrifuging in a J-20 or SS-35 rotor at 4°C for 30 min at 16,000 rpm.
6. Discard supernatant and save the pellet, which is *crude actomyosin*.
7. Gently dissolve the crude actomyosin in a small volume of KI-ATP buffer (0.6 *M* KI, 5 m*M* ATP, 5 m*M* dithiothreitol, 1 m*M* $MgCl_2$, 20 m*M* imidazole, pH 7) using a Dounce homogenizer.
8. Centrifuge out undissolved material at 16,000 rpm for 10 min at 4°C.

9. For ammonium sulfate fractionation, use cold saturated ammonium sulfate with 0.01 *M* EDTA, pH 7. Add ammonium sulfate slowly with stirring. After 10 min spin down the precipitate in a J-20 rotor at 16,000 rpm for 15 min at 4°C. Then measure the volume of the supernatant and use it for the next step.

	milliliter ammonium sulfate per liter
0–1.0 *M* ammonium sulfate	360
1.0–1.9 *M* ammonium sulfate	478 (enriched in myosin)
1.9–2.5 *M* ammonium sulfate	466 (enriched in myosin rod)

10. Dissolve precipitates in a small volume (~12 ml for a 450 ml agarose column) of cold KI-ATP buffer.

11. Clarify by centrifugation at 4°C for 30 min in a J-20 rotor at 16,000 rpm or in an ultracentrifuge at 100,000 *g*.

12. Apply either the myosin or myosin-rod-enriched fraction to a 4% agarose column behind a zone of KI-ATP buffer as described in Section V,A. The myosin will elute from the column with a partition coefficient of about 0.2–0.3.

This method with modifications will purify myosin from smooth muscle and many nonmuscle cells. The following are examples of modifications that improve results with specific tissues:

(a) For squid brain, extract with 0.6 *M* KCl rather than 0.9 *M* KCl to reduce lipid extraction (See and Metuzals, 1976).

(b) For macrophages, extract with a 0.34 *M* sucrose buffer rather than KCl to avoid rupture of lysosomes (Stossel and Pollard, 1973).

(c) For *Dictyostelium,* extract with 30% sucrose buffer and precipitate actomyosin by dialyzing out the sucrose at low ionic strength (Clarke and Spudich, 1974).

(d) For thyroid, extract with 0.3 *M* ammonium acetate buffer and substitute Blue-dextran Sepharose chromatography for actomyosin precipitation to remove thyroglobulin (Kobayashi *et al.*, 1977).

(e) For *Acanthamoeba* myosin-II extract with a 0.34 *M* sucrose buffer, fractionate on a DEAE-cellulose column (as described in the section on anion exchange chromatography), and precipitate actomyosin-II by hexokinase digestion of the ATP in the pooled actin and myosin fractions (Pollard *et al.*, 1978).

IX. Perspectives

The characterization of muscle myosin has now become quite sophisticated. The amino acid sequence will be completed in the near future by both direct methods (e.g., Elzinga *et al.*, 1980) and probably also by DNA-sequencing methods. The gross structure of myosin is known from electron microscopy (Lowey *et al.*, 1969) and a more detailed picture of the head is being worked out by image reconstruction of actin filaments decorated with subfragment-1 (Seymour and O'Brien, 1979). The mechanism of ATP hydrolysis has been characterized by extensive transient-state kinetic studies (reviewed in Taylor, 1979). In addition to pursuing these biochemical experiments in greater detail, it seems to me that substantial further progress will require the three-dimensional structure of the myosin head. Consequently, the formation of two-dimensional crystals for electron microscopic crystallography or three-dimensional crystals for X-ray crystallography would seem to have a very high priority. The improved purification methods reviewed here may be essential for making the homogenous subfragment-1 required for successful crystallization.

In contrast to muscle myosin, relatively little is known about the primary structure or catalytic mechanism of any cytoplasmic myosin (reviewed in Korn, 1978). In these ways our understanding of cytoplasmic myosins is likely to continue to lag behind muscle myosin, if for no other reason than that with a given effort and expense it is possible to purify more than 100 times as much muscle as cytoplasmic myosin. As discouraging as this might seem, it probably will not impede our understanding of cell motility. The reason is that the force-generating mechanisms of all myosins are likely to be similar on the submolecular level. Therefore, what we learn about such things in muscle can, most probably, be applied to nonmuscle systems.

On the other hand, there are major differences between muscle and nonmuscle cells in the ways in which myosin is organized and regulated. These are the issues that need to be investigated to learn how the universal actomyosin tension-generating system is used in various specialized ways to generate the wide spectrum of cellular movements. In the first place, decisive experiments like those in which Mabuchi and Okuno (1977) inhibited cytokinesis by injecting myosin antibodies are needed to demonstrate whether myosin is required for other cellular movements. Then much more biochemical and morphological information will be needed to understand how cells control the interaction of myosin with actin filaments, the assembly of myosin filaments, and the movement of myosin from one place to another within the cytoplasm during the cell cycle (Fujiwara and Pollard, 1976). One very promising lead is the growing evidence that reversible phosphorylation of myosin light chains (reviewed in Adelstein, 1980) and heavy chains (Maruta and Korn, 1977b; Collins and Korn, 1980; Kuczmarski and Spudich, 1980) can regulate actin–myosin interaction and perhaps even myosin filament assembly (Scholey *et al.*, 1980). The quality and quantity of cytoplasmic myosin that can be isolated from nonmuscle cells by the methods described here is probably adequate for these studies.

References

Adelman, M. R., and Taylor, E. W. (1969). *Biochemistry* **8,** 4976–4988.
Adelstein, R. S. (1980). *Fed. Proc., Fed. Am. Soc. Exp. Biol.* **39,** 1544–1546.
Adelstein, R. S., and Conti, M. A. (1975). *Nature (London)* **256,** 597–598.
Adelstein, R. S., Pollard, T. D., and Kuehl, W. M. (1971). *Proc. Natl. Acad. Sci. U.S.A.* **68,** 2703–2707.
Adelstein, R. S., Conti, M. A., Johnson, G., Pastan, I., and Pollard, T. D. (1972). *Proc. Natl. Acad. Sci. U.S.A.* **69,** 3693–3697.
Bernardi, G. (1971). *In* "Methods in Enzymology" (W. B. Jakoby, ed.), Vol. 22, pp. 325–339. Academic Press, New York.
Bottomley, R. C., and Trayer, I. P. (1975). *Biochem. J.* **149,** 365–379.
Burridge, K., and Bray, D. (1975). *J. Mol. Biol.* **99,** 1–14.
Clarke, M., and Spudich, J. A. (1974). *J. Mol. Biol.* **86,** 209–222.
Collins, J. H., and Korn, E. D. (1980). *J. Biol. Chem.* **255,** 8011–8014.
Dabrowska, R., Aromatorio, D., Sherry, J. M. F., and Hartshorne, D. J. (1977). *Biochem. Biophys. Res. Commun.* **78,** 1263–1272.
Elzinga, M., Behar, K., and Walton, G. (1980). *Fed. Proc., Fed. Am. Soc. Exp. Biol.* **39,** 2168.
Fairbanks, G., Steck, T. L., and Wallach, D. F. H. (1971). *Biochemistry* **10,** 2606–2617.
Fujiwara, K., and Pollard, T. D. (1976). *J. Cell Biol.* **71,** 848–875.
Gibson, W. (1974). *Virology* **62,** 319–336.
Gordon, D. J., Eisenberg, E., and Korn, E. D. (1976). *J. Biol. Chem.* **251,** 4778–4786.
Grantmont-Leblanc, A., and Gruda, J. (1977). *Can J. Biochem.* **55,** 949–957.
Kielley, W. W., and Bradley, L. (1956). *J. Biol. Chem.* **218,** 653–659.
Kobayashi, R., Goldman, R., Hartshorne, D., and Field, J. (1977). *J. Biol. Chem.* **252,** 8285–8291.
Korn, E. D. (1978). *Proc. Natl. Acad. Sci. U.S.A.* **75,** 588–599.
Kuczmarski, E. R., and Spudich, J. A. (1980). *J. Cell Biol.* **87,** 227a.
Laemmli, U. K. (1970). *Nature (London) New Biol.* **227,** 680–685.
Lamed, R., and Oplatka, A. (1974). *Biochemistry* **13,** 3137–3142.
Lamed, R., Levin, Y., and Oplatka, A. (1973). *Biochim. Biophys. Acta* **305,** 163–171.
Lowey, S., and Cohen, C. (1962). *J. Mol. Biol.* **4,** 293–308.
Lowey, S., Slayter, H. S., Weeds, A., and Baker, H. (1969). *J. Mol. Biol.* **42,** 1–29.
Mabuchi, I., and Okuno, M. (1977). *J. Cell Biol.* **74,** 251–263.
Margossian, S. S., and Lowey, S. (1978). *Biochemistry* **17,** 5431–5439.
Martin, J. B., and Doty, D. M. (1949). *Anal. Chem.* **21,** 965–967.
Maruta, H., and Korn, E. D. (1977a). *J. Biol. Chem.* **252,** 6501–6509.
Maruta, H., and Korn, E. D. (1977b). *J. Biol. Chem.* **252,** 8329–8332.
Maruta, H., Gadasi, H., Collins, J. H., and Korn, E. D. (1979). *J. Biol. Chem.* **254,** 3624–3630.
Moos, C. (1973). *Cold Spring Harbor Symp. Quant. Biol.* **37,** 137–143.
Nachmias, V. T. (1974). *J. Cell Biol.* **62,** 54–66.
Offer, G., Moos, C., and Starr, R. (1973). *J. Mol. Biol.* **74,** 653–676.
Ostlund, R. E., and Pastan, I. (1976). *Biochim. Biophys. Acta* **453,** 37–47.
Peterson, E. A. (1970). *In* "Laboratory Techniques in Biochemistry and Molecular Biology" (T. S. Work and E. Work, eds.), pp. 225–400. North-Holland Publ., Amsterdam.
Pollard, T. D., and Korn, E. D. (1973a). *J. Biol. Chem.* **248,** 4682–4690.
Pollard, T. D., and Korn, E. D. (1973b). *J. Biol. Chem.* **248,** 4691–4697.
Pollard, T. D., Thomas, S. M., and Niederman, R. (1974). *Anal. Biochem.* **60,** 258–266.
Pollard, T. D., Stafford, W. F., and Porter, M. E. (1978). *J. Biol. Chem.* **253,** 4798–4808.
Potter, J. D. (1974). *Arch. Biochem. Biophys.* **162,** 436–441.
Puszkin, S., and Berl, S. (1972). *Biochim. Biophys. Acta* **256,** 695–709.

Richards, E. G., Chung, C. S., Menzel, D. B., and Olcott, H. S. (1967). *Biochemistry* **6,** 528–540.
Scholey, J. M., Taylor, K. A., and Kendrick-Jones, J. (1980). *Nature* (*London*) **287,** 233–235.
See, Y. P., and Metuzals, J. (1976). *J. Biol. Chem.* **251,** 7682–7689.
Seymour, J., and O'Brien, E. J. (1979). *Nature* (*London*) **283,** 680–682.
Sobieszek, A., and Bremel, R. D. (1975). *Eur. J. Biochem.* **55,** 49–60.
Solaro, R. J., Pang, D. C., and Briggs, F. N. (1971). *Biochim. Biophys. Acta* **245,** 259–262.
Stephens, R. E. (1975). *Anal. Biochem.* **65,** 396–379.
Stossel, T. P., and Pollard, T. D. (1973). *J. Biol. Chem.* **248,** 8288–8294.
Szent-Györgyi, A. (1951). "The Chemistry of Muscle Contraction." Academic Press, New York.
Taylor, E. W. (1979). *CRC Crit. Rev. Biochem.* **6,** 103–164.
Toste, A. P., and Cooke, R. (1978). *Anal. Biochem.* **95,** 317–328.
Trayer, H. R., and Trayer, I. P. (1975). *FEBS Lett.* **54,** 291–296.
Trayer, H. R., Winstanley, M. A., and Trayer, I. P. (1977). *FEBS Lett.* **83,** 141–144.
Trayer, I. P., Trayer, H. R., Small, D. A., and Bottomley, R. C. (1974). *Biochem. J.* **139,** 609–623.
Wagner, P. D. (1977). *FEBS Lett.* **81,** 81–85.
Weeds, A. G., and Taylor, R. S. (1975). *Nature* (*London*) **257,** 54–56.
Winstanley, M. A., Trayer, H. R., and Trayer, I. P. (1977). *FEBS Lett.* **77,** 239–242.
Winstanley, M. A., Small, D. A. P., and Trayer, I. P. (1979). *Eur. J. Biochem.* **98,** 441–446.
Woodrum, D. T., Rich, S. A., and Pollard, T. D. (1975). *J. Cell Biol.* **67,** 231–237.
Yount, R. G., and Koshland, D. E., Jr. (1963). *J. Biol. Chem.* **238,** 1708.

Chapter 23

Preparation and Purification of Dynein

CHRISTOPHER W. BELL, CLARENCE FRASER, WINFIELD S. SALE, WEN-JING Y. TANG, AND I. R. GIBBONS

Pacific Biomedical Research Center
University of Hawaii
Honolulu, Hawaii

I. Introduction

On the basis of present knowledge, it appears that the great majority of motile processes in eukaryotes is caused by the action of one of two macromolecular

ISBN 0-12-564124-9

systems: the actomyosin system or the tubulin-dynein system. The systems appear to be generally similar inasmuch as, in both cases, the energy stored in the terminal phosphate of ATP is released by one protein (myosin or dynein) and utilized to perform work on a structural framework constructed of the other protein (actin or tubulin). This is about as far as the comparison can, with any degree of confidence, be taken at present, since although the structure and function of actomyosin are relatively well understood both as a complete system and as separated components, the same cannot be said of the tubulin–dynein system. This is partly because of the more recent identification of the proteins involved in ciliary and flagellar motility, and partly because of the relatively small quantities of material available for study. The latter problem is most acute with dynein, since tubulin comprises a large mass percentage of the proteins of cilia and flagella and is also available in larger quantities from other sources such as brain in an at least partly compatible form.

Dynein was first identified in ciliary axonemes of *Tetrahymena* as a high-molecular-weight protein with Mg-ATPase activity, which could be extracted from the axonemes by exposure to low ionic strength in the presence of EDTA (Gibbons, 1963). Selective extraction and recombination followed by electron microscopy indicated that dynein comprised part or all of the arms bridging the gap between the doublet microtubules of the axonemes (Gibbons, 1965a). These critiera—high molecular weight (sedimentation velocity up to 30S), possession of ATPase* activity, extractability with low-ionic-strength solutions in the presence of EDTA and relationship to the arms on doublet tubules—became the identifying characteristics of dynein from other sources. The introduction of polyacrylamide gel electrophoresis in the presence of sodium dodecyl sulfate led to the recognition that the polypeptide subunits of dynein were very large: Estimates of apparent molecular weight have ranged from 300,000 to 500,000 (e.g., Linck, 1973b; Burns and Pollard, 1974; Borisy *et al.*, 1975). Further investigation of dynein from a wide variety of sources has revealed multiple isoenzymic forms of the enzyme that are distinguishable by the different electrophoretic mobilities of their high-molecular-weight subunits. Extraction of axonemes with high concentrations of salt has also been found to release dynein, but whereas both the inner and outer arms of flagellar axonemes from *Chlamydomonas* are extractable in this manner (Piperno and Luck, 1979), only the outer arms of, for instance, see urchin sperm flagella are extracted (Gibbons and Fronk, 1972). These indications of both inter- and intraspecies differences in dynein are reinforced by reports of different forms of dynein with sedimentation coefficients varying from 10S to 30S, depending on the species and the conditions of sedimentation. Furthermore, the increasing resolution of sodium dodecyl sulfate–polyacrylamide gel electrophoresis has shown that the number of distinct bands in the region of the high-molecular-weight subunits is both large and

*Unless stated otherwise, "ATPase" is to be understood to mean "Mg-ATPase."

variable from species to species. At present the largest number of distinct high-molecular-weight chains identified in a single species is 10 from the flagella of *Chlamydomonas* (Piperno and Luck, 1979). At least some of the various high-molecular-weight chains in flagella of a given species have been shown to derive from multiple isoenzymic forms of dynein (Gibbons *et al.*, 1976; Ogawa and Gibbons, 1976). Further complexity has arisen in the recognition, in both *Chlamydomonas* and sea urchin sperm flagella, that two or more distinct ATPase may be associated with a single arm (Huang *et al.*, 1979; Tang *et al.*, 1982). It also appears that the dynein–tubulin system is not confined to cilia and flagella. There have been several reports of dynein-like ATPase activity in sea urchin egg cytoplasm (Miki, 1963; Weisenberg and Taylor, 1968; Mabuchi, 1973; Pratt *et al.*, 1980), and recently Dentler *et al.* (1980) demonstrated the presence of a dyneinlike ATPase in the ciliary membrane. These observations suggest that dynein may have a widespread role in cellular motility.

This structural and functional complexity of dynein serves to highlight the differences between this and the actomyosin system and to emphasize the difficulties inherent in obtaining a full understanding of flagellar and ciliary motility. Such understanding will depend on comprehensive enzymatic and physicochemical characterization of dynein, a process requiring large quantities of a homogeneous enzyme. Much of the remainder of this article describes in detail the preparation, storage, use, and properties of such a dynein ATPase from sea urchin sperm flagella, which is probably one of the best available sources for providing relatively large quantities of this enzyme. The methods described here relate specifically to our experience in preparing latent activity dynein-1 (LAD-1) from sperm flagella of the Hawaiian sea urchin *Tripneustes gratilla*. However, the general techniques appear to be applicable to sperm flagella of many other animals. The final section presents a brief comparative review of the methods that have been used for the preparation of dynein from various other sources and lists some of the salient properties of these dynein ATPases.

II. Preparation of Latent Activity Dynein-1

A. Collection and Storage of Sea Urchins

Sea urchins (*Tripneustes gratilla*) are collected weekly from sandy reef floor areas at depths of 3–5 m. They are returned to the laboratory in dry (i.e., not water-filled) buckets in order to prevent the widespread triggering of gamete shedding that occurs when one or more sea urchins shed into water in close contact with others; they are then stored in running seawater at ocean temperature. Ripe *Tripneustes* can be sexed by shaking them to and fro with the gonopore pointing down, which causes small amounts of gametes to appear at the gonopore: Eggs appear dark orange in such situations, whereas semen is off-white.

After carefully rinsing off the gonopores, males and females may be returned to separate tanks. Complete shedding of semen from male *Tripneustes* can be induced by injecting 0.5 *M* KCl into the body cavity. A speedier procedure is to remove the Aristotle lantern by cutting around the peristomatous membrane, empty out the body fluid, and fill the body cavity with 0.5 *M* KCl. In either case, the semen is collected by inverting the animal over a 50-ml beaker filled with seawater containing 0.1 m*M* EDTA. A single healthy male *Tripneustes* yields up to 40 ml of dense semen, although the average is 15–20 ml. If desired, semen may be collected in artificial seawater lacking certain constituents (e.g., Ca^{2+}). It may also be collected dry, without any overlying solution, but we have not found this to be of any particular advantage in the preparation of dynein.

Sea urchins may also be induced to shed gametes by electrical stimulation (Harvey, 1956; Osanai, 1975). Unlike injection with KCl, this method does not kill the animal and it may be returned to the storage tank, and with feeding it will become ripe again within a few weeks.

B. Preparation of Sperm Flagellar Axonemes

An important step in the preparation of LAD-1 from the sperm of *Tripneustes* is to isolate flagellar axonemes that are essentially free both of sperm heads and of surrounding membranes that would otherwise prevent direct access to the proteins of the axoneme. The removal of the heads is relatively straightforward and is described in detail later. The removal of membranes from flagellar and ciliary axonemes is a problem that has been approached in two ways. Membranes have been either solubilized by detergents, such as digitonin (Gibbons, 1963) or Triton X-100 (Gibbons *et al.*, 1970), or osmotically ruptured and fragmented by exposure of the intact organelle to high concentrations of glycerol or sucrose (Raff and Blum, 1969). Solubilization by detergents has the advantage of disrupting membranes on the molecular level and is generally more likely to effect their complete removal. However, it is often difficult to remove the detergent completely after membrane solubilization, even with extensive washing, since detergent molecules tend to bind to the protein structures of the axoneme.

We formerly used a method of axoneme preparation in which the membranes were solubilized by treatment with 1% w/v Triton X-100 (Gibbons and Fronk, 1972, 1979), but we have since found that LAD-1 prepared by this method sometimes has a partially activated ATPase level, variable from preparation to preparation and similar to but not as pronounced as that obtained when LAD-1 is incubated for 10–15 min at room temperature in the presence of 0.05% Triton X-100. We surmise that this raised level of ATPase activity is a side effect of the Triton in the membrane solubilization step. It has been known for some time

(Gibbons and Fronk, 1979) that incubation of soluble LAD-1 with Triton X-100 causes profound changes in the physical and chemical properties of the enzyme, not the least of which is a roughly 10-fold increase in ATPase activity over that of the latent form (see "Properties of LAD-1"). It is our present opinion that damage to LAD-1 during preparation is manifested by, among other things, increased basal ATPase activity and a consequent decrease in activation ratio (i.e., ratio of ATPase activity after exposure to 0.05% Triton X-100 to basal ATPase activity). By these criteria, solubilization of flagellar membranes by Triton X-100 is not the method of choice (Table II), and we no longer use it in the preparation of dynein.

The method now used to remove sperm flagella membranes involves osmotic rupture by exposure to 20% w/v sucrose. The detailed procedure for the preparation of axonemes follows. Compositions of buffers are given in Table I.

About 60 ml of semen are collected and are diluted to 300 ml with EDTA (0.1 m*M*) seawater. The diluted semen is centrifuged for 5 min at 30 g to remove sand and debris that was expelled from the urchins along with the semen. The semen supernatant is carefully decanted and centrifuged again at 3000 *g* for 5–10 min in order to pellet the sperm. The pelleted sperm are resuspended in 300 ml of cold (4°C or less) 20% w/v sucrose in glass-distilled water and are homogenized in a Dounce homogenizer with eight strokes of a tight pestle. This homogenization should be quite vigorous in order to ensure the complete separation of sperm heads and flagellar axonemes and to effect fragmentation of flagellar membranes. Twenty percent w/v sucrose is employed as a compromise between the more efficient membrane fragmentation but longer centrifugation times engendered by higher sucrose concentrations, and the much less efficient membrane

TABLE I

COMPOSITION OF BUFFERS USED IN THE PREPARATION AND ASSAY OF CRUDE LAD-1

Isolation buffer	High-salt buffer	ATPase assay buffer
0.1 *M* NaCl	0.6 *M* NaCl	0.1 *M* NaCl
5 m*M* Imidazole[a]-HCl, pH 7	5 m*M* Imidazole-HCl, pH 7	0.03 *M* Tris[b]-HCl, pH 8.1
4 m*M* $MgSO_4$	4 m*M* $MgSO_4$	2 m*M* $MgSO_4$
1 m*M* $CaCl_2$[c]	1 m*M* $CaCl_2$	0.1 m*M* EDTA
1 m*M* EDTA	1 m*M* EDTA	1 m*M* ATP
7 m*M* 2-Mercaptoethanol	7 m*M* 2-mercaptoethanol	
	1 m*M* dithiothreitol	

[a] Imidazole is recrystallized from 1 m*M* EDTA in 80% v/v ethanol.

[b] Tris-base is recrystallized first from 1 m*M* EDTA and then from 80% v/v methanol.

[c] Ca^{2+} is added because it appears to result in tighter axonemal pellets during preparation. However, preparation in the absence of Ca^{2+} yields crude LAD-1 with essentially identical characteristics.

fragmentation and concomitant low yields of dynein obtained with lower sucrose concentrations. This and all subsequent procedures are carried out at 0–4°C. The homogenate is then centrifuged at 3000 *g* for 7 min to pellet most of the sperm heads and membrane fragments. This pellet is generally not very tightly packed and care must be taken in order to prevent contamination of the supernate by pellet as the former is decanted. Contamination is usually easy to recognize as the pellet material has a yellow color, owing to mitochondrial membrane pigments. The decanted supernatant is centrifuged at about 27,000 *g* for 15 min. The resulting pellet should display a distinct stratification with a small, compact dark layer at the bottom comprising most of the remaining sperm heads and membrane fragments overlaid by a much thicker off-white layer comprising the axonemes. In resuspending the pellet, the darker bottom layer is discarded. If the pellet displays a large, diffuse yellow layer on the bottom, this indicates significant contamination of the axonemes by membranes and sperm heads, which should be removed by a second low-speed centrifugation (1500 *g* for 5 min) in isolation buffer.

The axonemal pellets are resuspended in isolation buffer (Table I) to a total volume of about 150 ml. Resuspension is achieved with four to five strokes of the loose-fitting pestle in a Dounce homogenizer. The resuspended axonemes are centrifuged at 12,000 *g* for 10 min, and the supernatant is discarded. The axonemal pellet is gently resuspended by homogenization in axoneme buffer to a volume of about 75 ml; any dark material at the bottom of the pellet is again discarded. At this point the concentration of axonemal protein is determined (we use the method of Lowry, calibrated with bovine serum albumin) and the total yield is calculated. The yield of axonemes from 60 ml of *Tripneustes* semen averages 400 mg when prepared by this method (Table II). Axonemes are generally used immediately after preparation, but they can be stored as a pellet in the

TABLE II

COMPARISON OF YIELDS AND OF PROPERTIES OF CRUDE LAD-1 PREPARED BY TWO DIFFERENT METHODS

Method of preparation[a]	mg Axonemes/ ml semen	mg Crude LAD-1[b]/ 100 mg axonemes	Latent activity[c]	Triton-activated activity[c]	Activation ratio
Triton X-100 (4)	3.1 ± 0.9	6.3 ± 1.2	0.62 ± 0.10	3.8 ± 0.8	6.1 ± 0.9
Sucrose (19)	6.4 ± 1.6	5.7 ± 0.9	0.26 ± 0.04	2.7 ± 0.4	10.4 ± 1.2

[a] Main entry refers to method of disrupting sperm flagellar membranes. Otherwise all steps in preparations were identical. Number in parentheses is the number of individual preparations from which data were calculated.

[b] Amount of LAD-1 obtained by single high-salt extract; protein measured by method of Lowry.

[c] Specific activity given as μmole P_i min^{-1} mg^{-1}.

cold with a small buffer overlay for at least 24 hr without noticeable deterioration.

C. Extraction of Crude LAD-1

Crude LAD-1 is extracted by resuspending the pelleted washed axonemes in high salt buffer at a protein concentration of 3 mg ml^{-1} (Table I). We have found that the lower the concentration at which axonemes are suspended in high-salt buffer the greater the total yield of crude LAD-1. Three mg ml^{-1} has been chosen as a compromise between maximizing yield and minimizing final volume and dilution of dynein. Axonemes are extracted for about 15 min at 0°C, before being centrifuged at 12,000 *g* for 15 min. The supernatant is decanted and clarified by centrifuging at 100,000 *g* for 15 min. The supernatant from this centrifugation constitutes the stock solution of crude LAD-1, and total protein yield is about 20–25 mg (Table II). If a greater quantity of dynein is required, it is most conveniently obtained by extracting the axonemes a second time with high-salt buffer, but this extract is generally contaminated to a greater degree by other heavy chains and by tubulin. Increasing the concentration of NaCl in high-salt buffer does not greatly increase the amount of dynein in each extraction, but there is a fairly sharp decrease in yield if the NaCl concentration is decreased.

The quality of the crude LAD-1 is assessed by measuring the ratio of its ATPase activity after activation by Triton X-100 (i.e., Triton-activated ATPase activity) to that before activation (i.e., latent ATPase activity). ATPase activity is assayed by incubating an aliquot of LAD-1 for 10–40 min (depending on amount added) at room temperature (23°C) in ATPase assay buffer (Table I) and determining inorganic phosphate according to the method of Fiske and Subbarow (1925). Dynein is activated by 10–15 min preincubation in the presence of 0.05% Triton; activation appears to take place within this time period equally well at either room temperature or 4°C. Typical specific activities are (1) latent: 0.25 μmole P_i min^{-1} mg^{-1} and (2) activated: 2.5 μmole P_i min^{-1} mg^{-1}, for an activation ratio of 10 (Table II). The composition of the LAD-1 may be determined by analyzing its polypeptide content by electrophoresis on sodium dodecyl sulfate–polyacrylamide gels. Typical gel pattern and band designations are shown in Fig. 1. Prominent components are the dynein 1 heavy chains A_α and A_β and the three intermediate and four light chains. There is also a variable amount of contamination by tubulin, probably deriving mainly from the central tubules of the axoneme, at least one of which tends to disintegrate during extraction with high salt buffer. Electron microscopic investigation of sectioned axonemes indicates that most of the dynein extracted by high salt buffer derives from the outer arms (Gibbons and Fronk, 1979). In *Tripneustes,* a single high-salt extraction at 3 mg ml^{-1} generally removes about 40% of the outer arms, and the second high-salt extraction removes approximately another 40%. A more complete ex-

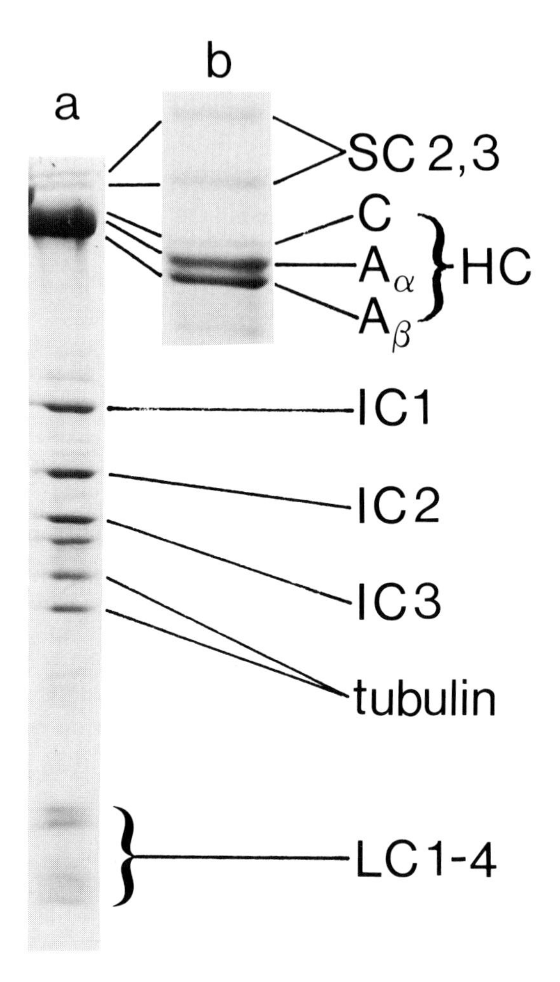

FIG. 1. Polypeptide composition of crude LAD-1 as revealed by polyacrylamide gel electrophoresis in the presence of sodium dodecyl sulfate. (a) 35-μg crude LAD-1 electrophoresed on a discontinuous buffer system (Laemmli, 1970) 5–15% w/v acrylamide gradient gel. (b) 10 μg of crude LAD-1 electrophoresed on a discontinuous buffer system 3–6% w/v acrylamide gradient gel, showing resolution of A_α and A_β heavy chains. SC, sky chain; HC, heavy chain, IC, intermediate chain; LC, light chain.

traction of outer arms is obtained with high-salt buffer using axonemes of some other species of sea urchin (Gibbons and Gibbons, 1973).

Crude LAD-1 may be stored in high-salt buffer at 4°C for about a week without a large change in ATPase activity. During this time the activation ratio of the dynein tends to decrease; the half-life for decay of an activation ratio of 10 is usually about one week. In the early stages this decay appears to be caused mainly by an increase in the latent activity of the dynein (i.e., a time-dependent activation), although the Triton-activated activity also decreases gradually. At room temperature the decay of activation ratio is greatly accelerated, with a half-life in the range of 2–4 hr. Thus the enzymatic activity of LAD-1 is relatively labile and it should be used as soon as possible. Physically, the enzyme appears to be more robust, with little change in sedimentation behavior and polypeptide content after two or three weeks storage at 4°C.

However, if the dynein is not to be used immediately, it can be stored frozen for longer periods. Solid sucrose is added to concentrated LAD-1 solution (see below) to a final concentration of about 10% w/v and allowed to dissolve slowly on ice for about 2 hr. The solution is then separated into 2-ml aliquots in Cryotubes (Vanguard International, Neptune, NJ) and frozen rapidly in liquid nitrogen for 5 min, followed by long-term storage at −80°C. We have recovered about 95% ATPase activity with roughly the original activation ratio after six months storage in this manner. Attempts to store the LAD-1/sucrose solution by lyophilization have resulted in a significant loss of activity.

Dynein is frequently required at concentrations greater than the 0.25 mg ml^{-1} yielded by high-salt extraction. Concentration can be carried out at 4°C in an Amicon ultrafiltration cell using a UM20 membrane. Concentrations up to 5 mg ml^{-1} can be attained, but at over 2 mg ml^{-1} nonspecific protein aggregation becomes an increasing problem. Although it has not been possible to prevent aggregation completely, it can be minimized by keeping the protein cold and maintaining a sufficient level of reducing agent (dithiothreitol or 2-mercaptoethanol). Aggregated material can be removed by centrifugation at 12,000 *g* for 5 min. A more delicate method of concentrating small volumes, which seems less prone to causing aggregation, is to concentrate the dynein in a dialysis membrane bag against dry Sephadex.

III. Further Purification of Dynein

Solutions of crude LAD-1 prepared as shown are pure enough for use in many basic enzymatic investigations and work involving recombination of dynein arms to extracted axonemes and reactivated sperm (Gibbons and Fronk, 1979; Gibbons and Gibbons, 1979). However, for many purposes, including detailed enzymological studies and physicochemical characterization (e.g., analytical ultracentrifugation), further purification is required.

A. Density Gradient Centrifugation

We routinely purify LAD-1 by zonal centrifugation on density gradients of sucrose or glycerol. Density gradients of 5–20% w/v sucrose or 8–30% w/v glycerol are made up in high-salt buffer. One ml crude LAD-1 solution, often concentrated to about 2–3 mg ml^{-1}, is layered on top of each gradient and centrifuged in an SW41 swinging bucket Beckman rotor at 35,000 rpm at 4°C for 15 hr. After fractionation, the approximate distribution of the protein in the gradient is determined by spotting an aliquot of each fraction onto Whatman 3 MM chromatography paper, drying the paper and staining it for 5–10 min with Coomassie Brilliant Blue (50 ml distilled water, 50 ml methanol, 10 ml glacial acetic acid, 0.05 g Coomassie Brilliant Blue R-250). After destaining (in 82.5 ml distilled water, 10 ml methanol, 7.5 ml glacial acetic acid), fractions containing protein can be seen as blue spots on the paper. This provides a rapid (~30–45 min) method of locating the major protein peaks in the gradient, but to be of most use it should be compared with the information derived from electrophoresis in sodium dodecyl sulfate–polyacrylamide slab gels. Such a gel of a complete fractionated sucrose gradient is shown in Fig. 2. The major peak, containing the A_α and A_β heavy chains and the intermediate and light chains, is the 21S form of LAD-1 (Gibbons and Fronk, 1979). In the 12–14S region, there is a faint second peak containing small amounts of the A chains and other dynein heavy chains. The sky chains and C chain tend to span the region between these two peaks.

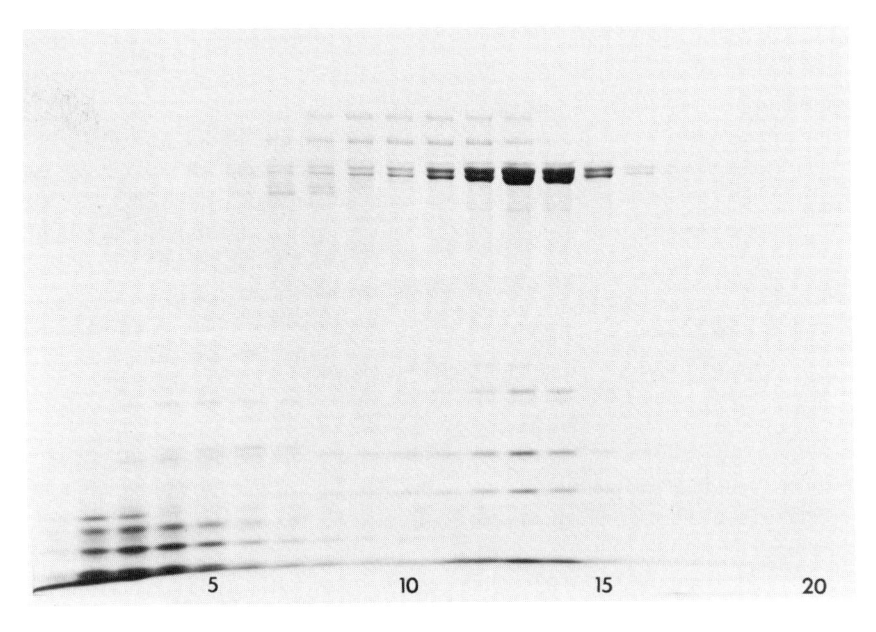

FIG. 2. Sucrose density gradient centrifugation of crude LAD-1. The 20 fractions from a 5–20% w/v sucrose density gradient were electrophoresed on a discontinuous buffer system 3–6% w/v acrylamide gradient slab gel, with the fraction from the top of the sucrose density gradient on the left. 21S LAD-1 peaks in fraction 13 (see numbers at the bottom of gel), the secondary peak of dynein heavy chains is in fractions 7 and 8, and tubulin peaks around fraction 3. The light chains of 21S LAD-1 (see Fig. 1) run with the dye front on this gel.

Tubulin and other small polypeptides are found mostly near the top of the gradient. Defining 21S LAD-1 as comprising the A_α and A_β chains, the intermediate chains and the light chains (Fig. 2), we find that it accounts for about 50–70% by stain intensity of the initial LAD-1 high-salt extract (Bell *et al.*, 1979). The purification of LAD-1 by density gradient centrifugation, although it produces a particle that appears monodisperse by sedimentation velocity and sedimentation equilibrium (Gibbons and Fronk, 1979), is done at the cost of reducing the ATPase activation ratio of this particle relative to that of initial crude LAD-1. The reason for this is uncertain, although it seems to stem primarily from an increase in the latent ATPase activity of the particle. Recombination of aliquots of all fractions of the sucrose gradient does not reinstate the original latent and Triton-activated ATPase activities, indicating that the change is not caused solely by the separation of some necessary component from the 21S LAD-1 particle. After it has been fractionated, the 21S LAD-1 may be stored at 4°C (in its sucrose gradient solution) for up to a week without significant further deterioration in activity. Under these conditions, aggregation appears to be insignificant, possibly because of the presence of ~15% w/v sucrose. However, if the sucrose is dialyzed out of purified 21S LAD-1, and especially if the protein is reconcentrated, aggregation again becomes a problem.

B. Other Purification Procedures

Gel filtration on cross-linked Sepharose Cl-4B in high-salt buffer may be used to purify 21S LAD-1. This method has the advantage of being adaptable to larger volumes but suffers in comparison with sucrose density gradient centrifugation in that the recovery of enzyme off the column is usually less than 50%.

Other purification procedures we have utilized are intended to separate particular polypeptides from the 21S LAD-1 particle. The method most often used involves dialyzing concentrated LAD-1 (2–3 mg ml^{-1}), originally in high-salt buffer, against a low-salt buffer containing EDTA (Table III). The concentrated, dialyzed LAD-1 is centrifuged at 12,000 *g* for 5 min to remove aggregated material and is then layered onto 5–20% sucrose density gradients made up in low-salt buffer. The gradients are centrifuged in an SW41 swinging bucket rotor at 35,000 rpm for 15 hr at 4°C. Sodium dodecyl sulfate polyacrylamide gels of samples from fractionated gradients (Fig. 3) show that the A_β and intermediate chains sediment at around 9–10S and are separated from the A_α chain, which is partially aggregated and sediments at velocities ranging from about 16 to 30S. The A_β chain and intermediate chain 1 appear to co-sediment as a unit, whereas intermediate chains 2 and 3 tend to spread to either side, most often the trailing side of this peak, indicating that in low salt they are not tightly associated with either the A_β or A_α heavy chains.

Another method leading to a similar separation involves chromatography on

TABLE III

COMPOSITION OF BUFFERS AND METHODS USED IN SEPARATION OF LAD-1 POLYPEPTIDES

Low-salt sucrose density gradient centrifugation	Phosphocellulose chromatography	Hydroxylapatite chromatography
5 m*M* Imidazole-HCl, pH 7 0.5 m*M* EDTA 14 m*M* 2-Mercaptoethanol LAD-1 is concentrated and dialyzed for 18–24 hr against two changes of 100 vol of low-salt buffer, then loaded onto 5–20% w/v sucrose density gradients prepared in same buffer.	5 m*M* Imidazole-HCl, pH 7 10 m*M* NaCl 14 m*M* 2-mercaptoethanol 0.5 × 2.0-cm phosphocellulose column is equilibrated in above buffer. LAD-1 is concentrated and dialyzed against buffer and then applied to column.	0.5 × 5-cm hydroxylapatite column is equilibrated to 0.01 *M* sodium phosphate, pH 6.9. LAD-1 (in original high-salt buffer) is applied to column. Column is eluted with a gradient of 0.01–0.5 *M* sodium phosphate, pH 6.9

phosphocellulose. Concentrated LAD-1 is dialyzed against column buffer (Table III) and 1 ml is applied to a 0.5 × 2.0-cm column of phosphocellulose equilibrated in the same buffer. The column is eluted with column buffer and the proteins in the flow-through peak are collected. With the best preparations, electrophoresis shows only the A_β chain and intermediate chain 1 in this peak. However, more often, all three intermediate chains emerge with the A_β chain. The reason for this variable performance is uncertain, and the procedure requires further development to improve its reproducibility.

Chromatography on hydroxylapatite as introduced by Ogawa (Ogawa and Mohri, 1972) for the purification of flagellar ATPase from the sperm of *Pseudocentrotus* can also be used to separate the major LAD-1 polypeptides. LAD-1 in high-salt buffer is loaded directly onto a 0.5 × 5-cm hydroxylapatite column equilibrated in 10 m*M* sodium phosphate, pH 6.9 (hydroxylapatite is prepared according to Bernardi, 1971). The protein is eluted with a gradient of sodium phosphate, pH 6.9, from 0.01 to 0.5 *M*. Electrophoresis shows that the first polypeptides to be eluted are contaminating tubulin. The next peak includes the A_β chain and intermediate chain 1, and following this are peaks containing the sky chains and intermediate chains 2 and 3, the C chain, and the A_α chain, respectively. Although tubulin and often the A_β chain and intermediate chain 1 are well separated from other peaks, there is considerable overlap among the remaining peaks even when shallow gradients are employed.

Ogawa used chromatography on hydroxylapatite to separate from a low-salt extract of *Tripneustes* sperm flagella, previously extracted with high salt, an ATPase other than that now known as LAD-1 (Ogawa and Gibbons, 1976). This ATPase was named dynein-2 and does not appear to be composed of A chain polypeptides.

Some of the properties of the separated A_α and A_β chains are given in the

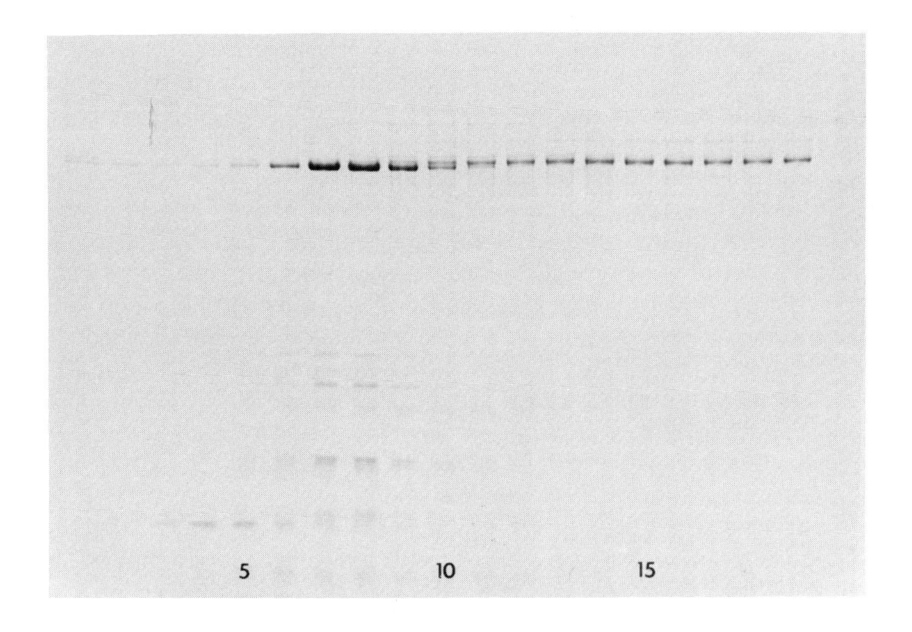

FIG. 3. Separation of the A_α and A_β heavy chains of 21S LAD-1 by density gradient centrifugation in low-salt buffer. The 19 fractions from a 5–20% w/v sucrose density gradient were electrophoresed on a discontinuous buffer system 3–6% w/v acrylamide gradient slab gel with the fraction from the top of the sucrose density gradient on the left. The A_β heavy chain and the intermediate chains peak in fractions 7 and 8, whereas the A_α chain spreads from about fraction 10 to fraction 19.

following section. However, it is of practical import to note here that the purified A_β chain appears to maintain its activity for extended periods and displays essentially no propensity for nonspecific aggregation in conditions ranging from 0.0 to 0.6 *M* NaCl and from pH 6 to 9. On the other hand, the A_α chain displays erratic ATPase activity, which is often difficult to distinguish from small quantities of contaminant A_β. The A_α chain also readily aggregates particularly at low salt concentration and low pH. It appears that the A_α chain may be the major factor in the aggregation of LAD-1.

IV. Selected Properties of LAD-1 and Its Subunits

A. LAD-1

A summary of the available data on enzymatic and physicochemical properties is given in Table IV. The sedimentation velocity and molecular weight yield a frictional ratio (f/f_0) of 1.9, which, as expected from electron microscopic evidence of outer-arm structure, suggests a relatively compact particle (cf. myosin,

f/f_0 = 3.5, myosin subfragment 1, f/f_0 = 1.4; calculated from data given in Lowey *et al.*, 1969). It should be noted that although we use density-gradient centrifugation for routine approximate measurement of sedimentation velocity, such measurements are best confirmed on the analytical ultracentrifuge (Beckman Spinco Model E). We have found that analytical centrifugation often displays more heterogeneity than zonal centrifugation, either because the presence of sucrose in the latter favors disaggregation or because the method is inherently less sensitive to heterogeneity. Densitometric scanning of sodium dodecyl sulfate polyacrylamide gels stained with Coomassie Brilliant Blue indicates that the A_α and A_β chains are present in equimolar quantities and that the intermediate chains are present in equimolar quantities relative to each other. The stoichiometry between the A chains and the intermediate chains is less certain, but it appears that there is probably 1 mole of intermediate chain 1 per mole of A_β chain. The staining of the light chains is, for reasons that are unclear, too variable to allow meaningful comparisons of this type.

The ATPase activity data in Table IV refer to the conditions of the standard ATPase assay buffer. Variation of NaCl concentration in the assay buffer (Gibbons and Fronk, 1979) causes a sigmoid increase in latent specific activity from 0.2 μmole P_i min^{-1} mg^{-1} at zero salt to a maximum about 3.5 μmole P_i min^{-1} mg^{-1} at 0.8 *M* NaCl, above which the activity levels out and then begins to decline. The effect of variation of NaCl concentration on Triton-activated ATPase activity is less dramatic, with the activity assayed in 0.5 *M* NaCl being about twice that assayed in zero NaCl. Substitution of Ca^{2+} for Mg^{2+} in the ATPase assay buffer does not significantly change the dependence of activity on NaCl concentration. However, the ATPase activity at any given NaCl concentration is less in the presence of Ca^{2+} than in the presence of Mg^{2+}, especially in the case of Triton-activated material, so that at concentrations above about 0.5 *M* NaCl, the "Triton-activated" Ca^{2+}-ATPase activity is actually less than the "latent" Ca^{2+}-ATPase activity (Gibbons and Fronk, 1979).

Occasionally a second kinetic component will appear at higher ATP concentrations (>100 μM) in a Lineweaver-Burk plot for latent LAD-1, giving a biphasic appearance similar to that noted by Takahashi and Tonomura (1978) in plots for 30S *Tetrahymena* dynein. In the case of *Tripneustes,* this biphasic appearance may be caused by the presence of a variable small amount of activated material in the latent preparation.

The specificity of dynein for ATP as substrate is fairly high, and most ATP-analogues as well as other nucleoside triphosphates (ITP, GTP, CTP, etc.) are hydrolyzed at less than 15% of the rate for ATP (Gibbons, 1966; Takahashi and Tonomura, 1978). Vanadate, the anionic form of vanadium (V), is a potent uncompetitive inhibitor of dynein ATPase with an apparent inhibition constant (K_i) of 100 n*M* or less when measured at low salt concentration (Gibbons *et al.*, 1978; Kobayashi *et al.*, 1978), and it has been used in attempts to probe the steps in

TABLE IV

SELECTED PROPERTIES OF LAD-1 AND ITS SUBUNITS

		Polypeptide composition[a]	$S^0_{20,w}$ (S)	MW[b]	f/f_0	K_m[c] (μM)	Specific[d] activity
LAD-1[e]		$A_\alpha \sim$330,000 $A_\beta \sim$320,000 IC1 122,000 IC2 90,000 IC3 76,000 LC1-4 24,000-14,000	21.2	1.25×10^6	1.9	latent, 1.3[f] activated, ~50	0.26 2.7
A_β	Low ionic strength	$A_\beta \sim$320,000 IC1 122,000	9.3	—	—	2.8	0.8
A_β	High ionic strength	$A_\beta \sim$320,000 IC1 122,000	14.3	—	—		
A_α		$A_\alpha \sim$330,000	—	—	—	7.3	0.3

[a] Polypeptide chain designation, followed by apparent molecular weight.
[b] Sedimentation equilibrium measurement.
[c] For ATP, derived from coupled assay measurements: coupled assay composition given in Gibbons *et al.* (1978).
[d] μmole P_i min^{-1} mg^{-1}, in 0.1 M NaCl, 30 mM Tris-HCl, pH 8.1, 2 mM $MgSO_4$, 0.1 mM EDTA, 1 mM ATP at 23°C.
[e] Crude LAD-1 used for enzymatic measurements.
[f] Lineweaver-Burk plots of latent LAD-1 activity sometimes show a second kinetic component at higher ATP concentrations: see text for details.

the mechanochemical cross-bridge cycle in which dynein produces sliding between flagellar tubules (Sale and Gibbons, 1979; Okuno, 1980).

B. Separated Subunits

A summary of available data is given in Table IV. The increase in sedimentation velocity of the A_β subunit from 9S to 14S between low-salt and high-salt buffers (Table IV) is possibly the result of a self-association of the monomeric subunit to form a dimer. If so, then this, coupled with the equimolarity of A_α and A_β chains in the 21S particle, would suggest that the latter comprises a heterotetramer containing two each of the A_α and A_β chains. However, the occasional presence of an intermediate 12S component suggests that the situation may be more complex. The present data are insufficient to choose between dimerization and/or a conformational change as the basis for the change in sedimentation rate. Regardless of whether the LAD-1 particle is considered to be a dimer or a tetramer of the A heavy chains, there is an apparent discrepancy between the molecular weight of LAD-1 derived from sedimentation equilibrium (Gibbons and Fronk, 1979) and the sum of apparent molecular weights (derived from electrophoretic mobilities in the presence of sodium dodecyl sulfate) of the probable polypeptide composition of LAD-1 (Bell *et al.*, 1979), which suggests that the results of one or both methods are inaccurate.

V. Preparation and Properties of Dynein ATPases from Other Sources

In this section the preparation and properties of dynein ATPase from other species will be briefly summarized from a comparative viewpoint. For more detailed treatment, reference should be made to the original work. Some of the data in this section are presented in tabulated form in Table V.

A. *Tetrahymena*

The cilia of *Tetrahymena* were the source from which the first well-defined axonemal ATPase was isolated and described by Gibbons (1963). The procedure developed by Gibbons has subsequently been used with only minor alteration by many workers (e.g., Blum, 1973; Mabuchi and Shimizu, 1974; Takahashi and Tonomura, 1978) and has also been applied to other species (see later discussion). Cilia are first detached from the cell bodies by one of a variety of methods that include treatment with ethanol and calcium (Gibbons, 1963, 1965a) or with glycerol (Gibbons, 1965b) or exposure to the local anesthetic dibucaine in the presence of calcium (Thompson *et al.*, 1974) and are then harvested by differential centrifugation. Prior to solubilization of dynein ATPase, the cilia must be demembranated with detergent, 0.5% w/v digitonin (Gibbons, 1963) or 1% w/v Triton X-100 (Mabuchi and Shimizu, 1974). Dynein ATPase is extracted from the demembranated cilia by dialysis against 0.1 m*M* EDTA, 1 m*M* Tris-thioglycolate, pH 8.3 (Gibbons, 1963, 1965a) or 0.1 m*M* EDTA, 1 m*M* dithiothreitol, 1 m*M* Tris-HCl, pH 8.3 (Takahashi and Tonomura, 1978) for approximately 18 hr at 0–4°C, which solubilizes about 90% of total axonemal ATPase. Raff and Blum (1969) used a somewhat different procedure involving extraction of dynein ATPase from glycerinated *Tetrahymena* cilia by incubation with 20 m*M* ATP in 20 m*M* imidazole-HCl buffer, pH 8.3, containing 2.5 m*M* $MgSO_4$. Warner *et al.* (1977) have solubilized dynein ATPase from Triton demembranated (0.2% w/v Triton X-100) *Tetrahymena* cilia by extraction in 0.5 *M* KCl, 2 m*M* HEPES, 4 m*M* $MgSO_4$, 0.5 m*M* EDTA, 1 m*M* dithiothreitol, and

0.1 m*M* ATP at pH 7.8. In all cases, differential centrifugation was used to separate the solubilized ATPase from insoluble axonemal residue.

Gibbons (1963) found that extraction of *Tetrahymena* axonemes with 1 m*M* Tris buffer, 0.1 m*M* EDTA, pH 8.0 solubilized most of the axonemal ATPase activity as well as removing both rows of arms from the doublet tubules, suggesting that most of the axonemal ATPase is localized in these arms. Studies with the analytical ultracentrifuge (Gibbons and Rowe, 1965) showed that the dynein ATPase existed in two forms sedimenting at 30S and 14S and suggested that the former was a polymer of the latter. However, Mabuchi and Shimizu (1974) reported that the 30S and 14S forms of dynein have electrophoretically distinct polypeptide subunits, suggesting that the two forms represent different isoenzymes rather than being related as monomer and polymer. The 30S form can rebind to extracted axonemes and restore the appearance of arms, but the 14S form does not possess this ability (Gibbons, 1965a). It is not possible to determine in a similar manner the location of the dynein ATPase extracted from *Tetrahymena* cilia by solutions containing ATP (Raff and Blum, 1969; Warner *et al.*, 1977; Warner and Zanetti, 1980), since in these cases the cilia disintegrate during extraction. In preparations from most species, such disintegration of ciliary and flagellar axonemes in the presence of ATP is seen only after mild proteolysis (Summers and Gibbons, 1971), suggesting that autoproteolysis may occur during the isolation of *Tetrahymena* cilia.

B. *Chlamydomonas*

Chlamydomonas possesses two main advantages over *Tetrahymena* as a source for the investigation of dynein ATPase. First, being an autotrophic organism, its proteins are less likely to be degraded by endogenous proteolytic enzymes during isolation. Second, a wide variety of paralyzed flagellar mutants that lack specific morphological components are available, and some of these have been used to investigate axonemal doublet microtubule arm structure (Lewin, 1954; Luck *et al.*, 1977; Huang *et al.*, 1979).

Flagella are generally isolated by pH shock or by the "STEEP-Ca" procedure (Witman *et al.*, 1972) followed by differential centrifugation. The isolated flagella may be demembranated by exposure to detergents, such as Nonidet P-40 at a final concentration of 0.5% w/v (Huang *et al.*, 1979). Dynein ATPase has been extracted from *Chlamydomonas* flagellar axonemes by an essentially unmodified Gibbons low-salt dialysis procedure (Watanabe and Flavin, 1976) and by extraction of axonemes with 0.5 *M* NaCl for 10 min at 4°C (Piperno and Luck, 1979). The latter method extracts virtually all the dynein ATPase from demembranated axonemes and leaves the doublet microtubules essentially free of both inner and outer arms (Piperno and Luck, 1979). One disadvantage engendered by the use of *Chlamydomonas* is that only relatively small quantities of flagellar proteins are available for biochemical studies. However, Luck and his co-workers (Piperno and Luck, 1979; Huang *et al.*, 1979) have alleviated this problem considerably by growing the organism in media containing $^{35}SO_4$ of high specific activity in order to produce highly labeled proteins.

Both low-salt dialysis and high-salt extraction release two major forms of dynein that sediment at 18S and 12S on sucrose density gradients made up in low-ionic-strength buffers (Watanabe and Flavin, 1976; Piperno and Luck, 1979). However, examination of outer- and inner-arm mutants strongly suggests that both these forms of dynein ATPase are localized in the outer arms (Huang *et al.*, 1979). Chromatography on hydroxylapatite coupled with sodium dodecyl sulfate–polyacrylamide gel electrophoresis shows that the two outer-arm dynein ATPases contain different high-molecular-weight polypeptides (apparent molecular weight 300,000–330,000), thus ruling out the possibility that the 18S particle is a dimer of the 12S particle. These two dyneins also differ in their content of intermediate and low-molecular-weight chains. The two dyneins have a similar specific ATPase activity, but it is not clear whether this is latent or activated (Piperno and Luck, 1979). The inner-arm dynein ATPase contains high-molecular-weight polypeptides that are different from those in both the 18S and 12S dyneins and that may form a particle of about 13S at low ionic strength (Huang *et al.*, 1979; Piperno and Luck, 1980).

C. Lamellibranch Gill Cilia

The gill cilia of lamellibranch mollusks have been used in the investigation of dynein ATPase. Linck has isolated gill cilia from *Aequipecten* by three different methods, using twice-concentrated seawater, 10% ethanol/10 m*M* CaCl, or 60% glycerol with apparently equivalent results (Linck, 1973a). After demembranation of the cilia in 1% w/v Triton X-100, low-salt dialysis of the ciliary axonemes for 48–60 hr extracted half of the ciliary ATPase (Linck, 1973b). The extracted dynein ATPase apparently corresponds to the outer doublet arms. The remaining ATPase, possibly representing the inner arms, is extractable in active form after brief digestion with trypsin (Linck, 1973a). In contrast, Linck (1973b) found that low-salt dialysis of sperm flagella of the same species extracted essentially all the dynein ATPase. Warner *et al.* (1977), using the mollusk *Unio*, isolated gill cilia by exposing excised gill tissue to a solution containing 20 m*M* dibucaine. After demembranation in 0.2% w/v Triton X-100, extraction of the ciliary axonemes in the same high-salt buffer that was used by this group in the study of *Tetrahymena* dynein (see above) resulted in the solubilization of most of the outer-arm dynein.

Both the low-salt-extracted and trypsin-released dynein ATPase isolated from gill cilia by Linck sediment in 10 m*M* Tris, pH 8, as 14S particles, and electrophoresis in the presence of sodium dodecyl sulfate reveals two large polypeptides with apparent molecular weights of 450,000 and 500,000 (Linck, 1973b).

However, Warner *et al.* (1977), using cross-linked bovine serum albumin as an electrophoretic standard, estimated the molecular weight of the large polypeptide subunits from *Unio* dynein to lie in the region of 300,000 to 460,000 and found up to six different chains.

D. Starfish Sperm

Mabuchi and co-workers have studied the flagellar dynein of starfish sperm (Mabuchi *et al.*, 1976). Sperm were obtained from the dissected testes of the starfish *Asterias*. The flagella were separated from the sperm heads by homogenization and then demembranated with 1% w/v Triton X-100. Dynein ATPase was extracted by dialysis for 18 hr at 0°C against either a low-salt buffer, 1 m*M* Tris-HCl (pH 7.8), 0.1 m*M* EDTA, 0.5 m*M* dithiothreitol, or a high-salt buffer, 0.6 *M* KCl, 10 m*M* Tris-HCl (pH 7.8) , 0.1 m*M* EDTA, 0.5 m*M* dithiothreitol. Low-salt dialysis solubilized over 90% of the axonemal ATPase activity, whereas slightly less than 50% was solubilized by high-salt dialysis.

These flagellar dyneins were examined in some detail by Mabuchi *et al.* (1976). The dynein ATPase extracted in high-salt sediments as a single peak at 20S in high-salt sucrose density gradients, whereas the low-salt dynein ATPase exhibits a single peak at about 12S when sedimented in low-salt sucrose density gradients. The 20S form is completely converted to the 12S form when dialyzed into low salt, and in the converse experiment there is partial conversion of the 12S form into the 20S form. Electron microscopic examination showed that extraction of axonemes with low salt appeared to remove both inner and outer arms, whereas extraction with high salt apparently removed only outer arms. On return to moderate ionic strength (20 m*M* Tris-HCl, pH 7.6, 2 m*M* $MgSO_4$, 0.5 m*M* dithiothreitol), the 20S dynein recombines with extracted flagella and the outer arms reappear on the doublet tubules. Sucrose density gradients run at this ionic strength suggest that the species of dynein that actually recombines is a 24S particle. Such recombination and restoration of arm structure could not be shown with the 12S low-salt dynein. Electrophoresis of whole axonemes in the presence of sodium dodecyl sulfate shows that five polypeptide chains migrate in the region expected of dynein heavy chains: the 20S and 24S dyneins contain only the upper two chains, whereas all five chains are found in the 12S dynein. Both the 12S and 20S dyneins show roughly similar changes in Mg^{2+} and Ca^{2+} ATPase activity in various conditions.

E. Sea Urchin Sperm

The sperm of many species of sea urchin have been used for the investigation of dynein ATPase, although few compete with *Tripneustes* for quantity and seasonal constancy of sperm production. In general, the techniques used to collect sperm and to isolate flagellar axonemes are similar to those described

TABLE V

SELECTED PROPERTIES OF DYNEIN ATPASE FROM VARIOUS SOURCES

Source	Preparation method[a]	Sed. Veloc.[b] (S)	Molecular weight	Polypeptide composition[c]	Activators	Specific activity[d]	K_m[e] (μM)	Refs.
Tetrahymena	Tris-EDTA	13 (A)	—	—	Mg^{2+}	1.3	—	1
		25	—	—	Mg^{2+}	0.67	—	
	Tris-EDTA	14 (A)	600,000	—	Mg^{2+},Ca^{2+},Mn^{2+}, Fe^{2+},Co^{2+},Ni^{2+}	3.5	35	2,3,4
		30	5.4×10^6	—		1.3	11	
	20 m*M* ATP	14 (D)	—	—	Mg^{2+},Ca^{2+},EDTA	0.5	—	5
		30	—	—	Mg^{2+},Ca^{2+}	2.0	—	
	Tris-EDTA	14 (D)	—	520,000	—	—	—	6
		30	—	560,000	—	—	—	
	0.5 *M* KCl	—	—	360,000	—	—	—	7
	Tris-EDTA	30(D)	—	—	Mg^{2+}	0.12	1†	8,9
Chlamydomonas	Tris-EDTA	12 (D)	—	—	Mg^{2+},Ca^{2+}	—	—	10
		18	—	—				
	0.5 *M* NaCl	12 (D)	—	315,000*	Mg^{2+},Ca^{2+}	6	—	11,12
		18	—	~320,000*		5.4	—	
		13(D)	—	>300,000*		4		
Gill cilia *Aequipecten*	Tris-EDTA	14(A)	—	500,000, 450,000	Mg^{2+}	—	—	13,14

Unio	0.5 *M* KCl	—	—	360,000, 320,000	—	—	—	7
Starfish sperm	Tris-EDTA	12 (D)	—	600,000	Mg^{2+},Ca^{2+}	0.8	23	15
	0.6 *M* KCl	20	—				16	
Sea urchin sperm								
Pseudocentrotus	Tris-EDTA	10 (A)	350,000	—	Mg^{2+},Ca^{2+}	0.38	160	16
	0.6 *M* KCl	25	—	—		—	—	
Hemicentrotus	Tris-EDTA	10,13(A)	—	—	Mg^{2+},Ca^{2+}	0.5	—	17
Pseudocentrotus	Tris-EDTA	—	—	—	Mg^{2+},Mn^{2+} (Ni^{2+},Ca^{2+})	2.5	50	18
Colobocentrotus	Tris-EDTA 0.5 *M* KCl	13(D)	—	[f]	Mg^{2+},Ca^{2+}	2.8	50	19,20
Tripneustes[g]	0.6 *M* NaCl	21(A)	1.25×10^6	~330,000*	Mg^{2+},Ca^{2+}	0.25	1†	21

[a] That is, method of preparation of dynein ATPase from isolated axonemes. Tris-EDTA = low-salt dialysis.

[b] Values derived from (A) analytical or (D) density gradient centrifugation.

[c] Apparent molecular weight of high-molecular-weight chains given; those marked ~ contain more than one heavy chain; those marked * also contain lower-molecular-weight chains.

[d] Mg^{2+}-ATPase, μmole P_i min^{-1} mg^{-1}; conditions generally 0.1 *M* salt, 1–4 m*M* Mg^{2+}, 1 m*M* ATP, pH 8, 20–25°C.

[e] For ATP, values marked † display second kinetic component.

[f] Although the apparent molecular weight of *Colobocentrotus* dynein heavy chains was originally estimated to be in the range of 500,000 (20), more recent work (Sale & Tang, unpublished results) has shown that these chains nearly comigrate with the heavy chains of *Tripneustes* dynein-1. Within a given electrophoretic system, dyneins from most sources nearly comigrate, so that the large scatter in apparent molecular weights of the heavy chains appears to be caused by differences between electrophoretic buffer systems and the lack of adequate high-molecular-weight standards.

[g] Refer to Table IV for a more detailed description.

References: 1. Gibbons (1963); 2. Gibbons (1965a); 3. Gibbons and Rowe (1965); 4. Gibbons (1966); 5. Raff and Blum (1969); 6. Mabuchi and Shimizu (1974); 7. Warner *et al.* (1977); 8. Takahashi and Tonomura (1978); 9. Shimizu *et al.* (1979); 10. Watanabe and Flavin (1976); 11. Piperno and Luck (1979); 12. Piperno and Luck (1980); 13. Linck (1973a); 14. Linck (1973b); 15. Mabuchi *et al.* (1976); 16. Mohri *et al.* (1969); 17. Hayashi and Higashi-Fujime (1972); 18. Ogawa and Mohri (1972); 19. Gibbons and Fronk (1972); 20. Kincaid *et al.* (1973); 21. Gibbons and Fronk (1979) and Tang *et al.* (1982).

earlier. Variations from these procedures include the use by Ogawa and Mohri (1972) of sonication rather than homogenization for separating the heads from the tails of sperm of the sea urchin, *Pseudocentrotus depressus,* and the use, in earlier work, of glycerol to disrupt flagellar membranes (Brokaw and Benedict, 1971; Ogawa and Mohri, 1972). Gibbons *et al.* (1970) introduced the use of Triton X-100 as a demembranating agent for sea urchin sperm flagella, and this has since been used widely. Various methods have been used to solubilize dynein ATPase from sea urchin sperm flagellar axonemes, but all are essentially variations of either the low-salt dialysis method originally described (Gibbons, 1963) for *Tetrahymena* ciliary axonemes (Mohri *et al.,* 1969; Hayashi and Higashi-Fujime, 1972; Ogawa and Mohri, 1972) or the high-salt extraction procedure (Gibbons, 1965a; Brokaw and Benedict, 1971; Gibbons and Fronk, 1972; Ogawa *et al.,* 1977).

Generally speaking, depending on the conditions under which they are isolated, the sperm flagella dynein ATPases from different species of sea urchin appear to be fairly similar in properties. Extraction of axonemes in low-salt solutions yields dynein ATPases that sediment at between 10 and 13S (*Pseudocentrotus* and other species, Mohri *et al.,* 1969; *Hemicentrotus,* Hayashi and Higashi-Fujime, 1972; *Colobocentrotus,* Gibbons and Fronk, 1972) and appears to solubilize most of the total axonemal ATPase. On the other hand, extraction of axonemes with high salt appears to solubilize specifically the outer arms (*Colobocentrotus,* Gibbons and Fronk, 1972; *Anthocidaris,* Ogawa *et al.,* 1977; *Tripneustes,* Gibbons and Fronk, 1979) and solubilizes only about half of the total axonemal ATPase activity. In all cases where they have been thus analyzed, dynein ATPases from sea urchin sperm flagella axonemes, regardless of extraction method, have been shown by electrophoresis in the presence of sodium dodecyl sulfate to contain the high-molecular-weight polypeptides customarily associated with dynein (Kincaid *et al.,* 1973; Ogawa and Gibbons, 1976; Gibbons and Fronk, 1979). Enzymatic characterization of dynein ATPase from sea urchin sperm flagella has been extensive; selected data are given in Table V.

Ogawa (1973) has prepared a well-defined fragment, Fragment A, by digesting with trypsin the dynein of *Hemicentrotus* sperm flagella. This fragment of molecular weight 380,000 retains high ATPase activity but has lost the ability to recombine with low-salt-extracted flagellar axonemes, thus suggesting a functional separation of doublet microtubule binding site and ATPase active site in dynein.

F. Trout Sperm

Ogawa and his colleagues have recently described the preparation of a dynein ATPase from the sperm of the rainbow trout, *Salmo gairdneri* (Ogawa *et al.,*

1980). Flagellar axonemes were prepared from the sperm in essentially the same way as from sea urchin sperm, and the dynein was extracted in high salt. The dynein appears to be localized in the outer arms. Physicochemical characterization of this dynein in the intact state has been limited so far to sodium dodecyl sulfate polyacrylamide gel electrophoresis, which shows the presence of high-molecular-weight chain(s) that approximately comigrate with the high-molecular-weight polypeptide subunits of sea urchin dynein-1. An immunological similarity between these two dyneins is also indicated.

One advantage of this organism is that a ripe male trout is relatively prolific, holding from 20 to 50 ml of sperm. However, Ogawa cautions that proteolytic enzyme(s) present in the trout spermatozoa lead to partial fragmentation of the dynein during purification.

VI. Summary

The following characteristics seem to be common to dyneins from all sources so far investigated and serve to distinguish dynein from myosin. First, dyneins are ATPase proteins that can be activated to a similar extent by either Mg^{2+} or Ca^{2+}, and in this respect they differ from myosin in which the Mg^{2+}-ATPase activity in the absence of actin is up to two orders of magnitude less than the Ca^{2+}-ATPase activity (Ebashi and Nonomura, 1973). Furthermore, dynein ATPases display a much greater substrate specificity for ATP than does myosin (Gibbons, 1966; Ogawa and Mohri, 1972). Second, all dyneins appear to exist in, or to be convertible to, active forms that sediment in the range 10–14S. No form of dynein displaying ATPase activity has been found that sediments at the 6S velocity of myosin. Third, all dyneins contain very large polypeptide subunits, with apparent molecular weights in the range of 300,000–500,000, which are electrophoretically distinct from the heavy chain of myosin. Finally, all ATPases with the preceding characteristics that have been isolated from cilia and flagella appear to be localized in the doublet microtubule arms.

Another property that is likely to assume greater importance in the future is the question of the functional capability of preparations of dynein as opposed to merely their ATPase activity. Gibbons and Gibbons (1979) found that only the latent form of LAD-1 had the ability to recombine functionally and increase the beat frequency of reactivated, dynein-depleted flagella of sea urchin sperm. Although activation of dynein ATPase from other sources by various chemical procedures has been reported (Blum and Hayes, 1974; Shimizu and Kimura, 1974; Blum and Hines, 1979; Mabuchi *et al.*, 1976), none has as yet been demonstrated to have functional activity. The chemical and physicochemical factors affecting latency and activation of LAD-1 and their relationship to the action of the functioning dynein arm are as yet unclear. However, it is necessary to postulate that a functional dynein ATPase, in order to perform work, must possess a rest state in which ATP turnover is significantly less than maximal, and it is possible that the latent form of LAD-1 represents the dynein arm in this rest state. Similarly, it is possible that one of the several activated forms of the enzyme that have been reported (Gibbons and Fronk, 1979) is analogous to the state of the dynein arm in which the energy of ATP hydrolysis is used to perform work, leading to the sliding of adjacent doublet tubules. The natural mechanisms controlling the cross-bridge cycle, which are bypassed or mimicked by the techniques used to activate LAD-1, are yet to be clarified.

Acknowledgments

This work has been supported in part by grants HD 06565 and HD 10002 from the National Institute of Child Health and Human Development, and by a Helen Hay Whitney Fellowship to W.S.S.

References

Bell, C. W., Fronk, E., and Gibbons, I. R. (1979). *J. Supramol. Struct.* **11,** 311–317.

Bernardi, G. (1971). *In* "Methods in Enzymology," (W. B. Jakoby, ed.), Vol. 22, pp. 325–339. Academic Press, New York.

Blum, J. J. (1973). *Arch. Biochem. Biophys.* **156,** 310–320.

Blum, J. J., and Hayes, A. (1974). *Biochemistry* **13,** 4290–4298.

Blum, J. J., and Hines, M. (1979). *Q. Rev. Biophys.* **12,** 103–180.

Borisy, G. G., Marcum, J. M., Olmsted, J. B., Murphy, D. A., and Johnson, K. A. (1975). *Ann. N.Y. Acad. Sci.* **253,** 107–132.

Brokaw, C. J., and Benedict, B. (1971). *Arch. Biochem. Biophys.* **142,** 91–100.

Burns, R. G., and Pollard, T. D. (1974). *FEBS Lett.* **40,** 274–280.

Dentler, W. L., Pratt, M. M., and Stephens, R. E. (1980). *J. Cell Biol.* **84,** 381–403.

Ebashi, S., and Nonomura, Y. (1973). *In* "The Structure and Function of Muscle" (G. H. Bourne, ed.), Vol. 3, pp. 285–362. Academic Press, New York.

Fiske, C. H.. and Subbarow, Y. (1925). *J. Biol. Chem.* **66,** 375–400.

Gibbons, B. H., and Gibbons, I. R. (1973). *J. Cell Sci.* **13,** 337–357.

Gibbons, B. H., and Gibbons, I. R. (1979). *J. Biol. Chem.* **254,** 197–201.

Gibbons, B. H., Fronk, E., and Gibbons, I. R. (1970). *J. Cell Biol.* **47,** 71a. (Abstr.)

Gibbons, I. R. (1963). *Proc. Natl. Acad. Sci. U.S.A.* **50,** 1002–1010.

Gibbons, I. R. (1965a). *Arch. Biol.* **76,** 317–352.

Gibbons, I. R. (1965b). *J. Cell Biol.* **26,** 707–712.

Gibbons, I. R. (1966). *J. Biol. Chem.* **241,** 5590–5596.

Gibbons, I. R., and Fronk, E. (1972). *J. Cell Biol.* **54,** 365–381.

Gibbons, I. R., and Fronk, E. (1979). *J. Biol. Chem.* **254,** 187–196.

Gibbons, I. R., and Rowe, A. J. (1965). *Science* **149,** 424–426.

Gibbons, I. R., Fronk, E., Gibbons, B. H., and Ogawa, K. (1976). *In* "Cell Motility" (R.

Goldman, T. Pollard, and J. Rosenbaum, eds.), pp. 915–932. Cold Spring Harbor Lab., Cold Spring Harbor, New York.
Gibbons, I. R., Cosson, M. P., Evans, J. E., Gibbons, B. H., Houck, B., Martinson, K. H., Sale, W. S., and Tang, W.-J. Y. (1978). *Proc. Natl. Acad. Sci. U.S.A.* **75,** 2220–2224.
Harvey, E. B. (1956). "The American Arbacia and Other Species of Sea Urchins." Princeton Univ. Press, Princeton, New Jersey.
Hayashi, M., and Higashi-Fujime, S. (1972). *Biochemistry* **11,** 2977–2982.
Huang, B., Piperno, G., and Luck, D. J. L. (1979). *J. Biol. Chem.* **254,** 3091–3099.
Kincaid, H. L., Gibbons, B. H., and Gibbons, I. R. (1973). *J. Supramol. Struct.* **1,** 461–470.
Kobayashi, T., Martensen, T., Nath, J., and Flavin, M. (1978). *Biophys. Biochem. Res. Commun.* **81,** 1313–1318.
Laemmli, U. K. (1970). *Nature (London)* **227,** 680–685.
Lewin, R. A. (1954). *J. Gen. Microbiol.* **11,** 358–363.
Linck, R. W. (1973a). *J. Cell Sci.* **12,** 345–367.
Linck, R. W. (1973b). *J. Cell Sci.* **12,** 951–981.
Lowey, S., Slayter, H. S., Weeds, A. G., and Baker, H. (1969). *J. Mol. Biol.* **42,** 1–29.
Luck, D. J. L., Piperno, G., Ramanis, Z., and Huang, B. (1977). *Proc. Natl. Acad. Sci. U.S.A.* **74,** 3456–3460.
Mabuchi, I. (1973). *Biochim. Biophys. Acta* **297,** 317–332.
Mabuchi, I., and Shimizu, T. (1974). *J. Biochem. (Tokyo)* **76,** 991–999.
Mabuchi, I., Shimizu, T., and Mabuchi, Y. (1976). *Arch. Biochem. Biophys.* **176,** 564–576.
Miki, T. (1963). *Exp. Cell Res.* **29,** 92–101.
Mohri, H., Hasegawa, S., Yamamoto, M., and Murakami, S. (1969). *Sci. Pap. Coll. Gen. Educ., Univ. Tokyo* **19,** 195–217.
Ogawa, K. (1973). *Biochim. Biophys. Acta* **293,** 514–524.
Ogawa, K., and Gibbons, I. R. (1976). *J. Biol. Chem.* **251,** 5793–5801.
Ogawa, K., and Mohri, H. (1972). *Biochim. Biophys. Acta* **256,** 142–155.
Ogawa, K., Mohri, T., and Mohri, H. (1977). *Proc. Natl. Acad. Sci. U.S.A.* **74,** 5006–5010.
Ogawa, K., Negishi, S., and Obika, M. (1980). *Arch. Biochem. Biophys.* **203,** 196–203.
Okuno, M. (1980). *J. Cell Biol.* **85,** 712–715.
Osanai, K. (1975). *In* "The Sea Urchin Embryo" (G. Czihak and R. Peter, eds.), pp. 26–40. Springer-Verlag, Berlin and New York.
Piperno, G., and Luck, D. J. L. (1979). *J. Biol. Chem.* **254,** 3084–3090.
Piperno, G., and Luck, D. J. L. (1980). *J. Cell Biol.* **87,** 36a. (Abstr.)
Pratt, M. M., Otter, T., and Salmon, E. D. (1980). *J. Cell Biol.* **86,** 738–745.
Raff, E. C., and Blum, J. J. (1969). *J. Biol. Chem.* **244,** 366–376.
Sale, W. S., and Gibbons, I. R. (1979). *J. Cell Biol.* **82,** 291–298.
Shimizu, T., and Kimura, I. (1974). *J. Biochem. (Tokyo)* **76,** 1001–1008.
Shimizu, T., Kimura, I., Murofushi, H., and Sakai, H. (1979). *FEBS Lett.* **108,** 215–218.
Summers, K. E., and Gibbons, I. R. (1971). *Proc. Natl. Acad. Sci. U.S.A.* **68,** 3092–3096.
Takahashi, M., and Tonomura, Y. (1978). *J. Biochem. (Tokyo)* **84,** 1339–1355.
Tang, W.-J. Y., Bell, C. W., Sale, W. S., and Gibbons, I. R. (1982). *J. Biol. Chem.* (in press).
Thompson, G. A., Baugh, L. C., and Walker, L. F. (1974). *J. Cell Biol.* **61,** 253–257.
Warner, F. D., and Zanetti, N. C. (1980). *J. Cell Biol.* **86,** 436–445.
Warner, F. D., Mitchell, D. R., and Perkins, C. R. (1977). *J. Mol. Biol.* **114,** 367–384.
Watanabe, T., and Flavin, M. (1976). *J. Biol. Chem.* **251,** 182–192.
Weisenberg, R., and Taylor, E. W. (1968). *Exp. Cell Res.* **53,** 372–384.
Witman, G. B., Carlson, K., Berliner, J., and Rosenbaum, J. L. (1972). *J. Cell Biol.* **54,** 507–539.

"cleaner" in appearance following negative staining. Exact subunit composition and subunit stoichiometric information of the IF can be determined by gel electrophoresis or isoelectric focusing and quantitated by densitometry (Steinert *et al.*, 1976).

Keratin IF subunits of hair and other related "hard" keratins do not reassemble *in vitro* under any known conditions, although highly ordered "protofilamentous" particles are formed (Dobb *et al.*, 1973).

3. Minimal Subunit Composition

Single keratin IF subunits do not assemble *in vitro* into filamentous structures. However, most combinations of two subunits of large and small molecular weight do under the standard conditions described. Highest yields are obtained when purified subunits dissolved in SDS solution are mixed prior to removal of SDS for filament assembly (Steinert *et al.*, 1976, 1981a). The reason for this is unclear, but may indicate that the subunits begin to associate in a structurally relevant manner as soon as some SDS is removed. Thus keratin IF assembled *in vitro* from subunits derived from the epidermis or other epithelial tissues, such as the esophagus (Milstone, 1981), are obligate copolymers.

4. Isolation of Filaggrin

Filaggrin is a prominent structural protein of the stratum corneum of mammalian epidermis (up to 30% of the total dry weight mass), and it or its immunological analogue may be present in many other types of epithelial tissues (R. D. Goldman and P. M. Steinert, unpublished observations; Steinert *et al.*, 1981b). This protein was previously termed stratum corneum basic protein (Dale, 1977) or histidine-rich protein II (Ball *et al.*, 1978). It is synthesized in large amounts in the granular layer of mammalian epidermis as a highly phosphorylated precursor of acidic pI where it accumulates as keratohyalin granules, but it is dephosphorylated at terminal stages of differentiation to become highly cationic in charge (Lonsdale-Eccles *et al.*, 1980). This cationic protein is of interest because of its capacity to aggregate *in vitro* all types of IF so far examined into highly oriented parallel arrays or macrofibrils (Steinert *et al.*, 1981b). It is thus an important epithelial cell-type IF-associated protein.

Mouse or rat stratum corneum is washed in PBSa to remove soluble proteins and extracted with a buffer of 8 *M* urea, 0.1 *M* Tris-HCl (pH 7.4), 0.1 *M* 2-mercaptoethanol, 0.5 m*M* PMSF, with homogenization, which releases both the keratin IF subunits and the filaggrin. The filaggrin is recovered in an enriched form by exclusion chromatography on DEAE-cellulose, to which it does not bind. It may then be further purified by preparative SDS gel electrophoresis (Steinert *et al.*, 1981b). Macrofibril formation with an IF type *in vitro* occurs simply on mixing the IF with filaggrin dissolved in a buffer compatible with the IF type. At total protein concentrations above about 100 μg/ml, macrofibril formation occurs within 5 sec.

B. Vimentin IF

1. CHARACTERISTICS

A wide variety of mesenchymal tissues and cells derived from them or various transformed cell lines maintained in culture contain vimentin IF (up to 5–10% of the total cellular mass), whose properties differentiate them from the IF of more specialized cell types. Unlike the IF of other cells, the organizational and assembly state of vimentin IF can be altered during different stages of cellular activity or treatment with drugs. Exposure of BHK-21 or neuroblastoma cells to colcemid, for example, results in the rapid withdrawal of the IF from their cytoplasmic arrays and their accumulation in a perinuclear location as a birefringent juxtanuclear cap (Goldman and Follet, 1970; Starger and Goldman, 1977; Starger *et al.*, 1978; Solomon, 1980). A similar accumulation of the IF occurs during normal spreading of the cells onto a substrate upon plating, or on spreading of the daughter cells following a mitotic event (Starger *et al.*, 1978).

2. ISOLATION FROM FIBROBLASTS

Methods for the isolation of vimentin IF have been developed that take advantage of the cells' capacity to concentrate the IF. The following procedures were adopted for the isolation of vimentin IF from BHK-21 fibroblasts (Starger *et al.*, 1978) and are applicable to most other fibroblastic cells grown in culture. BHK-21/c13 cells were replated following trypsinization and allowed to spread for 45–60 min. They were washed with 6 m*M* Na^+/K^+ phosphate buffer (pH 7.4), 3 m*M* KCl, 0.17 *M* NaCl (PBSa), harvested with a rubber policeman, and collected by centrifugation at 750 *g* for 3 min. All subsequent steps were done at 4°C to minimize degradation of the IF proteins by proteases. The cells were lysed with buffer (2.5 ml/100 mm dish) containing 0.6 *M* KCl, 1% Triton X-100, 10 m*M* $MgCl_2$, 0.5 m*M* PMSF, 1 mg/ml TAME in PBSa and homogenized. The viscosity due to the released chromatin was reduced by addition of DNAse 1 (0.3–0.5 mg/ml) and incubated for 1 min. These procedures dissolve the cellular constituents and most of the microtubules and microfilaments, but leave the IF cap intact. The caps were then harvested by centrifugation (1600 *g* for 5 min) and washed three times by centrifugation with PBSa containing 5 m*M* EDTA and 0.1 m*M* PMSF.

Some cells such as CHO Pro^-5 cells do not cap in the same way (Cabral and

Gottesman, 1979). However, their IF can be harvested in high yield although in a less pure form.

The washed pellets of IF may then be utilized for electron microscopy, SDS gel electrophoresis by standard procedures, or IF reassembly *in vitro*. In most types of fibroblastic cells grown in culture, the isolated IF consist of a major protein (50–70% of the total), vimentin, of M_r ~55,000 (range of 54–58,000, depending on the exact gel system used), as well as many other proteins in smaller amounts, including tubulin, actin, and their associated proteins (Cabral *et al.*, 1981).

Further purification of the vimentin can be achieved by disassembly and reassembly of the IF *in vitro* using conditions in which these other proteins do not assemble (Starger *et al.*, 1978; Steinert *et al.*, 1978, 1981a; Zackroff and Goldman, 1979; Cabral *et al.*, 1981). The pelleted IF of BHK-21 or CHO cells were resuspended (1–2 mg/ml total protein) in 6 m*M* Na^+/K^+ phosphate [or 5 m*M* Tris-HCl] buffer (pH 7.4), 1 m*M* dithiothreitol, 1 m*M* EGTA, and 1 mg/ml TAME (vimentin disassembly buffer), homogenized, and dialysed against 1000 vol of the buffer for 16 hr (overnight) at 4°C. This procedure dissociates the vimentin IF into their constituent protofilaments. Centrifugation steps at 40,000 *g* for 15 min and 250,000 *g* for 1 hr were utilized to clarify the solution. The supernatant was carefully withdrawn to avoid contamination with the superficial layer of lipid. Analysis of the supernatant by gel electrophoresis showed that more than 75% of the vimentin had remained in solution. Upon addition of 2.5 *M* KCl to a final concentration of 0.17 *M* (this is now vimentin IF assembly buffer), the vimentin reassembled within 1 hr at 23°C or 6 hr at 4°C and were then pelleted at 100,000 *g* for 1 hr. Analysis of the pellet by two-dimensional gel electrophoresis by the method of O'Farrell (1975) or Douglas *et al.* (1979) revealed significant purification of the vimentin, since many of the contaminating proteins did not sediment. The vimentin was then resuspended (0.5–1 mg/ml) and purified by two further cycles of disassembly–reassembly as described above.

Pellets or caps of IF may also be dissociated in an 8 *M* urea solution such as keratin IF disassembly buffer (see Section II,A,2), and reassembled by dialysis against 1000 vol of vimentin IF assembly buffer (Cabral *et al.*, 1981). This method has several advantages: (a) Urea solutions offer a more quantitative solubilization of the IF pellets; (b) the IF subunits are not irreversibly denatured by this urea solution (although care must be used to avoid modification by contaminating cyanate), whereas tubulin, actin, and many other associated proteins are insoluble in urea and thus removed during recycling; and (c) contaminating proteases that may damage the IF subunits are also denatured by the urea solutions.

Some minor amounts of proteins nevertheless still coassemble with the vimentin. These may be vimentin IF associated proteins that are important in the *in vivo* regulation of the IF. Note that these proteins are usually lost when the urea procedures of above are utilized. BHK-21 cells are of considerable interest, however, because their IF contain significant amounts of two other more basic proteins of slightly lower molecular weight that have now been unequivocally identified as the α and β subunits of desmin, the IF type of muscle cells (Gard *et al.*, 1979; Tuszynski *et al.*, 1979; Steinert *et al.*, 1981a) (see also Chapter 13 of this volume for details of muscle cell IF). Thus desmin and vimentin polymerize *in vitro* under the same conditions, and may in fact copolymerize into the same IF (Steinert *et al.*, 1981a).

Homogeneous vimentin of CHO and BHK-21 cells and the two desmin subunits of BHK-21 cells have been isolated by preparative polyacrylamide gel electrophoresis in the presence of SDS on a Uniphor device (Cabral *et al.*, 1981; Steinert *et al.*, 1981a). Following removal of the SDS by ion-pair extraction (see Section II,A,2), the proteins were redissolved in keratin IF disassembly buffer (0.5–1 mg/ml) and reassembled by dialysis against vimentin IF assembly buffer. The single proteins vimentin, α-desmin, and β-desmin form homopolymer IF in high yields. Such IF, like keratin IF, are uniformly 8–10 nm in diameter, tubular in cross section, and many micrometers long and have a "smooth" appearance as judged by negative staining (Fig. 1a) (Steinert *et al.*, 1981a).

C. Neural IF

1. Invertebrate

Neurofilaments are the quantitatively major protein of worm giant axons and relatively simple neural tissues such as squid, snail, and clam brains and axoplasms (10–20% of the total tissue).

Neural IF from these tissues can be isolated by taking advantage of their unique high-salt solubility and capacity to reassemble *in vitro* in solutions of reduced ionic strength, again by utilizing conditions in which the major contaminating protein, tubulin, does not reassemble *in vitro*. The following procedures were established for squid brain (optic lobe) and axon neural IF (Zackroff and Goldman, 1980), but are equally applicable with only minor modifications to other tissues, such as the clam *Spisula solidissima* and various snails (e.g., *Busycon* sp.).

Optic lobes were dissected from freshly killed squid and freed of as much connective tissue as possible. The tissue was rinsed, minced [1 ml/8 lobes (or 2 ml/g wet weight for clam or snail tissue)], and homogenized in a buffer of 0.25 *M* MES (pH 6.6), 1.0 *M* KCl, 5 m*M* EGTA, 2 m*M* dithiothreitol, and 0.5 m*M* PMSF (invertebrate neural IF disassembly buffer). The homogenate was then dialysed against 200 vol of this buffer for 5–6 hr and centrifuged at 250,000 *g* for

90 min at 4°C. This step removed most of the microtubules. The clear supernatant was brought to 20°C and diluted with 9 vol of a buffer of 0.25 *M* MES (pH 6.6), 5 m*M* EGTA, 0.5 m*M* PMSF, and 2 m*M* dithiothreitol (this is now invertebrate neural IF assembly buffer). Reassembly of long IF occurred within 60 min. The IF were pelleted (150,000 *g* for 30 min at 20°C), resuspended (4–8 mg/ml) by homogenization in disassembly buffer, dialysed, and recycled as above. Alternatively, the pellets of the once reassembled IF were redissolved in an 8 *M* urea buffer such as keratin IF disassembly buffer (see Section II,A,2) and clarified by centrifugation. The urea was removed by chromatography on a 5 × 1-cm column of Sephadex G-25 equilibrated in invertebrate IF assembly buffer and IF were reassembled as described earlier.

Two cycles of disassembly-reassembly yielded about 0.5 mg of neural IF from 30 squid optic lobes. The IF appeared after negative staining to be 8–10 nm wide and many micrometers long and had "smooth" walls. Many IF appeared to emanate from densely staining bodies, which may serve as nucleation centers for assembly *in vitro* and *in vivo* as well. These structures were lost when urea was used in the *in vitro* cycling steps. Interestingly, unlike all other IF types we have studied, these neural IF could assembly from very low protein concentrations (~10 μg/ml), at ionic strengths ranging from 0.01 to about 0.3 mol/liter and at temperatures ranging between 0 to 37°C.

Analysis of the squid optic lobe neural IF by SDS gel electrophoresis revealed a major protein of M_r ~ 60,000 (70% of the total protein) and smaller (3–6%) amounts of proteins of M_r about 74,000, 100,000, and 200,000 (Zackroff and Goldman, 1980). Squid axoplasm and clam and snail neural tissues contained IF subunits of similar molecular weight but in different relative molar amounts. The subunits of squid optic lobes have been separated by preparative SDS gel electrophoresis, and following removal of the SDS by ion-pair extraction (see Section II,A,2), those of M_r 60,000, 74,000, and 200,000 were capable of forming homopolymer IF *in vitro* with the same morphological characteristics as those containing all subunits (R. V. Zackroff and P. M. Steinert, unpublished observations).

2. Mammalian

Mammalian neural IF are quite different from those of invertebrate species, and, accordingly, different methods have been developed for their isolation. The established procedure for the isolation of microtubules from vertebrate neural tissue was based on polymerization cycles *in vitro* at 37°C and depolymerization at 4°C (Shelanski *et al.*, 1973). The cold nondepolymerizable fraction obtained during these cycles consists of neural (Delacourte *et al.*, 1977, 1980; Berkowitz *et al.*, 1977) and glial (Bignami and Dahl, 1977) IF. The fraction consists of α- and β-tubulin (~60%), glial IF protein (M_r ~ 50,000, which comigrates with β-tubulin; ~10%), and the neurofilament triplet subunits of M_r about 220,000, 150,000, and 70,000 (~20%) (Thorpe *et al.*, 1979; Runge *et al.*, 1979; Shelanski and Liem, 1979). The subsequent isolation and *in vitro* characterization of glial IF from these preparations has been described in detail by Rueger *et al.* (1979).

We describe here a procedure based on that of Delacourte *et al.* (1980) for the isolation and purification of the neural IF triplet subunits from bovine spinal cord. The same procedure is applicable to other vertebrate animals and tissues such as the cerebral hemisphere and brain stem and is predicated on selectively removing the microtubules by cycles of disassembly and reassembly *in vitro*. Bovine spinal cord, obtained within a few minutes of slaughter, was homogenized (2 ml/g wet weight) at 4°C in a buffer of 0.1 *M* MES (pH 6.6), 5 m*M* EGTA, 1 m*M* PMSF and centrifuged at 100,000 *g* for 30 min at 4°C. The pellet was then resuspended in the MES buffer (1 ml/g of original tissue) at 4°C by gentle homogenization and recentrifuged at 100,000 *g*. This pellet consisted of a "loose" translucent upper layer and a more compact lower layer. The upper layer was carefully removed and found by electron microscopy to contain neural IF and a significantly reduced amount of microtubules, most of which had been depolymerised in the cold centrifugation steps.

Two subsequent purification methods may then be utilised (Zackroff *et al.*, 1981, 1982). The first takes advantage of the observation that vertebrate neural IF disassemble in a buffer of low ionic strength ($I \sim 0.02$ mole/liter) and reassemble at higher ionic strength ($I \sim 0.15$ mole/liter). The neural IF pellet was resuspended (1–2 mg/ml) by homogenization in disassembly buffer of 5 m*M* Tris (pH 8.6), 2 m*M* dithiothreitol, 0.125 m*M* EGTA, 0.1 m*M* PMSF, dialysed against two changes of 1000 vol of this buffer at 4°C overnight and clarified by centrifugation at 250,000 *g* for 90 min. Reassembly was induced by the addition of 0.1 vol of 0.25 *M* imidazole buffer (pH 6.8–7.0), 1.5 *M* KCl, 50 m*M* $MgSO_4$ within 1 hr at 37°C. The IF were harvested by centrifugation at 100,000 *g* for 1 hr and recycled exactly as above. In an alternative procedure, neural IF pellets were dissolved (3–5 mg/ml) in a buffer of 8 *M* urea, 20 m*M* MES (pH 6.6), 1 m*M* EGTA, 1 m*M* PMSF, 0.1 m*M* $MgCl_2$, 0.25 *M* 2-mercaptoethanol, centrifuged at 250,000 *g* for 1 hr, and then dialyzed overnight at 23°C against 1000 vol of 25 m*M* imidazole buffer (pH 7.1), 0.15 *M* KCl, 5 m*M* $MgSO_4$, 2 m*M* dithiothreitol, 0.125 m*M* EGTA, and 0.2 m*M* PMSF. The IF were collected by centrifugation at 100,000 *g* for 1 hr and recycled. A notable advantage of this procedure was that the tubulin was insoluble in the urea solution and thus removed more completely than in the low-high-salt cycling procedure.

Analysis of the IF by SDS-gel electrophoresis revealed that the neural IF triplet subunits had been greatly enriched. By negative staining the IF appeared to be ~10 nm wide and many micrometers long. In contrast to all other IF types we have studied *in vitro*, these neural IF had a "rough" wall appearance (Fig. 1b). The three triplet subunits have been separated by preparative SDS gel elec-

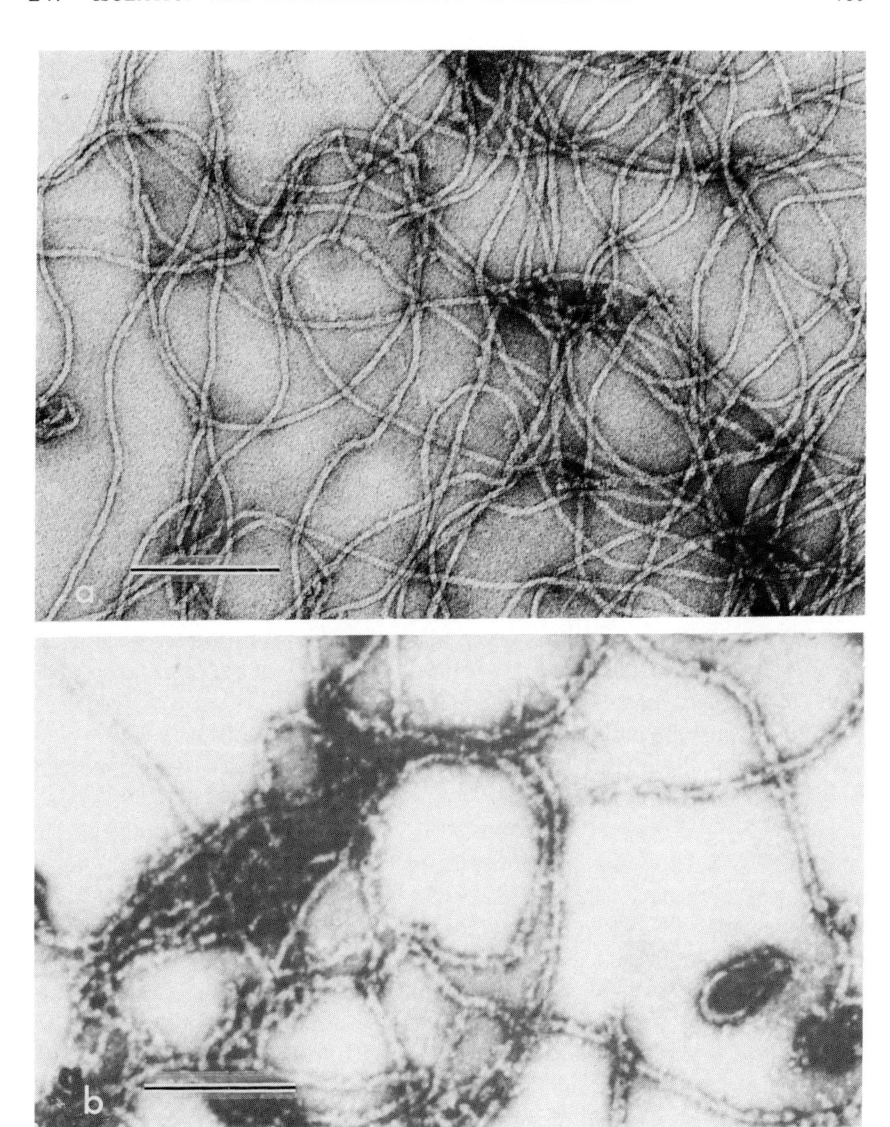

FIG. 1. Morphology of IF reconstituted *in vitro*. (a) Homopolymer vimentin IF that had been isolated and purified from CHO cells (Cabral *et al.*, 1981). Note that the long IF adopt a curvilinear appearance, are uniformly 8–10 nm wide, and have "smooth" walls or edges. (b) Neural IF three times cycled with urea from bovine spinal cord. This preparation contained negligible amounts of tubulin. Note that the long IF are also curvilinear and ~10 nm wide, but have a "rough" wall appearance. This may be due to the attachment of the higher molecular weight subunits of the neural IF triplet, which in effect, serve as associated IF proteins to the IF formed from the M_r 70,000 subunit. Negatively stained with uranyl acetate. ×103,000; bar = 0.2 μm.

trophoresis and individually characterized. The $M_r \sim 70{,}000$ subunit was found to be capable of assembly into homopolymer IF *in vitro*, and such IF had a "smooth" wall appearance (Zackroff *et al.*, 1981, 1982). The higher-molecular-weight subunits did not assemble by themselves, but did coassemble with the M_r 70,000 subunit to form long IF with a "rough" wall appearance. These findings suggest that the higher-molecular-weight triplet subunits are M_r 70,000 IF-associated proteins, a conclusion supported by recent antibody decoration studies of neural IF (Willard and Simon, 1981).

D. Isolation of IF from Complex Multicomponent Systems: HeLa Cells

HeLa cells, derived from a human cervical adenocarcinoma, have been shown by immunofluorescence techniques to contain IF of the keratin and vimentin types (Franke *et al.*, 1979; Osborn *et al.*, 1980). The presence of two types of IF within a single cell type may be a common feature of transformed cells, not only those of epithelial origin. The procedures given here for HeLa cells (Whitman *et al.*, 1980) should be of general applicability to other complex multicomponent systems. In principle, conditions should be chosen that maximize the yield of IF and take advantage of the known properties of the individual IF types suspected of being present in the cell type of interest.

HeLa cells may be harvested from culture, lysed, and the IF pellets collected as described for fibroblasts (see Section II,B,2). The IF pellets were dissolved (1–2 mg/ml) in a buffer of 8 *M* urea, 10 m*M* Tris-HCl (pH 7.4), 1 m*M* dithiothreitol, and 0.1 m*M* PMSF within 2 hr, and clarified by centrifugation at 250,000 *g* for 1 hr. Reassembly of some of the IF occurred on removal of the urea by dialysis against 1000 vol of 6 m*M* Na^+/K^+ phosphate (or 5 m*M* Tris-HCl) buffer (pH 7.4), 1 m*M* dithiothreitol, and 0.1 m*M* PMSF at 23°C for 16 hr (overnight), but optimal reassembly occurred in this buffer containing 0.1 *M* KCl. Such IF could then be pelleted, redissolved in the urea buffer, and reassembled. As judged by polyacrylamide gel electrophoresis, the subunit composition of the IF was essentially unchanged by cycling in this way, except that several high molecular weight proteins, possible candidates for HeLa-cell-IF-associated proteins, were lost. The major subunits have M_r values ($\times 10^{-3}$) of 43, 54, 55, 60, 65, 70. The tendency of the HeLa cell IF to form lateral aggregates in solution in especially high total protein concentration (>1 mg/ml), suggests the presence of IF-associating proteins, perhaps like epidermal filaggrin (see Section II,A,4).

The major IF subunits can be individually isolated in the following manner, initially developed for epidermal keratin IF subunit separation (Steinert and Idler, 1975) and should be of general applicability to other complex multisubunit IF systems. The subunits are applied to a DEAE-cellulose column equilibrated in a urea buffer and fractionated into simpler mixtures by stepwise gradient elution with increasing salt (KCl) concentrations. Subsequently, the subunits of these mixtures can be purified by other techniques, such as preparative polycrylamide gel electrophoresis in SDS, and their homogeneity confirmed on one- or two-dimensional gels. Subunits purified in this way are then amenable to individual characterization with respect to their chemical, structural, and *in vitro* reassembly properties. In this way, it was demonstrated (M. Aynardi-Whitman, unpublished observations) that the HeLa cell IF subunit of $M_r \sim 55{,}000$ was very similar to fibroblastic (BHK-21) vimentin: it could be reassembled in vimentin assembly buffer (see Section II,B,2) into homopolymer IF; it cross-reacted with antivimentin antibodies; it had a similar one-dimensional peptide map and a similar, although not identical, two-dimensional tryptic peptide map; and its amino acid composition was similar to that of vimentin. The other HeLa cell IF subunits were more typically keratin-like; combinations such as the 43,000 and 54,000 subunits could form IF *in vitro* in keratin IF assembly buffer (see Section II,A,2); they were immunologically related to epidermal keratin IF; and their amino acid contents were similar to epidermal keratin IF subunits.

E. *In Vitro* Solution Properties of IF

1. Properties

The solution properties of the various types of IF that have been studied *in vitro* are summarized in Table I. The various classes of IF differ in their subunit sizes and complexities and in the optimal ionic strength conditions for their assembly and stability in solution. Most IF types disassemble at moderate extremes of pH or in certain specific ionic conditions into soluble protofilaments, and in solutions containing denaturants into soluble subunits. All reassemble upon return to optimal conditions and/or the removal of the denaturant. In no case has a requirement for high-energy phosphates, metal ions, prosthetic groups, or accessory proteins been identified for assembly *in vitro*. (In certain cases such as neural IF, Mg^{2+} afford higher yields of subunits in the IF.) Most IF types, with the exception of keratin IF, form homopolymer IF *in vitro*. In those tissues in which more than one subunit is present, homologous copolymer IF can be formed *in vitro* as well. The subtle differences in the solution properties of the IF types identified so far should be useful for further studies and aid in the identification and characteristization of IF-associated proteins.

2. Copolymerization

We have recently found that IF subunits of certain types can form heterologous copolymer IF *in vitro* with subunits of other types (Steinert and Goldman, 1980;

TABLE I

In Vitro SOLUTION PROPERTIES OF INTERMEDIATE FILAMENTS[a]

	Filament type				
Parameter	Epithelia (keratin)	Muscle (desmin)	Mesenchymal (vimentin)	Glial[b]	Neural (neurofilaments)
Subunit composition	Very heterogeneous	Limited variations	Limited variations	Conserved(?)	Heterogeneous
Range of M_r	45–70,000	54,000	55,000	50,000	60–220,000
In vitro assembly conditions	$I < 0.01$	$I \sim 0.17$	$I \sim 0.17$	$I \sim 0.17$	$I > 0.01$ (invertebrate) $I \sim 0.2$ (vertebrate)
	pH 7–8	pH 7–8	pH 7–8	pH 7	pH 6.8–7.1
Critical concentration of assembly (μg/ml)	<50	<50	<50	Not known	<10 (invertebrate) >1000 (vertebrate)
In vitro disassembly conditions	pH <3, >9.5	pH <3,>9.5	pH <3,>9.5	Not known	pH<3, >9.5 I>0.5 (invertebrate)
	No value of I denaturant only	$I < 0.02$ Denaturant	$I < 0.05$ Denaturant	$I < 0.02$ Denaturant	$I < 0.02$ (vertebrate) Denaturant
Solution stability	$I < 0.01$	$I > 0.02$–2	$I > 0.05$–2	Not known	$I < 0.35$ (invertebrate) $I \sim 0.1$–0.2 (vertebrate)
	pH > 6.5, < 9.5	pH > 6.5, < 9.5	pH > 6.5, < 9.5	Not known	pH >6.8, <7.3
Minimal subunit composition	Obligate copolymers	Homopolymer	Homopolymer	Homopolymer	Homopolymer (invertebrate) Homopolymer (low M_r vert.) "Copolymer" (high M_r vert.)

[a] See Section IIA, C for details.

[b] Glial IF data are taken from Rueger *et al*. (1979).

Steinert *et al.*, 1982a). The experimental proof of the copolymerization of two different subunits into the same IF is predicated on the exploitation of a difference in the properties of the two subunits by using conditions in which one subunit alone does not form an IF. Since it seems that only a few combinations of subunits can be induced to form heterologous copolymer IF *in vitro* (Steinert *et al.*, 1982), the phenomenon may not be of general significance, and, with few exceptions, may not be of physiological significance. It does suggest, however, that certain subunits have closely similar structures. The experiments were assayed simply by visualizing the products of a heterologous mixture of subunits in an electron microscope after negative staining. Quantitative stoichiometric subunit information (Steinert *et al.*, 1976) and isolation of stable coiled-coil α-helical segments (see Section III,B,3) from the resulting IF were also done.

a. Keratin–Keratin Combinations. Mixtures of single subunits purified to homogeneity from bovine and mouse epidermis formed IF under standard keratin IF conditions (Steinert *et al.*, 1982). The rationale used in this case was that single keratin subunits do not assemble into IF.

b. Vimentin–Desmin Combinations. Homopolymer vimentin and desmin IF disassemble in solutions of $I \sim 0.05$ mole/liter and ~ 0.02 mole/liter, respectively (Steinert *et al.*, 1981a) (Table I). When vimentin and desmin denatured in SDS solution were mixed, freed of SDS, and allowed to assemble *in vitro,* the resulting IF disassembled at intermediate ionic strength values. Similarly, mixtures of preformed homopolymer protofilaments of vimentin and desmin formed apparent copolymer IF of intermediate solubility properties. The copolymerization of these two types of subunits may be of physiological significance since: (i) the two subunits coexist in certain cells such as BHK-21 fibroblasts (Gard *et al.*, 1979; Tuszynski *et al.*, 1979; Steinert *et al.*, 1981a); (ii) indirect immunofluorescence staining of such cells with specific vimentin and desmin antisera displayed overlapping cytoplasmic arrays of IF (Frank *et al.*, 1980); and (iii) native IF isolated from IF caps of BHK-21 cells have similar intermediate solubility properties (Steinert *et al.*, 1981a). In addition, partial amino acid sequencing analyses of vimentin and desmin have revealed significant homologies (Geisler and Weber, 1981).

c. Vimentin–Keratin combinations. Mixtures of a pure mouse epidermal keratin IF subunit and hamster vimentin formed IF in low yield in keratin IF assembly buffer (Steinert *et al.*, 1982). Under these conditions *in vitro,* the keratin subunit alone did not form an ordered structure and the vimentin alone formed protofilaments. Copolymerization of keratin and vimentin IF subunits may not be of physiological significance, since, for example, in HeLa cells where these two types coexist (see Section II,D), indirect immunofluorescence studies have shown distinctly different cytoplasmic arrays of keratin and vimentin IF (Osborn *et al.*, 1980).

III. Chemical and Structural Features of IF

A. Chemical Characterization

None of the IF subunit types so far studied contain carbohydrate or other prosthetic groups, but they are variably phosphorylated (mostly through *O*-phosphoserine linkages) (Gilmartin *et al.*, 1980; Steinert *et al.*, 1982b). When keratin IF subunits are examined by two-dimensional gel electrophoresis, for example, a marked degree of apparent heterogeneity is observed (Sun and Green, 1978; Gilmartin *et al.*, 1980) because of the presence of phosphate isomers. In addition, some apparent heterogeneity of IF subunits may arise because of their tendency to aggregate in the isoelectric-focusing dimension where inclusion of a strong denaturant such as SDS is not possible. In general, however, IF subunits can be examined by all routine protein chemical techniques, with the precaution that aggregation may frequently pose technical problems. The subunits may be stored frozen (−70°C) indefinitely in SDS solution containing 1 m*M* dithiothreitol and 0.1 m*M* PMSF, and recovered in native form following removal of the SDS by ion-pair extraction (see Section II,A,2). Freeze-thawing of subunits or intact IF in aqueous solutions in the absence of a denaturant and lyophilization of the subunits usually result in irreversible denaturation. IF subunits should not be stored in urea solutions.

IF subunits contain generally similar amino acids compositions, although keratin subunits contain somewhat higher contents of serine, glutamate, and glycine (Starger *et al.*, 1978). IF subunits may be cleaved at their tryptophan residues with *o*-iodosobenzoate or *N*-bromosuccinimide and at methionine residues with CNBr (Steinert *et al.*, 1980c). All IF subunits so far studied have *N*-acetylated aminotermini (Steinert and Idler, 1975; Steinert *et al.*, 1980c), which will pose problems for chemical sequencing. Cleveland *et al.* (1977)-type limited proteolytic digestions have also proved useful in comparative studies of IF subunits, but care should be made in interpretation of maps because certain bands may arise from structurally related regions that do not necessarily indicate close sequence homologies. In contrast, autoradiography or fluorography of two-dimensional tryptic peptide maps of subunits labeled *in vivo* or *in vitro* avoid this problem and afford a more sensitive basis for comparison of subunits (Shih *et al.*, 1978; Cabral *et al.*, 1981; Steinert *et al.*, 1981a).

B. Physicochemical Characterization

1. X-Ray Diffraction

One feature common to all IF so far examined is their α-type X-ray diffraction pattern (Steinert *et al.*, 1978), so termed because of a characteristic meridional arc at 5.15 Å. This arises because of the presence within the IF of regions in which α-helices on adjacent subunits form a supercoil or coiled coil. This property can be examined simply from a high-speed pellet of IF (minimum of 0.25 mg). The disk of pelleted IF is aligned approximately perpendicular to the direction of sedimentation. Upon removal from the tube, a portion of the pellet can be teased along a diagonal and stretched into a fiber between two wire supports and allowed to dry in a humid atmosphere. This results in further alignment of the IF within the fiber as judged by its birefringence. This is quite suitable for the detection of the most frequently repeated structural (α-helical) feature of the IF by high- (wide) angle X-ray diffraction in, for example, a Norelco microcamera using CuK_α X-radiation (γ = 1.54 Å) (Steinert *et al.*, 1976, 1978). Higher orders of IF structure are not discernible by this method, however.

2. Estimation of α-Helix Contents

As mentioned, one of the characteristic features of IF subunits is their relatively high content of α-helix (40–60%). This can be estimated by optical rotatory dispersion or circular dichroism in equipment such as a Carey or Jasco spectropolarimeter. The protein of interest is equilibrated (5–50 μg/ml) in a non-amine-based buffer (such as PBSa or 0.05 *M* sodium borate). The exact protein concentration can be measured by routine protein assay or more accurately by amino acid analysis if this is possible. In the case of circular dichroism, amino acid content data are required in order to calculate the mean residue molecular weight. Refractive index data of the solutions can be measured in a pycnometer. Proteins with high α-helix contents yield large negative extrema at about 222 and 208 nm (circular dichroism) and 233 nm (optical rotatory dispersion) (Van Holde, 1971; Greenfield and Fasman, 1969). Assignment of α-helix contents depend on the rotations afforded by model protein systems whose helicities are known from other physicochemical data. Thus in view of these implicit assumptions, only estimates (usually within $\pm$10%) of the α-helix content are possible.

3. Isolation of Common α-Helical Components of IF

The coiled-coil α-helical regions of IF are more resistant to proteolytic cleavage than the other neighboring regions of non-α-helix. Thus as was shown for myosin (Lowey *et al.*, 1969), α-helix-enriched fragments can be released from intact IF or their protofilaments by limited proteolytic digestion. The exact conditions of protein concentration, pH, time, and enzyme for optimal release of such particles varies with the nature of the IF examined. For example, α-helix-enriched particles were obtained from bovine epidermal keratin IF protofilaments within 2–10 min of digestion with 2% (w/w) trypsin in 0.05 *M* sodium borate (pH 9.2) (Steinert, 1978a). Different types of trypsin were more effective with inner root sheath keratin IF (Steinert, 1978b). Chymotrypsin offered improved yields in the digestion of vimentin and desmin IF (Steinert *et al.*, 1980c; P. M. Steinert, unpublished observations) and wool keratin protofilaments (Crewthar *et al.*, 1976). Subsequently, the particles can be isolated and separated by column chromatography (Crewther and Dowling, 1971; Steinert, 1978a,b). Properties such as molecular weight, size, shape, α-helix content, and subunit structure can then be examined by standard physicochemical techniques.

C. Kinetics of IF Assembly *in Vitro*

A more quantitative approach to the study of the structure and possible biological function of fibrous proteins in general and IF in particular has been to follow their ordered assembly *in vitro* by turbidometric means. The long IF scatter coherent light much more than soluble subunits or protofilaments. Accordingly, the assembly of the IF can be monitored while in progress by measurement of the increase in turbidity. Theoretical models for such measurements have been advanced (Oosawa and Higashi, 1967; and see earlier chapters in this volume on tubulin and actin assembly). Preliminary kinetic studies of the assembly of epidermal keratin IF (Steinert *et al.*, 1976; Steinert, 1977), BHK-21 cell IF (Zackroff and Goldman, 1979), and squid neural IF (Zackroff and Goldman, 1980) have been published.

The following general procedures are meant to represent possible experimental conditions and can be modified as required within the constraints of the IF system of interest and equipment to be utilized. IF subunits, preferably homogeneous, are dissolved in a urea buffer (see urea disassembly buffers used in Section II) and rapidly desalted on a calibrated 5 × 1-cm column of Sephadex G-25 (e.g., Pharmacia PD-10 column) equilibrated in the salt solution compatible with assembly of the IF type. This step takes about 1 min. The solution of reequilibrated IF subunits is then transferred to a spectrophotometer set at 300–500 nm to measure turbidity. A sigmoidal curve was observed (Steinert *et al.*, 1976; Steinert, 1977), revealing a slow initial step, possibly corresponding to the formation of small oligomers (''nuclei'') followed by a more rapid assembly rate. In an alternative procedure, assembly may be initiated from the protofilamentous forms of the IF. For example, vimentin IF dissociated in buffers of low ionic strength rapidly reassemble to IF when the ionic strength is adjusted to near 0.17 *M* with 2 *M* KCl (Zackroff and Goldman, 1979; Steinert *et al.*, 1981a). Squid neural IF protofilaments in solution of high ionic strength rapidly reassemble on reduction of the ionic strength by dilution (Zackroff and Goldman, 1980). Most IF types disassemble at pH ~9.5 to their protofilaments (Table I) and rapidly reassemble when the pH is lowered to 7–8 with a small volume of a concentrated buffer (P. M. Steinert, unpublished observations). In each of these cases, the reassembly follows apparent zero-order kinetics.

Centrifugal analysis of these reactions has provided information on the critical concentrations required for *in vitro* assembly. It would seem that monitoring of the assembly process by negative staining and electron microscopy should provide useful information on IF structure and assembly *in vivo*. Further studies using more sophisticated analyses such as stop-flow and temperature-jump techniques could afford more detailed information.

Buffers used

Keratin IF disassembly buffer:
  8 *M* urea, 0.1 *M* Tris-HCl (pH 7.6), 25 m*M* 2-mercaptoethanol
Keratin IF assembly buffer:
  5 m*M* Tris-HCl (pH 7.6), 25 m*M* 2-mercaptoethanol, 1 m*M* EGTA, 0.1 m*M* PMSF
Vimentin or desmin IF disassembly buffer:
  5 m*M* Tris-HCl [or 6 m*M* Na^+/K^+ phosphate] (pH 7.4), 1 m*M* dithiothreitol, 1 m*M* EGTA, 1 mg/ml TAME
Vimentin or desmin IF assembly buffer:
  5 m*M* Tris-HCl [or 6 m*M* Na^+/K^+ phosphate] (pH 7.4), 1 m*M* dithiothreitol, 1 m*M* EGTA, 0.17 *M* KCl, 1 mg/ml TAME
Invertebrate neural IF disassembly buffer:
  0.25 *M* MES (pH 6.6), 1.0 *M* KCl, 5 m*M* EGTA, 2 m*M* dithiothreitol, 0.5 m*M* PMSF
Invertebrate neural IF assembly buffer:
  0.25 *M* MES (pH 6.6), 0.1 *M* KCl, 5 m*M* EGTA, 2 m*M* dithiothreitol, 0.5 m*M* PMSF
Vertebrate neural IF disassembly buffers:
  Either 5 m*M* Tris (pH 8.6), 2 m*M* dithiothreitol, 0.125 m*M* EGTA, 0.1 m*M* PMSF
  Or 8 *M* urea, 20 m*M* MES (pH 6.6), 1 m*M* EGTA, 0.1 m*M* $MgCl_2$, 0.25 *M* 2-mercaptoethanol, 1 m*M* PMSF
Vertebrate neural IF assembly buffer:
  25 m*M* imidazole (pH 6.8–7.0), 0.15 *M* KCl, 2 m*M* dithiothreitol, 5 m*M* $MgSO_4$, 0.125 m*M* EGTA, 0.1 m*M* PMSF

References

Ball, R. D., Walker, G. K., and Bernstein, I. A. (1978). *J. Biol. Chem.* **254,** 5861–5868.
Berkowitz, S. A., Katagiri, J., Binder, H. K., and Williams, R. C. (1977). *Biochemistry* **16,** 5610–5617.
Bignami, A., and Dahl, D. (1977). *J. Histochem. Cytochem.* **25,** 466–471.

Cabral, F., and Gottesman, M. M. (1979). *J. Biol. Chem.* **254,** 6203–6206.

Cabral, F., Gottesman, M. M., Zimmerman, S. B., and Steinert, P. M. (1981). *J. Biol. Chem.* **256,** 1428–1431.

Cleveland, D., Fischer, S., Kirschner, M., and Laemmli, U. (1977). *J. Biol. Chem.* **252,** 1102–1106.

Crewther, W. G., and Dowling, L. M. (1971). *Appl. Polymer Symp.* **18,** 1–20.

Crewther, W. G., Fraser, R. D. B., Lennox, F. G., and Lindley, H. (1965). *Adv. Protein Chem.* **20,** 191–346.

Crewther, W. G., Dowling, L. M., Gough, K. H., Inglis, A. S., McKern, N. M., Sparrow, L. G., and Woods, E. F. (1976). *Proc.-Int. Wolltextil-Forschungskonf., 5th, Aachen, 1975* **2,** 233–242.

Dale, B. A. (1977). *Biochim. Biophys. Acta* **491,** 193–204.

Delacourte, A., Plancot, M.-T., Han, K. K., Hildebrand, F., and Biserte, G. (1977). *FEBS Lett.* **77,** 41–46.

Delacourte, A., Filliatreau, G., Boutteau, F., Biserte, G., and Schrevel, J. (1980). *Biochem. J.* **191,** 543–546.

Dobb, M. G., Millward, G. R., and Crewther, W. G. (1973). *J. Text. Res.* **64,** 374–385.

Douglas, M., Finkelstein, D., and Butow, R. (1979) *In* "Methods in Enzymology," (S. Fleischer and L. Packer, eds.), Vol. 56, pp. 58–66. Academic Press, New York.

Frank, E. D., Tuszynski, G. P., and Warren, L. (1980). *J. Cell Biol.* **87,** 183a.

Franke, W. W., Schmid, E., Weber, K., and Osborn, M. (1979). *Exp. Cell Res.* **118,** 95–109.

Fraser, R. D. B., MacRae, T. P., and Rogers, G. E. (1972). "Keratins, Their Composition, Structure and Biosynthesis." Thomas, Springfield, Illinois.

Fuchs, E., and Green, H. (1978). *Cell* **15,** 887–897.

Gard, D. L., Bell, P. B., and Lazarides, E. (1979). *Proc. Natl. Acad. Sci. U.S.A.* **76,** 3894–3898.

Geisler, G., and Weber, K. (1981). *Proc. Natl. Acad. Sci. U.S.A.* **78,** 4120–4123.

Gilmartin, M. E., Culbertson, V. B., and Freedberg, I. M. (1980). *J. Invest. Dermatol.* **75,** 211–216.

Goldman, R. D., and Follet, E. (1970). *Science* **169,** 286–288.

Greenfield, N., and Fasman, G. D. (1969). *Biochemistry* **8,** 4108–4116.

Huang, L.-Y., Stern, I. B., Clagett, J. A., and Chi, E.-Y. (1975). *Biochemistry* **14,** 3573–3580.

Lazarides, E. (1980). *Nature (London)* **283,** 249–256.

Lee, L. D., Fleming, B. F., Waitkus, R. F., and Baden, H. P. (1975). *Biochim. Biophys. Acta* **412,** 82–90.

Lee, L. D., Kubilus, J., and Baden, H. P. (1979). *Biochem. J.* **177,** 187–196.

Lonsdale-Eccles, J. D., Haugen, J. A., and Dale, B. A. (1980). *J. Biol. Chem.* **255,** 2235–2238.

Lowey, S., Slayter, H. S., Weeds, A. G., and Baker, H. (1969). *J. Mol. Biol.* **42,** 1–29.

Matoltsy, A. G. (1965). *In* "The Biology of the Skin and Hair Growth" (A. G. Lyne and B. F. Short, eds.), pp. 291–305. Angus & Robertson, Sydney, Australia.

Milstone, L. (1981). *J. Cell Biol.* **88,** 317–322.

O'Farrell, P. H. (1975). *J. Biol. Chem.* **250,** 4007–4021.

Oosawa, F., and Higashi, S. (1967). *Prog. Theor. Biol.* **1,** 79–164.

Osborn, M., Franke, W. W., and Weber, K. (1980). *Exp. Cell Res.* **125,** 37–46.

Rheinwald, J. G., and Green, H. (1975). *Cell* **6,** 331–343.

Rueger, D. C., Huston, J. S., Dahl, D., and Bignami, A. (1979). *J. Mol. Biol.* **135,** 53–68.

Runge, M. S., Hewgley, P. B., Puett, D., and Williams, D. C. (1979). *Proc. Natl. Acad. Sci. U.S.A.* **76,** 2561–2565.

Shelanski, M. L., and Liem, R. K. H. (1979). *J. Neurochem.* **33,** 5–13.

Shelanski, M. L., Gaskin, F., and Cantor, C. R. (1973). *Proc. Natl. Acad. Sci. U.S.A.* **70,** 765–768.

Shih, T. Y., Williams, D. R., Weeks, M. O., Maryak, J. M., Vass, N. C., and Scolnick, E. M. (1978). *J. Virol.* **27,** 45–55.

Solomon, F. (1980). *Cell* **21,** 333–338.
Starger, J. M., and Goldman, R. D. (1977). *Proc. Natl. Acad. Sci. U.S.A.* **74,** 2422–2426.
Starger, J. M., Brown, W. E., Goldman, A. E., and Goldman, R. D. (1978). *J. Cell Biol.* **78,** 93–109.
Steinert, P. M. (1975). *Biochem. J.* **149,** 39–48.
Steinert, P. M. (1977). *In* "Biochemistry of Cutaneous Epidermal Differentiation" (M. Seiji and I. A. Bernstein, eds.), pp. 444–464. Tokyo Univ. Press, Tokyo.
Steinert, P. M. (1978a). *J. Mol. Biol.* **123,** 49–70.
Steinert, P. M. (1978b). *Biochemistry* **17,** 5045–5052.
Steinert, P. M., and Goldman, R. D. (1980). *J. Cell Biol.* **87,** 180a.
Steinert, P. M., and Idler, W. W. (1975). *Biochem. J.* **151,** 603–614.
Steinert, P. M., and Idler, W. W. (1979). *Biochemistry* **18,** 5664–5669.
Steinert, P. M., and Yuspa, S. H. (1978). *Science* **200,** 1491–1493.
Steinert, P. M., Idler, W. W., and Zimmerman, S. B. (1976). *J. Mol. Biol.* **108,** 547–567.
Steinert, P. M., Zimmerman, S. B., Starger, J. M., and Goldman, R. D. (1978). *Proc. Natl. Acad. Sci. U.S.A.* **75,** 6098–6101.
Steinert, P. M., Idler, W. W., Poirier, M. G., Katoh, Y., Stoner, G. A., and Yuspa, S. H. (1979). *Biochim. Biophys. Acta* **577,** 11–21.
Steinert, P. M. Idler, W. W., and Wantz, M. L. (1980a). *Biochem. J.* **187,** 913–916.
Steinert, P. M., Peck, G. L., and Idler, W. W. (1980b). *In* "Biochemistry of Normal and Abnormal Epidermal Differentiation" (I. A. Bernstein and M. Seiji, eds.), pp. 391–406. Tokyo Univ. Press, Tokyo.
Steinert, P. M., Idler, W. W., and Goldman, R. D. (1980c). *Proc. Natl. Acad. Sci. U.S.A.* **77,** 4534–4538.
Steinert, P. M., Idler, W. W., Cabral, F., Gottesman, M. M., and Goldman, R. D. (1981a). *Proc. Natl. Acad. Sci. U.S.A.* **78,** 3692–3696.
Steinert, P. M., Cantieri, J. C., Teller, D. C., Lonsdale-Eccles, J. D., and Dale, B. A. (1981b). *Proc. Natl. Acad. Sci. U.S.A.* **78,** 4097–4101.
Steinert, P. M., Idler, W. W., Aynardi-Whitman, M., Zackroff, R. V., and Goldman, R. D. (1982a). *Cold Spring Harbor Symp. Quant. Biol.* **46** (in press).
Steinert, P. M., Wantz, M. L., and Idler, W. W. (1982b). *Biochemistry* **21,** 177–183.
Sun, T.-T., and Green, H. (1978). *J. Biol. Chem.* **253,** 2053–2060.
Thorpe, R., Delacourte, A., Ayers, M., Bullock, C., and Anderton, B. H. (1979). *Biochem. J.* **181,** 275–284.
Tuszynski, G. P., Frank, E. D., Damsky, C. H., Buck, C. A., and Warren, L. (1979). *J. Biol. Chem.* **254,** 6138–6143.
Van Holde, K. E. (19781). "Physical Biochemistry." Prentice-Hall, Englewood Cliffs, New Jersey.
Whitman, M., Steinert, P. M., Zimmerman, S. B., and Goldman, R. D. (1980). *J. Cell Biol.* **87,** 176a.
Willard, M., and Simon, C. (1981). *J. Cell Biol.* **89,** 198–205.
Zackroff, R. V., and Goldman, R. D. (1979). *Proc. Natl. Acad. Sci. U.S.A.* **76,** 6226–6230.
Zackroff, R. V., and Goldman, R. D. (1980). *Science* **208,** 1152–1155.
Zackroff, R. V., Steinert, P. M., Aynardi-Whitman, M., and Goldman, R. D. (1981). *Cell Surf. Rev.* **7** (in press).
Zackroff, R. V., Idler, W. W., Steinert, P. M., and Goldman, R. D. (1982). *Proc. Natl. Acad. Sci. U.S.A.* **79** (in press).

Index

A

V

X

CONTENTS OF RECENT VOLUMES

(Volumes I–XX edited by David M. Prescott)

Volume X

Volume XI

Volume XII

Volume XIII

Volume XIV

Volume XV

Volume XVI

Part A. *Isolation of Nuclei and Preparation of Chromatin. I*

Volume XVII

Volume XVIII

Volume XIX

Volume XX

Volume 21A

Normal Human Tissue and Cell Culture A. Respiratory, Cardiovascular, and Integumentary Systems

Volume 21B

Normal Human Tissue and Cell Culture B. Endocrine, Urogenital, and Gastrointestinal Systems

Volume 22

Three-Dimensional Ultrastructure in Biology

Volume 23

Basic Mechanisms of Cellular Secretion